TREE AUTOMATA AND LANGUAGES

STUDIES IN COMPUTER SCIENCE AND ARTIFICIAL INTELLIGENCE

10

Editors:

R.B. Banerji
Saint Joseph's University
Philadelphia

M. Nivat
Université Paris VII
Paris

M. Wirsing
Universität Passau
Passau

NORTH-HOLLAND – AMSTERDAM • LONDON • NEW YORK • TOKYO

TREE AUTOMATA AND LANGUAGES

edited by

Maurice NIVAT

LITP
Université Paris VII
Paris, France

Andreas PODELSKI

DIGITAL
Paris Research Laboratory
Centre de Recherche de Paris
Rueil-Malmaison, France

1992

NORTH-HOLLAND – AMSTERDAM • LONDON • NEW YORK • TOKYO

ELSEVIER SCIENCE PUBLISHERS B.V.
Sara Burgerhartstraat 25
P.O. Box 211, 1000 AE Amsterdam, The Netherlands

Library of Congress Cataloging-in-Publication Data

Tree automata and languages / edited by Maurice Nivat, Andreas
  Podelski.
       p.   cm. -- (Studies in computer science and artificial
  intelligence ; 10)
    Includes bibliographical references.
    ISBN 0-444-89026-2 (alk. paper)
    1. Sequential machine theory. 2. Trees (Graph theory)
  3. Programming languages (Electronic computers)   I. Nivat, M.
  II. Podelski, Andreas. III. Series.
  QA267.5.S4T74   1992
  005.13'1--dc20
                                                      92-17937
                                                      CIP

ISBN: 0 444 89026 2

Transferred to digital printing 2005

FOREWORD

The workshop on Tree Languages and Tree Automata was the starting point of the present volume and was made possible by the ESPRIT – BRA working group ASMICS (Algebraic and Synthetic Methods in Computer Science), which came into existence in September 1989 and hosts approximately 15 research teams from France, Germany, Italy, Belgium, and Portugal. Among the main themes of its research are the combinatorial and algebraic properties of words and trees, sets of words and sets of trees and the main devices to define such sets, which we call automata.

We were surprised by the very positive, even enthusiastic reaction to our suggestion of a workshop dedicated to trees, tree automata and tree languages from all the researchers we wrote to. We did not limit ourselves to just ASMICS members but included researchers from another ESPRIT working group "Graph Grammars" and from outside the CCE. (The invitation to participate was extended solely to European researchers.) Eventually, thirty-five researchers attended the 3 day meeting held in Le Touquet, Northern France in June 1990. The meeting was both pleasant and fruitful with a friendly atmosphere and a rich number of scientific exchanges between the participants.

It so happens that the theory of tree languages, which was founded in the late sixties and still active in the seventies, was much less active during the eighties: and there is now undoubtedly a simultaneous revival in several countries, with a number of significant results proved in the past five years. (A large proportion of them appear in the present volume.) The organizers of the workshop, who are also the editors of this volume suggested that the authors should write comprehensive half-survey papers. This collection would be a useful tool for everyone who is interested in the theory of tree languages and would cover most of the recent questions which are not treated in the (very few) rather old standard books on the subject. As one knows, trees appear naturally in many chapters of computer science and each new property is likely to result in improvement of some computational solution of a real problem (in handling logical formulae, data structures, programming languages on systems, algorithms etc.). The point of view we have adopted is to put an emphasis on the properties themselves and their rigorous mathematical exposition rather than on the multiple possible applications. We believe that this book will be a source book of useful concepts and methods which may be applied successfully in many situations: its philosophy is very close to the whole philosophy of the ESPRIT Basic Research Actions and to that of the European Association for Theoretical Computer Science.

TABLE OF CONTENTS

Tree Automata and Languages
M. Nivat and A. Podelski (editors)
1992 Elsevier Science Publishers B.V.

BINARY TREE CODES

Maurice NIVAT
L.I.T.P.
University Paris 7

O. INTRODUCTION

We deal with labeled binary trees which need not be complete : a node can have only one son, and we consider as different the trees in which the root has only one left son and one right son respectively.[6,7]

We define a decomposition of a tree in elements of a given set of trees X and we say that X is a tree code iff no tree can have two different decomposition in element of X . The notion of tree code extends naturally the notion of word codes if we write the words in a set of words U as trees in which each node has only one left son, the set $\hat{U}$ of such trees is a tree code iff U is a word code.

The main result of the present paper is an algorithm to decide whether a given finite set of finite trees is a tree code. This algorithm extends naturally the well known algorithm of Sardinas and Paterson for word codes.[1, 2, 3, 4, 5]

Trees and bushes over A

We denote D the set $\{1,2\}$, D^* the free monoid generated by D and ε the empty word.

The set A is a finite alphabet of labels.

A tree t over A is a partial mapping of D^* into A whose domain dom(t) is closed by left factor ie satisfies the condition

$$\forall\ f,g \in D^* : fg \in \text{dom}\ (t) \Rightarrow f \in \text{dom}\ (t)$$

We associate to every subset E of D^* which is closed by left factor a rooted tree in the classical sense of graph theory : the nodes of the tree are the elements of E and the set of arcs is the set $\{(f,fi) \mid i \in D, f\ i \in$

E}. The root ε is the only node which is the extremity of no arc. The set of leaves $Max(E) = \{f \in E \mid fD \cap E = \varnothing\}$ is the set of nodes which are the origin of no arc.

According with a traditional terminology the node f is the father of its sons $f1, f2$ which we shall call left son and right son. In what follows we distinguish the two trees $\{\varepsilon,1\}$ and $\{\varepsilon,2\}$ in which the root has only one son but this only son is a left son in $\{\varepsilon,1\}$ and a right son in $\{\varepsilon, 2\}$. A tree is finite iff its domain is finite and the set of finite trees over A is denoted $A^{\#}$. In this paper we deal only with finite trees and thus call simply a tree a finite tree.

Trees are represented by drawings in the usual most way, with the root at the top and the leaves at the bottom.

A (finite) bush b over A is a partial mapping of D^* into A (whose domain is finite).

We denote Ω the empty tree with $\mathrm{dom}\,(\Omega) = \varnothing$.

The punctual tree a is the tree whose domain is restricted to the root and the root is labeled by a.

Fundamental operations on trees [cf. 5, 6]

A fundamental relation on the set of bushes over A is the order relation of inclusion denoted by the usual symbol $\subseteq$

$$b \subseteq b' \Leftrightarrow \mathrm{dom}\,(b) \subseteq \mathrm{dom}\,(b') \text{ and } b' \mid \mathrm{dom}\,(b) = b$$

The mapping $b' \mid \mathrm{dom}\,(b)$ is the restriction of b' to $\mathrm{dom}\,(b)$ ie the mapping whose domain is $\mathrm{dom}\,(b)$ and whose value at each, point of $\mathrm{dom}\,(b)$ is equal to the value of b' at that point.

In the special case of trees if $t \subseteq t'$ we say that t is an initial subtree of t'.

The two bushes b_1 and b_2 are said to be compatible iff
$$b_1 \mid (\mathrm{dom}\,(b_1) \cap \mathrm{dom}\,(b_2)) = b_2 \mid (\mathrm{dom}\,(b_1) \cap \mathrm{dom}\,(b_2))$$

The union $b_1 \oplus b_2$ is defined iff b_1 and b_2 are compatible and given in that case by

$$\mathrm{dom}\,(b_1 \oplus \beta_2) = \mathrm{dom}\,(b_1) \cup \mathrm{dom}\,(b_2)$$
$$\forall\, f \in \mathrm{dom}\,(b_1) \qquad (b_1 \oplus b_2)\,(f) = b_1\,(f)$$
$$\forall\, f \in \mathrm{dom}\,(b_2) \qquad (b_1 \oplus b_2)\,(f) = b_2\,(f)$$

The union $b_1 \oplus b_2$ is the smallest bush which is greater than b_1 and b_2, when bushes are ordered by inclusion.

The bushes b_1 and b_2 are disjoint iff dom $(b_1) \cap$ dom $(b_2) = \emptyset$.

Obviously two disjoint bushes are always compatible and these union exists : we denote it $b_1 + b_2$ rather than $b_1 \oplus b_2$.

The intersection $b_1 \cap b_2$ is also defined iff b_1 and b_2 are compatible and is given by $b_1 \cap b_2 = b_1 \mid ($dom $(b_1) \cap$ dom $(b_2)) = b_2 \mid$ (dom $(b_1) \cap$ dom (b_2)). It should be noted that $b_1 \cap b_2$ is not in general the greatest bush which is smaller than b_1 and b_2.

Translations of trees and bushes

Let B be a bush and f be a word in D.

The bush fb has f dom $(b) = \{fg \mid g \in$ dom $(b)\}$ as domain and for all $g \in$ dom (fb) we have $fb \, (g) = b \, (f^{-1}g)$.

We use the standard notation $f^{-1}g$ which is defined for all $f, g \in D^*$ by

$f^{-1}g$ is undefined if $g \notin fD^*$;
$f^{-1}g$ is the unique g' such that $g = fg'$ otherwise

The translation ft of a tree t by $f \neq \varepsilon$ is not a tree but a bush.

The bush $f^{-1}b$ has f^{-1} dom $(b) = \{g \in D^* \mid fg \in$ dom $(b)\}$ as domain and for all $g \in$ dom $(f^{-1}b)$ we have $(f^{-1}b) \, (g) = b \, (fg)$.

If t is a tree, $f^{-1}t$ is a tree since dom $(f^{-1}t)$ is closed by left factor and if $f \in$ dom(t) it is called the terminal subtree of t rooted in f. Clearly if $f \notin$ dom (t) the set f^{-1} dom (t) is empty and we write $f^{-1}t = \Omega$.

An another notation is $b \setminus E$ to denote the bush $b \mid ($dom $(b) \setminus E)$. If t is a tree and f a word in dom (t) ie a node of t, $t \setminus fD^*$ denotes the initial subtree of t which remains when one has deleted the terminal subtree rooted in f, including f.

Our notations allow us to state

Lemma I.1

For all tree $t \in A^{\#}$, for all $f \in D^*$ are has
$$t = t \backslash f\, D^* + f\, f^{-1} t$$

Proof

dom $(t) = ($dom $(t) \backslash f\, D^*) \cup (f\, D^* \cap$ dom $(t))$ implies

dom $(t) =$ dom $(t \backslash f D^*) \cup f(f^{-1}$ dom $(t))$

The two sets on the right are disjoint and one checks immediately the equality of the labels. $\square$

We shall denote TST (t) the set of terminal subtrees of t and TST (X) the set of all terminal subtrees of elements of X if X is a subset of $A^{\#}$. A terminal subtree of t is proper iff it is different from t. We denote $PTST$ $(t) = TST$ $(t) \backslash \{t\}$ and $PTST$ $(X) = U$ $\{PTST$ $(t) \mid t \in X\}$.

Border of a tree

If $t \in A^{\#}$ is a tree we call border of t and denote B (t) the set B $(t) =$ dom $(t)\, D \backslash$ dom (t).

Thus is the set of nodes $f\, i$ whose father f belongs to D but do not belong to D themselves. By definition B $(\Omega) = \{\varepsilon\}$.

We can state the fundamental

Lemma I.2

For all $t_1, t_2 \in A^{\#}$ if $t_1 \subseteq t_2$ one has
$$t_2 = t_1 + \Sigma\, \{f\, f^{-1}\, t_2 \mid f \in B\, (t_1)\}$$

Proof

Clearly t_1 is obtained from t_2 by deleting the terminal subtrees of t_2 which are rooted in B $(t_1) \cap$ dom (t_2) ie the highest nodes which are in dom (t_2) but not in dom (t_1). If f_1 and f_2 are two such nodes the bushes $f_1\, f_1^{-1} t_2$ and $f_2 f_2^{-1}\, t_2$ are disjoint and we can reconstruct t_2 as the

disjoint union above. $\square$

Let X be a set of trees and t be a tree. We define the product tX as the set of all trees of the form

$$t + f_1 x_1 + f_2 x_2 \ldots + f_k r_k \text{ where}$$
$$-1 \le k \le card\ (B\ (t))$$
$$- f_1,\ldots,f_k \text{ are distinct elements of } B\ (t)$$
$$- x_1,\ldots,x_k \text{ are elements of } X$$

The sum $t + f_1 x_1 + f_2 x_2 \ldots + f_k r_k$ is well defined since t is disjoint from all the bushes $f_i x_i$ and two bushes $f_i x_i$, $f_j x_j$ and two bushes $f_i x_i$, $f_j x_j$ where f_i and f_j are distinct elements of $B(t)$ are disjoint.

If Y and X are sets of trees we define YX as
$$YX = \cup \{tX \mid t \in Y\}$$

We need be careful at that point for this product is not associative. One has $X (YZ \subseteq (XY)Z$ but the equality does not usually hold, as proved by the exemple :

$$X = \{< \varepsilon,1,2 >\} \qquad Y = \{ < \varepsilon,1,2 >\} \quad Z = \{< \varepsilon,1 >\}$$

One has $< \varepsilon,1,11,12,2 > \in XY$ and thus $< \varepsilon,1,11,21,2,21 > \in$ $(XY)Z$ but $YZ = < \varepsilon,1,11,2 >$, $<\varepsilon,1,2,21 >$, $<\varepsilon,1,11,2,21 >\}$ and $<$ $\varepsilon,1,11,21,2,21 > \notin X(YZ)$.

The following figures show the difficulty.

All the elements of X(YZ) are of the form

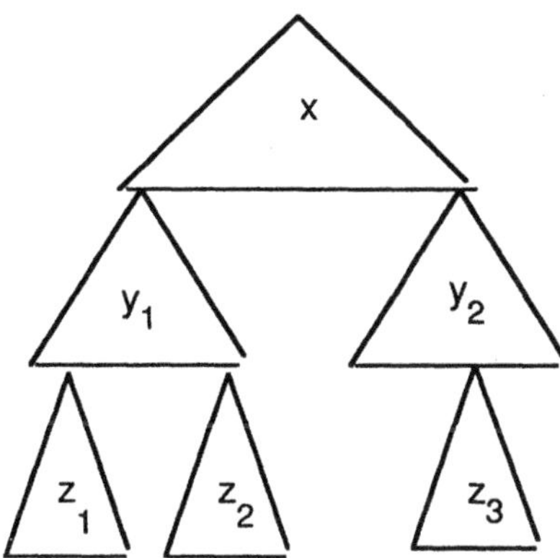

And (XY)Z contains elements of a different form :

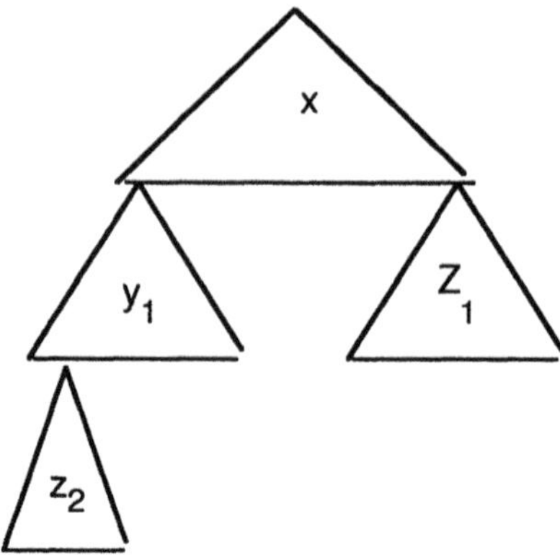

Clearly the elements of the second form are in $X(YZ)$ if but only if Y contains the empty tree Ω.

We write

$$X^3 = (XX)X \quad \text{and more generally}$$
$$X^{n+1} = (X^n)\, X$$

We denote $X^\#$ the union $X^\# = \{\Omega\} \cup X \cup X^2 \cup \ldots \cup X^n \cup \ldots$ Every tree t in $X^\#$ has a decomposition in elements of X which we can write as a sum

$$t = x_0 + \Sigma \,\{f_i x_i \mid f_i \in M\}$$

where the x_i's are elements of X and the set M has the following property. One can write $M = \{\,f_1,\ldots,f_k\,\}$ in such a way that $f_1 \in B(x_0)$

and for all $j = 2,\ldots,k \ \ f_j \in B\left(x_0 + \displaystyle\sum_{i=1}^{i=j-1} f_i x_i\right).$

We call decomposition of a tree t in elements of X any representation of t as a disjoint union of bushes which are translations of elements of X.

If $t = x_0 + \Sigma \,\{f_i x_i \mid f_i \in M\}$ is a decomposition the initial component of the decomposition is x_0 ant the terminal components are the bushes $f_i x_i$ for $f_i \in \mathrm{Max}(M)$.

Each element $f \in M$ is called a node of the decomposition which indeed can be identified with M.

The following lemmas are quite obvious but essential.

Lemma I.3

If $t = x_0 + \Sigma \{f_i x_i \mid f_i \in M\}$ is a decomposition of t in elements of X and if $f \in M$ one can write

$$f^{-1}t = \Sigma \{f^{-1}f_i x_i \mid f_i \in fD \cap M\}$$
$$t \setminus fD = x_0 + \Sigma \{f_i x_i \mid f_i \in M \setminus fD\}$$

Thus a decomposition with M as a set of nodes gives immediately for every node $f \in M$ a decomposition of $f^{-1}(t)$ and $t \setminus fD$.

Lemma I.4

If $t = x_0 + \Sigma \{f_i x_i \mid f_i \in M\}$ and $\bar{t} = \bar{x}_0 + \Sigma \{g_j \, \bar{x}_j \mid g_j \in \bar{M}\}$ are two decompositions of t and $\bar{t}$ in elements of X and if $h \in B$ (t).

$$t + h\bar{t} = x_0 + \Sigma \{f_i x_i \mid i \in M\} + \Sigma \{hg_j \, \bar{x}_j \mid j \in \bar{M}\}$$

is a decomposition of $t + h\bar{t}$ in elements of X.

An other notation will be useful in this paper.

For all subsets X, Y of $A^{\#}$ we denote
$$XY^{-1} = \{t \in A^{\#} \mid tY \cap X \neq \varnothing\}$$

This notation needs also be used carefully for in general X and $XY^{-1}Y$ are different. The set XY^{-1} is empty if $TST(X) \cap Y = \varnothing$ and thus $XY^{-1}Y$ can be empty when X is not empty. We have $X \subseteq XY^{-1}Y$ if either $\Omega \in Y$ or $X \subseteq Y$ for t $\Omega = \{t\}$ and if $t \in X$, one has $\Omega \in tX^{-1}$ and thus $t \in tX^{-1}X$.

II. TREE CODES

A subset X of $A+\# \setminus \Omega$ is a tree code iff every tree t has at most one decomposition in elements of X.

We can immediately exhibit tree codes

1 - The set of all punctual trees is a tree code.

For every tree t can be written as

$$t = \Sigma \; \{f \; t(f) \mid f \in \text{dom } (t)\}$$

and clearly there cannot be two distinct such decompositions.

This fact justifies the notation $A^{\#}$ to designate the set of all trees.

2 - Let $\mathcal{U} \subseteq A^+$ be a (word) code ie a subset of A^+ such that every word $v \in A^*$ has at most one factorisation in elements of $\mathcal{U}$ of the form $v = u_1...u_k$ with $u_1,...,u_h \in \mathcal{U}$

We associate to each element u of $\mathcal{U}$ the tree $\hat{u}$ given by dom ($\hat{u}$) $= \{\varepsilon,1,11,...,1^{\ell(u)-1}\}$.

$\hat{u} \, (i^k) =$ the h+1 th letter of u read from left to right.

It is easy to see that $\hat{\mathcal{U}} = \{\hat{u} \mid u \in \mathcal{U}\}$ is a tree-code.

Say that a tree t has right depth k iff $k = \max\{n \mid 1*2^{n-1} \cap \text{dom}(t) \neq \varnothing\}$. A tree with right depth 1 is a tree whose domain is contained in $\{1\}^*$. Assume such a tree t has two decompositions over $\hat{\mathcal{U}}$.

$$t = u_1 + 1^{\ell(u_1)} u_2 +...+ 1^{\ell(u_1)+...+\ell(u_i)}u +...+ 1^{\ell(u_1)+...+\ell(u_h-1)}u_k$$

$$t = \bar{u}_1 + 1^{\ell(\bar{u}_1)} \bar{u}_2 +...+ 1^{\ell(\bar{u}_1)+...+\ell(\bar{u}_{m-1})} \bar{u}_m$$

Clearly one has $u_1...u_h = \bar{u}_1 ... \bar{u}_m$ and this contradicts the fact that $\mathcal{U}$ is a code.

Consider now a tree with right depth $n+1$ and assume it has two decompositions. We can isolate the subtree rooted in 2^n and write each decomposition as the sum of one decomposition of $t \setminus 2^n D$ and the translation by 2^n of a decomposition of $(2^n)^{-1}t$. But since $(2^n)^{-1}t$ has right depth one it has at most one decomposition in elements of $\hat{\mathcal{U}}$, and also $t \setminus 2^n D$ by induction. $\square$

We can establish the following

Lemma II.1

If the set of trees with two different decomposition in elements of X is not empty then there exists a tree t with two different decompositions which have two different initial components ;

Proof

Take t with two different decompositions such that card (dom (t)) is minimal.

Assume $t = x_0 + \Sigma\{f_i x_i \mid i \in M\}$ and $t = x_0 + \Sigma\{g_j \bar{x}_j \mid j \in \bar{M}\}$ are two different decompositions with the same initial component x_0. For every $f \in B(x_0) \cap$ dom (t) one can find by lemma I.3 two decompositions of $f^{-1}t$ namely $\Sigma\{f_i x_i \mid f_i \in M \cap f D\}$ and

$$\Sigma \{g_j \bar{x}_j \mid g_j \in \bar{M} \cap f D\}$$

If for all $f \in B(x_0) \cap$ dom (t) these two decompositions were equal then the two decompositions of t would be equal. Thus for at least one f these two decompositions are different and this contradicts the minimality of card (dom (t)). $\square$

In order to state the fundamental property on which relies the rest of this paper we need introduce two definitions.

The difference $t_1 - t_2$ of two trees is defined iff t_1 and t_2 are compatible and when they are compatible it is defined as the set
$$t_1 - t_2 = \{f^{-1}t_1 \mid f \in B(t_2) \cap \text{dom } (t_1)\}$$
The symetric difference $t_1 \,\Delta\, t_2$ is equal to the union
$$t_1 \,\Delta\, t_2 = (t_1 - t_2) \cup (t_2 - t_1)$$
Obviously $t_1 \,\Delta\, t_2$ is defined iff t_1 and t_2 are compatible and $t_1 \,\Delta\, t_2$ is empty iff the two trees are equal.

Remark II.1

One has for all compatible $t, \bar{t}$
$$t \cup \bar{t} = (t \cap \bar{t}) + \Sigma_i \{f_i f_i^{-1} t \mid f_i \in B(\bar{t}) \cap \text{dom } (t)\}$$
$$+ \Sigma_j \{\bar{f}_j \bar{f}_j^{-1} \bar{t} \mid \bar{f}_j \in B(t) \cap \text{dom } (\bar{t})\}$$

and $t \,\Delta\, \bar{t} = \cup \{ f_i^{-1} t , \bar{f}_j^{-1} \bar{t} \}$.

Let $Q(n,X)$ be the set of trees t such that for some $k, \ell \in \mathbb{N}$, $k + \ell \le n$ one has $t\, X^k \cap X^\ell \ne \varnothing$.

We denote $Q(X)$ the union of all $Q(n,X)$.

Remark that $Q(0,X) = \{a\}$, $Q(1,X) = X$, $Q(2,X) = X^2 \cup XX^{-1}$.

Lemma II.2

There exists a tree t with two different decompositions in elements of X iff there exists two different elements x_0 and $\bar{x}_0$ of X such that

$$x_0 \, \Delta \, \bar{x}_0 \subseteq Q(X)$$

Proof

We can take t with two different decompositions which have two different initial components x_0 and $\bar{x}_0$. Consider $s \in x_0 - \bar{x}_0$ if $x_0 - \bar{x}_0$ is non empty. We can write $s = f^{-1} x_0$ for some $f \in B(\bar{x}_0) \cap \mathrm{dom}(x_0)$.

The conditions $f \in B(\bar{x}_0)$ and $t \in \bar{x}_0 X^{\#}$ imply $f^{-1}t \in X^{\#}$. Call $g_1,\dots,g_k$ the elements of $B(s)$: the trees $(fg_1)^{-1}t,\dots, (fg_h)^{-1}t$ are rooted at the border of x_0 and thus they all belong to $X^{\#}$. For all i one has

$$(fg_i)^{-1}t = g_i^{-1}\, f^{-1}t \quad \text{and we can write} \quad f^{-1}t = \sum_{i=1}^{i=k} g_i\, g_i^{-1}\, f^{-1}t$$

proving that $f^{-1}t \in sX^{\#}$.

The intersection $sX^{\#} \cap X^{\#}$ contains $f^{-1}t$ and is non empty. A similar proof holds for $s \in \bar{x}_0 - x_0$.

Conversely let us write

$$x_0 \cup \bar{x}_0 = (x_0 \cap \bar{x}_0) + \Sigma \{f_i s_i \mid f_i \in B(\bar{x}_0) \cap \mathrm{dom}(x_0)\}$$
$$+ \Sigma \{\bar{f}_j\, \bar{s}_j \mid \bar{f}_j \in B(x_0) \cap \mathrm{dom}(\bar{x}_0)\}$$

and assume that for all $s \in x_0 \Delta \bar{x}_0$ one has $sX^{\#} \cap X^{\#} \ne \varnothing$.

This means that there exists for all $s_i \in x_0 \Delta \bar{x}_0$ a tree $t_i \in X^{\#}$ which can be written $t_i = s_i \Sigma \{g_{i\ell}\, t_{i\ell} \mid g_{i\ell} \in B(s_i)\}$ where $t_{i\ell} \in X^{\#}$.

We can consider now the tree t given by
$$t = (x_0 \cap \bar{x}_0) + \Sigma \{f_i s_i \mid f_i \in B(\bar{x}_0) \cap \mathrm{dom}(x_0)\}$$

$$+ \Sigma \{ \overline{f}_j \, \overline{s}_j \mid \overline{f}_j \in B(\overline{x}_0) \cap \mathrm{dom}(\overline{x}_0) \}$$
$$+ \Sigma \{ f_i \, g_{i\ell} \, t_{i\ell} \mid g_{i\ell} \in B(s_i) \}$$
$$+ \Sigma \{ \overline{f}_j \, \overline{g}_{j\ell} \, \overline{t}_{j\ell} \mid \overline{g}_{j\ell} \in B(\overline{s}_j) \}$$

If we merge terms in two different ways we obtain the two different decompositions

$$t = x_0 + \Sigma \{ f_i \, g_{i\ell} \, t_{i\ell} \} + \Sigma \{ \overline{f}_j \, \overline{t}_j \}$$
$$t = \overline{x}_0 + \Sigma \{ \overline{f}_j \, \overline{g}_{j\ell} \, \overline{t}_{j\ell} \} + \Sigma \{ f_i t_i \} \qquad \square$$

Obviously we can restrict the condition to lemma II.2 to become $x_0 \, \Delta \, \overline{x}_0 \in Q(X) \cap TST(X)$ since for all $x_0, \overline{x}_0 \; x_0 \, \Delta \, \overline{x}_0 \subseteq TST(X)$.

We now give an algorithm to compute $Q(X) \cap PTST(X)$ which we denote $Res(X)$. We denote $Res(n,X)$ the set $Q(n,X) \cap TST(X)$.

Algorithm

We compute first the following sequences of subsets of $TST(X)$
$$\overline{S}_0 = X \cup \{\Omega\}$$
$$S_1 = \cup \{ x \, \Delta \, x' \mid x, x' \in X \}, \quad \overline{S}_1 = S_1 \cup \overline{S}_0$$
$$S_{n+1} = \cup \{ s \, \Delta \, x \mid s \in S_n \setminus \overline{S}_{n-1} \}, \quad \overline{S}_{n+1} = S_{n+1} \cup \overline{S}_n$$

It is clear that for all $n \; S_n \subseteq TST(X)$ and thus since $PTST(X)$ is finite the sequence $\overline{S}_n$ is stationary. For some n_0 we have $\overline{S}_{n_0} = \underset{n \geq 0}{\cup} \overline{S}_n$.

In a second phase we compute the following sequences of subsets of.
$$\overline{S} = \underset{n \geq 0}{\cup} \overline{S}_n.$$
$$\overline{T}_0 = X \cup \{\Omega\}$$
$$T_1 = \{ s \in \overline{S} \mid \exists \, x \in X \setminus \{s\} : s \, \Delta \, x \subseteq \overline{T}_0 \}$$
$$\overline{T}_1 = T_1 \cup \overline{T}_0$$
$$T_{n+1} = \{ s \in \overline{S} \setminus \cup \{ T_m \mid m < n \} \mid \exists \, x \in X \setminus \{s\} \; s \, \Delta \, x \subseteq \overline{T}_n \}$$
$$\overline{T}_{n+1} = T_{n+1} \cup \overline{T}_n$$

Clearly the increasing sequence $\overline{T}_n$ is also stationary ant there exists an m_0 such that $\overline{T}_{m_0} = \underset{m \geq 0}{\cup} \overline{T}_m = \overline{T}$.

This algorithm computes all the residues.

Property II.1 $Res(X) = \overline{T}$.

Let us prove first $\overline{T} \subseteq Res(X)$. We prove it by induction. $\overline{T}_0 \subseteq Res(1,X)$. Assume $\overline{T}_n \subseteq Res(X)$.

Consider $s \in T_{n+1}$. There exists an $x \in X \setminus \{s\}$ such that $s \Delta x \subseteq \overline{T}_n$ and by induction $s \Delta x \in Res(X)$.

By the same construction as lemme II.2 we can exhibit a tree t in $sX^{\#} \cap X^{\#}$ thus proving that $s \in Res(X)$.

Conversely consider $s \in Res(X)$ ie a terminal subtree s fo some $x \in X$ such that $sX^{\#} \cap X^{\#} \neq \varnothing$.

We need prove that $s \in \overline{T}_n$ for some n.

We consider $s \in Res(0,X) = \{\Omega\}$. Clearly $s \in \overline{T}_0$.

Assume $Res(n,X) \subseteq \overline{T}_n$ and consider $s \in Res(n+1,X)$.

The tree s can be in X^{n+1} in which case we can write
$s = x_0 + \Sigma\{f_i t_i \mid f_i \in B(x_0)\}$ with for all i

$$t_i \in \cup \{ X^{\ell} \mid \ell \leq n\}$$
$$\cup \{ X^{\ell} \mid \ell \leq n\}$$

Then $s \Delta x_0 \subseteq \underset{\ell \leq n}{\cup} X^{\ell}$ which implies $s \Delta x_0 \subseteq Res(n,X)$ and by induction $s \Delta x_0 \subseteq \overline{T}_n$. By construction then $s \in T_{n+1}$.

Consider now $s \in Res(n+1,X) \setminus X^{n+1}$. Certainly for some $k,l \geq 1$ one has $sX^k \cap X^{\ell} \neq O$. We can take a tree t in $sX^k \cap x^{\ell-1}$ for some $x \neq s$ and this tree can be written as

- $t = s + \Sigma\{f_i t_i \mid f_i \in B(s)\}$ where for all i $t_i \in \cup \{X^m \mid m \leq k)$
- $t = x + \Sigma\{g_j \mathsf{T}_j \mid g_j \in B(x)\}$ where for all j $t_j \in \cup \{X^m \mid m \leq \ell)$

Let us now consider $s_i = f_i^{-1} s$ an element of s-x. We have $f_i \in B(x) \cap dom(s)$ and we can write

$$t_i = s_i + \Sigma\{gh\, \mathsf{T}_h \mid gh \in B(x) \cap fD^*\}$$

Since $t_i \in \cup \{X^j \mid j \leq k\}$ and $\mathsf{T}_h \in \cup \{X^j \mid j \leq \ell)$ we have

$s_i X^{\ell'} \cap X^{k'} \neq \varnothing$ for some k', ℓ' with $k' \leq k$ and $\ell' \leq \ell$. If we consider $\bar{s}_i \in x\text{-}s$ we prove in a similar way that

$s_i X^{k'} \cap X^{\ell'} \neq \varnothing$ for some k', ℓ' with $k' \leq k$ and $\ell' \leq \ell$. Thus $s \; \Delta \; x \subseteq \cup \operatorname{Res} \mid m \leq n\}$ which is by induction contained in T_n. By construction then $s \in T_{n+1}$. $\square$

We have established that

Property II.2

X is not a code iff there exists two different x, x' in X such that $x \; \Delta \; x' \subseteq T$.
Since $X \subseteq \bar{S}$ we also have

Property II.3

X is a code iff for all $n \geq 1$ $T_n \cap X \neq \varnothing$.

And a few remarks can be made

Remark II.2

The set S_1 is empty iff for all $x, x' \in X$ $x \neq x'$ implies that $x \; \Delta \; x'$ is undefined.

Obviously $S_1 = \varnothing$ implies $S_n = \varnothing$ for all $n > 1$ and thus $\bar{S} = X \cup \{\Omega\}$. Also $T_n = \varnothing$ for all $n > 1$ and thus X is a code.

The subsets X of $A^{\#}$ such that $\forall x, x' \in X$ $x \neq x' \Rightarrow x \; \Delta \; x'$ is undefined are called prefix sets of trees and we can state the following.

Property II.4 Every prefix set of trees is a tree code.

On a one letter alphabet a prefix set can contain at most one element since any two treeson a one letter alphabet are compatible.

Remark II.3

If no tree in X is a proper terminal subtree of an other element of X then the set X is called suffix. And we can easily prove the following

Property II.5   Every suffix set of trees is a tree code

If X is suffix the set T_1 is empty for all the elements of $s \ \Delta \ x$ where $s \neq x$ are non empty terminal subtrees of s or x, and since s is itself a terminal subtree of some tree in X we have $s \ \Delta \ x \subseteq PTST(X) \setminus \{\Omega\}$. If X is suffix $s \ \Delta \ x \subseteq T_0 = X \cup \{\Omega\}$ is impossible whence $T_1 = \varnothing$ and for all $n \geq 1$ the set T_n is also empty. $\square$

Remark II.4

If we apply our algorithm to set $\mathfrak{U}$ of trees with right depth one corresponding to some finite set $\mathfrak{U}$ of words in A^* we obtain exactly the well-known algorithm of Sardinas and Paterson to cheeck whether a finite set of words is a (word) code [4].

Exemple

$$x_1 = \qquad = \ <\varepsilon,1,12,2>$$

$$x_2 = \qquad = \ <\varepsilon,1,\ 222,\ 222,\ 2221,\ 2222>$$

$$x_3 = \qquad <\varepsilon,\ 2,\ 21,\ 22,\ 222>$$

We compute the symetric differences

$$x_1\ \Delta\ x_2 = \{ \cdots \} = \{x_4\ ,x_6\ \}$$

$$x_1\ \Delta\ x_3 = \{ \cdots \} = \{x_4\ ,x_5\}$$

$$x_1\ \Delta\ x_3 = \{ \cdots \} = \{x_4\ ,x_5\}$$

Denoting $x_4 = \ \cdot \qquad x_5 = \qquad x_6 =$

We compute all the symetric differences

$$x_4\ \Delta\ x_1 = \{x_4,\ x_5\}$$

$$x_4\ \Delta\ x_2 = \{x_4\ ,\ x_7\} \ \text{with}\ x_7 =$$

$x_4 \Delta x_3 = \{x_8\}$ with $x_8 = $

$x_5 \Delta x_1 = \{x_5\}$
$x_5 \Delta x_2 = \{x_4, x_6\}$
$x_5 \Delta x_3 = \{x_4, x_5\}$
$x_6 \Delta x_1 = \{x_4, x_5\}$
$x_6 \Delta x_2 = \{x_4, x_1\}$
$x_6 \Delta x_3 = \{x_4\}$
$x_7 \Delta x_1 = \{x_1, x_5\}$
$x_7 \Delta x_2 = \{x_4, x_5\}$
$x_7 \Delta x_3 = \{x_4\}$
$x_8 \Delta x_1 = \{x_4\}$
$x_8 \Delta x_2 = \{x_1\}$
$x_8 \Delta x_3 = \{x_4\}$

Thus $\overline{S} = x_1,...,x_8\}$

Then $T_1 = \{x_8\}$ since $x_8 \Delta x_2 = \{x_1\} \subseteq X$

$T_1 = X \cup \{x_8\}$

$T_2 = \{x_4\}$ since $x_4 \Delta x_3 = \{x_8\} \subseteq T_1$

$T_2 = X \cup \{x_4, x_8\}$

$T_3 = \{x_6, x_7\}$ since $x_6 \Delta x_3 = \{x_4\} \subseteq T_2$

and $x_7 \Delta x_3 = \{x_4\} \subseteq T_2$

$T_3 = X \cup \{x_4, x_6, x_7, x_8\}$

$T_4 = \{x_1, x_2, x_5\}$ we knows at that point that X is not a code

$T_4 = X \cup \{x_4, x_5, x_6, x_7, x_8\}$

$T_5 = \{x_3\}$

$T = T_5 = \overline{S}$

What is interesting is to build from the information given by the algorithm a tree with two different decompositions

x_8 is a residue of order 1

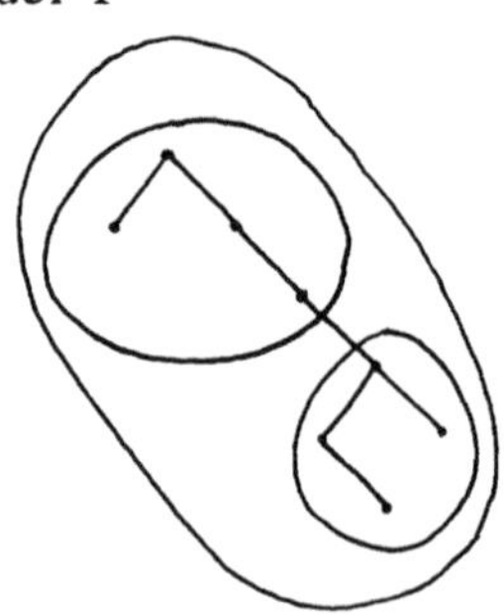

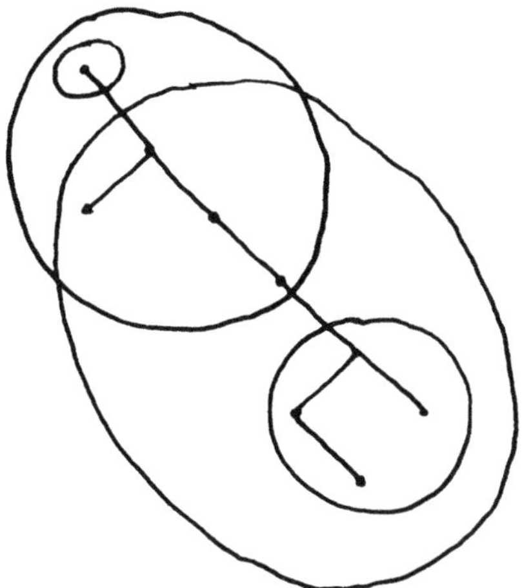

x_4 is a residue of order 2

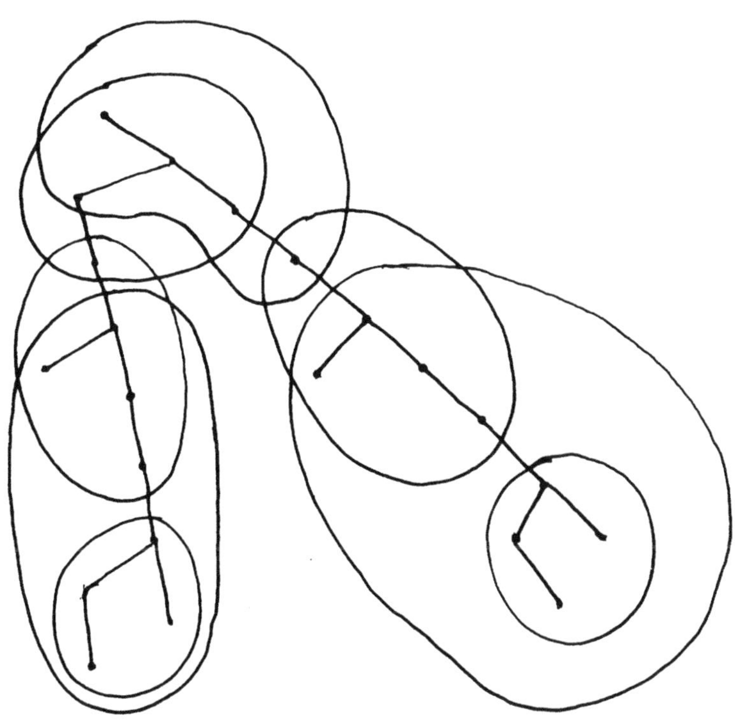

x_6 is a residue of order 3

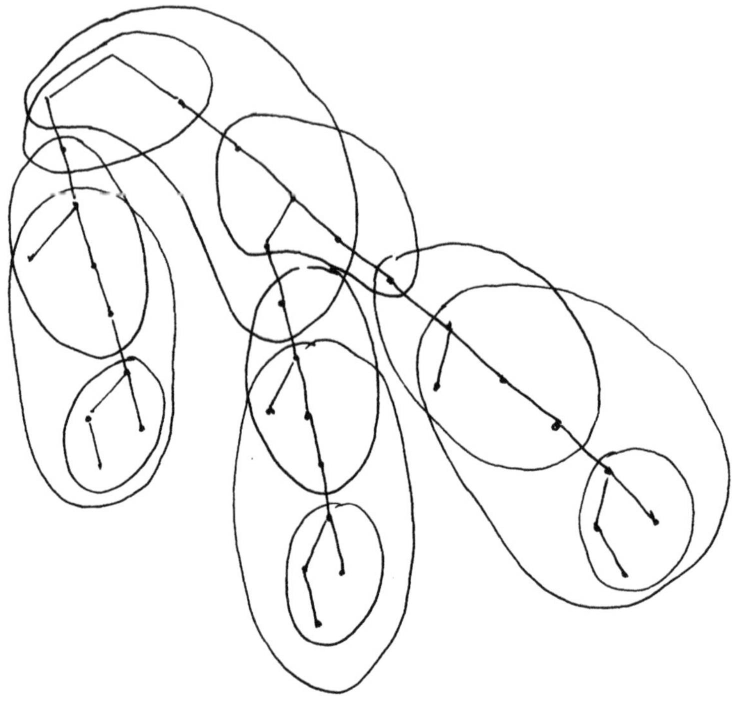

and eventually x_1 is a residue of order 4

This gives us the smallest tree with two different decompositions in elements of X

$$x_1 + 122\, x_2 + 22\, x_3 + 22212\, x_3 + 222122222\, x_1 + 222222\, x_2 =$$
$$x_2 + 12\, x_3 + 122222\, x_1 + 222122\, x_2 + 22222\, x_3 + 222222222\, x_1 .$$

Conclusion

The possibility of deciding whether a finite set of finite trees is a code make tree codes handable. It is tempting to entend to tree codes the study of standard classes of word codes such as prefix, suffix and maximal codes. First steps towards this goal appear in a paper by P. Aigrain and M. Nivat in the present volume. The algebraic method which consists in relating properties of (word) codes to properties of their syntactic monoïd may be also extended if one has a proper definition of syntactic monoïds of sets of trees. Such a definition is proposed in [4].

Bibliography

[1]   J. Berstel et D. Perrin, Theory of Codes, Academic Press (1985).

[2]   P. Franchi-Zanettacci, An extension to trees of the Sardinas and Patterson algorithm, Inf. Proc. Letters 14 (1982), 168-173.

[3]   M. Nivat, Eléments de la théorie générale des codes in E. Caianello (ed.), Automata theory, Academic Press (1966).

[4]   M. Nivat and A. Podelski, Tree monoïds and recognizability of sets of finite trees, in H. Aït-Kaci and M. Nivat (eds), Resolution of equations in algebraic structures, Academic Press 1989.

[5]   A. Sardinas and C. Patterson, A necessary and sufficient condition for the unique decomposition of coded messages, IRE Intern. Conv. Rec. 8 (1953), 104-108.

[6]   J. Thatcher, Tree automata, an informal survey in A. Aho (ed.), Currents in the theory of computing, Prentice Hall (1973).

[7]   W. Thomas, Logical aspects in the theory of tree languages, in B. Courcelle (ed.), Proceedings ninth colloquium on Trees in Algebra and Programming, Cambridge University Press, 1984.

Tree Automata and Languages
M. Nivat and A. Podelski (editors)
© 1992 Elsevier Science Publishers B.V. All rights reserved.

Suffix, prefix and maximal tree codes

Philippe Aigrain[a] and Maurice Nivat[b]

[a]Institut de Recherche en Informatique de Toulouse, Université Paul Sabatier, 118, route de Narbonne, 31062 Toulouse Cedex, France[§]

[b]L.I.T.P., Université Paris 7, 2, place Jussieu, 75251 Paris Cedex 05, France

Abstract

We give a characterization of suffix maximal tree codes, and show that every suffix maximal set is a maximal tree code. We give a sufficient condition for a prefix maximal set of trees to be a maximal tree code. We show that every maximal tree code is complete and that every suffix maximal tree code is self-complete. Various other properties of prefix and suffix maximal tree codes are given.

I. INTRODUCTION AND DEFINITIONS

We use definitions and notations from [Nivat90] most of them first introduced in [Nivat86] and [Nivat86b].

A *graded alphabet* Σ is a set of letters endowed with an arity function: $\alpha: \Sigma \rightarrow N$

A subset E of the free monoid N^* generated by N is the *domain of a tree* <u>iff</u> it is closed by left factor, i.e. it satisfies $\forall f,g \in N^*$, $fg \in E \Rightarrow f \in E$
We also write $E = LF(E)$.

The elements of E are called *nodes*. For all $f \in N^*$ and $i \in N$ such that $fi \in E$, we say that the node f is the *parent* of the node fi and that fi is the *ith child* of the node f. The set of children of a node f is thus $fN \cap E$. NB: we consider trees in a geometrical meaning: a node with a single first child and a node with a single nth child are different objects.

A *tree* t is the pair (E, σ), formed by the domain E of t and the labelling σ which maps E into the graded alphabet Σ of labels in such a way that for all $f \in E$ and for all i such that $fi \in E$:

$$i \leq \alpha(\sigma(f))$$

Such a labelling is said to be *arity-compatible*. A tree such that:

$$\forall f \in E, \; card(fN \cap E) = \alpha(\sigma(f)) \text{ or } card(fN \cap E) = 0$$

is said to be *complete*.

[§]Research conducted under support from the French Ministère de la Culture, de la Communication, des Grands Travaux et du Bicentenaire.

                    P. Aigrain and M. Nivat

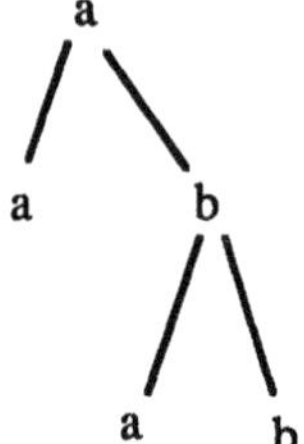

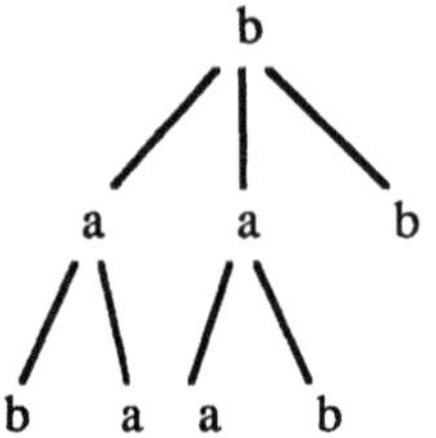

A tree on the graded alphabet
$\{a\, ,\, b\}$. Arity of a is 2 and arity
of b is 3.

A complete tree on the
same alphabet

Figure 1: Examples of trees

A node with no children ($card(fN \cap E) = 0$) is called a leaf. If $t = (E\, ,\, \sigma)$ the domain E will also be denoted as $dom(t)$, and the labelling $\sigma(f)$ as $t(f)$.

We write as usual, for all $f \in N^*$ and $E \subset N^*$:

$$fE = \{\ fg\ /\ g \in E\ \}$$
$$f^{-1}E = \{\ g\ /\ fg \in E\ \}$$

If f is a node of a tree $t = (E\, ,\, \sigma)$ the tree $f^{-1}t$ is defined as the tree whose domain is $f^{-1}E$ and whose labelling $f^{-1}\sigma$ is given by:

$$f^{-1}\sigma(\ g\ ) = \sigma(\ fg)$$

The tree $f^{-1}t$ has a clear geometrical interpretation: it is the terminal subtree of t rooted in f. If f is not in E we write $f^{-1}t = \Omega$ where Ω is the empty tree whose domain is $\varnothing$.

Let us call *a bush* on the graded alphabet Σ any pair $(E\, ,\, \sigma)$ where E is a (not necessarily left-factor closed) subset of N^* and σ is an arity-compatible labelling of E. For a tree $t = (E\, ,\, \sigma)$ and $f \in N^*$, the bush ft is defined by its domain fE and the labelling $f\sigma$ given by :

$$f\sigma(\ g\ ) = \sigma(f^{-1}\ g\ )$$

Two bushes $(E\, ,\, \sigma)$ and $(E'\, ,\, \sigma')$ are said to be compatible <u>iff</u> the restrictions of the two labellings σ and σ' to $E \cap E'$ are identical. By generalization two trees are said to be compatible if they are compatible as bushes. We can define the *compatible union* of two bushes:

$$(E\, ,\, \sigma)\ \oplus\ (E'\, ,\, \sigma') = \Omega\ \ if\, (E\, ,\, \sigma)\ and\ (E'\, ,\, \sigma')\ are\ not\ compatible$$

Otherwise the domain of $(E\, ,\, \sigma)\ \oplus\ (E'\, ,\, \sigma')$ is the union $E \cup E'$ and the labelling is given by:

$\sigma''(f) = \sigma(f)$ if $f \in E$
$\sigma''(f) = \sigma'(f)$ if $f \in E'$.

Two bushes are disjoint <u>iff</u> the intersection of their domain is empty, and obviously two disjoint bushes are compatible. In the case where (E, σ) and (E', σ') are disjoint, the compatible union is denoted:

$$(E, \sigma) + (E', \sigma')$$

The difference $t - t'$ of two trees is defined <u>iff</u> they are compatible as:

$$t - t' = \{ f^{-1} t \mid f \in B(t') \cap dom(t) \}$$

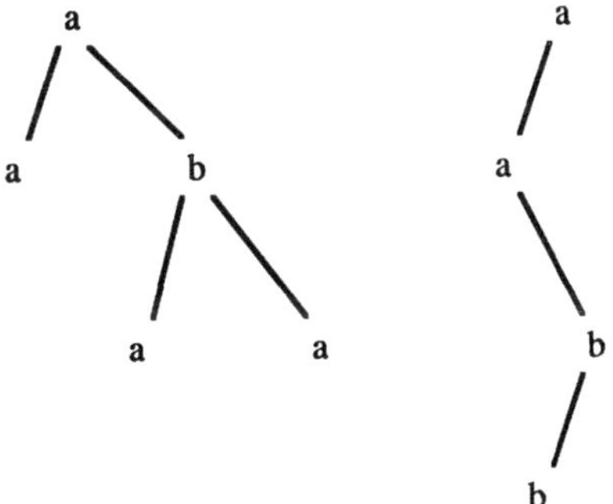

Two trees on the $\{a, b\}$ alphabet. Both letters have arity 2.

The symetric difference of the two trees is the set of two trees:

Figure 2: Symetric difference of two trees

The symetric difference $t \Delta t'$ of two compatible trees t and t' is the union:

$$t \Delta t' = (t - t') \cup (t' - t)$$

A restriction of a tree $t = (E, \sigma)$ to a subset E' of N^* is a bush denoted as t / E' and defined by:

- its domain $dom(t / E') = E \cap E'$
- its labelling $\sigma' = \sigma / E \cap E'$

The set of all trees on an alphabet Σ will be denoted $\Sigma^{\#}$. Classically, we will denote:
$$\Sigma^+ = \Sigma^{\#} - \Omega$$

The trees with a single node are called *punctual trees*. Trees on alphabets of letters of arity *1* are words in the usual theory of codes meaning. We will speak on *non-word alphabets* and *non-word trees* when at least one letter in the alphabet is of arity greater than *1*.
We also denote:

$$\forall f \in E , t \in \Sigma^{\#}, \quad t \setminus f = t \mid (E \setminus fN^*)$$

With this notation we can write the fundamental identity:

$$\forall f \in E, t \in \Sigma^{\#}, \quad t = t \setminus f + f f^{-1} t$$

which expresses the fact that a tree can be identified to the union of the bush rooted in any node *f* of its domain and of the restricted tree obtained by deleting this bush.
Let us say that t' is an initial subtree of t and denote it by $t' \leq t$ iff:

$$dom(t) \ C \ dom(t') \quad \text{and} \quad t \mid dom(t') = t'$$

The *border B(t)* of a tree $t = (E, \sigma)$ is the set of words of N^*:

$$\{ fn \mid f \in E, n \leq \alpha(last(f)), fn \notin E \}$$

where *last(f)* is the last letter of a word *f*. By definition $B(\Omega) = \{ \varepsilon \}$.
We have the property [Nivat86]:

I.1. Property: For all $t, t' \in \Sigma^{\#}$ such that $t' \leq t$: $\quad t = t' + \Sigma \ \{ f f^{-1} \ t \mid f \in B(t) \}$

This property expresses the fact that a tree is the bush union of any of its initial subtrees and of all bushes rooted on the border of this initial subtree.
The *height h(t)* of a tree $t = (E, \sigma)$ is defined as:

$$h(t) = \max_{f \in E} \ (length(f)) \ - 1$$

If X is a family of trees of $\Sigma^{\#}$, and t is a tree of $\Sigma^{\#}$, a *decomposition* of t on X is a set of couples (f_i , x_i) such that:

$$\forall i, x_i \in X$$
$$t = \Sigma_i f_i x_i \quad \text{(the use of } \Sigma \text{ indicates a disjoint union)}$$

A family X of trees of $\Sigma^{\#}$ is a tree code iff every tree in $\Sigma^{\#}$ admits at most one decomposition on X. The set of trees of $\Sigma^{\#}$ which admit a decomposition on X will be noted $X^{\#}$. A justification for this notation is the fact that $\Sigma^{\#}$ is the set of trees coverable by the punctual trees on Σ. If all letters of Σ have arity *1*, $\Sigma^{\#}$ is the monoid Σ^* and $X^{\#}$ is the free submonoid X^* generated by X. We will note:

$$\Sigma^* \setminus X = \{ t \mid t \in \Sigma^* \text{ and } t \notin X \}$$

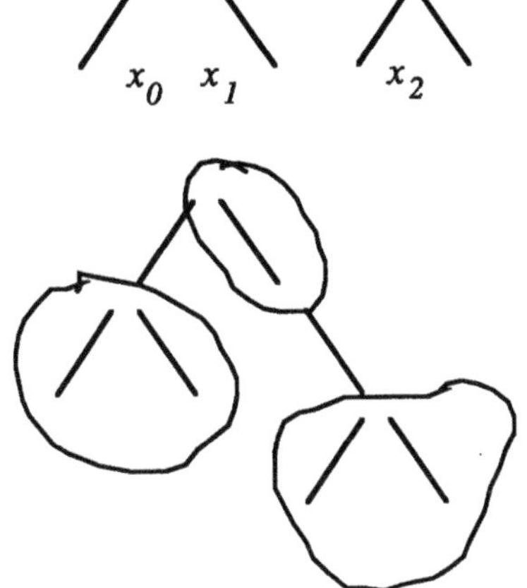

A family X of trees on a arity 2 one-letter alphabet (the labelling can be omitted since there is only one letter in the alphabet)

A tree t on the same alphabet, and a decomposition of this tree on X:

$$t = x_1 + (1\,x_2) + (22\,x_2)$$

Figure 3: Example of decomposition on a family of trees.

A set X of trees of $\Sigma^{\#}$ is a *maximal code* <u>iff</u> it is a code and $\forall t \in \Sigma^{\#} \setminus X, X \cup \{t\}$ is not a code.

If X and Y are two sets of trees on the same alphabet, we note:

$$X\ Y = \{\,t\,/\,t = x + \sum_{i \in I} f_i\,y_i,\ x \in X,\ card(I) \geq 1,\ \forall i,\ y_i \in Y,\ f_i \in B(x)\,\}$$

This operation is not generally associative, but two important remarks can be made:

- If $\Omega \in Y,\ (X\ Y)\ Z = X\ (Y\ Z)$
- $\forall\ X,Y,Z,\ X\ (Y\ Z)\ C\ (X\ Y)\ Z$

If X, Y, Z, are sets of trees on the same alphabet, we note:

$$X^{-1}\ Y = \{\,t\,/\,(X\,\{t\}) \cap Y \neq \phi\,\}$$
$$X\ Y^{-1} = \{\,t\,/\,(\{t\}\,Y) \cap X \neq \phi\,\}$$
$$X^{-1}\ Y\ Z^{-1} = \{\,t\,/\,((X\,\{t\})\,Z) \cap Y \neq \phi\,\}$$

$\Sigma^{\#-1}\ X$ (resp. $\Sigma^{+-1}\ X$) is the set of *terminal subtrees of* X (resp. proper terminal subtrees of X). $X\ \Sigma^{\#-1}$ (resp. $X\ \Sigma^{+-1}$) is the set of *initial subtrees of* X (resp. proper initial subtrees of X). For simplicity, we will denote $\Sigma^{\#-1}\ \{x\}$ as $TST(x)$, $\Sigma^{+-1}\ \{x\}$ as $PTST(x)$, $\{x\}\ \Sigma^{\#-1}$ as $IST(x)$ and $\{x\}\ \Sigma^{+-1}$ as $PIST(x)$.

A set X of trees of $\Sigma^{\#}$ will be said to be suffix <u>iff</u> no tree of X is a subtree of another tree of X. If all letters of Σ have arity 1, this definition is the classical definition of suffixity for words. A set X of trees of $\Sigma^{\#}$ will be said to be prefix <u>iff</u> $\forall x, y \in X,\ x \neq y \Rightarrow x$ and y are incompatible.

Once again, if all letters of Σ have arity *1*, we find back the classical definition of prefixity for words. As one could anticipate, when considering trees, there is a breaking of symmetry between prefixity and suffixity. This will be most notable when discussing maximal codes. [Nivat86] and [Nivat86b] show that any suffix (resp. prefix) set of trees is a tree code, and give two algorithms for codicity of any set of trees, which generalize the Sardinas-Patterson algorithms for words. The notion of a *suffix residue*, used in the "suffix" algorithm, is of special interest to our subject:

If X is a set of trees of $\Sigma^{\#}$ and $x \in X$:

$$RES_x(X) = PIST(x) \cap X^{\#} \, X^{\#-1}$$

By generalization $RES(X) = \underset{x \in X}{\cup} \; RES_x(X).$

A suffix (resp. prefix) set X of $\Sigma^{\#}$ is a maximal suffix (resp. prefix) set <u>iff</u>:

$$\forall t \in \Sigma^{\#} \setminus X , \; X \cup \{\, t \,\} \text{ is not suffix (resp. prefix)}$$

A suffix set X of $\Sigma^{\#}$ is *full suffix set* <u>iff</u>:

$$\forall t \in \Sigma^{\#} \setminus TST(X) , \; PTST(t) \cap X \neq \varnothing$$

We shall use the notations:

TC	The family of tree codes
MTC	The family of maximal tree codes
SC	The family of suffix codes
MSS	The family of maximal suffix sets
FSS	The family of full suffix sets
MSC	The family of maximal codes which are suffix (MSC = MTC $\cap$ SC)
PC	The family of prefix codes
MPS	The family of maximal prefix sets
MPC	The family of maximal codes which are prefix (MPC = MTC $\cap$ PC)

The notations FTC, FMTC, FSC, FMSS, FFSS, FMSC, FPC, FMPS and FMPC stand for the intersection of the above family with the family of finite subsets of $\Sigma^{\#}$.

II. COMPLETENESS AND MAXIMALITY

We say that a tree t' is a *factor* of a tree t <u>iff</u> t' is an initial subtree of a terminal subtree of t. In other words t' is a factor of t <u>iff</u>:

$$t \in \Sigma^{\#} \, t' \, \Sigma^{\#} \text{ or equivalently } t' \in \Sigma^{\#-1} \, t \, \Sigma^{\#-1}$$

The set of factors of t is denoted $F(t)$ and if X is a subset of $\Sigma^{\#}$ we denote:

$$F(X) = \cup \; \{\, F(t) \mid t \in X \,\}$$

If X is a subset of $\Sigma^{\#}$, a tree t is said to be completable in X <u>iff</u>:

$t \in F(X)$ or equivalently $t \in \Sigma^{\#-1} X \Sigma^{\#-1}$

The set X is *dense* in $\Sigma^{\#}$ <u>iff</u> $F(X) = \Sigma^{\#}$. It is *thin* in $\Sigma^{\#}$ <u>iff</u> it is not dense. The set X is *complete* in $\Sigma^{\#}$ <u>iff</u> $X^{\#}$ is dense in $\Sigma^{\#}$ i.e. <u>iff</u>:

$$\Sigma^{\#} = \Sigma^{\#-1} X^{\#} \Sigma^{\#-1}$$

A complete tree code is a tree code which is a complete subset of $\Sigma^{\#}$. Fundamental results of the theory of word codes state that a maximal word code is complete and that a complete thin code is maximal. We will retrieve only the first of these results for tree codes.

II.1. Property: A finite complete tree code is not necessarily maximal

Proof: Let us consider the case of a code X formed by a single tree t on a one-letter alphabet $\Sigma = \{ a \}$. For all n, we define the trees (of $X^{\#}$):

$t_0 = t$
For $n > 0$, $t_n = t_{n-1} + \Sigma \{ f t \mid f \in B(t_{n-1}) \}$

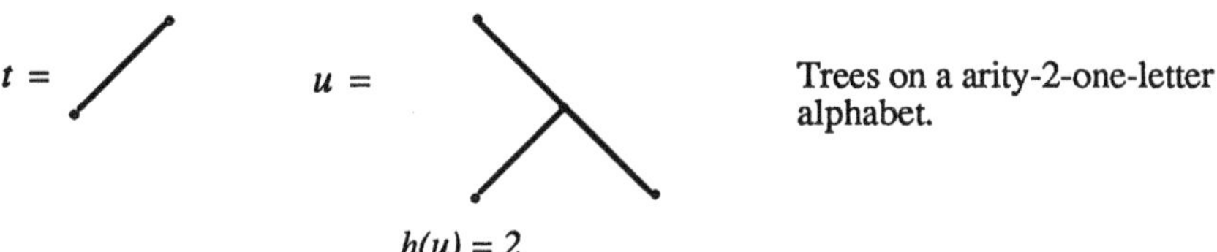

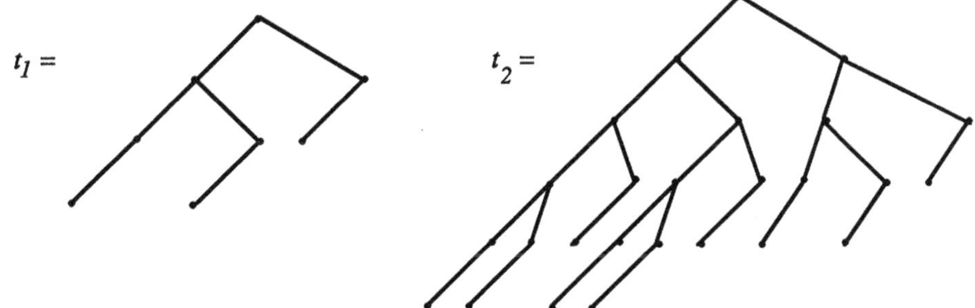

u is an initial subtree of $a^{(2)}$ which is an initial subtree of t_2

Figure 4: Example of construction of t_k

Let u be any tree of $\Sigma^{\#}$. We can show that for all n greater that some k, u is an initial subtree of t_n. Let us denote $a^{(k)}$ the tree whose domain is:

$$dom(a^{(k)}) = \{ f \mid f \in \{ 1, .. , \alpha(a) \}^{*} , l(f) \leq k \}$$

and for which $\forall f \subset dom(a^{(k)})$, $\sigma(f) = a$. If $k = h(u)$, we have:

$$u \ C \ a^{(k)} \ C \ t_k$$

Since u is an inital subtree of t_k, property I.1. gives us an obvious completion of u.

$$[\ ]$$

II.2. Theorem: A maximal tree code is complete

Proof: Let us first remark that the theorem is trivial for one-letter alphabets: we have seen that a single tree code on a one-letter alphabet is complete. We will know take care of the case of alphabets with 2 or more letters. Let us say that a tree t is *self-overlapping* <u>iff</u>:

$$\exists f \in dom(t) \setminus \varepsilon, \text{ such that the union } t \oplus ft \text{ is defined } (t \text{ and } ft \text{ are compatible})$$

We know that $t \oplus ft$ is defined <u>iff</u> t and ft coincide on $dom(t)$ and $f\,dom(t)$. This is equivalent to say that:

$$\forall g \in dom(t) \cap f\,dom(t), \ \sigma(g) = \sigma(f^{-1}\,g)$$

A similar notion for word codes is that of an *unbordered* word [Berstel-Perrin85].
A word $u \in A^{+}$ is *unbordered* <u>iff:</u>

$$FG(u) \cap FD(u) = \{ \varepsilon \}$$

Let us now prove a related lemma:

II.3. Lemma: Let $\Sigma = \{ a , b \}$. Let us consider the trees $a^{(n)}$ and $b^{(n)}$ defined as in II.1.. Then, for all trees $t \in \beta \Sigma^{\#}$ (where $\beta = a$ or $\beta = b$) the tree:

$$t = t + \Sigma \{ g \underline{\beta}^{(n)} \mid g \in B(t) \}$$

is non self-overlapping ($\underline{\beta} = a$ if $\beta = b$ and $\underline{\beta} = b$ if $\beta = a$).

Proof of the lemma: Assume that for some $f \in dom(\underline{t})$, the union $\underline{t} \oplus f\underline{t}$ is defined. Clearly f cannot be of length greater than $n+1$. For then surely $f \in g \underline{\beta}^{(n)}$ for some $g \in B(t)$ and this implies that the labelling of f in $\underline{t}$ is $\underline{\beta}$ when:

$$\underline{t}(f) = \underline{t}(\varepsilon) = \underline{t}(\varepsilon) = \beta.$$

Thus, one can assume $f \in dom(t) \setminus \varepsilon$. Certainly for some integer $p \geq 1$:

$f^p \in dom(t)$ and $f^{p+1} \in dom(\underline{t}) \setminus dom(t)$

Since $\underline{t} \oplus f\underline{t}$ is defined, $\underline{t} \oplus f\underline{t} \oplus f^2\underline{t}$ is also defined for $\underline{t}$ and $f\underline{t}$ coincide on $dom(\underline{t}) \cap f\, dom(\underline{t})$ and thus $f\underline{t}$ and $f^2\underline{t}$ coincide on $dom(f\underline{t}) \cap f^2\, dom(\underline{t})$. Clearly:

$(dom(f\underline{t}) \cap f^2\, dom(\underline{t}))\ \subset\ (dom(\underline{t}) \cap f\, dom(\underline{t}))$

and thus $f^2\underline{t}$ coïncides with $\underline{t}$ on $dom(\underline{t}) \cap f^2\, dom(\underline{t})$ leading to the fact that:

$\underline{t}(f^2) = (f^2\underline{t})(\varepsilon) = \underline{t}(\varepsilon).$

By induction we have $\underline{t}(f^p) = \underline{t}(\varepsilon)$ for all $p \geq 1$. But by construction $\underline{t}(f^{p+1}) = \underline{b}$ if $\underline{t}(\varepsilon) = b$ and we have a contradiction.

$$[\ ]$$

Proof of the theorem: If X is not complete, there exists a tree t which is not completable in $X^{\#}$. By Lemma II.3. we can assume that t is non self-overlapping: indeed if t is self-overlapping we replace t by $\underline{t}$ which is not completable in $X^{\#}$ if t is not. If X is a tree code, then $X \cup \{t\}$ is also a tree code. Let us assume that $X \cup \{t\}$ is not a tree code. Then, there exists trees with two different decompositions in elements of $X \cup \{t\}$. If s is such a tree, at least one of the two decompositions contains an element equal to t and we can find a tree s with two different decompositions which has the smallest possible size. Let us write these two decompositions:

$$s = \Sigma\ \{f\, \beta_f\ /\ f \in M1\} = \Sigma\ \{g\, \gamma_f\ /\ g \in M2\}$$

Let us consider then $f_0 \in M$ such that $\beta_{f_0} = t$ and let us assume $f_0 \neq \varepsilon$. The node f_0 cannot be in $M2$ for $f \in M2$ implies the existence of two decompositions in elements of $X \cup \{t\}$ for $s \setminus_{f_0} N^*$ and $f_0^{-1} s$ namely:

$$s \setminus_{f_0} N^* = \beta_{f_0} + \Sigma\ \{f\, \beta_f\ /\ f \in M1 \setminus_{f_0} N^*\} = \gamma_{f_0} + \Sigma\ \{g\, \gamma_g\ /\ g \in M2 \setminus_{f_0} N^*\}$$
$$f_0^{-1} s = \Sigma\ \{f\, \beta_f\ /\ f \in M1 \cap f_0 N^*\} = \Sigma\ \{g\, \gamma_g\ /\ g \in M2 \cap f_0 N^*\}$$

Then either $s \setminus_{f_0} N^*$ or $f_0^{-1} s$ have two different decompositions and a size strictly smaller than the size of s. Since f_0 cannot be in $M2$, we have:

$$f_0 \in g_0(\, dom(\gamma_{g_0}) \setminus \varepsilon\,)$$

and since t is non self-overlapping, we are sure that $\gamma_{g_0} \neq t$.

Thus, if we look at $g_0^{-1} s$, we are in the following situation:

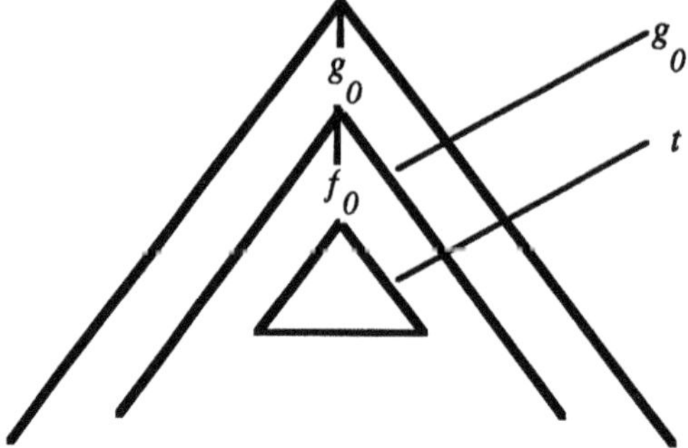

None of the γ_g with $g \in M2 \cap f_0\, dom(t)$ can be equal to t since t is non-self-overlapping. Thus, if we delete in the decomposition of $g_0^{-1}\, s$ all the components γ_g which are equal to t or lie below a component equal to t we get a tree which is in $X^\#$ and contains t as a factor.

The case $f_0 = \varepsilon$ is treated in the same way. If β_ε is also equal to t, then one of the trees $h^{-1}\, s$ where $h \in B(t)$ has two different decompositions in elements of $X \cup \{\, t\, \}$ and a size which is strictly smaller than the size of s. Thus $\beta_\varepsilon = x$ is different from t. We can then also delete in the second decomposition of s all the components equal to t and get a tree in $X^\#$ which contains t as a factor.

$$[\ ]$$

A tree code X of $\Sigma^\#$ is said to be:

> *self-complete* <u>iff</u>　　　　　　　$\Sigma^\# = X^{\#-1}\, X^\#\, X^{\#-1}$
>
> *prefix-self-complete* <u>iff</u>　　　$\Sigma^\# = X^{\#-1}\, X^\#$
>
> *suffix-self-complete* <u>iff</u>　　　$\Sigma^\# = X^\#\, X^{\#-1}$

The difference between self-completeness and completeness is that if X is a self-complete code of $\Sigma^\#$, every tree t of $\Sigma^\#$ can be completed in $X^\#$ not only by trees of $\Sigma^\#$ but by trees of $X^\#$. This is of course a very strong property. As we will see, self-completeness is the key notion for the study of maximal tree codes. A first and simple result is that:

II.4. Theorem: A self-complete code is maximal

Proof: This is in fact trivial, since if X is a self-complete code of $\Sigma^\#$, and t is a tree of $\Sigma^\# \setminus X$, by definition of self-complete:

$$(X^\#\, t\, X^\#) \cap X^\# \neq \varnothing$$

which means that there is a tree with a double decomposition on $X \cap \{\, t\, \}$. The two factorizations are distinct since t is a component of one of them and not of the other.

For word codes a fundamental result states that any thin complete code is maximal. As we have seen, this is not true for non-word tree codes. For word codes self-completeness is related to synchronization: a thin prefix self-complete code is synchronizing [Schutzenberger65]. There are non self-complete maximal word codes, for instance bi-prefix maximal codes, and other codes built from them [Perrin77]. We will see that there are no non-trivial (non-reduced to sets of punctual trees) bi-prefix maximal non-word tree codes.

III. MAXIMAL SUFFIX SETS AND SUFFIX MAXIMAL CODES

III.1. Definition: If X is a set of trees of $\Sigma^{\#}$ the *suffix expansion* of X is the set:

$$SE(X) = \{ t \mid t \in \Sigma^{\#}, PTST(t) \subset X, t \notin X \}$$

i.e. the set of trees which are not in X, but have all their proper terminal subtrees in X.

III.2. Property: A set X of trees of $\Sigma^{\#}$ is *full suffix* <u>iff</u> it is suffix and $X = SE(PTST(X))$

Proof:
- If X is suffix, $x \in X \Rightarrow x \in SE(PTST(X))$
- If $\exists x \in SE(PTST(X))$ and $x \notin X$:

$$PTST(x) \subset PTST(X) \Rightarrow X \cap PTST(x) = \varnothing$$
$x \notin PTST(X) \Rightarrow x \notin TST(X)$ since $x \notin X$
Thus if $X \neq SE(PTST(X))$, X is not full suffix.

$$[\,]$$

III.3. Property: A set X of trees of $\Sigma^{\#}$ is full suffix <u>iff</u> it is a maximal suffix set:
$MSS = FSS$.

Proof:
- If X is not full suffix, by property III.4.:

$$\exists t \in SE(PTST (X)), t \notin X$$

t is not in X and is not in $PTST(X)$ by definition of the suffix expansion. Thus t is not a terminal subtree of a tree of X. But no tree of X can be a terminal subtree of t since every terminal subtree of t is in $PTST(X)$ and X is suffix. Thus:

$$X \cup \{ t \} \text{ is suffix.}$$

- If X is a full suffix set, it is trivial that it is also a maximal suffix set, since every tree of $\Sigma^{\#}$ either has a terminal subtree in X or is a terminal subtree of a tree of X.

$$[\,]$$

III.4. Example: The set $H_n(\Sigma)$ of all trees of $\Sigma^{\#}$ of height n is a full suffix set and thus a maximal suffix set.

This follows from the fact that:

$$PTST(H_n) = \bigcup_{k < n} H_k$$

and:

$$X = SE\left(\bigcup_{k < n} H_k\right)$$

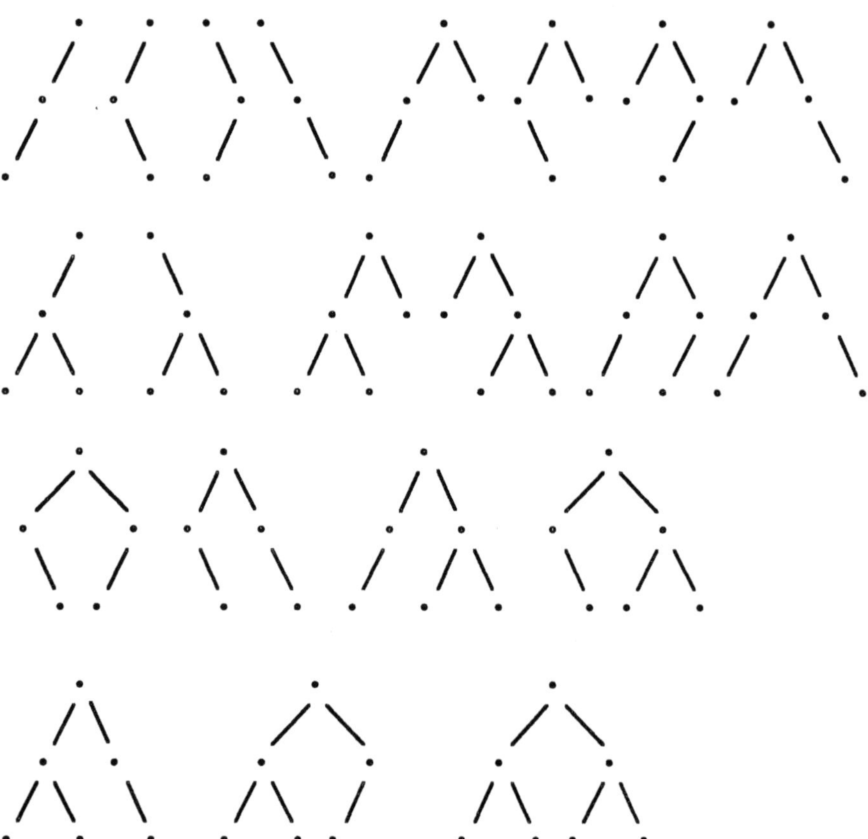

Figure 5: The set $H_2(\{\bullet\})$ of all trees of height 2 on an arity-2-one-letter alphabet

Remark: The number of elements of the sets H_n grows very fast with n. For arity-2-one-letter alphabets it is given by:

$$|H_0| = 1$$

$$| H_{n+1} | = 2 | H_n | + 2 | H_n | \sum_{k=1}^{k=n-1} | H_k | + | H_n |^2$$

For instance, $| H_3 | = 651$ and $| H_4 | = 457\,653$

III.5. Property: If X is a finite maximal suffix set of non-word trees of $\Sigma^{\#}$ $(X \in FMSS)$ and $t \in PTST(X)$, then: $\exists\, x_1 , x_2 \in X, \exists\, f \in B(x_2)$ such that $x_1 = x_2 + f\,t$

Proof: Let us assume that $t \in PTST(X)$. We are going to build a sequence of pairs of trees:

$< t_1(0), t_2(0) >$
$\ldots\ldots\ldots$
$< t_1(n), t_2(n) >$

such that:

(1) $\forall\, i \geq 0, \exists\, f_i \in B(\, t_2(i)\,)$ such that $t_1(i) = t_2(i) + f_i\, t$
(2) $\forall\, i \geq 0,\; h(\, t_1(i+1)\,) = h(\, t_2(i+1)\,) = 1 + h(\, t_1(i)\,)$
(3) $\forall\, i \geq 0,\; t_1(i), t_2(i) \in PTST(X) \cup X$

Let a be a letter such that $\alpha(a) > 1$ (there exists such a letter since Σ is a non-word alphabet). Let us have:

$t_1(0) = a + 1\, t + 2\, t$
$t_2(0) = a + 2\, t$
$f_0 = 1$

$t_1(0)$ and $t_2(0)$ have all their proper terminal subtrees in $PTST(X)$ since $t \in PTST(X)$. Thus they either are themselves in $PTST(X)$ or are in $X = SE(\, PTST(X)\,)$.

We choose $t_1(i + 1)$ and $t_2(i + 1)$ from $t_1(i)$ and $t_2(i)$ in the following manner:

- If $t_1(i)$ and $t_2(i)$ are not in $L(X)$, then they are in X and we have found the x_1 and x_2 searched for.
- If $t_2(i) \in PTST(X)$, then let us have:

$t_1(i + 1) = a + 1\, t + 2\, t_2(i)$
$t_2(i + 1) = a + 2\, t_2(i)$
$f_{i+1} = 1$

- If $t_2(i) \in X$ and $t_1(i) \in PTST(X)$, let us have:

$t_1(i + 1) = a + 1\, t + 2\, t_1(i)$
$t_2(i + 1) = a + 2\, t_1(i)$
$f_{i+1} = 1$

The fact that condition *(3)* is true $\forall\, i$ comes from $t_1(i)$ and $t_2(i)$ having all their proper terminal subtrees in $PTST(X)$.

If the first case never occurs, then X is infinite.

$$[\,]$$

III.6. Lemma: If X is a finite maximal suffix set of non-word trees of $\Sigma^{\#}$ and t is a tree such that $X \cap TST(t) = \varnothing$ then: $\exists x_1 , x_2 \in X, \exists f \in B(x_2)$ such that $x_1 = x_2 + f\,t$

Proof: This follows immediately from properties III.2. and III.5. This lemma is important not only for the following theorem, but also because it is most useful to prove maximality of a class of composed codes (results to appear in [Aigrain91]).

III.7. Theorem: A finite maximal suffix set of finite trees is a maximal tree code. Thus $FMSS = FMSC \subset MTC$.

Proof: Let X be a finite maximal suffix set of non-word trees of $\Sigma^{\#}$. By lemma III.6., we have:

$$\Sigma^{\#} = X^{-1} \; X^{\#}$$

which is stronger that X being self-complete. Thus X is a maximal tree code.

$$[\,]$$

Remark: We do not need the condition "non-word" in the theorem, since it is already known that it is true for word codes.
We now propose a simple characterization of finite maximal suffix tree codes.

III.8. Definition: A set X of trees is *Terminal Subtree Closed* (TSC) <u>iff</u>:
$\forall x \in X, y \in TST(x) \Rightarrow y \in X$

Remark: For words, this is usually stated as being *Right Factor Closed*.

III.9. Property: If X and Y are two Terminal Subtree Closed sets of finite trees of $\Sigma^{\#}$, $SE(X) = SE(Y) \Rightarrow X = Y$

Proof: Let us suppose (without loss of generality) that:

$$SE(X) = SE(Y) \text{ and } \exists x \in X, x \notin Y$$

Then let a be any letter in the alphabet Σ and let:

$$t_0 = a + 1\,x$$

t_0 cannot be in Y, since Y is Terminal Subtree Closed and x is not in Y. By definition of the suffix expansion, either $t_0 \in X$ or $t_0 \in SE(X)$
- If $t_0 \in X$, we repeat the process with $t_1 = a + t_0 \cdot$
- If $t_0 \in SE(X)$, $t_0 \notin SE(Y)$ since t_0 has a proper terminal subtree which is not in Y, in contradiction with the hypothesis.
Thus either we have a contradiction or X is infinite.

$$[\,]$$

III.10. Property: If X is a set of trees, $PTST(X)$ is Terminal Subtree Closed

Proof: This is trivial.

By property III.2. we know that a set of trees is maximal suffix <u>iff</u> it is equal to the suffix expansion of its set of proper terminal subtrees. By property III.10. we know that this set of proper terminal subtrees is Terminal Subtree Closed. Finally, by property III.9. we know that no two different finite Terminal Subtree Closed sets can generate the same finite maximal suffix set of trees by suffix expansion. Thus, there is a bijection between the family FMSS (= FMSC) of all finite maximal suffix sets of trees and the family of all finite Terminal Subtree Closed sets of trees. This is the equivalent for trees of the result associating to every suffix maximal word code a complete tree.

IV. PREFIX MAXIMAL CODES

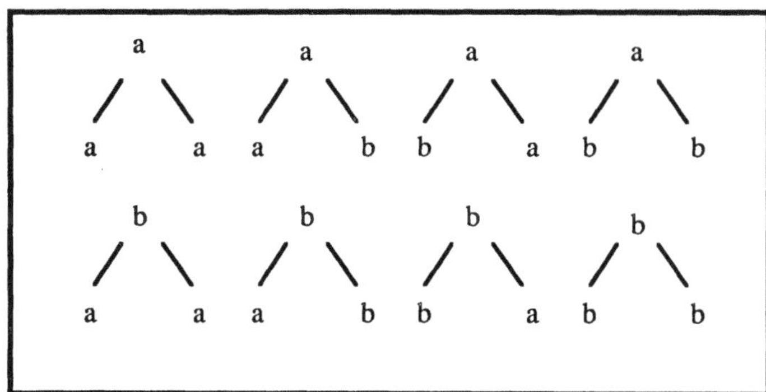

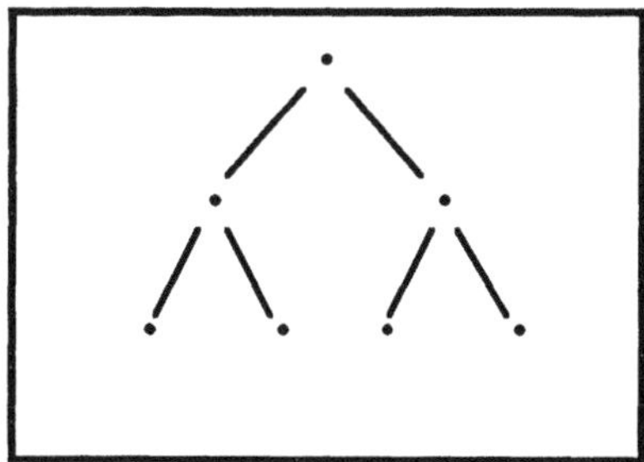

Figure 6: The sets $\{a,b\}$ [1] and $\{\bullet\}$ [2] where a, b and $\bullet$ are arity-2 letters

For trees, the symmetry between suffix and prefix properties is broken. This is most notable when discussing maximality.

IV.1. Property: Not every maximal prefix set is a prefix maximal code
(MPS C MPC, MPS ≠ MPC)

Proof: It is trivial that a prefix code is maximal only if it is a maximal prefix set. To show that this condition is not sufficient, let us denote $\Sigma^{(k)}$ the family of complete homogeneous-length trees of height k on the alphabet Σ, defined by:

$$\Sigma^{(k)} = \{ t \mid t \in \Sigma^{\#}, \forall f \in dom(t), f \text{ is a leaf} \Rightarrow l(f) = k \}$$

$\Sigma^{(k)}$ is a maximal prefix set. Let t be a tree of $\Sigma^{\#} \setminus \Sigma^{(k)}$. Let $t^{(k)}$ be the tree whose domain is:

$$dom(t^{(k)}) = \{ f \mid f \in dom(t), l(f) \leq k \}$$

and whose labelling is compatible with the labelling of t. $t^{(k)} \in \Sigma^{(k)}$. Thus t is compatible with a tree of $\Sigma^{(k)}$. But $\Sigma^{(k)}$ is also a suffix set ($\Sigma^{(k)}$ is bi-prefix). $\Sigma^{(k)}$ is not a maximal suffix set, though, since it is strictly included in $H_k(\Sigma)$.

[ ]

The main problem when studying prefix maximal codes is thus to state a necessary and sufficient condition for a prefix maximal set to be a maximal tree code. We here give a sufficient condition, which we conjecture to also be necessary:

IV.2. Property: A self-complete maximal prefix set is a maximal code

Proof: This is trivial since every self-complete set of trees is a maximal code by theorem II.4.

The following properties are useful complements to property IV.2.

IV.3. Property: A prefix set is self-complete <u>iff</u> it is suffix-self-complete

Proof: Let X be a prefix self-complete code of $\Sigma^{\#}$. Let t be any tree of $\Sigma^{\#} \setminus X$. Since X is self-complete, we know that:

$$(X^{\#} t X^{\#}) \cap X^{\#} \neq \varnothing$$

Let t' be a tree of this non-empty intersection whose size is minimal. Let us write down the two decompositions of t':

$$t' = \Sigma \{ f \beta_f \mid f \in M1 \} = \Sigma \{ g \gamma_f \mid g \in M2 \}$$

with:

$$\forall f \in M1, \ \beta_f \in X \text{ or } \beta_f = t$$
$$\forall g \in M2, \ \gamma_g \in X$$

Let us assume that $\beta_\varepsilon \in X$. Since X is prefix (no two distinct elements of X can be compatible), we have $\gamma_\varepsilon = \beta_\varepsilon$. But then all elements of $t' \Delta \beta_\varepsilon$ are in $(X^{\#} t X^{\#}) \cap X^{\#}$ in contradiction with t' being of minimal size. Thus, $\beta_\varepsilon = t$ and:

$$(t X^{\#}) \cap X^{\#} \neq \varnothing$$

[ ]

IV.4. Property: A maximal prefix set of $\Sigma^{\#}$ is suffix-self-complete <u>iff</u> $PTST(X) \subset X^{\#} X^{\#-1}$

Proof: The fact that the condition is necessary is trivial, since X being suffix self-complete is defined by:

$$\Sigma^{\#} = X^{\#} X^{\#-1}$$

Let us assume now that $PTST(X) \subset X^{\#} X^{\#-1}$. We will show that X is suffix-self complete by induction on the height of trees.

0) Let t be a punctual tree of $\Sigma^{\#} \setminus X$. Since X is a maximal prefix set, there is an element x of X which is compatible with t. Let us consider the elements of $t \Delta x$. Since t is a punctual tree, they are all proper terminal subtrees of x. Thus:

$$t \in X^{\#} X^{\#-1}$$

n) Let us assume that all elements of $\Sigma^{\#}$ of height smaller than n are in $X^{\#} X^{\#-1}$. Let t be a tree of height n. Since X is a maximal prefix set there is an element x of X compatible with t. The elements of $t \Delta x$ are either proper terminal subtrees of x or their height is smaller than n. Thus we have again:

$$t \in X^{\#} X^{\#-1}$$

$$[\ ]$$

IV.5. Example: Let $\Sigma = \{ a , b \}$, with $\alpha(a) = \alpha(b) = 2$, and let X be the maximal prefix set composed of:

- All complete homogeneous-length trees of height 2 whose labelling of ε is a.

- All complete homogeneous-length trees of height 1 whose labelling of ε is b.
X is a maximal code.

Proof: The fact that X is a maximal prefix set is trivial (see IV.1.). $PTST(X)$ is composed of:
- The punctual trees labelled a and b which can be prolongated in $X^{\#}$ as shown below:

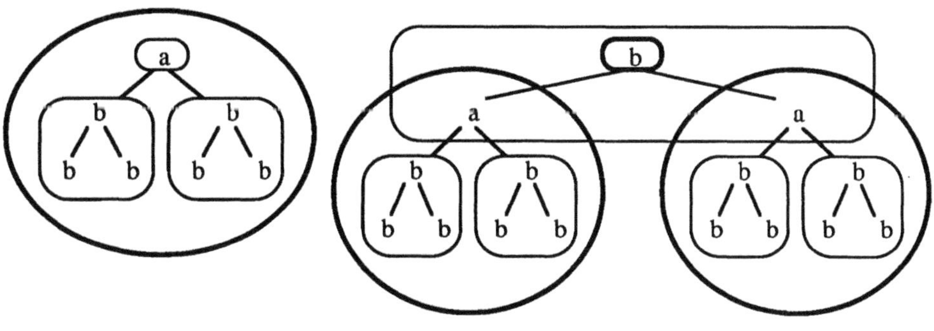

- The set of homogenous-length complete trees of height 1, which are in X if the labelling of ε is b and can be prolongated in $X^{\#}$ as shown below if the labelling of ε is a:

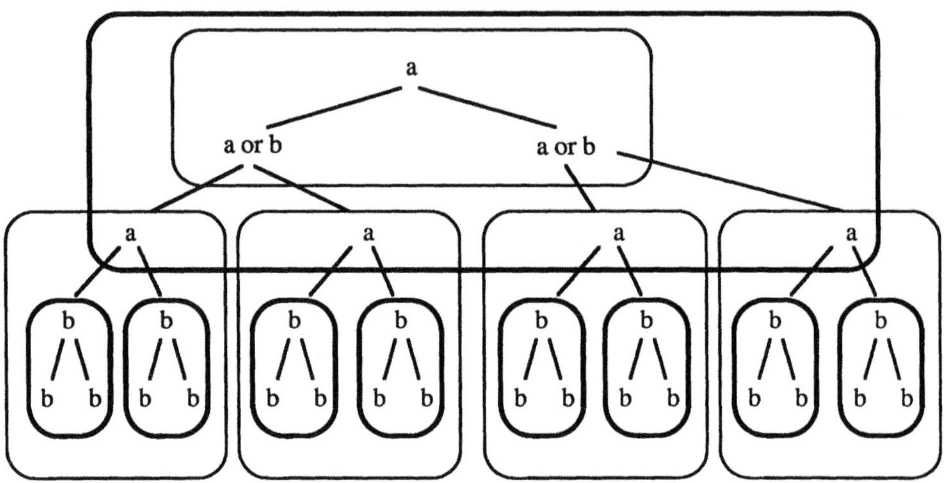

IV.6. Property: The only finite biprefix maximal code of $\Sigma^{\#}$ is the set of punctual trees

Proof: Let X be a finite bi-prefix maximal code of $\Sigma^{\#}$ of height $n > 1$. Since X is suffix, X is a maximal suffix set. If there is punctual tree t on Σ which is not in X then $TST(t) \cap X = \varnothing$. Thus, by Lemma III.6., there exists trees x and y in X such that:

$$x = y + f t \quad (f \in B(y))$$
which means that X is not prefix and we have a contradiction. If all punctual trees belong to X, we also have a contradiction, since X is prefix and of height greater than 1.

$$[\ ]$$

V. CONCLUSION

The theory of suffix maximal codes we have presented is complete. Two results need to be achieved in the theory of prefix maximal codes:
- Give a better characterization of prefix maximal sets.
- Prove or disprove our

Conjecture 1: Every non-word maximal tree code which is prefix is self-complete.

REFERENCES

[Aigrain91]   Philippe Aigrain, "Composed Maximal Tree Codes and Self-Completeness", submitted for publication.
[Berstel-Perrin85]   Jean Berstel, Dominique Perrin, "Theory of Codes", Academic Press, N.Y., 1985.

[Nivat86]   Maurice Nivat, "Eléments d'une théorie générale des codes d'arbres: I",
                 Rapport de Recherche n°86-61, L.I.T.P., Université Paris 7, 1986.
[Nivat86b]   Maurice Nivat, "Eléments d'une théorie générale des codes d'arbres:
                 Deuxième Partie: Codes Suffixes", Rapport de Recherche n°86-70,
                 L.I.T.P., Université Paris 7, 1986.
[Nivat90]   Maurice Nivat, "Binary Tree Codes", to appear in this book.
[Perrin77]   Dominique Perrin, "Codes asynchrones", Bull. Soc. Math., t. 105, 1977,
                 p. 385-404.
[Schutzenberger65] Marcel Schutzenberger, "Codes à longueur variable", Royan, 1965
                 republished in D. Perrin, ed., "Théorie des Codes", Actes de la 7ème
                 Ecole de Printemps d'Informatique Théorique, L.I.T.P., Jougne, 1979.

Tree Automata and Languages
M. Nivat and A. Podelski (editors)
1992 Elsevier Science Publishers B.V.

A Monoid Approach to Tree Automata

Andreas Podelski

LITP, Université Paris 7
2, pl. Jussieu, 75251 Paris Cedex 05
podelski@litp.ibp.fr

We survey a number of results which are based on a novel imbedding of sets of trees into a free monoid structure. With every set L of trees one associates the set $\hat{L}$ of all those trees which, when their only occurrence of a variable is substituted by a constant, become a tree of L. The standard concepts of tree automata, such as the pumping lemma, determinization, minimization and equivalence testing, can be treated in a concise way and, moreover, be fully formally based on the theory of automata on words. With every tree automaton A one associates an automaton $\hat{A}$ on words on an alphabet which depends on the states of A. This approach becomes useful also for the investigation of structural properties of sets of trees, such as aperiodicity (threshold-, but not modulo-counting of subtrees), since the deep theory of varieties of word languages can be directly applied.

1 Introduction

It is well-known that the theory of tree automata arises as a straightforward extension of the theory of word automata when words are viewed as elements of unary term algebras. If, however, words are viewed as elements of the free monoid (over a finite alphabet), it is not evident of what the extension would consist.

One possibility (thoroughly investigated in the article of Steinby in this volume) is to consider the monoid as a (non-free) algebra over the function symbols "·" (binary, for the product) and "1" (for the neutral element) and all letters of the alphabet as nullary symbols, and whose equations are the monoid axioms. The extension considered in this article is a completely different one.

As underlying algebraic structure we use again a freely generated monoid. Its (countably infinite) basis Γ_Σ consists of all trees with exactly one occurrence of a variable (the auxiliary substitution or concatenation symbol) at depth one, that is, as left or right son of the root. The free monoid generated by Γ_Σ, the so-called *tree monoid* $S_\Sigma = \Gamma_\Sigma^\star$, consists of all trees with exactly one occurrence of the variable (at any depth k, $k \geq 0$). These trees have been used by various authors; they are called pointed trees [9], special trees [19], or trees with a handle [8]. The fact that the underlying structure forms a free monoid has been pointed out for the first time in [16]. Also, the idea of embedding the theory of recognizability for tree algebras fully into the theory of recognizability for free monoids is an original one from [9]. This idea is realized as follows:

With every set L of trees one associates the set $\hat{L}$ of all special trees which, when their only occurrence of a variable is substituted by a constant, become a tree of L. The central theorem in the embedding approach is: The set L of trees is recognizable (*i.e.*, in the tree algebra T_Σ) if and only if the set $\hat{L}$ is recognizable (*i.e.*, in the free monoid $\Gamma_\Sigma^\star$).

Although the theory of recognizability in the free monoid $\Gamma^\star$, for some alphabet Γ, is usually treated under the overall assumption that the generating basis Γ is finite, none of the proofs actually needs this assumption. Hence, the theorems generally extend to the case of Γ being infinite.

This is the case, for example, for the pumping lemma. Its formulation and proof in the tree case become concise and elegant if we use the tree monoid: A tree t of height k corresponds to a special tree $s \in \Gamma_\Sigma^k$, *i.e.*, an element of the free monoid of length k. Thus, one simply applies the standard pumping lemma on the automaton $\hat{A}$ on words over Γ_Σ which is directly deduced from a tree automaton A, with the same number of states. The correspondence between t and s means precisely: s becomes t when the occurrence of the variable is replaced by a constant a, which is written $t = s[a]$. The pumping lemma for words says: If k is greater than the number of states, then s can be decomposed, $s = s_1 \cdot s_2 \cdot s_3$, such that $s \in \hat{L}$ iff $s_1 \cdot s_2{}^n \cdot s_3 \in \hat{L}$, which gives the result for trees: $t \in L$ iff $s_1 \cdot s_2{}^n \cdot s_3[a] \in L$, for any $n \geq 0$.

An automaton is called finite if the set Q of its states is finite—regardless of the cardinality of the alphabet. The automaton is no longer a finite device if the alphabet is infinite in the sense that its transition table $\delta : \Gamma \times Q \mapsto Q$ is infinite. Observe, however, that only finitely many letters can be distinguished by the automaton; thus, we obtain a finite alphabet by identifying letters which induce the same transitions of the automaton. Indeed, algorithms on tree automata A can be obtained by first calculating this finite alphabet and then executing the standard (*i.e.*, the word case) algorithm on the automaton $\hat{A}$ on words over the finite alphabet.

In the next section we recall the dual views: free monoid *vs.* unary term algebra—of words in the theory of finite word automata, much in the same way this has already been done by Thatcher and Wright in [18]. Whereas their conclusion was the necessity of introducing an infinite auxiliary alphabet of substitution variables, we will show that one can stick to the case of just one variable. Thus, along the lines of this section, we will try to provide the intuition for the notions defined in the sections to follow, in particular for the "$\hat{L}$–construction."

In Section 3 we introduce trees and the tree monoid, always in the framework of term algebras. In previous articles, trees were defined as functions from a tree domain (which is a prefix-closed subset of a free monoid) into the set Σ of symbols labeling the nodes. The presentation of the tree monoid theory in the term algebra framework is an asset of this article.

In Section 4 we recall the notions of tree automata and algebra–recognizability and then show in Section 5 that latter is equivalent to monoid recognizability. Thus, not the duality presented in Section 2 is new; what is new is to pursue the theory of recognizability of tree languages also in the free monoid.

In Section 6 we consider regular tree representations which are very different from the known regular tree expressions. Here the infinity of the base generating the tree monoid plays a role for the first time in the presentation, and we come upon effectivity and decidability questions. We obtain a new version of the Theorem of Kleene for trees, and, by appropiately restricting the regular tree representations, a normal form theorem for the known regular tree expressions. The latter is important mainly when we want to characterize recognizable aperiodic tree languages. We are able to to extend the Theorem of Schützenberger for regular tree representations (which, as shown by Thomas (1984), does not hold for regular tree expressions) and hereby give the only known characterization of aperiodic tree languages. Finally, in Section 7, we discuss the linear aspects of algorithms concerning tree automata, meaning that already existing algorithms for the word case apply to the tree case as well. In particular, we show how the syntactic monoid of a tree language can be effectively computed and give a small example indicating its importance, namely for a theory of varieties of tree languages.

2 Words and special words

In this section we present the duality of the theory of recognizability of word languages: either in the free monoid or in the unary term algebra. This point of view is not new; *cf.*, for example, [18]. What is new is to pursue the theory of recognizability of tree languages also in the free monoid.

In the following Σ always denotes a finite set (of letters of words, in this section; of labels of tree nodes, in the other sections). When words (or strings) over Σ are considered, one usually uses either of the two following algebraic structures (but *cf.* also Courcelle (1989)).

(I) The set Σ is the base of the free monoid $(\Sigma^*, \cdot, \lambda)$ over Σ. Its elements are the words, its product is the concatenation of words, and λ is its neutral element, the "empty word."

(II) The set Σ is the set of unary operator symbols, λ is the only constant symbol, and the words are the (ground) terms in the free (or initial) Σ'–algebra (or Σ'–magma), where $\Sigma' = \Sigma \uplus \{\lambda\}$. The set T_Σ of these terms is defined inductively as follows: $\lambda \in T_\Sigma$, and if $f \in \Sigma$ and $t \in T_\Sigma$, then $ft = f(t) \in T_\Sigma$.

Thus the word $w = fgh$, written here as an element of the free monoid over $\{f, g, h\}$, is the element $t_w = fgh\lambda$, also written $f(g(h(\lambda)))$, of the free algebra over $\{f, g, h\} \cup \{\lambda\}$.

Generally, given an algebraic structure, a subset L of its domain is said to be *recognizable* if there exists a congruence (*i.e.*, an equivalence relation compatible with this structure) which has a finite index (*i.e.*, a finite number of equivalence classes) and which saturates L (*i.e.*, L is union of some of its equivalence classes).

The notions of recognizability in (I) and in (II) coincide. However, congruences in the two cases have different meanings. Every finite state automaton recognizing L defines congruences on words of the two kinds. Always, its 'algebra-congruence' classes are its states, and its 'monoid-congruence' classes are its transition functions.

We define the monoid S_Σ of all polynomials (or derived terms) on T_Σ, *i.e.*, the set of all non-ground terms over Σ. They have exactly one occcurrence of a variable, since the operator symbols in Σ are unary. They are built up from the identity function, denoted x, and the functions $f(x)$, for $f \in \Sigma$. We note that S_Σ is a free monoid, over the base $\Gamma_\Sigma = \{f(x) \mid f \in \Sigma\}$. To every word, say $w = fgh \in \Sigma^*$ (*i.e.*, $t_w = fgh\lambda = f(g(h(\lambda))) \in T_\Sigma$), corresponds uniquely the "special word" $s_w = fghx = f(g(h(x))) \in S_\Sigma$; $s_w[\lambda] = t_w$.

A ground term t_w gets evaluated by a word automaton to a value (*i.e.*, a state $q = t^A \in Q$) and a function s_w in x gets evaluated to a function mapping states onto states (called the transition function associated with w, $s_w{}^A : Q \mapsto Q$).

Two words w and w' are said to be Nerode congruent iff t_w and t'_w are congruent (*i.e.*, in the unary algebra); they are syntacticly congruent if s_w and $s_{w'}$ are congruent (*i.e.*, in the free monoid). Let A be the minimal automaton of L. Then the two congruences above mean that w and w', as terms t_w and $t_{w'}$, are evaluated to the same state, or, respectively, that, as special words s_w and $s_{w'}$, they are evaluated to the same transition function.

That is, the syntactic congruence of L is a congruence on the free monoid Σ^*, whereas the Nerode congruence of L is an algebra congruence defined on T_Σ. The Nerode congruence saturates the set of all terms $\{w\lambda \mid w \in L\}$. It can be defined as an equivalence relation on Σ^*. (Here it is a so-called left-congruence,*i.e.*, a left invariant equivalence relation.) The syntactic congruence can not be defined on T_Σ. However, we can define its corresponding notion over derived terms, *i.e.*, on S_Σ. The set that the syntactic congruence of L then saturates is the subset $\hat{L} = \{wx \mid w \in L\}$ of S_Σ, or, sticking with the algebra notion, $\hat{L} = \{s_w \mid s_w[\lambda] \in L\}$.

Words as unary ground terms t_w and words as unary derived terms s_w correspond to each other bijectively: this is why the dual algebraic settings of words can deliberately be confused in most presentations of the theory of finite automata.

3 Trees and special trees

The step from words and word automata to trees and tree automata is a straightforward generalization in the formalism of universal algebra from unary Σ–algebras (*i.e.*, Σ contains only unary operator symbols, besides a dummy constant symbol λ) to Σ–algebras where Σ is an arbitrary ranked alphabet. This means that Σ is of the form $\Sigma = \Sigma_m \uplus \ldots \uplus \Sigma_0$ where the index i of Σ_i indicates the rank of the operator symbol $f \in \Sigma_i$. Trees are the elements of the domain of the free algebra over Σ (ground terms). Before we give formal definitions, we will introduce a restriction which simplifies the notation considerably, but is inessential in the sense that all statements (and proofs) to follow are valid for the unrestricted case as well (*i.e.*, can be reformulated in a straightforward way). This reduction from arbitray to binary trees is standard in computer science (*cf.* e.g. Downey et al. (1980), Courcelle (1989)).

We assume a finite set Σ of labels for interior as well as leaf nodes. We set $\Sigma_2 = \Sigma$ and $\Sigma_0 = \{\Omega\}$ where Ω is a new constant symbol (to be read as "empty tree"), and abuse notation by noting the ranked alphabet $\Sigma_2 \uplus \Sigma_0$ again Σ, *i.e.*, $\Sigma = Sigma \uplus \{\Omega\}$.

An *algebra* $\mathcal{A}$ over Σ is a structure $\mathcal{A} = (Q, (f^{\mathcal{A}})_{f \in \Sigma \uplus \{\Omega\}})$ consisting of a domain Q and a family of functions $f^{\mathcal{A}}$ for every function symbol $f \in \Sigma$. Each of those symbols is interpreted by $\mathcal{A}$ according to its rank, *i.e.*, the arity of $f^{\mathcal{A}}$ is the rank of f: $f^{\mathcal{A}} : D \times D \mapsto D$ is a binary function, and $\Omega^{\mathcal{A}} \in D$ is a constant, denoted from now on q_0. We will also denote the Σ-algebra $\mathcal{A} = (Q, (f^{\mathcal{A}})_{f \in \Sigma})$ (which is, to be precise, a $(\Sigma \uplus \{\Omega\})$-algebra.

(Binary finite labeled) *trees* over Σ are the ground terms of the free algebra over Σ, called the tree algebra or tree magma, which is $\mathcal{T}_\Sigma = (T_\Sigma, (f^{\mathcal{T}})_{f \in \Sigma \uplus \{\Omega\}})$. Thus, the set of trees over Σ is the smallest set T_Σ containing Ω, and, for any $f \in \Sigma$ and two trees $t_1, t_2 \in T_\Sigma$, the tree $f(t_1, t_2)$.

The *interpretation* (or: valuation) of a tree t by an algebra $\mathcal{A}$ over Σ is defined inductively: $t = \Omega$ and $t^{\mathcal{A}} = q_0$, or $t = f(t_1, t_2)$ and $t^{\mathcal{A}} = f^{\mathcal{A}}(t_1{}^{\mathcal{A}}, t_2{}^{\mathcal{A}})$. (The interpretation in the free algebra is simply $\Omega^{\mathcal{T}} = \Omega$ and $f^{\mathcal{T}}(t_1, t_2) = f(t_1, t_2)$.)

Note that in this formalism every label in the given alphabet Σ denotes a binary operator symbol. The symbol Ω is introduced solely in order to simplify the formal representation. In the standard representation of a tree t as a graph, its nodes are labeled by symbols of Σ or by Ω. We call the nodes labeled Ω the *border nodes* of t; they don't have descendants. The *leaves* of the tree t are the nodes whose *both* sons are labeled Ω. All the other nodes are interior nodes, labeled with a letter from Σ. The picture of the graphical representation of a tree usually shows only nodes labeled by a symbol of Σ; the border nodes are implicit.

In other frameworks (*cf.* e.g. Thomas (1984)) a word is considered a special case of a tree, namely one which is built up solely from unary operator symbols $f \in \Sigma_1$. Here, however, words are embedded as trees where the right son of every node is Ω. Thus, the word fgh corresponds to the tree $f(g(h(\Omega, \Omega), \Omega)\Omega)$.

Let us point out a consequence of our framework on the reformulation of an unsolved conjecture. A tree language is anti-chain definable (*cf.*.[19]) if it can be defined by a weak monadic second-order formula over two successors whose set variables range only over anti-chains. An anti-chain is a set of nodes such that no two of them lie on the same path. In the formulation adapted to this framework, the conjecture, first postulated in [6], states that every recognizable tree language is anti-chain definable. This conjecture, as surprising as it seems at a first glance, has resisted quite a number of counter attacks, by Niwiński, Heuter, and others (*cf.*. [6]. — A set of trees over solely unary function symbols is anti-chain definable iff it is first-order definable, which is the case iff it is aperiodic (see Section 6). Thus, every non-aperiodic, recognizable word language, formalized as a unary-tree language, provides a counter-example to the conjecture. In our formalism, we do no longer know of any example of a tree language which is *not* anti-chain definable.

We define *pointed trees* over Σ to be the derived terms over Σ in one variable x in exactly one occurence. They are called "special trees" in [19] and "trees with a handle" in [8]. Thus, pointed trees denote unary functions which can be applied on ground terms. For example, x denotes the identity, and thus, for all $t \in T_\Sigma$, $x[t] = t$. In particular, every pointed tree applied on a tree is a tree,

$$\forall s \in S_\Sigma \forall t \in T_\Sigma : s[t] \in T_\Sigma.$$

In the presentation as graphs, pointed trees are trees where one (and only one) border node is labeled x (instead of Ω, like all other leaves). Or, in the concise presentation omitting the border nodes, we would have the tree with a "pointer" from a node of the tree to a son which is a node outside the tree. Note that the other son might be a node of the tree. Then $s[t]$ denotes the graph obtained from s by attaching the root of the tree t onto the pointed node.

Clearly, a (unique) tree t corresponds to a given pointed tree s in the sense that $t = s[\Omega]$ can be obtained from a pointed tree s by substituting Ω for x. Vice versa, if $t \neq \Omega$, several different pointed trees s can be obtained from t by replacing one label Ω through x; *i.e.*, such that $s[\Omega] = t$.

The set S_Σ of pointed trees can be defined inductively as follows: $x \in S_\Sigma$, and if $f \in \Sigma$, $s_1, s_2 \in S_\Sigma$, then $f(s_1, s_2[\Omega])$, $f(s_1[\Omega], s_2) \in S_\Sigma$.

We define the concatenation of two pointed trees $s\, s' = s \cdot s'$ as the ordinary composition of (unary) functions. This corresponds to the substitution $s\, s' = s[s']$ of the node labeled x of s through s'. (One also could use the inductive definition of S_Σ from above to define

$x\,s' = x[s'] = s'$, etc..) Clearly, this operation is associative and has a neutral element: the pointed tree denoted x. Thus, we obtain a monoid $(S_\Sigma, \cdot, x)$, in the following referred to as the tree monoid (over Σ).

Since we can decompose $f(s_1, s_2[\Omega]) = f(x, s_2[\Omega])\,s_1$ and, symmetrically, $f(s_1[\Omega], s_2) = f(s_1[\Omega], x)\,s_2$, we see from the inductive definition of S_Σ that every pointed tree can be decomposed into pointed trees of the form $f(x,t)$ or $f(t,x)$. Graphically speaking, this decomposition works by "cutting off" the subtrees at each node on the path which is denoted by the leaf labeled x. It is not difficult to show that this decomposition is unique so that we obtain:

Remark 1 *The tree monoid* $(S_\Sigma, \cdot, x)$ *of pointed trees over* Σ *is a free monoid. It is infinitely generated by the basis:*

$$\Gamma_\Sigma = \{f(x,t),\, f(t,x) \mid f \in \Sigma,\, t \in T_\Sigma\};$$

i.e., $S_\Sigma = \Gamma_\Sigma{}^\star$.

Thus, a pointed *base* tree consist of the root, a node labeled x which is one of its sons, and a subtree, which can be any tree over Σ (and which is rooted in the other son of the root).

The *height* $h(t)$ of a tree $t \in T_\Sigma$ can be defined as the maximal length of a pointed tree, *i.e.*, a word $s \in \Gamma_\Sigma{}^\star$. That is,

$$h(t) = max(\{k \mid \exists s, s_1, \ldots, s_k \in \Gamma_\Sigma : s[\Omega] = t,\ s = s_1 \cdot \ldots \cdot s_k\}).$$

The pointed base trees correspond to the 'elementary translations' of the term algebra T_Σ used by Gegsec and Steinby [5]; their algebraic structure, however, is different: Two elementary translations s and s' are composed via an operator $f \in \Sigma$ to the unary function $f(s, s')$, which has now two occurences of x and is no longer a pointed tree (called a translation in [5]).

4 Recognizability in the tree algebra

A (finite root-to-frontier) tree automaton A over Σ is a quadruple:

$$A = (Q, (f^A)_{f \in \Sigma}, q_0, Q_{fin})$$

such that Q is a finite set, $f^A : Q \times Q \mapsto Q$ is a binary function, for each $f \in \Sigma$, q_0 is an element of Q, and Q_{fin} is a subset of Q, $Q_{fin} \subseteq Q$.

In other words, if we set $\Omega^A = q_0$, then $A = (Q, (f^A)_{f \in \Sigma \uplus \{\Omega\}})$ is an algebra over Σ (*i.e.*, a $\Sigma \uplus \{\Omega\}$–algebra). Vice versa, given an algebra A over Σ whose domain Q is finite, and a subset Q_{fin} of its domain, then (A, Q_{fin}) constitutes a tree automaton.

We call the elements of the domain Q the values or the states, the elements of the subset Q_{fin} the final states, and $q_0 = \Omega^{\mathcal{A}}$ the initial state of the tree automaton A. Sometimes it is convenient to write the functions $(f^{\mathcal{A}})_{f \in \Sigma}$ bundled together into one function $\lambda : \Sigma \times Q \times Q \mapsto Q$.

The *valuation* of a tree $t \in T_\Sigma$ by the tree automaton A with underlying algebra $\mathcal{A}$ (*i.e.*, A can be written in the form $A = (\mathcal{A}, Q_{fin})$), is the interpretation of the term t by the underlying algebra $\mathcal{A}$. We recall that this means that the state $t^{\mathcal{A}} \in Q$ is given inductively through $\Omega^{\mathcal{A}} = q_0$ and:

$$(f(t_1, t_2)^{\mathcal{A}} = f^{\mathcal{A}}(t_1{}^{\mathcal{A}}, t_2^{\mathcal{A}}).$$

We say that the set L of trees over Σ, $L \subseteq T_\Sigma$, is the tree language *accepted* by the tree automaton A if exactly the trees in L are evaluated to a final state,

$$L = \mathcal{L}(A) \quad \text{iff} \quad L = \{t \in T_\Sigma \mid t^{\mathcal{A}} \in Q_{fin}\}.$$

In particular, L is then *recognized* by the algebra $\mathcal{A}$ which underlies A. Generally, a tree language L is recognized by the algebra $\mathcal{A} = (Q, (f^{\mathcal{A}})_{f \in \Sigma})$ if there exists a subset $Q' \subseteq Q$ of its domain such that $L = \{t \in T_\Sigma \mid t^{\mathcal{A}} \in Q'\}$.

Obviously equivalent conditions for the *recognizability* of a tree language $L \subseteq T_\Sigma$ are that:

— there exists a homomorphism ϕ from the free Σ–algebra T_Σ onto a Σ–algebra A with finite domain such that $L = \phi^{-1}(\phi(L))$;

— there exists a congruence $\approx$ on the tree algebra T_Σ which has a finite index (*i.e.*, number of congruence classes) and saturates L, *i.e.*, L is the union of some of the congruence classes, or $L = \bigcup_{t \in L}[t]_\approx$.

5 Monoid-Recognizability

By definition a subset S of a free monoid $(\Gamma^\star, \cdot, 1)$ is recognizable if

— there exists an automaton $A = (Q, \delta, q_0, Q_{fin})$ on words over Γ which accepts S. Here the transition function of A is $\delta : \Gamma \times Q \mapsto Q$. If we denote $\delta : \Gamma^\star \times Q \mapsto Q$ its extension defined by $\delta(\lambda, q) = q$, $\delta(w \cdot a, q) = \delta(a, \delta(w, q))$ for $w \in \Gamma^\star, a \in \Gamma$, then $w \in L$ iff $\delta(w, q) \in Q_{fin}$;

— there exists a homomorphism ϕ from the monoid $\Gamma^\star$ onto a finite monoid such that $S = \phi^{-1}(\phi(S))$; or, obviously equivalently,

— there exists a congruence $\equiv$ on the monoid $\Gamma^\star$ which has finite index and saturates S, *i.e.*, $S = \bigcup_{s \in S}[s]_\equiv$.

The connection between tree languages $L \subseteq T_\Sigma$ and subsets $S \subseteq S_\Sigma$ of the tree monoid is established as follows. Given a set $L \subseteq T_\Sigma$ of trees over Σ, we define the set $\hat{L} \subseteq S_\Sigma$ of pointed trees by:

$$\hat{L} = \{s \mid s[\Omega] \in L\}.$$

Thus $\hat{L}$ is the set of all pointed trees s which we can obtain from trees t of L by replacing the label Ω of one (and only one) leaf of t through x. Observe that the embedding of sets L of trees into sets of pointed trees $\hat{L}$ is injective, but not surjective. In fact, a set S of pointed trees corresponds to a set of trees (in the sense that $S = \hat{L}$ for some $L \subseteq T_\Sigma$) if and only if it is invariant against moving the label x along all nodes on the border, formally, iff S is saturated by the following equivalence relation.

$$s \sim_{[\Omega]} s' \quad \text{iff} \quad s[\Omega] = s'[\Omega]$$

In this case we will simply say that S is *border-invariant*. Observe that $\sim_{[\Omega]}$ is generally a left– but not a right congruence on S_Σ (*i.e.*, $s \sim_{[\Omega]} s'$ implies for any s'' that $s'' s \sim_{[\Omega]} s'' s'$, but not $s\, s'' \sim_{[\Omega]} s'\, s''$).

We note already here that, given any monoid congruence $\equiv$ on S_Σ, the left-congruence $\sim_{[\Omega]}$ does not generally refine it (even if $\equiv$ saturates a border-invariant set, *i.e.*, of the form $\hat{L}$). In fact, the only congruence that is refined is the one which identifies *all* elements:

Remark 2 *If $\equiv$ is any congruence on the monoid Γ^* which is refined by the equivalence relation $\sim_{[\Omega]}$ (which means that $s \sim_{[\Omega]} s'$ implies $s \equiv s'$), then $\equiv$ is the trivial equivalence relation: $s \equiv s'$ for all pointed trees s and s'.*

Theorem 1 *[9] A set of trees $L \subseteq T_\Sigma$ is algebra–recognizable if and only if the set of pointed trees $\hat{L} \subseteq S_\Sigma$ is monoid–recognizable.*

The theorem thus says that L is saturated by a congruence on the free algebra T of finite index if and only if $\hat{L}$ is saturated by a congruence on the free monoid $S_\Sigma = \Gamma_\Sigma{}^*$ of finite index.

Given a tree automaton $A = (Q, (f^A)_{f \in \Sigma}, q_0, Q_{fin})$, say $A = (\mathcal{A}, Q_{fin})$, accepting L (or simply a finite algebra $\mathcal{A}$ recognizing L), we can define an automaton

$$\hat{A} = (Q, \hat{\delta}, q_0, Q_{fin})$$

on words over Γ_Σ, by defining its transition function $\hat{\delta} : \Gamma_\Sigma \times Q \mapsto Q$ as follows: $\hat{\delta}(f(x,t),q) = f^A(q,t^A)$ and $\hat{\delta}(f(t,x),q) = f^A(t^A,q)$. Clearly, $\hat{A}$ accepts the word language $\hat{L}$.

Whereas trees, as ground terms, are evaluated to constant values (states) by the algebra $\mathcal{A}$, pointed trees $s \in S_\Sigma$, as derived terms or unary functions, are evaluated to unary functions on the domain Q, $s^A : Q \mapsto Q$. They can be defined as the *transition function* associated with the pointed tree s of the automaton $\hat{A}$. Thus, if $s \in \Gamma_\Sigma$, then s^A is given by $s^A(q) = \hat{\delta}(s,q)$ for $q \in Q$, and if s is of the form $s = s_1 \cdot \ldots \cdot s_n$ with $s_1, \ldots, s_n \in \Gamma_\Sigma$ for some $n \geq 0$, then s^A is given by composition of the functions $s_1^A, \ldots, s_n^A$. If $n = 0$, then $s = x$ and s^A is the identity function. Using the inductive definition of S_Σ, $s^A(q)$ can be calculated as follows: $x^{\mathcal{A}}(q) = q$, and $(f(s_1, s_2[\Omega]))^{\mathcal{A}}(q) = f^{\mathcal{A}}(s_1{}^{\mathcal{A}}(q), s_2{}^{\mathcal{A}}(q_0))$, and, symmetrically, $(f(s_1[\Omega], s_2))^{\mathcal{A}}(q) = f^{\mathcal{A}}(s_1{}^{\mathcal{A}}(q_0), s_2{}^{\mathcal{A}}(q))$.

The following observation relates the valuation of a tree and the corresponding pointed tree by a tree automaton A, *i.e.*, by its underlying algebra $\mathcal{A}$

Remark 3 *For all pointed trees $s \in S_\Sigma$ and all trees $t \in T_\Sigma$ over T_Σ,*

$$if \ \ s[\Omega] = t \ \ then \ \ s^{\mathcal{A}}(q_0) = t^{\mathcal{A}}.$$

The automaton $\hat{A}$ corresponding to the tree automaton A is useful for technical purposes. Known properties and algorithms for automata translate, in a *formal* way, to those for tree automata. Let us illustrate this at a small example, the proof of the pumping lemma.

Lemma 1 (Pumping Lemma) *If a tree t of height k is contained in a tree language L which is accepted by a tree automaton A whose number of states is smaller than k, then t can be decomposed into $t = s_1 \cdot s_2 \cdot s_3[\Omega]$ such that $s_1 \cdot s_2{}^n \cdot s_3[\Omega] \in L$, for any $n \geq 0$.*

Proof: A tree t of height k corresponds to a special tree $s \in \Gamma_\Sigma{}^k$, *i.e.*, a word of the free monoid of length k (*cf.* the definition of height at the end of Section 3). Thus, one simply applies the standard pumping lemma on the automaton $\hat{A}$ on words over Γ_Σ, with the same number of states. The pumping lemma for words says: If k is greater than the number of states, then s can be decomposed, $s = s_1 \cdot s_2 \cdot s_3$, such that $s \in \hat{L}$ iff $s_1 \cdot s_2{}^n \cdot s_3 \in \hat{L}$. This yields the result for trees: $t \in L$ iff $s_1 \cdot s_2{}^n \cdot s_3[\Omega] \in L$, for any $n \geq 0$.

Given a tree automaton $A = (Q, (f^{\mathcal{A}})_{f \in \Sigma}, q_0, Q_{fin})$, say $A = (\mathcal{A}, Q_{fin})$, accepting L (or simply a finite algebra $\mathcal{A}$ recognizing L), we can define the congruence $\equiv_A$ (also noted $\equiv_{\mathcal{A}}$) on the tree monoid S_Σ by:

$$s_1 \equiv_A s_2 \ \ \text{iff} \ \ s_1^{\mathcal{A}} = s_2^{\mathcal{A}}.$$

That is, pointed trees s are congruent iff the associated transition functions $s^{\mathcal{A}} : Q \mapsto Q$ are identical. Clearly, the congruence $\equiv_A$ has a finite index (bound by the cardinality of Q^Q) and saturates $\hat{L}$, since $\hat{L} = \{s \mid s^{\mathcal{A}}(q_0) \in Q_{fin}\}$.

The only-if direction in the theorem above is easily established, using either the automaton or the congruence characterization of recognizability.

It has already been observed (by Thomas [19], who also was the first to consider S_Σ) that, if the tree automaton A is finite, we obtain a congruence on S_Σ of finite index. The introduction of the set $\hat{L}$ and the saturation property of the congruence, *i.e.*, the notion of monoid-recognizability of a tree language L, are new. The fact that this implies the algebra-recognizability of L, was discovered by Nivat and Podelski [9]. Another proof of this fact in a different setting was done by Heuter [6].

We will not investigate the general recognizability of (not necessarily border-invariant) sets of pointed trees any further here. We only note that there are uncountably many recognizable, and only countably many border-invariant recognizable sets of pointed trees.

The border-invariance of the recognizable set of pointed trees which is saturated by a monoid congruence is essential in the inference of recognizability of the associated tree language. The same is true when structural properties other than recognizability, such as aperiodicity, are inferred.

6 Regular Tree Representations and Aperiodicity

In this section we first formulate an extension of the Theorem of Kleene from the word to the tree case which is different from the one in [18] (*cf.*also [5]). We obtain this extension by extending the Theorem of Kleene for recognizable/regular subsets over a free monoid from the case of a finite base to the case of an infinite base, and by furthermore applying *Theorem 1.*

We call a *regular representation* of a tree language $L \subseteq T_\Sigma$ a regular expression of the free monoid over Γ_Σ which denotes $\hat{L}$. We recall that the regular expressions of a free monoid over the (possibly infinite) base Γ are obtained by composing arbitray subsets of Γ using the Boolean operations, the concatenation and the star-operator (the two latter canonically extended from elements to sets). In other words, they constitute the closure of the powerset of the base Γ against these operations.

Theorem 2 *[16, 10] A tree language $L \subseteq T_\Sigma$ is recognizable if and only if it can be denoted by regular representation.*

Thus, it is possible to construct a regular representation with sets of pointed base trees (*i.e.*, subsets of Γ_Σ) which are given by properties of arbitray complexity (a priori even undecidable). The (somewhat surprising) fact the denoted tree language lies still in the family of recognizable tree languages relies on the border-invariance of $\hat{L}$. Intuitively, this property ensures that the regular representation of L accounts for the decomposition along every path of a tree in the denoted tree language.

The theorem also says that it is sufficient to use regularity very much like in the word case, namely with only one product (concatenation) and one star-operator.

The subsets of pointed base trees which are the constituents of the regular representations can always be chosen in the form $\{f(x,t) \mid t \in T\}$ or $\{f(t,x) \mid t \in T\} \subseteq \Gamma_\Sigma$ for some set $T \subseteq T_\Sigma$ of trees. We abbreviate sets of this form by $f(x,T)$, or $f(T,x)$, respectively.

We can show that, for a recognizable tree language L, we can find a representation such that the sets T occurring in constituents of the above forms are recognizable. Now, if we use the regular tree expressions of Thatcher and Wright [18] in order to denote these sets

T (which, as we know from [18], is always possible), then the regular tree representations are exactly regular tree expressions in the sense of [18]—in a particular "path-oriented" form.

The following questions arise from these results (pointed out by W. Thomas). How weak can the expressibility for predicates defining these sets T in the constituents of the regular tree representation be so that still all recognizable tree languages are denotable? E.g., is first order definability sufficient? (If not: what is the family of tree languages obtained?) — For which power of expressibility (e.g. recursive predicates?) for defining these sets T is the problem still decidable whether the expression over the subsets of Γ_Σ is a border-invariant set of pointed trees, *i.e.*, is actually a regular tree representation?

We next investigate the structural property *aperiodicity* whose extension from word to tree languages was considered first by Thomas [19]. In our framework we may reformulate the definition of this property in a simple way: A tree language L is called aperiodic if the word language $\hat{L}$ is aperiodic. This means that the following (sharpened pumping property) holds.

$$\exists n \geq 1 \, \forall s, s', s'' \in S_\Sigma : \; s' \, s^n \, s'' \in \hat{L} \text{ iff } s' \, s^{n+1} \, s'' \in \hat{L}.$$

A tree language L has a *star-free regular representation* if $\hat{L}$ can be denoted by a star-free regular expression, *i.e.*, one without the star-operator (but with the complement). In other words, $\hat{L}$ lies in the closure of the powerset of the base Γ_Σ against the Boolean operations and the concatenation of pointed trees (the latter canonically extended from elements to sets).

Thanks to our monoid framework, we are able to apply the deep Theorem of Schützenberger for recognizable subsets over a free monoid, which can easily be extended from the case of a finite base to the case of an infinite base. By also applying *Theorem 1*, we obtain the only (up to now) known characterization of aperiodic tree languages.

Theorem 3 *[16, 10] A recognizable tree language L is aperiodic if and only if it has a star-free regular representation.*

[16, 10] Again we can choose the constituents of the regular representations to be sets of the form $f(x, T)$, or $f(T, x)$. Given an aperiodic tree language L, we can now find a regular representation such that the sets T occurring in constituents are not only recognizable, but also aperiodic.

Moreover, we can use the regular tree expressions of Thatcher and Wright [18] in order to denote these sets T and then obtain regular tree expressions in the sense of [18] in "path-oriented" form "with star-free paths." Without formalizing this form we note that the converse is true as well: If a tree language is denoted by such a regular expression, then it is aperiodic.—The open problem here is: Can the sets T occurring in the regular representations always be expressed by regular tree expressions as in [18], but *without* star-operators? A positive answer to this question would solve the open problem whether every aperiodic tree language L can be expressed by regular tree expressions as in [18] without star-operators (which is equivalent to: L is anti-chain definable). We know that the converse is not true [19]. Every first-order definable tree language is aperiodic, but again the converse does not hold [7].

7 Minimization and syntactic monoids

Given a tree automaton $A = (Q, (f^A)_{f \in \Sigma}, q_0, Q_{fin})$ accepting the tree language L, we calculate the minimal (deterministic frontier-to-root) tree automaton A^L of L as follows.

(1) For each state $q \in Q$ we find a tree t_q whose evaluation by A yields the state q, *i.e.*, $t_q{}^A = q$. We form the subset of Γ_Σ given by:

$$\Gamma_{\Sigma Q} = \{s \mid s = f(x, t_q) \text{ or } s = f(t_q, x), \ f \in \Sigma, \ q \in Q\}.$$

The size of $\Gamma_{\Sigma Q}$ is $2 \cdot |\Sigma| \cdot |Q|$.

(2) We minimize the automaton $\hat{A} = (Q, \hat{\delta}, q_0, Q_{fin})$ on words over $\Gamma_{\Sigma Q}$, which is defined as in Section 5, but with its transition function $\hat{\delta}$ restricted to $\Gamma_{\Sigma Q}$ as the alphabet.

(3) The minimal tree automaton A^L of L is obtained from A by identititifying states q and q' whenever they were identitified by the minimization of $\hat{A}$.

Since the transition tables of $\hat{A}$ and A have the same size (which we will denote $|A|$ and call the size of the tree automaton—which is not necessarily proportional to $|Q|^2$, but possibly $|A| << |Q|^2$), and the known algorithms on finite automata on words can be applied on $\hat{A}$, we obtain:

Theorem 4 *The minimization of tree automata A can be done in $O(|A| \cdot log|Q|)$.*

We mention another complexity result using the monoid approach.

Theorem 5 *[14] The equivalence of two tree automata A_1 and A_2 can be decided in $O(n \cdot \alpha(n))$, i.e., in time* almost linear *in the size of the tree automata $n = |A_1| + |A_2|$.*

It is well known by the work of Brainerd [1] and others that the states of A^L are the classes of the Nerode congruence $\approx_L$ of L. This (algebra!) congruence is defined as the smallest congruence on the tree algebra $\mathcal{T}_\Sigma$ saturating L. If we define $\sim_L$ to be the smallest left-congruence on the tree monoid saturating $\hat{L}$, then we have a bijection:

$$\mathcal{T}_{\Sigma / \approx_L} \leftrightarrow S_{\Sigma / \sim_L}; \quad [t]_{\approx_L} \leftrightarrow [s]_{\sim_L} \text{ if } s[\Omega] = t.$$

Observe that this is not an isomorphism: the set on the right-hand side has neither an algebra nor a monoid structure.

With respect to Remark 2 it is interesting to note that $\sim_{[\Omega]}$ refines the left-congruence $\sim_L$ (*i.e.*, $s \sim_{[\Omega]} s'$ implies $s \sim_L s'$)..

We define the *syntactic monoid of a tree language* L, denoted $\mathcal{M}(L)$, to be the syntactic monoid of the subset $\hat{L}$ in the free monoid $S_\Sigma = \Gamma_\Sigma{}^\star$. Thus, $\mathcal{M}(L) = S_{\Sigma/\equiv_L}$ is the quotient of the tree monoid with the syntactic congruence of $\hat{L}$ which we simply denote $\equiv_L$. Thus, $\equiv_L$ is defined as the smallest congruence on S_Σ which saturates $\hat{L}$, *i.e.*, $s_1 \equiv_L s_2$ iff

$$\forall s', s'' \in S_\Sigma : \ s' s_1 s'' \in \hat{L} \ \text{iff} \ s' s_2 s'' \in \hat{L}.$$

It is now easy to see that the transition monoid of the minimal tree automaton A^L is identical to the syntactical monoid of L,

$$\mathcal{M}(A^L) = \mathcal{M}(L).$$

This is an identity in the algebraic sense, *i.e.*, the two monoids above (with the product canonically induced from S_Σ) are isomorphic.

We will next study how to compute the *transition monoid of a tree automaton*. Then, together with the above identity, we have obtained a way to effectively compute the syntactic monoid of a tree language.

Given any (deterministic finite frontier-to-root) tree automaton A (with, say, the underlying finite algebra $\mathcal{A}$), its transition monoid is defined as $\mathcal{M}(A) = \{s^{\mathcal{A}} \mid s \in S_\Sigma\}$, together with functional composition as product and the identity on Q a the neutral element. Clearly, $\mathcal{M}(A)$ is the transition monoid of the automaton $\hat{A}$ on words over Γ_Σ.

Since $\mathcal{M}(A)$ is the homomorphic image of $S_\Sigma = \Gamma_\Sigma{}^\star$ under the mapping $s \mapsto s^{\mathcal{A}}$, it is generated by the set $\Gamma_\Sigma{}^{\mathcal{A}} = \{s^{\mathcal{A}} \mid s \in \Gamma_\Sigma\}$. Clearly, if we construct the finite alphabet $\Gamma_{\Sigma Q}$ (of size $\leq 2{\cdot}|\Sigma|{\cdot}|Q|$) as described as above, then we have:

$$\Gamma_\Sigma{}^{\mathcal{A}} = \Gamma_{\Sigma Q}{}^{\mathcal{A}}.$$

Thus we obtain

Remark 4 *The transition monoid of a tree automaton $\mathcal{M}(A)$ is the transition monoid of the automaton $\hat{A}$ on words over $\Gamma_{\Sigma Q}$, i.e., with its transition function $\hat{\delta}$ restricted to $\Gamma_{\Sigma Q}$ as the alphabet.*

Thus, for the calculation of the syntactic monoid of a tree language the already existing programs, e.g. MONOID, implemented at the LITP, or AMORE, implemented at the University of Aachen, can be used.

We define an *M-variety of tree languages* a family $\mathcal{V}(T_\Sigma)$ (indexed by the alphabet Σ) of tree languages L such that the family of associated word languages $\hat{L}$ constitutes exactly the set of the border-invariant languages of an M-variety $\mathcal{V}(\Gamma_\Sigma)$ of word languages (in the sense of [4]; *cf.* also [15]); *i.e.*, for all alphabets Σ,

$$\mathcal{V}(T_\Sigma) = \{L \subseteq T_\Sigma \mid \hat{L} \in \mathcal{V}(\Gamma_\Sigma)\}.$$

Applying the Stream Theorem of Eilenberg [4], there exists a unique M-variety of monoids $\underline{V}$ such that, for any alphabet Σ and any tree language $L \subseteq T_\Sigma$, $L \in \mathcal{V}(\Sigma)$ if and only if $\mathcal{M}(L) \in \underline{V}$. An S-variety of tree languages is defined in the analogous way, replacing monoids by semi-groups; the syntactic semi-group $\mathcal{S}(L)$ of a tree language L is $\Gamma_{\Sigma/\equiv_L}^+$, with $\Gamma_\Sigma^+ = (\Gamma_\Sigma^* - \{x\})$.

Given a structural property, the problem is to determine whether the tree languages which are characterized by this property form an M-variety (or an S-variety). This step generally uses the border-invariance of the set $\hat{L}$ in a non-forward way. *If* this is the case, then the deep and already well-founded theory of varieties of word languages (*cf.* [4, 15]) can be applied. With this method, it has been possible to characterize the varieties of finite/cofinite, definite, aperiodic, recognizable, label-counting (modulo or threshold) tree languages by exactly the same semigroup/monoid varieties as in the word case. This method yields simple decision tools for structural properties of tree languages. Whether these exist for local testability, or for piecewise testability, are still open problems which, however, we hope to solve applying the methods outlined above.

References

[1] Brainerd, W.S. "The Minimization of tree automata" *Information and Control* **13** (1968) 484-491.

[2] Courcelles, B. "Tree Automata and Recognizability" in: Aït-Kaci, H., and Nivat, M., eds., *Resolution of Equations in Algebraic Structures*, Vol. 1 (Academic Press, London, 1989).

[3] Downey, P.J., Sethi, R., Tarjan, R.E. "Variations on the Common Subexpression Problem". *J. ACM* **25** 4 (1980) 758-771.

[4] Eilenberg, S., *Automata, Languages and Machines, vol. B.* (Academic Press, New York, 1976).

[5] F. Gecsec, M. Steinby. *Tree Automata.* (Akademiai Kiado, Budapest, 1984).

[6] Heuter, U. *Zur Klassifizierung regulärer Baumsprachen.* Dissertation, University of Aachen (1989), Aachen, FRG.

[7] Heuter, U. "First-order properties of trees, star-free expressions and aperiodicity". *RAIRO Informatique* 25 (1991), pp. 125-145.

[8] Muller, D.E. *Notes on the Theory of Automata.* Script, University of Illinois, Urbana-Champain (1987).

[9] Nivat, M., and Podelski, A. "Tree Monoids and Recognizable Sets of Trees", in: Aït-Kaci, H., and Nivat, M., eds., *Resolution of Equations in Algebraic Structures,* Vol. 1 (Academic Press, London, 1989); Technical Report 87-43 LITP, Université Paris 7 (1987).

[10] Nivat, M., and Podelski, A. "Regular Tree Representations, the Theorems of Kleene and Schützenberger". submitted for publication.

[11] N. Nivat and A. Podelski, Definite tree languages (cont'd), *Bulletin of the EATCS* **38**, 1989, 186–190.

[12] Nivat, M., and Podelski, A. "Another Variation of the Common Subexpression Problem". to appear in *Discrete Mathematics.*

[13] Nivat, M., and Podelski, A. "Minimal ascending and descending tree automata". Rapport LITP N° 89-39, submitted for publication.

[14] Nivat, M., and Podelski, A. "Regular Tree Expressions". Rapport LITP N° 89-40, submitted for publication.

[15] Pin, J.E., *Varieties of formal languages.* (Plenum Press, New York, 1986)

[16] Podelski, A. *Tree Automata and Tree Monoids.* Dissertation. Université Paris 7 (1989), Paris, France.

[17] Rhodes, J. *Private Communication.* Berkeley, Ca. (1989).

[18] Thatcher, J.W., Wright, J.B. "Generalized Finite Automata Theory with an Application to a Decision Problem of Second Order Logic" *Math.Syst. Th.* **2** (1968).

[19] W. Thomas. "Logical aspects in the study of tree languages", in: Courcelle, B., ed., *Proc. Ninth Coll. Trees in Algebra and in Programming,* (Cambridge Univers. Press, Cambridge, 1984) 31-49.

Tree Automata and Languages
M. Nivat and A. Podelski (editors)
© 1992 Elsevier Science Publishers B.V. All rights reserved.

A THEORY OF TREE LANGUAGE VARIETIES

Magnus Steinby

Department of Mathematics, University of Turku
SF-20500 Turku, Finland

1. INTRODUCTION

Syntactic monoids and semigroups have played a central role in
the theory of regular languages ever since the discoveries of
M.P. Schützenberger in the mid-sixties, and S. Eilenberg's
theory of varieties is a general framework for the study of
families of regular languages characterizable by them (*cf*
[10,19,20,25]). Of course, tree language counterparts of these
notions would be highly desirable. Although in the seventies
several ways to define transition monoids of tree automata
were suggested (*cf* [13], p. 123, for references), it was only
much later that Thomas [32,33] introduced syntactic congruences and monoids of tree languages. These have been studied
further by Nivat and Podelski [22,23].
Here we shall present an alternative approach to the classification of regular tree languages, originally put forward in
[28], which is based on syntactic algebras rather than syntactic monoids. This idea ensues quite naturally form Mezei's and
Wright's [21] general notion of recognizable subsets of algebras if one regards regular string languages as recognizable
subsets of free monoids (or semigroups) and regular tree
languages as recognizable subsets of term algebras (*cf* also
[29]). In fact, [28] was formulated so that both varieties of
string languages and varieties of tree languages can be seen
as special cases of the same general concept. Here the theory
is, however, presented explicitly for tree languages. On the
other hand, it will be extended into another direction by
introducing varieties of congruences. Hence, we shall obtain
an extended form of Eilenberg's variety theorem which establishes triple bijections between varieties of tree languages,
varieties of finite algebras and varieties of congruences. The
varieties of congruences arise naturally from the study of
concrete examples; many varieties of tree languages are
defined most easily using some suitable congruences of term
algebras. In the case of string languages varieties of
congruences have been studied and utilized by Thérien [30,31].
A general notion of varieties of congruences (practically the
same as ours) has recently been considered by Almeida [1]. The
reader interested in the algebraic aspects of the topic is
urged to consult this paper for an alternative generalized

variety theorem as well as for many further results concerning varieties of finite algebras. The emphasis of this paper is on the general properties of syntactic algebras and the three types of varieties mentioned above, but to show the applicability of the theory, examples of tree language varieties are also considered. However, much remains to be done even with these rather simple varieties, and other families of tree languages can be hoped to fit into the scheme.
It has not been possible to make the presentation totally self-contained, but a working knowledge of basic universal algebra must be assumed. This can perhaps be acquired most conveniently by studying Chapter II of [7] or the first six chapters of [18]. On the other hand, I have tried to increase the readability of the paper and its tutorial usefulness by including even some proofs and explanations which somebody undoubtedly will find superfluous. As regards tree languages, the formalism follows closely that used in [13,28,29].
We shall often refer to propositions and lemmas simply by their numbers.

2. PRELIMINARIES

In what follows, Σ is always a *ranked alphabet*, *ie* a finite set of symbols each of which has a nonnegative integer *arity* or *rank*. For any $m \geq 0$, Σ_m denotes the set of m-ary symbols in Σ. We call Σ *nontrivial* if $\Sigma \neq \Sigma_0$.

In a Σ-*algebra* $\mathcal{A} = (A,\Sigma)$, A is the nonempty set of *elements* of $\mathcal{A}$ and each $f \in \Sigma_m$ is realized as an operation $f^{\mathcal{A}} \colon A^m \longrightarrow A$ $(m \geq 0)$. If $\mathcal{A} = (A,\Sigma)$ and $\mathcal{B} = (B,\Sigma)$ are Σ-algebras, then

(1) $\mathcal{A} \subseteq \mathcal{B}$ means that $\mathcal{A}$ is isomorphic to a subalgebra of $\mathcal{B}$,

(2) $\mathcal{A} \twoheadleftarrow \mathcal{B}$ means that $\mathcal{A}$ is an epimorphic image of $\mathcal{B}$, and

(3) $\mathcal{A} < \mathcal{B}$ means that $\mathcal{B}$ *covers* $\mathcal{A}$, *ie* $\mathcal{A}$ is an epimorphic image of a subalgebra of $\mathcal{B}$.

The letters X and Y will denote ordinary (finite) alphabets disjoint from Σ.The set $T(\Sigma,X)$ of all Σ-*terms over* X is the smallest set T for which $X \subseteq T$ and $ft_1...t_m \in T$ whenever $m \geq 0$, $f \in \Sigma_m$ and $t_1,...,t_m \in T$. The elements of $T(\Sigma,X)$ are regarded as formal representations of labelled left-to-right ordered finite trees, and hence we shall call them ΣX-*trees* (or just *trees*). Since Σ is always fixed, we write $T(X)$ instead of $T(\Sigma,X)$. Subsets of $T(X)$ are called ΣX-*tree languages* or just *tree languages*.

The *height* and the *root* of a ΣX-tree t, denoted respectively by *hg*(t) and *root*(t), are defined as follows:

(1) $hg(t) = 0$ and *root*$(t) = t$ for every t in $X \cup \Sigma_0$;

(2) $hg(t) = \max(hg(t_1),\ldots,hg(t_m)) + 1$ and *root*$(t) = f$, if t is of the form $ft_1\ldots t_m$ $(m > 0)$.

The *ΣX-term algebra* $\mathcal{T}(\Sigma,X) = (T(X),\Sigma)$, which we shall denote simply by $\mathcal{T}(X)$, is defined so that for any f in Σ_m and $t_1,\ldots,t_m$ in $T(X)$, $f^{\mathcal{T}(X)}(t_1,\ldots,t_m) = ft_1\ldots t_m$.

It is well-known that any mapping $\alpha\colon X \longrightarrow A$ into a Σ-algebra $\mathcal{A}$ has a unique extension to a morphism $\alpha^{\mathcal{A}}\colon \mathcal{T}(X) \longrightarrow \mathcal{A}$.

A *ΣX-recognizer* $\mathbf{A} = (\mathcal{A},\alpha,F)$ consists of a finite Σ-algebra $\mathcal{A} = (A,\Sigma)$, an *initial assignment* $\alpha\colon X \longrightarrow A$, and a set F $(\subseteq A)$ of *final states*. The *tree language recognized* by $\mathbf{A}$ is $T(\mathbf{A}) = \{t \in T(X)\colon t\alpha^{\mathcal{A}} \in F\}$. A ΣX-tree language T is *recognizable*, if $T = T(\mathbf{A})$ for some ΣX-recognizer $\mathbf{A}$. The set of all recognizable ΣX-tree languages is denoted by $\mathcal{R}ec(\Sigma,X)$ or by $\mathcal{R}ec(X)$.

Finally, a few conventions concerning relations and mappings:

If θ $(\subseteq A{\times}B)$ is a relation from A to B, we express $(a,b) \in \theta$ by writing $a\,\theta\,b$ or $a \equiv b\,(\theta)$. For any $a \in A$ and any $L \subseteq A$, let $a\theta = \{b \in B : a\,\theta\,b\}$ and $L\theta = \bigcup(a\theta : a \in L)$. The *product* of two relations θ $(\subseteq A{\times}B)$ and ρ $(\subseteq B{\times}C)$ is the relation $\theta{\circ}\rho = \{(a,c)\colon (\exists b{\in}B)\ a\,\theta\,b,\ b\,\rho\,c\}$ $(\subseteq A{\times}C)$. The *converse* of θ is the relation $\theta^{-1} = \{(b,a)\colon a\,\theta\,b\}$. These notations are also applied to mappings when these are regarded as special relations.

If θ is an equivalence relation of A, then the θ-class of an element a $(\in A)$ is denoted by a/θ, and the quotient set modulo θ is $A/\theta = \{a/\theta : a \in A\}$. A subset L of A is said to be *saturated* by θ, if L is the union of some θ-classes.

If $\varphi\colon A \longrightarrow B$ is a mapping and θ is an equivalence on B, then $\varphi{\circ}\theta{\circ}\varphi^{-1}$ is an equivalence on A such that if $a,a' \in A$, then $a \equiv a'\ (\varphi{\circ}\theta{\circ}\varphi^{-1})$ iff $a\varphi\,\theta\,a'\varphi$.

3. SYNTACTIC CONGRUENCES AND SYNTACTIC ALGEBRAS

Two words s,t $(\in X^*)$ are *syntactically equivalent* with respect to a language L $(\subseteq X^*)$ if they appear in L in exactly the same *contexts* u_v, where $u,v \in X^*$. Such a context u_v corresponds to a *translation* $X^* \longrightarrow X^*$, $s \longmapsto usv$ of the free monoid X^*. Hence, s and t are syntactically equivalent with respect to L

exactly in case $p(s) \in L$ iff $p(t) \in L$, for every translation p of X^*. The generalization goes as follows:

The set $\mathrm{Tr}(\mathcal{A})$ of *translations* [8] of a Σ-algebra $\mathcal{A}$ is the smallest set of mappings $A \longrightarrow A$ which contains the identity mapping 1_A of A, all *elementary translations*

$$p(\xi) = f^{\mathcal{A}}(a_1, \ldots, a_{i-1}, \xi, a_{i+1}, \ldots, a_m) \quad (m > 0,\ f \in \Sigma_m,\ a_j \in A),$$

and is closed under composition.

The *syntactic congruence* $\approx_L$ of a subset L $(\subseteq A)$ of a Σ-algebra $\mathcal{A}$ is defined by the condition

$$a \approx_L b \quad \text{iff} \quad (\forall p \in \mathrm{Tr}(\mathcal{A}))(p(a) \in L \Leftrightarrow p(b) \in L) \qquad (a, b \in A).$$

It is easy to see that $\approx_L$ has even in the general case the following fundamental property of syntactic congruences.

Proposition 3.1 The syntactic congruence $\approx_L$ of a subset L of a Σ-algebra $\mathcal{A}$ is the greatest congruence of $\mathcal{A}$ which saturates L.
$\qquad\qquad\square$

The *syntactic algebra* $\mathcal{A}/L = (A/L, \Sigma)$ of a subset L of $\mathcal{A}$ is the quotient algebra $\mathcal{A}/\approx_L$, and the canonical morphism

$$\varphi_L : \mathcal{A} \longrightarrow \mathcal{A}/L, \quad a \longmapsto a/L$$

is called the *syntactic morphism* of L; here a/L denotes the $\approx_L$-class of the element a $(\in A)$ and $A/L = \{a/L : a \in A\}$ is the quotient set modulo $\approx_L$.

We shall need the following technical lemma.

Lemma 3.2 Let $\varphi : \mathcal{A} \longrightarrow \mathcal{B}$ be a morphism of Σ-algebras. For every p in $\mathrm{Tr}(\mathcal{A})$ there exists a p_φ in $\mathrm{Tr}(\mathcal{B})$ such that $p(a)\varphi = p_\varphi(a\varphi)$, for every a in A. If φ is surjective, then there is for every q in $\mathrm{Tr}(\mathcal{B})$ a p in $\mathrm{Tr}(\mathcal{A})$ such that $q = p_\varphi$.

Proof. It suffices to note that for an elementary translation

$$p(\xi) = f^{\mathcal{A}}(a_1, \ldots, a_{i-1}, \xi, a_{i+1}, \ldots, a_m),$$

we may set

$$p_\varphi(\xi) = f^{\mathcal{B}}(a_1\varphi, \ldots, a_{i-1}\varphi, \xi, a_{i+1}\varphi, \ldots, a_m\varphi).$$
$\qquad\qquad\square$

Proposition 3.3 Let $\mathcal{A}$ and $\mathcal{B}$ be any Σ-algebras.

(a) For any L ($\subseteq$ A), $\approx_{A-L} = \approx_L$.

(b) For any K,L ($\subseteq$ A), $\approx_K \cap \approx_L \subseteq \approx_{K\cap L}$.

(c) For any L ($\subseteq$ A) and any p in $\mathrm{Tr}(\mathcal{A})$, $\approx_L \subseteq \approx_{p^{-1}(L)}$.

(d) For any morphism $\varphi\colon \mathcal{A} \longrightarrow \mathcal{B}$ and any subset L of B, $\varphi\circ\approx_L\circ\varphi^{-1} \subseteq \approx_{L\varphi^{-1}}$, and equality holds if φ is surjective.

Proof. The claims (a) and (b) are trivial. For (c) it suffices to observe that

$$a \approx_L b \implies (\forall q \in \mathrm{Tr}(\mathcal{A}))(p(q(a)) \in L \Leftrightarrow p(q(b)) \in L) \implies$$

$$(\forall q \in \mathrm{Tr}(\mathcal{A}))(q(a) \in p^{-1}(L) \Leftrightarrow q(b) \in p^{-1}(L)) \implies a \approx_{p^{-1}(L)} b.$$

In the proof of (d) we use 3.2:

$$a \equiv a' \ (\varphi\circ\approx_L\circ\varphi^{-1}) \ \Leftrightarrow\ a\varphi \approx_L a'\varphi$$

$$\implies (\forall p \in \mathrm{Tr}(\mathcal{A}))(p_\varphi(a\varphi) \in L \Leftrightarrow p_\varphi(a'\varphi) \in L)$$

$$\Leftrightarrow (\forall p \in \mathrm{Tr}(\mathcal{A}))(p(a) \in L\varphi^{-1} \Leftrightarrow p(a') \in L\varphi^{-1})$$

$$\Leftrightarrow a \approx_{L\varphi^{-1}} a'.$$

If φ is surjective, the only '$\implies$' also becomes an equivalence, which proves the last claim of (d). $\qquad\square$

Now the corresponding facts for syntactic algebras:

Proposition 3.4 Let $\mathcal{A}$ and $\mathcal{B}$ be Σ-algebras.

(a) For any L ($\subseteq$ A), $\mathcal{A}/A-L = \mathcal{A}/L$.

(b) For any K,L ($\subseteq$ A), $\mathcal{A}/K\cap L \subseteq \mathcal{A}/K \times \mathcal{A}/L$.

(c) For any L ($\subseteq$ A) and any p in $\mathrm{Tr}(\mathcal{A})$, $\mathcal{A}/p^{-1}(L) \twoheadleftarrow \mathcal{A}/L$.

(d) For any morphism $\varphi\colon \mathcal{A} \longrightarrow \mathcal{B}$ and any subset L of B, $\mathcal{A}/L\varphi^{-1} < \mathcal{B}/L$. If φ is an epimorphism, then $\mathcal{A}/L\varphi^{-1} \cong \mathcal{B}/L$.

Proof. The claims (a), (b) and (c) follow directly from the corresponding parts of 3.3. For example, 3.3(b) implies that

$$\varphi\colon A/K\cap L \longrightarrow A/K \times A/L, \quad a/K\cap L \longmapsto (a/K, a/L)$$

is an embedding of $\mathcal{A}/K\cap L$ into $\mathcal{A}/K \times \mathcal{A}/L$. For proving (d) consider first the case where φ is an epimorphism. By using 3.2 one easily shows that $\psi\colon A/L\varphi^{-1} \longrightarrow B/L, \ a/L\varphi^{-1} \longmapsto a\varphi/L$ is well-defined and injective. Obviously, ψ is surjective, and a straightforward computation shows that it is a morphism from $\mathcal{A}/L\varphi^{-1}$ to $\mathcal{B}/L$. Hence $\mathcal{A}/L\varphi^{-1} \cong \mathcal{B}/L$. Consider any morphism

$\varphi: \mathcal{A} \longrightarrow \mathcal{B}$. Let $\mathcal{C} = (C, \Sigma)$ be the subalgebra of $\mathcal{B}/L$ obtained as the image of $\mathcal{A}$ under the morphism $\varphi\varphi_L: \mathcal{A} \longrightarrow \mathcal{B}/L$. Then

$$\gamma: \mathcal{A} \longrightarrow \mathcal{C}, \quad a \longmapsto a\varphi\varphi_L$$

is an epimorphism, so by the first part of the proof, $\mathcal{A}/(L\varphi^{-1}\gamma)\gamma \cong \mathcal{C}/L\varphi^{-1}\gamma$. But one easily shows that $(L\varphi^{-1}\gamma)\gamma^{-1} = L\varphi^{-1}$. Since $\mathcal{C} \subseteq \mathcal{B}/L$, this means that $\mathcal{A}/L\varphi^{-1} < \mathcal{B}/L$. $\quad\square$

Lemma 3.5 If $\varphi: \mathcal{A} \longrightarrow \mathcal{B}$ is a morphism and $L \subseteq B$, then
$$\varphi \circ \approx_L \circ \varphi^{-1} = \bigcap (\approx_{p^{-1}(L)} \varphi^{-1}: p \in \mathrm{Tr}(\mathcal{B})).$$

Proof. Let ρ be the righthand side of the claimed equality. The parts (c) and (d) of 3.3 imply that for any p in $\mathrm{Tr}(\mathcal{A})$,

$$\varphi \circ \approx_L \circ \varphi^{-1} \subseteq \varphi \circ \approx_{p^{-1}(L)} \circ \varphi^{-1} \subseteq \approx_{p^{-1}(L)} \varphi^{-1}.$$

Hence $\varphi \circ \approx_L \circ \varphi^{-1} \subseteq \rho$. The converse inclusion is given by the following chain of implications in which $a, a' \in A$, and p and q range over $\mathrm{Tr}(\mathcal{B})$:

$$a \rho a' \implies (\forall p)(a\varphi \approx_{p^{-1}(L)} a'\varphi)$$
$$\implies (\forall p, q)(q(a\varphi) \in p^{-1}(L) \leftrightarrow q(a'\varphi) \in p^{-1}(L))$$
$$\implies (\forall p, q)(p(q(a\varphi)) \in L \leftrightarrow p(q(a'\varphi)) \in L)$$
$$\implies a\varphi \approx_L a'\varphi \implies a \, \varphi \circ \approx_L \circ \varphi^{-1} \, a'.$$

$\quad\square$

Let us call an algebra *syntactic*, if it is isomorphic to the syntactic algebra of a subset of some algebra. A subset of an algebra $\mathcal{A}$ is said to be *disjunctive*, if its syntactic congruence is the diagonal relation δ_A of A.

Proposition 3.6 An algebra is syntactic iff it has a disjunctive subset.

Proof. Consider any syntactic algebra $\mathcal{A}/L$. We claim that $K = \{a/L: a \in L\}$ is disjunctive in $\mathcal{A}/L$. Let $\varphi: \mathcal{A} \longrightarrow \mathcal{A}/L$ be the syntactic morphism. If $a/L \approx_K a'/L$ for some a and a' in A, then for every p in $\mathrm{Tr}(\mathcal{A})$,

$$p(a) \in L \leftrightarrow p(a)\varphi = p_\varphi(a/L) \in K \leftrightarrow p_\varphi(a'/L) = p(a')\varphi \in K$$
$$\leftrightarrow p(a') \in L,$$

and hence a/L = a'/L. This proves the claim. The converse part is trivial; if L is disjunctive in $\mathcal{A}$, then $\mathcal{A} \cong \mathcal{A}/L$. $\quad\square$

The following proposition generalizes a result by Schein [26].

Proposition 3.7 Every subdirectly irreducible algebra is syntactic.

Proof. Let $\mathcal{A}$ be any algebra. It is obvious that for any a and b in A, $a \approx_{\{a\}} b$ implies a = b. Hence $\bigcap (\approx_{\{a\}}: a \in A) = \delta_A$. If $\mathcal{A}$ is subdirectly irreducible, this implies by Birkhoff's subdirect decomposition theorem that $\approx_{\{a\}} = \delta_A$ for at least one a in A. Therefore $\mathcal{A}$ is syntactic by 3.6. $\quad\square$

By the *syntactic congruence* $\approx_T$ of a ΣX-tree language T we mean its syntactic congruence as a subset of the term algebra $\mathcal{T}(X)$. For the syntactic algebra $\mathcal{T}(X)/T$ we use the notation SA(T). The next two propositions both depend on the fact that $\mathcal{T}(X)$ is freely generated by X over the class of all Σ-algebras.

Proposition 3.8 A finitely generated algebra is syntactic iff it is isomorphic to the syntactic algebra of a tree language. More precisely: if the Σ-algebra $\mathcal{A}$ is generated by a subset G such that $|G| \leq |X|$, then $\mathcal{A}$ is syntactic iff $\mathcal{A} \cong SA(T)$ for some ΣX-tree language T.

Proof. Suppose $\mathcal{A}$ is a syntactic algebra with D as a disjunctive subset. If G generates $\mathcal{A}$ and $|G| \leq |X|$, there is an epimorphism $\varphi: \mathcal{T}(X) \longrightarrow \mathcal{A}$, and by 3.4(d), $\mathcal{A} \cong \mathcal{A}/D \cong SA(D\varphi^{-1})$. The converse part is even more obvious. $\quad\square$

Proposition 3.9 For every Σ-algebra $\mathcal{A}$ and every ΣX-tree language T, the following two conditions are equivalent:
(1) $T = P\varphi^{-1}$ for some morphism $\varphi: \mathcal{T}(X) \longrightarrow \mathcal{A}$ and some $P \subseteq A$;
(2) $SA(T) < \mathcal{A}$.

Proof. The implication (1) $\Rightarrow$ (2) follows from 3.4(d). On the other hand, $SA(T) < \mathcal{A}$ means that there is a subalgebra $\mathcal{B}$ of $\mathcal{A}$ with an epimorphism $\psi: \mathcal{B} \longrightarrow SA(T)$. Since $\mathcal{T}(X)$ is free, there exists a morphism $\eta: \mathcal{T}(X) \longrightarrow \mathcal{B}$ such that $\eta\psi = \varphi_T$. Let P =

$T\varphi_T\psi^{-1}$ ($\subseteq A$). If we denote by φ the morphism from $\mathcal{T}(X)$ to $\mathcal{A}$ such that $t\varphi = t\eta$ for every t in T(X), then $P\varphi^{-1} = T\varphi_T\psi^{-1}\varphi^{-1} = T\varphi_T\varphi_T^{-1} = T$.

$\square$

In 3.9 (1) means that $\mathcal{A}$ can be turned into a (possibly infinite) ΣX-recognizer of T. Hence the proposition actually says that, for every tree language T, SA(T) gives the minimal recognizer of T; for infinite recognizers minimal with respect to the covering relation <. In particular, we also get

Corollary 3.10 A tree language T is recognizable iff SA(T) is finite.

$\square$

Finally, let us note that the translations of the term algebra $\mathcal{T}(X)$ are mappings

$p* : T(X) \longrightarrow T(X), \quad t \longmapsto p(t),$

where p is a $\Sigma(X\cup\{\xi\})$-tree in which ξ ($\notin \Sigma\cup X$) appears exactly once as the label of a leaf, and $p(t)$ is the ΣX-tree obtained from p by substituting t for ξ. Following [32,33] such trees p are called *special ΣX-trees*, and we denote their set by Sp(X). Also, we denote $Tr(\mathcal{T}(X))$ by Tr(X). Since $p \longmapsto p*$ ($p \in Sp(X)$) defines a natural bijection between Sp(X) and Tr(X), we shall write simply p for $p*$, too.

In $\mathcal{T}(X)$ syntactic equivalence with respect to some ΣX-tree language T means, exactly as in the string case, interchangeability in all 'contexts':

$s \approx_T t$ iff $(\forall p \in Sp(\Sigma))(p(s) \in T \Leftrightarrow p(t) \in T).$

Moreover, the inverses of translations are the natural counterparts of the quotient operations applied to string languages, and we have the following useful fact (which holds also more generally, *cf* [29,1]).

Lemma 3.11 If $T \in \mathcal{R}ec(X)$, then $\{p^{-1}(T): p \in Tr(X)\}$ is finite.

Proof. By 3.3(c), $\approx_T$ saturates every $p^{-1}(T)$. This implies the lemma since $\approx_T$ has but finitely many equivalence classes.

$\square$

4. VARIETIES OF FINITE ALGEBRAS

A *variety* of Σ-algebras is a class of Σ-algebras closed under the operations of forming subalgebras, epimorphic images and direct products. Analogously, a *variety of finite Σ-algebras* (here abbreviated Σ-VFA or just VFA) is a nonempty class **K** of finite Σ-algebras such that

(1) $\mathcal{A} \subseteq \mathcal{B}$, $\mathcal{B} \in$ **K** implies $\mathcal{A} \in$ **K**,

(2) $\mathcal{A} \twoheadleftarrow \mathcal{B}$, $\mathcal{B} \in$ **K** implies $\mathcal{A} \in$ **K**, and

(3) $\mathcal{A}, \mathcal{B} \in$ **K** implies $\mathcal{A} \times \mathcal{B} \in$ **K**.

Conditions (1) and (2) can be replaced by the single condition

(4) $\mathcal{A} < \mathcal{B}$, $\mathcal{B} \in$ **K** implies $\mathcal{A} \in$ **K**.

We denote the class of all Σ-VFAs by VFA(Σ).

Varieties of finite monoids were introduced by Eilenberg and Schützenberger [11] who call them pseudovarieties. These in connection with varieties of rational languages are discussed extensively in [10,19,25]. The general theory of VFAs has been studied in universal algebra (*cf* [1,2,3,4], for example]).

It is obvious that the intersection of any family of Σ-VFAs is a Σ-VFA. Therefore every class **K** of finite Σ-algebras is contained in a smallest Σ-VFA, the VFA *generated* by **K**, which we denote by VFA(**K**). Tarski's description of the variety generated by a given class of algebras has the following analogue. A proof is obtained by adapting the original proof to finite algebras and finite products. Note that $n = 0$ is allowed in order to capture the trivial algebras in case **K** is empty.

Proposition 4.1 If **K** is class of finite Σ-algebras and $\mathcal{A}$ is a finite Σ- algebra, then

$\mathcal{A} \in$ VFA(**K**) iff $\mathcal{A} < \mathcal{A}_1 \times \ldots \times \mathcal{A}_n$ for some $n \geq 0$ and $\mathcal{A}_1, \ldots, \mathcal{A}_n \in$ **K**.

□

A modification of Birkhoff's subdirect decomposition theorem shows that any VFA is generated by its subdirectly irreducible members. Hence 3.7 yields the following useful fact.

Proposition 4.2 Every VFA is generated by syntactic algebras.

□

Recall that a nonempty subset S of a partially ordered set is *directed*, if for any $a, b \in S$ there is a $c \in S$ such that $a, b \leq c$. In view of 4.1 it is obvious that the union of any directed set of Σ-VFAs is again a Σ-VFA. Together with the closure of VFA(Σ) under intersections this observation yields

Proposition 4.3 (VFA(Σ), ⊆) is an algebraic lattice in which $\inf(S) = \bigcap S$ and $\sup(S) = \text{VFA}(\bigcup S)$ whenever $S \subseteq \text{VFA}(\Sigma)$. □

5. VARIETIES OF TREE LANGUAGES

A mapping V which assigns to every X a nonempty subset $V(X)$ of $\mathcal{R}ec(X)$ is called a *family of regular Σ-tree languages*, and we write $V = \{V(X)\}$ with the understanding that X ranges over all alphabets. Such a family V is a *variety of Σ-tree languages* (abbreviated Σ-VTL or VTL), if the following four conditions hold for all alphabets X and Y:

(1) $T \in V(X)$ implies $T(X) - T \in V(X)$;

(2) $T, U \in V(X)$ implies $T \cap U \in V(X)$;

(3) $T \in V(X)$, $p \in \text{Tr}(X)$ implies $p^{-1}(T) \in V(X)$;

(4) if $\varphi: \mathcal{J}(X) \longrightarrow \mathcal{J}(Y)$ is a morphism and $T \in V(Y)$, then $T\varphi^{-1} \in V(X)$.

This definition parallels Eilenberg's [10] definitions of * - varieties and + - varieties with inverse translations in the role of quotient operations. Note, however, that in a Σ-VTL V every $V(X)$ is nonempty. Condition (3) could obviously be restricted to elementary translations.

The union $\bigcup (V_i : i \in I)$ and the intersection $\bigcap (V_i : i \in I)$ of a class of families $V_i = \{V_i(X)\}$ of regular Σ-tree languages $(i \in I)$ are defined as the families $U = \{U(X)\}$ and $V = \{V(X)\}$, respectively, such that for any X, $U(X) = \bigcup (V_i(X) : i \in I)$ and $V(X) = \bigcap (V_i(X) : i \in I)$. The inclusion relation is defined by componentwise inclusion: $U \subseteq V$ iff $U(X) \subseteq V(X)$ for every X.

The intersection of any class of Σ-VTLs is again a Σ-VTL, and hence the VTL VTL(V) *generated* by a family V of regular Σ-tree languages is defined as $\bigcap (U : V \subseteq U, U \in \text{VTL}(\Sigma))$.

It is also easy to see that the union of any directed family of Σ-VTLs is again a Σ-VTL. Hence we have the following fact.

Proposition 5.1 $(\mathrm{VTL}(\Sigma),\subseteq)$ is an algebraic lattice in which $\inf(\mathcal{F}) = \bigcap\mathcal{F}$ and $\sup(\mathcal{F}) = \mathrm{VTL}(\bigcup\mathcal{F})$ $(\mathcal{F} \subseteq \mathrm{VTL}(\Sigma))$. □

The least element in this lattice is the trivial Σ-VTL $\mathcal{T}\textit{riv}$ defined so that for any X, $\mathcal{T}\textit{riv}(\mathrm{X}) = \{\varnothing,\mathrm{T}(\mathrm{X})\}$. The greatest element is $\mathcal{R}ec = \{\mathcal{R}ec(\mathrm{X})\}$.
The following observation will be useful.

Lemma 5.2 If V is a Σ-VTL, then for every X and any T in $V(\mathrm{X})$, every $\approx$ - class is also in $V(\mathrm{X})$.

Proof. For any t in $\mathrm{T}(\mathrm{X})$,

$$t/T = \bigcap (p^{-1}(T): p\in\mathrm{Tr}(\mathrm{X}), p(t)\in T) - \bigcup (p^{-1}(T): p\in\mathrm{Tr}(\mathrm{X}), p(t)\notin T).$$

Since there are finitely many different sets $p^{-1}(T)$, each of which is in $V(\mathrm{X})$, this shows that $t/T \in V(\mathrm{X})$. □

6. VARIETIES OF FINITE CONGRUENCES

There are some formal differences, but the following notion of a variety of congruences is, when adapted to free monoids, equivalent to the concept used by Thérien [30,31], and it is practically the same as that studied by Almeida [1].
Let FCon(X) denote the set of all *finite congruences* of $\mathcal{T}(\mathrm{X})$, *ie* the congruence with a finite number of congruence classes. It is clear that FCon(X) forms a filter of the congruence lattice (Con $\mathcal{T}(\mathrm{X})$, $\subseteq$) of $\mathcal{T}(\mathrm{X})$. By a Σ-*class of finite congruences* we mean a mapping Γ which associates with each alphabet X a subset $\Gamma(\mathrm{X})$ of FCon(X), and we write then $\Gamma = \{\Gamma(\mathrm{X})\}$. Such a class Γ is called a Σ-*variety of finite congruences* (a Σ-VFC or VFC, for short), if for all X and Y
(1) $\Gamma(\mathrm{X})$ is a filter of Con $\mathcal{T}(\mathrm{X})$, and
(2) $\varphi\circ\theta\circ\varphi^{-1}\in \Gamma(\mathrm{X})$, for any morphism $\varphi:\mathcal{T}(\mathrm{X}) \longrightarrow \mathcal{T}(\mathrm{Y})$ and any θ in $\Gamma(\mathrm{Y})$.
The set of all Σ-VFCs is denoted by VFC(Σ).
The inclusion relation, intersections and unions of Σ-classes of finite congruences are defined in the natural componentwise way. As the intersection of any family of Σ-VFCs is obviously

again a Σ-VFC, the Σ-VFC *generated by any Σ-class* $\Gamma = \{\Gamma(X)\}$ can be defined as the Σ-VFC $VFC(\Sigma) = \bigcap (\Delta: \Gamma \subseteq \Delta, \Delta \in VFC(\Sigma))$. In any lattice, the union of a directed family of filters is a filter. This implies easily that the union of any directed family of Σ-VFCs is a Σ-VFC. Hence the following proposition.

Proposition 6.1 $(VFC(\Sigma),\subseteq)$ is an algebraic lattice in which $\inf(\mathcal{G}) = \bigcap \mathcal{G}$ and $\sup(\mathcal{G}) = VFC(\bigcup \mathcal{G})$ whenever $\mathcal{G} \subseteq VFC(\Sigma)$. $\qquad\square$

The least element of $VFC(\Sigma)$ is $I = \{\{\iota_X\}\}$, where $\iota_X = T(X)^2$, and the greatest element is $\Omega = \{FCon(X)\}$.

The following fact was noted for semigroups by Schein [26] and generally in [28]. The proof is straightforward.

Lemma 6.2 Every congruence θ of $\mathcal{T}(X)$ is the intersection of the syntactic congruence of some ΣX-tree languages. In fact, $\theta = \bigcap (\approx_{t/\theta}: t \in T(X))$. $\qquad\square$

In any lattice $(L,\leq)$, the filter generated by a subset H ($\neq \emptyset$) is $\{a \in L: (\exists G) \inf(G) \leq a, G \subseteq H, G \text{ finite}\}$. Any θ in $FCon(X)$ is the meet of the syntactic congruences of finitely many recognizable ΣX-tree languages; each t/θ is recognizable since it is saturated by θ. From these observations and 6.2 we get

Lemma 6.3 Every Σ-VFC is generated by syntactic congruences.

$\qquad\square$

Lemma 6.2 gives also the following fact related to 3.7.

Lemma 6.4 Every finite meet-irreducible congruence of $\mathcal{T}(X)$ is the syntactic congruence of a recognizable ΣX-tree language. $\qquad\square$

Many VFCs are of the special type described in the following proposition. For any element a of a lattice L, the *principal filter generated* by a is $[a) = \{x \in L: a \leq x\}$. Accordingly, these VFCs will be called *principal VFCs*.

Proposition 6.5 Suppose we are given for each alphabet X a congruence $\theta_X \in FCon(X)$. Then $\Gamma = \{[\theta_X)\}$ is a Σ-VFC iff $\theta_X \subseteq \varphi \circ \theta_Y \circ \varphi^{-1}$ for all X and Y and every morphism $\varphi: \mathcal{T}(X) \longrightarrow \mathcal{T}(Y)$. $\qquad\square$

7. THE VARIETY THEOREM

We shall now show that the lattices $(VTL(\Sigma), \subseteq)$, $(VFA(\Sigma), \subseteq)$ and $(VFC(\Sigma), \subseteq)$ are all isomorphic. First we define the mappings which, when restricted to varieties, give these isomorphisms.
a. For any class K of finite Σ-algebras, let K^t be the class of regular Σ-tree languages and K^c be the Σ-class of finite congruences such that for any X, $K^t(X) = \{T \subseteq T(X): SA(T) \in K\}$ and $K^c(X) = \{\theta \in FCon(X): \mathcal{T}(X)/\theta \in K\}$.
b. For any family V of regular Σ-tree languages, let $V^a = VFA(\{SA(T): T \in V(X)$ for some $X\})$, and let V^c be the Σ-class of finite congruences such that for every alphabet X, $V^c(X) = [\{\approx_T: T \in V(X)\}]$.
c. For any Σ-class Γ of finite congruences, let Γ^a be the Σ-VFA generated by all algebras $\mathcal{T}(X)/\theta$, where X is any alphabet and $\theta \in \Gamma(X)$. The family Γ^t of regular Σ-tree languages is defined so that for every X, $\Gamma^t(X) = \{T \subseteq T(X): \approx_T \in \Gamma(X)\}$.
First we observe the obvious isotonicity of these mappings.

Lemma 7.1
(a) If K and L are classes of finite Σ-algebras such that $K \subseteq L$, then $K^t \subseteq L^t$ and $K^c \subseteq L^c$.
(b) If U and V are families of regular Σ-tree languages such that $U \subseteq V$, then $U^a \subseteq V^a$ and $U^c \subseteq V^c$.
(c) If Γ and Δ are Σ-classes of finite congruences such that $\Gamma \subseteq \Delta$, then $\Gamma^a \subseteq \Delta^a$ and $\Gamma^t \subseteq \Delta^t$. □

Lemma 7.2
(a) $K \in VFA(\Sigma)$ implies $K^t \in VTL(\Sigma)$ and $K^c \in VFC(\Sigma)$.
(b) $V \in VTL(\Sigma)$ implies $V^a \in VFA(\Sigma)$ and $V^c \in VFC(\Sigma)$.
(c) $\Gamma \in VFC(\Sigma)$ implies $\Gamma^a \in VFA(\Sigma)$ and $\Gamma^t \in VTL(\Sigma)$.

Proof. (a) If $K \in VFA(\Sigma)$, then $K^t \in VTL(\Sigma)$ by 3.4. Let us show that $K^c \in VFC(\Sigma)$. If $\theta, \rho \in FCon(X)$, $\theta \subseteq \rho$ and $\theta \in K^c(X)$, then $\mathcal{T}(X)/\rho \twoheadleftarrow \mathcal{T}(X)/\theta$ implies $\mathcal{T}(X)/\rho \in K$ and hence $\rho \in K^c(X)$. Suppose now that $\theta, \rho \in K^c(X)$. Then $\mathcal{T}(X)/\theta, \mathcal{T}(X)/\rho \in K$ and $\mathcal{T}(X)/\theta\cap\rho < \mathcal{T}(X)/\theta \times \mathcal{T}(X)/\rho$ together imply $\theta\cap\rho \in K^c(X)$. Since obviously $\varnothing \neq K^c(X) \subseteq FCon(X)$, these facts imply that $K^c(X)$ is a filter of $Con\ \mathcal{J}(X)$ contained in $FCon(X)$.

Let $\varphi: \mathcal{J}(X) \longrightarrow \mathcal{J}(Y)$ be a morphism, and suppose $\theta \in K^c(Y)$. Let $\rho = \varphi \circ \theta \circ \varphi^{-1}$. It is easy to see that $\eta: T(X)/\rho \longrightarrow T(Y)/\theta$, $s/\rho \longmapsto s\varphi/\theta$ defines a monomorphism from $\mathcal{J}(X)/\rho$ into $\mathcal{J}(Y)/\theta$. Since $\mathcal{J}(Y)/\theta \in K$, this implies $\mathcal{J}(X)/\rho \in K$. Hence $\rho \in K^c(X)$.

(b) Let $V \in VTL(\Sigma)$. By definition, $V^a \in VFA(\Sigma)$. The definition of V^c already stipulates that $V^c(X)$ is a filter in $FCon(X)$. Let $\varphi: \mathcal{J}(X) \longrightarrow \mathcal{J}(Y)$ be a morphism and suppose $\theta \in V^c(Y)$. Then

$$\approx_{T_1} \cap \ \ldots \ \cap \approx_{T_n}\ \subseteq \theta \qquad\qquad (*)$$

for some $n \geq 1$ and some $T_1, \ldots, T_n$ in $V(Y)$. For each i,

$$\varphi \circ \approx_{T_i} \circ \varphi^{-1} = \bigcap\ (\approx_{p^{-1}(T_i)\varphi^{-1}}\ :\ p \in Tr(Y))$$

by 3.5. Since V is a VTL, every $p^{-1}(T_i)\varphi^{-1}$ is in $V(X)$, and by 3.11 there are only finitely many such sets. Hence every $\varphi \circ \approx_{T_i} \circ \varphi^{-1}$ is in $V^c(X)$. Now $(*)$ implies $\varphi \circ \theta \circ \varphi^{-1} \in V^c(X)$.

(c) For any Σ-VFC Γ, $\Gamma^a \in VFA(\Sigma)$ by definition. The fact $\Gamma^t \in VTL(\Sigma)$ follows directly from Proposition 3.3. $\qquad\qquad\square$

The strategy for proving the isomorphism of the three lattices will be as follows. First we show (as is [28]) that $K \longmapsto K^t$ and $V \longmapsto V^a$ define mutually inverse bijections between $VFA(\Sigma)$ and $VTL(\Sigma)$; we already know that they are isotone. Then we show that $K \longmapsto K^c$ and $\Gamma \longmapsto \Gamma^a$ form a similar pair of bijections between $VFA(\Sigma)$ and $VFC(\Sigma)$. The third isomorphism pair is derived indirectly by first showing that all these mappings compose as one would expect: $K^{ct} = K^t$, $V^{ac} = V^c$ etc.

Proposition 7.3 The lattices $(VFA(\Sigma), \subseteq)$ and $(VTL(\Sigma), \subseteq)$ are isomorphic: (a) $K^{ta} = K$, for every Σ-VFA K, and (b) $V^{at} = V$, for every Σ-VTL V.

Proof. It suffices to prove (a) and (b).

(a) The VFA K^{ta} is generated by the algebras $SA(T)$, where $T \in K^t(X)$ for some X. But every such algebra is also in K. Hence, $K^{ta} \subseteq K$. On the other hand, let $\mathcal{A}$ be any syntactic algebra in K. By 3.8, $\mathcal{A} \cong SA(T)$ for some ΣX-tree language T, if X is sufficiently large. Then $T \in K^t(X)$ and hence $\mathcal{A} \in K^{ta}$. This gives the inclusion $K \subseteq K^{ta}$.

(b) The inclusion $V \subseteq V^{at}$ is obvious: for any T in $V(X)$, $SA(T) \in V^a$ implies $T \in V^{at}(X)$. Suppose now that $T \in V^{at}(X)$. Then $SA(T) \in V^a$ which means that

$$SA(T) < SA(T_1) \times \ldots \times SA(T_k), \tag{1}$$

for some $k \geq 1$, $X_1, \ldots, X_k$ and $T_1 \in V(X_1), \ldots, T_k \in V(X_k)$. The syntactic morphisms

$$\varphi_i: \mathcal{T}(X_i) \longrightarrow SA(T_i) \qquad\qquad (i = 1, \ldots, k)$$

define a morphism $\eta = \varphi_1 \times \ldots \times \varphi_k$,

$$\eta: \mathcal{T}(X_1) \times \ldots \times \mathcal{T}(X_k) \longrightarrow SA(T_1) \times \ldots \times SA(T_k),$$

such that for every i, $1 \leq i \leq k$, $\eta \pi_i = \tau_i \varphi_i$, where

$$\pi_i: SA(T_1) \times \ldots \times SA(T_k) \longrightarrow SA(T_i)$$

and

$$\tau_i: T(X_1) \times \ldots \times T(X_k) \longrightarrow T(X_i)$$

are the *i*th projections. From (1) and 3.9 it follows that we have a morphism $\varphi: \mathcal{T}(X) \longrightarrow SA(T_1) \times \ldots \times SA(T_k)$ and a subset H of $SA(T_1) \times \ldots \times SA(T_k)$ such that $T = H\varphi^{-1}$. Because $\mathcal{T}(X)$ is freely generated by X and η is surjective, we may define a morphism $\psi: \mathcal{T}(X) \longrightarrow \mathcal{T}(X_1) \times \ldots \times \mathcal{T}(X_k)$ such that $\psi\eta = \varphi$. Then

$$\varphi \pi_i = \psi \tau_i \varphi_i \tag{2}$$

for every $i = 1, \ldots, n$. Since H is finite, T is expressible as the union of finitely many sets $h\varphi^{-1}$ with $h = (h_1, \ldots, h_k)$ in $SA(T_1) \times \ldots \times SA(T_k)$. By using (2) we get

$$h\varphi^{-1} = \bigcap (h_i(\varphi\pi_i)^{-1}: 1 \leq i \leq k) = \bigcap (h_i\varphi_i^{-1}(\psi\tau_i)^{-1}: 1 \leq i \leq k).$$

By Lemma 5.2, each $h_i\varphi_i^{-1}$ is in $V(X_i)$. Since V is a VTL, this implies that also each $h_i\varphi_i^{-1}(\psi\tau_i)^{-1}$ is in $V(X)$. Hence $h\varphi^{-1} \in V(X)$ and $T \in V(X)$.

$$\square$$

In order to facilitate the proof of the corresponding result for algebras and congruences, we first present a lemma.

Lemma 7.4 For any Σ-VFC Γ and any Σ-algebra $\mathcal{A}$, $\mathcal{A} \in \Gamma^a$ iff there exist an X and an epimorphism $\varphi: \mathcal{T}(X) \longrightarrow \mathcal{A}$ such that ker $\varphi \in \Gamma(X)$.

Proof. If $\mathcal{A} \in \Gamma^a$, then $\mathcal{A} < \mathcal{T}(X_1)/\theta_1 \times \ldots \times \mathcal{T}(X_k)/\theta_k$ for some $k \geq 1$, $X_1, \ldots, X_k$ and $\theta_1 \in \Gamma(X_1), \ldots, \theta_k \in \Gamma(X_k)$. Hence we have an algebra $\mathcal{B}$, an epimorphism $\eta: \mathcal{B} \longrightarrow \mathcal{A}$ and a monomorphism

$$\varphi: \mathcal{B} \longrightarrow \mathcal{T}(X_1)/\theta_1 \times \ldots \times \mathcal{T}(X_k)/\theta_k.$$

This $\mathcal{B}$ is finite and there is an epimorphism $\psi: \mathcal{T}(X) \longrightarrow \mathcal{B}$, for some X. Let the epimorphism

$$\beta: \mathcal{T}(X_1) \times \ldots \times \mathcal{T}(X_k) \longrightarrow \mathcal{T}(X_1)/\theta_1 \times \ldots \times \mathcal{T}(X_k)/\theta_k,$$

defined by $\beta: (t_1, \ldots, t_k) \longmapsto (t_1/\theta_1, \ldots, t_k/\theta_k)$. Furthermore, let $\pi_i: T(X_1) \times \ldots \times T(X_k) \longrightarrow T(X_i)$ be th *ith* projection $(i = 1, \ldots, k)$. We can also define a morphism

$$\gamma: \mathcal{T}(X) \longrightarrow \mathcal{T}(X_1) \times \ldots \times \mathcal{T}(X_k)$$

so that $\gamma\beta = \psi\varphi$. Then $\psi\eta: \mathcal{T}(X) \longrightarrow \mathcal{A}$ is an epimorphism, and

$$\ker\psi\eta \supseteq \ker\psi = \ker\psi\varphi = \ker\gamma\beta = \bigcap (\gamma\pi_i \cdot \theta_i \cdot (\gamma\pi_i)^{-1}: 1 \leq i \leq k)$$

shows that ker $\psi\eta \in \Gamma(X)$. The converse implication is obvious. $\quad\square$

Proposition 7.5 The lattices $(\text{VFA}(\Sigma), \subseteq)$ and $(\text{VFC}(\Sigma), \subseteq)$ are isomorphic: (a) $\mathbf{K}^{ca} = \mathbf{K}$, for every Σ-VFA $\mathbf{K}$, and (b) $\Gamma^{ac} = \Gamma$, for every Σ-VFC Γ.

Proof. Again it suffices to prove (a) and (b).

(a) Let $\mathcal{A}$ be an Σ-algebra. By 7.4, $\mathcal{A} \in \mathbf{K}^{ca}$ iff there exist an X and an epimorphism $\varphi: \mathcal{T}(X) \longrightarrow \mathcal{A}$ such that ker $\varphi \in \mathbf{K}^c(X)$. But it is obvious that this is the case exactly in case $\mathcal{A} \in \mathbf{K}$.

(b) Let $\Gamma \in \text{VFC}(\Sigma)$. Consider any X and any $\theta \in \text{FCon}(X)$. If $\theta \in \Gamma(X)$, then $\mathcal{T}(X)/\theta \in \Gamma^a$, which implies $\theta \in \Gamma^{ac}(X)$. Assume now that $\theta \in \Gamma^{ac}(X)$. By 7.4 there exists for some Y an epimorphism $\psi: \mathcal{T}(Y) \longrightarrow \mathcal{T}(X)/\theta$ such that ker ψ is in $\Gamma(Y)$. Since ψ is surjective, we may choose for every x in X a ΣY-tree t_x such that $t_x\psi = x/\theta$. If $\varphi: \mathcal{T}(X) \longrightarrow \mathcal{T}(Y)$ is the morphism such that $x\varphi = t_x$ for every x in X, then $\varphi\psi = \theta''$, where θ'' is the canonical morphism from $\mathcal{T}(X)$ to $\mathcal{T}(X)/\theta$. Hence $\theta = \ker \theta'' = \ker \varphi\psi = \varphi \circ (\ker \psi) \circ \varphi^{-1} \in \Gamma(X)$. $\quad\square$

Propositions 7.3 and 7.5 already imply the isomorphism of $(\mathrm{VTL}(\Sigma), \subseteq)$ and $(\mathrm{VFC}(\Sigma), \subseteq)$, but that it is defined by the mappings $V \longmapsto V^c$ and $\Gamma \longmapsto \Gamma^t$, we show using the following composition laws.

Proposition 7.6
(a) $\mathbf{K}^{ct} = \mathbf{K}^t$, for every Σ-VFA $\mathbf{K}$.
(b) $V^{ac} = V^c$, for every Σ-VTL V.
(c) $\Gamma^{ta} = \Gamma^a$, for every Σ-VFC Γ.

Proof. (a) For any ΣX-tree language T,

$$T \in \mathbf{K}^t(X) \quad \text{iff} \quad SA(T) \in \mathbf{K} \quad \text{iff} \quad \approx_T \in \mathbf{K}^c(X) \quad \text{iff} \quad T \in \mathbf{K}^{ct}(X).$$

(b) If $\theta \in V^c(X)$, then $\approx_{T_1} \cap \ldots \cap \approx_{T_k} \subseteq \theta$ for some $k \geq 1$ and $T_1, \ldots, T_k$ in $V(X)$. Thus $\mathcal{F}(X)/\theta < SA(T_1) \times \ldots \times SA(T_k)$ which implies $\mathcal{F}(X)/\theta \in V^a$, and hence $\theta \in V^{ac}(X)$.

If $\theta \in V^{ac}(X)$, then $\mathcal{F}(X)/\theta < SA(T_1) \times \ldots \times SA(T_k)$, for some $k \geq 1$, $X_1, \ldots, X_k$ and $T_1 \in V(X_1), \ldots, T_k \in V(X_k)$. Hence, for some algebra $\mathcal{B}$, we have an epimorphism $\psi \colon \mathcal{B} \longrightarrow \mathcal{F}(X)/\theta$ and a monomorphism $\eta \colon \mathcal{B} \longrightarrow SA(T_1) \times \ldots \times SA(T_k)$. Moreover, $\mathcal{B}$ may be chosen so that there is an epimorphism $\varphi \colon \mathcal{F}(X) \longrightarrow \mathcal{B}$ for which $\varphi \psi = \theta$". Again, let π_i be the *i*th projection from $\mathcal{F}(X_1) \times \ldots \times \mathcal{F}(X_k)$ onto $\mathcal{F}(X_i)$, and define

$$\pi \colon \mathcal{F}(X_1) \times \ldots \times \mathcal{F}(X_k) \longrightarrow SA(T_1) \times \ldots \times SA(T_k)$$

by $(t_1, \ldots, t_k)\pi = (t_1/T_1, \ldots, t_k/T_k)$. Then π is an epimorphism and we can define a morphism $\gamma \colon \mathcal{F}(X) \longrightarrow \mathcal{F}(X_1) \times \ldots \times \mathcal{F}(X_k)$ such that $\gamma\pi = \varphi\eta$. Then $\theta = \ker \varphi\psi \supseteq \ker \varphi\eta = \ker \gamma\pi$. Since $\ker \gamma\pi = \bigcap (\gamma\pi_i \circ \approx_{T_i} \circ (\gamma\pi_i)^{-1} \colon 1 \leq i \leq k)$, this implies $\theta \in V^c(X)$.

(c) A Σ-algebra $\mathcal{A}$ is in Γ^a iff $\mathcal{A} < \mathcal{F}(X_1)/\theta_1 \times \ldots \times \mathcal{F}(X_k)/\theta_k$ for some $k \geq 1$, $X_1, \ldots, X_k$ and $\theta_1 \in \Gamma(X_1), \ldots, \theta_k \in \Gamma(X_k)$. Since every $\Gamma(X)$ is generated by syntactic congruences, we may assume that each θ_i is the syntactic congruence of some $T_i \ (\subseteq T(X_i))$. Then $T_i \in \Gamma^t(X_i)$, and it can now be seen that $\mathcal{A} \in \Gamma^a$ iff $\mathcal{A} \in \Gamma^{ta}$. □

For any Σ-VTL V, 7.6(b) applied to V, 7.6(a) applied to V^a, and 7.3(b) yield: $V^{ct} = V^{act} = V^{at} = V$. Similarly, one can show using 7.6(b), 7.6(c) and 7.5(b) that for any Σ-VFC Γ, $\Gamma^{tc} = \Gamma^{tac} = \Gamma^{ac} = \Gamma$. Hence we have

Proposition 7.7 The lattices $(\mathrm{VTL}(\Sigma),\subseteq)$ and $(\mathrm{VFC}(\Sigma),\subseteq)$ are isomorphic: (a) $V^{ct} = V$, for every Σ-VTL V, and (b) $\Gamma^{tc} = \Gamma$, for every Σ-VFC Γ.

$\square$

These results also yield the remaining composition laws:

Corollary 7.8
(a) $K^{tc} = K^c$, for every Σ-VFA **K**.
(b) $V^{ca} = V^a$, for every Σ-VTL V.
(c) $\Gamma^{at} = \Gamma^t$, for every Σ-VFC Γ.

$\square$

The results of this section can be summed up as follows:

The extended variety theorem for tree languages
The mappings
$$\mathrm{VFA}(\Sigma) \longrightarrow \mathrm{VTL}(\Sigma), \quad K \longmapsto K^t; \quad \mathrm{VTL}(\Sigma) \longrightarrow \mathrm{VFA}(\Sigma), \quad V \longmapsto V^a,$$
the mappings
$$\mathrm{VFA}(\Sigma) \longrightarrow \mathrm{VFC}(\Sigma), \quad K \longmapsto K^c; \quad \mathrm{VFC}(\Sigma) \longrightarrow \mathrm{VFA}(\Sigma), \quad \Gamma \longmapsto \Gamma^a,$$
and the mappings
$$\mathrm{VTL}(\Sigma) \longrightarrow \mathrm{VFC}(\Sigma), \quad V \longmapsto V^c; \quad \mathrm{VFC}(\Sigma) \longrightarrow \mathrm{VTL}(\Sigma), \quad \Gamma \longmapsto \Gamma^t$$
form three pairs of mutually inverse isomorphisms between the lattices $(\mathrm{VFA}(\Sigma),\subseteq)$, $(\mathrm{VTL}(\Sigma),\subseteq)$ and $(\mathrm{VFC}(\Sigma),\subseteq)$. Moreover, $K^t = K^{ct}$, $K^c = K^{tc}$, $V^a = V^{ca}$, $V^c = V^{ac}$, $\Gamma^t = \Gamma^{at}$ and $\Gamma^a = \Gamma^{ta}$ whenever $K \in \mathrm{VFA}(\Sigma)$, $V \in \mathrm{VTL}(\Sigma)$ and $\Gamma \in \mathrm{VFC}(\Sigma)$.

$\square$

8. EXAMPLES OF VARIETIES OF TREE LANGUAGES

In this section we consider some simple examples of families of tree languages which form varieties as defined above.

a. The varieties $\mathcal{R}ec$, $\mathcal{T}riv$ and $\mathcal{N}il$
The two extreme examples of Σ-VTLs are the variety $\mathcal{R}ec = \{\mathcal{R}ec(X)\}$ of all recognizable Σ-tree languages and the trivial variety $\mathcal{T}riv = \{\{\emptyset, T(X)\}\}$. $\mathcal{R}ec^a$ contains all finite Σ-algebras

while $\mathcal{T}riv^a$ consists of the trivial Σ-algebras only. For every X, $\mathcal{R}ec^C(X) = \text{FCon}(X)$ and $\mathcal{T}riv^C(X) = \{\iota_X\}$.

For each X, let $\mathcal{N}il(X)$ consist of all finite and all cofinite ΣX-tree languages. For strings, the corresponding pseudo-variety is formed by the nilpotent semigroups [10]. Nilpotence has been defined for Σ-algebras in [12] in a natural way: a finite Σ-algebra $\mathcal{A} = (A,\Sigma)$ is *nilpotent*, if there exist an *absorbing state* a_0 in A and an $n > 0$ such that $t\alpha^{\mathcal{A}} = a_0$ for every ΣX-tree t of height $\geq n$ and every $\alpha: X \longrightarrow A$. The smallest n for which this holds (if it exists) is called the *degree of nilpotency* of $\mathcal{A}$. Let **Nil** be the class of all nilpotent Σ-algebras.

Proposition 8.1 $\mathcal{N}il$ is a Σ-VTL and $\mathcal{N}il^a = $ **Nil**.

Proof. It is clear that $\mathcal{N}il(X)$ is closed under the Boolean set operations. Also, if $p \in \text{Tr}(X)$ and $T \in \mathcal{N}il(X)$, then $p^{-1}(T) = \{s: p(s) \in T\}$ is finite or cofinite according to whether T is finite or cofinite. Similarly, $T\varphi^{-1} \in \mathcal{N}il(X)$, for any morphism $\varphi: \mathcal{T}(X) \longrightarrow \mathcal{T}(Y)$ and T in $\mathcal{T}(Y)$. Hence $\mathcal{N}il$ is a Σ-VTL.

Let $\mathcal{A}$ be nilpotent of degree n with a_0 as the absorbing state. Clearly, every subalgebra of $\mathcal{A}$ is nilpotent of degree $\leq n$ with a_0 as the absorbing state. Let $\varphi: \mathcal{A} \longrightarrow \mathcal{C}$ be an epimorphism. For any mapping $\gamma: X \longrightarrow C$, we may define a mapping $\alpha: X \longrightarrow A$ so that $\alpha\varphi = \gamma$. Then for any ΣX-tree t of height $\geq n$, $t\gamma^{\mathcal{C}} = t\alpha^{\mathcal{A}}\varphi = a_0\varphi$. Hence $\mathcal{C}$ is nilpotent of degree $\leq n$. Finally, if $\mathcal{A}$ is nilpotent of degree n and $\mathcal{B}$ is nilpotent of degree m, then it is obvious that $\mathcal{A} \times \mathcal{B}$ is nilpotent of degree $\max(n,m)$. Hence, **Nil** is a Σ-VFA.

Let $\mathcal{A}$ be a syntactic nilpotent Σ-algebra of degree n with absorbing state a_0 and a disjunctive subset D. Choose X so that there is an epimorphism φ from $\mathcal{T}(X)$ onto $\mathcal{A}$. Then $T = D\varphi^{-1}$ is cofinite or finite depending on whether or not $a_0 \in D$. Hence $\mathcal{A} \cong \mathcal{A}/D \cong \text{SA}(T) \in \mathcal{N}il^a$. This shows that **Nil** $\subseteq \mathcal{N}il^a$. For the converse inclusion, consider a finite ΣX-tree language T. We may assume that $T \neq \emptyset$. Let $n = \max(hg(t): t \in T)$. Define a Σ-algebra $\mathcal{A} = (A,\Sigma)$ as follows. Let

$A = \{s \in T(X) : hg(s) \leq n\} \cup \{a_0\}$ $(a_0 \notin T(X))$.

For c in Σ_0, set $c^{\mathcal{A}} = c$. If $m > 0$, $f \in \Sigma_m$ and $a_1, \ldots, a_m \in A$, define

$$f^{\mathcal{A}}(a_1, \ldots, a_m) = \begin{cases} fa_1 \ldots a_m, & \text{if } fa_1 \ldots a_m \in A; \\ a_0 & \text{otherwise.} \end{cases}$$

It is now obvious that for any $\alpha : X \longrightarrow A$ and any ΣX-tree t of height $> n$, $t\alpha^{\mathcal{A}} = a_0$. Hence $\mathcal{A}$ is nilpotent. Choose $\alpha : X \longrightarrow A$ so that $x\alpha = x$ for every x in X. Then for every ΣX-tree t,

$$t\alpha^{\mathcal{A}} = \begin{cases} t, & \text{if } hg(t) \leq n; \\ a_0 & \text{otherwise.} \end{cases}$$

This means that T is recognized by $\mathbf{A} = (\mathcal{A}, \alpha, F)$, where $F = A \cap T$. Therefore $SA(T) < \mathcal{A}$, which implies $SA(T) = SA(T(X)-T) \in$ **Nil**. Hence $\mathcal{N}il^{\alpha} \subseteq$ **Nil**.   □

b. Definite tree languages

A string language L is called *weakly k-definite* [24], if a word of length $\geq k$ is in L iff its suffix of length k is in L. The following extension to trees is essentially the one studied by Heuter [15,16].

The *k-root* $r_k(t)$ $(k \geq 0)$ of a ΣX-tree t is defined as follows:

0. For every t, $r_0(t) = \varepsilon$, where ε is a special symbol which represents an empty root segment.

1. For every t, $r_1(t) = root(t)$.

2. Let $k \geq 2$. If $hg(t) < k$, then $r_k(t) = t$. If $hg(t) \geq k$ and $t = ft_1 \ldots t_m$, then $r_k(t) = fr_{k-1}(t_1) \ldots r_{k-1}(t_m)$.

For example, if $t = fgfcxc$ ($f \in \Sigma_2$, $g \in \Sigma_1$, $c \in \Sigma_0$, $x \in X$), then $r_0(t) = \varepsilon$, $r_1(t) = f$, $r_2(t) = fgc$, $r_3(t) = fgfc$ and $r_k(t) = t$ for all $k \geq 4$.

Let $k \geq 0$ and $T \subseteq T(X)$. We call T *weakly k-definite*, if for all $s, t \in T(X)$, $t \in T$ whenever $r_k(s) = r_k(t)$ and $s \in T$. Let $\mathcal{D}(k) = \{\mathcal{D}(k,X)\}$, where $\mathcal{D}(k,X)$ is the set of all weakly k-definite ΣX-tree languages. The family of all *definite Σ-tree languages* is $\mathcal{D} = \{\mathcal{D}(X)\}$, where $\mathcal{D}(X) = \bigcup(\mathcal{D}(k,X) : k \geq 0)$ for every X. Clearly, $\mathcal{D}(0) \subseteq \mathcal{D}(1) \subseteq \ldots \subseteq \mathcal{D}(k) \subseteq \ldots \subseteq \mathcal{D} \subseteq \mathcal{R}ec$, and all inclusions are proper if Σ is nontrivial.

Proposition 8.2 For any $k \geq 0$, $\mathcal{D}(k)$ is a Σ-VTL, and so is $\mathcal{D}$.

Proof. It would be easy to verify directly that each $\mathcal{D}(k)$ satisfies the four defining conditions of a Σ-VTL, but the result follows also readily from the fact that $\mathcal{D}(k) = \Gamma^t$, where Γ is the principal Σ-VFC generated by the congruences θ_X defined by: $s \; \theta_X \; t$ iff $r_k(s) = r_k(t)$ $(s,t \in T(X))$.
For this the following three observations are needed:
(1) $\theta_X \in \mathrm{FCon}(X)$, for every X;
(2) $\theta_X \subseteq \varphi \circ \theta_X \circ \varphi^{-1}$, for every morphism $\varphi \colon \mathcal{T}(X) \longrightarrow \mathcal{T}(Y)$ (this holds since $r_k(s) = r_k(t)$ implies $r_k(s\varphi) = r_k(t\varphi)$);
(3) $T \in \mathcal{D}(k,X)$ iff T is saturated by θ_X. $\square$

c. Reverse definite tree languages

The membership of a sufficiently long word in a reverse definite language [6] is determined by its prefix of a given length. In trees the natural counterparts of prefixes are subtrees. However, a tree may have many subtrees of a given height and a subtree may appear several times in it. Moreover, subtrees can be at different 'depths' in a tree. Therefore reverse definite tree languages could be defined in several natural ways. Our choice is again essentially the notion studied by Heuter [15].
If $k \geq 0$ and $t \in T(X)$, let $S_k(t)$ denote the set of all subtrees of t of height $< k$. Hence, $S_0(t) = \emptyset$ and $S_1(t)$ consists of the leaves of t. We call a ΣX-tree language *weakly reverse* k-*definite*, if for any ΣX-trees s and t,

$$S_k(s) = S_k(t), \; s \in T \implies t \in T.$$

Let $\mathcal{DR}(k) = \{\mathcal{DR}(k,X)\}$ be the family of all weakly reverse k-definite Σ-tree languages. The union of these families is the family $\mathcal{RD} = \{\mathcal{RD}(X)\}$ of all *reverse definite Σ-tree languages*. Obviously, $\mathcal{RD}(0) \subseteq \mathcal{RD}(1) \subseteq \ldots \subseteq \mathcal{RD}(k) \subseteq \ldots . \subseteq \mathcal{RD} \subseteq \mathcal{Rec}$,   and the inclusions are proper whenever Σ is nontrivial.
We may again use congruences for showing that these families are VTLs. Given $k \geq 0$ and X, define the relation θ_X so that

$$s \; \theta_X \; t \quad \text{iff} \quad S_k(s) = S_k(t) \qquad\qquad (s,t \in T(X)).$$

Obviously, $\theta_X \in \text{FCon}(X)$. If $\varphi: \mathcal{T}(X) \longrightarrow \mathcal{T}(Y)$ is a morphism and $s,t \in T(X)$, $s \; \theta_X \; t$ implies $s\varphi \; \theta_Y \; t\varphi$. Hence, $\theta_X \subseteq \varphi \circ \theta_Y \circ \varphi^{-1}$. Finally, it is clear that for any $T \; (\subseteq T(X))$, $T \in \mathcal{RD}(k,X)$ iff T is saturated by θ_X. All this means that $\mathcal{DR}(k)$ is the Σ-VTL corresponding to the principal VFC $\Gamma = \{[\theta_X]\}$. Hence:

Proposition 8.3   Every $\mathcal{RD}(k)$ is a Σ-VTL $(k \geq 0)$, and so is $\mathcal{RD}$.

□

d. Generalized definite tree languages

Our definition of generalized definite tree languages is in essence that of Heuter [15,17], but similarly as in [27] we introduce a double hierarchy of families by allowing the root heights and the subtree heights to vary independently.

Let $h,k \geq 0$. The relation $\gamma(h,k,X)$ on $T(X)$ is defined so that for any ΣX- trees s and t,

$$s \; \gamma(h,k,X) \; t \quad \text{iff} \quad S_h(s) = S_h(t) \text{ and } r_k(s) = r_k(t).$$

It is clear that $\gamma(h,k,X) \in \text{FCon}(X)$. A ΣX-tree language is called *weakly h,k-definite*, if it is saturated by $\gamma(h,k,X)$. The set of these tree languages is denoted by $\mathcal{GD}(k,X)$, and let $\mathcal{GD}(h,k) = \{\mathcal{GD}(h,k,X)\}$. For any X, let

$$\mathcal{GD}(X) \;\; = \;\; \bigcup_{h \geq 0} \bigcup_{k \geq 0} \mathcal{GD}(h,k,X).$$

Then $\mathcal{GD} = \{\mathcal{GD}(X)\}$ is the family of *generalized definite Σ-tree languages*. The following facts are obvious.

Lemma 8.4   For all $h,k \geq 0$,

(a) $\mathcal{GD}(h,k) \subseteq \mathcal{GD}(h+1,k) \cap \mathcal{GD}(h,k+1)$;

(b) $\mathcal{GD}(h,k) \subseteq \mathcal{GD} \subseteq \mathcal{Rec}$;

(c) $\mathcal{GD}(h,0) = \mathcal{RD}(h)$ and $\mathcal{GD}(0,k) = \mathcal{D}(k)$.

If Σ is nontrivial, then all inclusions are proper.

□

Again one can prove that each $\mathcal{GD}(h,k)$ is a Σ-VTL by showing that $\mathcal{GD}(h,k) = \Gamma(h,k)^t$, where $\Gamma(h,k)$ is the principal Σ-VFC defined by the congruences $\gamma(h,k,X)$. As the union of a directed set of Σ-VTLs, $\mathcal{GD}$ is also a Σ-VTL.

Proposition 8.5 For every pair $h,k \geq 0$, $\mathcal{GD}(h,k)$ is a Σ-VTL, and so is $\mathcal{GD}$.

$\square$

e. Local tree languages

The words of a local string language [5,9] are determined by their first letters, their last letters and the pairs of consecutive letters appearing in them. For defining a local tree language one has to specify the root symbols and the 'forks' (which also determine the possible leaves) allowed.

The set $fork(t)$ of *forks* of a ΣX-tree t is defined as follows:

(1) $fork(t) = \emptyset$, for all t in $X \cup \Sigma_0$;

(2) for $t = ft_1 \ldots t_m$ $(m > 0)$,

$$fork(t) = \bigcup (fork(t_i) : 1 \leq i \leq m) \cup \{f(root(t_1), \ldots, root(t_m))\}.$$

Let $fork(\Sigma, X)$ denote the (finite) set of all forks possible in a ΣX-tree. A ΣX-tree language T is *local in the strict sense*, if there are two sets F $(\subseteq fork(\Sigma, X))$ and R $(\subseteq \Sigma \cup X)$ such that for any ΣX-tree t, $t \in T$ iff $fork(t) \subseteq F$ and $root(t) \in R$. Then we write $T = SL(F,R)$.

The family $\mathcal{SLoc} = \{\mathcal{SLoc}(X)\}$ of these tree languages is not a Σ-VTL since the sets $\mathcal{SLoc}(X)$ are not closed under complements. Let us define the family of *local Σ-tree languages* $\mathcal{Loc} = \{\mathcal{Loc}(X)\}$ so that for every X, $\mathcal{Loc}(X)$ is the Boolean closure of $\mathcal{SLoc}(X)$.

Proposition 8.6 $\mathcal{Loc}$ is the Σ-VTL generated by $\mathcal{SLoc}$.

Proof. Obviously, if $\mathcal{Loc}$ is a Σ-VTL, then it equals $VTL(\mathcal{SLoc})$. The definition of $\mathcal{Loc}$ also implies directly that each $\mathcal{Loc}(X)$ is closed under complements and intersections. Since inverse translations and inverse morphisms commute with the Boolean operations, it suffices to consider these operations applied to local tree languages in the strict sense. Let $T = SL(F,R) \in \mathcal{SLoc}(X)$. We claim that for any special ΣX-tree p, $p^{-1}(T) \in \mathcal{SLoc}(X)$. The case $p = \xi$ is obvious. Let $p = fp_1 \ldots p_m$. If $f \notin R$, then $p^{-1}(T) = \emptyset$ $(\in \mathcal{SLoc}(X))$, so suppose $f \in R$. Let $fork'(p)$ be the set of all forks of p in which ξ does not appear. We may assume that $fork'(p) \subseteq F$, since otherwise $p^{-1}(T)$ is again empty. Let $g(\ldots, \xi, \ldots)$ be the fork of p in which ξ appears.

Now, for any t in T(X),

$p(t) \in T$ iff $g(...,root(t),...) \in F$ and $fork(t) \subseteq F$.

Hence $p^{-1}(T) = SL(F,R')$ with $R' = \{d \in \Sigma \cup X: g(...,d,...) \in F\}$. A similar analysis shows that $T\varphi^{-1} \in \mathscr{SLoc}(X)$, for every morphism $\varphi: \mathscr{T}(X) \longrightarrow \mathscr{T}(Y)$ and any T in $\mathscr{SLoc}(Y)$.

□

References

1 Almeida, J.: On pseudovarieties, varieties of languages, filters of congruences, pseudoidentities and related topics. *Algebra Universalis* 27 (1990), 333-350.

2 Ash, C.J.: Pseudovarieties, generalized varieties and similarly described classes. *J. Algebra* 92 (1985), 104-115.

3 Baldwin, J.T.; J. Berman: Varieties and finite closure conditions. *Colloq. Math.* 35 (1976), 15-20.

4 Banaschewski, B.: The Birkhoff Theorem for varieties of finite algebras. *Algebra Universalis* 17 (1983), 360-368.

5 Berstel, J.: *Transductions and context-free languages.* B.G. Teubner, Stuttgart 1979.

6 Brzozowski, J.A.: Canonical regular expressions and state graphs for definite events. *Mathematical Theory of Automata.* Proc. Symp. Math. Theory of Automata, Microwave Research Inst. Symp. Ser. 12, Brooklyn, New York 1963, 529-561.

7 Burris, S.; H.P. Sankappanavar: *A course in universal algebra.* Springer-Verlag, New York 1981.

8 Cohn, P.M.: *Universal algebra* (2. rev. ed.). D. Reidel Publ. Company, Dordrecht 1981.

9 Eilenberg, S.: *Automata, languages, and machines, Vol. A.* Academic Press, New York 1974.

10 Eilenberg, S.: *Automata, languages, and machines, Vol. B.* Academic Press, New York 1976.

11 Eilenberg, S.; M.P. Schützenberger: On pseudovarieties. *Advances in Math.* 19 (1976), 413-418.

12 Gécseg, F.; B. Imreh: On a special class of tree automata. *Automata, Languages and Programming Systems* (Proc. Conf., Salgógarján 1988). Dep. Math., K. Marx University of Economics, Budapest 1988, 141-152.

13 Gécseg, F.; M. Steinby: *Tree automata.* Akadémiai Kiadó, Budapest 1984.

14 Ginzburg, A.: About some properties of definite, reverse-definite and related automata. *IEEE Trans. Elecronic Computers* EC-15 (1966), 809-810.

15 Heuter, U.: *Zur Klassifizierung regulärer Baumsprachen.* Dissertation, Technical University of Aachen, Faculty of Natural Sci., Aachen 1989.

16 Heuter, U.: Definite tree languages. *Bull. EATCS* 35 (1988), 137-144.

17   Heuter, U.: Generalized definite tree languages. *Mathem. Found. Comput. Sci.* (Proc. Symp., Porabka-Kozubnik, Poland 1989). Lect. Notes in Comput. Sci. 379, Springer-Verlag, Berlin 1989, 270-280.

18   Ihringer, Th.: *Allgemeine Algebra.* B.G. Teubner, Stuttgart 1988.

19   Lallement, G.: *Semigroups and combinatorial applications.* John Wiley & Sons, New York 1979.

20   McNaughton, R.; S. Papert: *Counter-free automata.* The M.I.T. Press, Cambridge, Mass. 1971.

21   Mezei, J; J.B. Wright: Algebraic automata and context-free sets. *Information and Control* 11 (1967), 3-29.

22   Nivat, M.; A. Podelski: Tree monoids and recognizability of sets of finite trees. *Resolution of equations in algebraic structures, Vol.1* (eds. H. Aït-Kaci and M. Nivat), Academic Press, Boston 1989, 351-367.

23   Nivat, M.; A. Podelski: Definite tree automata (cont'd). *Bull. EATCS* 38 (1989), 186-190.

24   Perles, M.; M.O. Rabin; E. Shamir: The theory of definite automata. *IEEE Trans. Electronic Computers* EC-15 (1963), 233-243.

25   Pin, J.E.: *Varieties of formal languages.* North Oxford Academic Publ., London 1986.

26   Schein, B.M.: Homomorphisms and subdirect decompositions of semigroups. *Pacific J. Math.* 17 (1966), 529-547.

27   Steinby, M.: On definite automata and related systems. *Ann. Acad. Sci. Fenn.*, Ser. A I, 444 (1969).

28   Steinby, M.: Syntactic algebras and varieties of recognizable sets. *Les arbres en algebre et en programmation, 4eme Coll. Lille* (Proc. Coll., Lille 1979), University of Lille, Lille 1979, 226-240.

29   Steinby, M.: Some algebraic aspects of recognizability. *Fundamentals of computation theory* (Proc. Conf., Szeged 1981). Lect. Notes in Comput. Sci. 117, Springer-Verlag, Berlin 1981, 360-372.

30   Thérien, D.: *Classification of regular languages by congruences.* Rep. CS-80-19, Univ. Waterloo, Dept. Comput. Sci., Waterloo, Ontario 1980.

31   Thérien, D.: Recognizable languages and congruences. *Semigroup Forum* 23 (1981), 371-373.

32   Thomas, W.: On noncounting tree languages. *Grundl. Theor. Informatik* (Proc. 1. Intern. Workshop, Paderborn 1982), 234-242.

33   Thomas, W.: Logical aspects in the study of tree languages. *Ninth colloquium on trees in algebra and in programming* (ed. B. Courcelle), Cambridge University Press, Cambridge 1984, 270-280.

Tree Automata and Languages
M. Nivat and A. Podelski (editors)
1992 Elsevier Science Publishers B.V.

Interpretability and Tree Automata: a simple way to solve Algorithmic Problems on Graphs closely related to Trees

Detlef Seese

FB11–Praktische Informatik, Universität Duisburg,
Postfach 10 15 03, D–W–4100 Duisburg 1, (Germany)

1 Introduction

The main goal of this article is to point out the strong connection between recent results in complexity theory and results on decidability of monadic second order theories.

Both sets of results employ the following three concepts:

- tiling problems, in order to deduce high complexity or undecidability,

- tree automata, in order to deduce polynomial time complexity or decidability,

- the interpretability method, in order to transfer complexity or decidability results from trees to other structures.

This article constitutes an attempt to explain why high complexity and undecidability correlate with the fact that the considered structures "contain" large grids, and why polynomial or linear time complexity and decidability correlate with the fact that the structures "resemble" trees.

Many algorithmic problems in graph theory are NP–hard and (at least until now) have no solutions in polynomial time. A great number of them remain NP–hard even if they are restricted to smaller and simpler classes of graphs (such as the class of planar graphs of maximum degree ≤ 4). But if the considered class of graphs is very simple, as for instance the class of trees, or the class of series–parallel graphs, then some of the NP–hard problems have solutions in polynomial or even in linear time. This suggests the problem of determining the structural reason for this complexity gap.

A similar change can be observed if monadic second order theories are investigated with respect to decidability. It seems that decidability is correlated with the similarity of the considered structures to trees.

The present paper tries to shed some light into both areas.

Using tiling problems it is pointed out that the containment or definability of large grids is one of the reasons for high complexity or undecidability.

This together with the fundamental results of N. Robertson and P.D. Seymour concerning the structure of graphs avoiding a given minor (see [RoSey84], [RoSey86a]) implies that one can concentrate the attention in the first step to graphs which are similar to trees, or more exactly have universally bounded tree width (see [RoSey84]).

For the description of the algorithmic problems we use an extension of the very powerful monadic second order logic and denote such problems as EMS–problems. Using a variant of the machinery of interpretations it can be proved that each EMS problem over a class K of graphs of universally bounded tree width can be reduced to another EMS problem on a class of binary trees. In comparison: Each decidable monadic theory of graphs has only models of universally bounded tree width and hence is interpretable in a suitable monadic theory of trees. EMS problems for binary trees can be easily solved in polynomial time by an adaptation of automata–theoretical methods which showed their value already in proofs on the decidability of monadic theories of trees.

It is a goal of this survey, which is based on material presented in [ArLaSe89] and [Se86], [Se87a] to demonstrate the usefulness of a fusion of ideas from logic, graph theory and computer science.

The article is organized as follows: After fixing the notation from graph theory in Section 2, the notions of tree width and graph minors are introduced. In Section 3 tilings and generalizations of them are introduced. In Section 4, applications of tilings on the complexity investigations are given. Graph properties which can be definded in monadic second order logic are introduced on Section 5. Section 6 is devoted to interpretability as a simple method to transform problems to more simple ones. Tree automata are introduced in section 7. In the following section some conclusions about ways to deduce the existence of polynomial time algorithms are drawn. In Section 9 we study local properties. Monadic second order theories are investigated with respect to decidability in Section 10. The article concludes with some remarks and problems.

The notation from mathematical logic in this article is standard and can be found e.g. in [Sh67]. For a given logic L and a class K of structures for L, the L–theory of K, denoted as $Th_L(K)$, is the set of all L sentences which are true in K. The *monadic second order logic* (see further Section 5) results from elementary logic by adding to the corresponding elementary language a sequence of quantifiable set variables and allowing new atomic formulas $t \in X$, where t is an elementary term and X is a set variable. Here the set variables range over all subsets of the domain of the structures for L and the intended interpretation of $\in$ is the membership relation. The *weak monadic second order logic* is defined in the same way, but for the interpretation of the set variables are allowed finite sets only. $Th_2(K)$ denotes the monadic second order theory of K and $Th_{w2}(K)$ denotes the weak monadic second order theory of K.

A theory is *decidable* if there is an algorithm deciding for each sentence Φ of the corresponding language whether $\Phi \in T$ or not. In the other case T is defined to be *undecidable*. For further informations concerning decidability the reader is refered to [Ra77]. A profound introduction into the contemporary world of monadic second order theories can be found in [Gu85]. Here we shall follow an idea from descriptive complexity theory (see [Im88]) to desribe algorithmic problems using special languages. Assume that a property P for a class K of structures is described by a formula Φ of a language L, i.e. $P(G)$ holds if and only if $G \models \Phi$ holds for all structures $G \in K$. This approach has the advantage that one can try to use results on the decidability of $Th_L(K)$ to deduce results on the complexity of the problem "does hold $P(G)$?" for $G \in K$. Here we will show that this approach is useful for graphs closely related to trees when the problems are formalized in a certain extension of monadic second order logic.

2 The structure of graphs avoiding planar minors

The usual graph–theoretic terminology will be standard and can be found in any classical textbook on graph theory, e.g. [Ha69]. Throughout the paper all considered graphs will be simple (i.e. have no loops or multiple edges) and finite. Exceptions of this rule will be stated explicitly. A graph $H = (V(H), E(H))$ with vertex set $V(H)$ and edge set $E(H)$ is said to be a *subgraph* of a graph $G = (V(H), E(G))$ if

$$V(H) \subseteq V(G) \text{ and } E(H) \subseteq E(G).$$

Assume that G is a graph and that A is a subset of the set of its vertices, then $G|A$ denotes the subgraph of G *induced* by A, i.e. the graph (A, E'), where E' contains all edges whose both endvertices (the vertices incident with the considered edge) are in A.

The *length* of a path in a graph is the number of its edges. The *distance* of two vertices is the length of the shortest path between them.

Assume that r is a natural number and a is a vertex of a graph G. Then $N_r(G, a)$ denotes the r–*neighbourhood* of a in G, i.e. the set

$$\{b : \ b \in V(G) \text{ and the distance of } b \text{ and } a \text{ is } \leq r\}.$$

Sometimes $N_r(G, a)$ will denote also the graph $G|N_r(G, a)$. In case that $r = 1$ we shall write also $N(a)$ instead of $N_1(G, a)$. If G is a graph and e is an edge of G with endvertices a and b, then a new graph H results by identifying a and b in such a way, that the resulting new vertex a' is incident to all those edges (others than e) which were originally incident to a or b in G, and multiple edges which would result in this way are substituted by a single representative.

This operation is called *contracting* the edge e. If a graph H can be obtained from G by a succession of such edge contractions, then G is *contractible* to H. H is called a *minor* of G if there is a subgraph of G which is contractible to H (see [RoSey83], [RoSey84]). H is a proper minor of G if H is isomorphic to a minor of G, but H is not isomorphic to G. For a natural number $n \geq 2$ we define the $n \times n$ *grid* Q_n, as follows:

$$V(Q_n) := \{(i, j) : 0 \leq i < n \text{ and } 0 \leq j < n\},$$
$$E(Q_n) := \{\{(i, j), (i', j')\} : |i - j'| + |j - j'| = 1$$
$$\text{and } 0 \leq i, i', j, j' < n\}.$$

It is easy to verify the following fact by an induction on the size of the considered graphs.

(1) **Fact** (see [RoSey83], [RoSey84]): If H is planar, then there is an n such that H is isomorphic to a minor of the $n \times n$–grid Q_n.

One of the most well–known results concerning minors is the characterization of planar graphs as those graphs having no minor isomorphic to K_5 or $K_{3,3}$. K_5 and $K_{3,3}$ are then called also minimal forbidden minors for the property to be not planar.

A graph H is said to be a *minimal forbidden minor* for a property P of graphs if H has not the property P but each graph H' which is isomorphic to a proper minor of G has property P. A property P is *closed under minor taking* if with each graph G with property P also each graph H isomorphic to a minor of G has property P.

One of the famous results of N. Robertson an P.D. Seymour (see [RoSey89]) is their proof of the conjecture of Wagner:

Each property of graphs, which is closed under minor taking has up to isomorphism a finite number of minimal forbidden minors.

To prove this conjecture N. Robertson and P.D. Seymour devoted a fundamental series of papers (see e.g. [RoSey83], [RoSey84], [RoSey86a], [RoSey88]) to investigate the structure of graphs avoiding a given minor. The structure theory of N. Robertson and P.D. Seymour has many interesting consequences as well in graph theory itself (see [Thom]) as in theoretical computer science (see [FeLan86], [FeLan88]). One of their results will be a key for our investigations. N. Robertson and P.D. Seymour definded in [RoSey84]:

A *tree decomposition* of a graph G is a pair (T, X), where T is a tree and $X = (X_t : t \in V(T))$ is a family of subsets of $V(G)$, with the following properties:

1. $\bigcup(X_t, : t \in V(T)) = V(G)$,

2. for every edge e of G there exists $t \in V(T)$ such that e has both ends in X_t,

3. for $t, t', t'' \in V(T)$; if t' is on a path of T between t and t'', then $X_t \bigcup X_{t''} \subseteq X_{t'}$.

The *width* $w((T, X))$ of the tree–decomposition (T, X) is $max(|X_t| - 1 : t \in V(T))$. The graph G has *tree width* m if m is minimum such that G has a tree–decomposition of width m'. The tree width of G is denoted by $w(G)$.

In [RoSey84] are given the following examples:

Trees and forests have tree width 1 (or 0 in degenerated cases) and series–parallel graphs have tree width ≤ 2. For $n \geq 1$ the complete graph K_n has tree width $n - 1$ and Q_n has tree width n.

A collection of classes of graphs of universally bounded tree width is given in the following table.

Classes of graphs	tree width
trees, forests	≤ 1
Almost trees with parameter k	$\leq k + 1$
Graphs with bandwidth $\leq k$	$\leq k$
Graphs with cutwidth $\leq k$	$\leq k$
Series parallel graphs	≤ 2
Outerplanar graphs	≤ 2
Halin graphs	≤ 3
k–outerplanar graphs	$\leq 3^k - 1$
k–bounded tree–partitie graphs	$\leq 2k - 1$
Chordal graphs with max. cliquesize k	$\leq k - 1$
Undirected path graphs with max. cliquesize k	$\leq k - 1$
Interval graphs with max. cliquesize k	$\leq k - 1$
Circular are graphs with max. cliquesize k	$\leq 2k - 1$
partial k–trees	$\leq k$

These classes, many of them have importance in applications and complexity investigations, were gathered from [Bo86], [Bo88], [Sch88] One of the main results presented in this section is:

(5) Theorem ([RoSey86b]): For every finite planar graph H there is a number n_H such that every finite graph G with no minor isomorphic to H has tree width $\leq n_H$.

Recently a short proof of this result and several improvements of it were found by N. Roberson, P.D. Seymour and R. Thomas. They proved e.g. the following variant of Theorem 5.

(5') Theorem ([RoSeyTh89a]: Assume that n is a natural number and that G is a graph without a minor isomorphic to Q_n . Then the tree width of G is $\leq 20^{2n^5}$.

A quite simple subcase of (5) was shown in [Se79], where in a different terminology it was proved that a planar graph with maximum degree ≤ 3, which has no minor isomorphic to the graph $(\{1,2,3,4,5\}, \{\{1,2\},\{2,3\},\{3,4\},\{1,4\},\{1,5\},\{2,5\},$
$\{3,5\},\{4,5\}\})$ can be generated by very simple operations. There was also conjectured a similar result for planar graphs with no minor isomorphic to a slight variant of Q_n (for $n \geq 3$). Theorem (5) holds also for infinite graphs G.

(6) Theorem ([Th]): For every finite planar graph H there is a natural number n_H such that every (possibly infinite) graph G with no minor isomorphic to H has tree width $\leq n_H$.

(7) Theorem ([Th]): For an arbitrary graph G it holds the following equation:
tree width$(G) = \sup($tree width(G'): G' is a finite subgraph of $G)$.

A short proof of this result was found in [Thom]. It is interesting to note that Theorem 6 and a special case of Theorem 7 can be deduced from the usual Compactness–theorem of mathematical logic via a structure result concerning pseudotree decompositions which are a generalization of the usual tree–decomposition (see [Se87]), where instead of a tree in the tree–decomposition in the pseudotree decomposition a partial ordering is used which has the property that each of its initial segments is a linear ordering. Similar ideas were developed in [RoSeyTh89].

(8) Lemma ([ArCorPr87a]): For every natural number m, a tree–decomposition of width m for a graph G can be found in polynomial time, if one exists.

What is actually proved in [ArCorPr87a] is that a simple graph can be embedded in a partial k–tree in time $O(n^{k+2})$. But from such an embedding of the underlying simple graph of G it is easy to produce a tree decomposition [Sch86], [Sch89]. This tree decomposition can be found in such a way that the size of the tree is $\leq n - m$.

A purely existential proof that there is an $O(|V(G)|^2)$ algorithm to decide whether the tree width of G is $\leq k$ for every fixed k was given in [RoSey86] [RoSey86b]. Recently [BoKl90] proved

(9) Theorem([BoKl90]): For every natural number m, there exists an algorithm, that given a graph $G = (V, E)$, determines whether the tree–width of G is at most m, and if so, finds a tree–decomposition of G with tree–width at most m in time $O(|V| \cdot log^2|V|)$.

The algorithm to prove this theorem is given explicitely. This result can be improved in case that one is not interested to find the tree decomposition.

(10) Theorem([ArCouPrSe90]): For every natural number m, there exists an algorithm, that given a graph $G = (V, E)$, determines whether the tree–width of G is at most m in time $O(|V|)$.

The proof of the following result is based on a similar result from [La90].

(11) Theorem([BoKl90]): For each m there exists an algorithm, that uses $O(log^3|V|)$ time on a CRCW PRAM with $O(|V|)$ processors, that given graph $G = (V, E)$, determines whether the tree–width of G is at most m, and if so finds a tree–decomposition of G with tree–width at most m.

In case that there is no fixed upper bound for the tree width the problem becomes NP–complete.

(12) Theorem([ArCorPr87a]): If k is a part of the input, the decision problem "is tree width of $G \leq k$" is NP–complete.

We shall need also the following notion, which was introduced in [Se86].

Let $n \geq 1$ be a natural number. A graph G is called *n–bounded tree partite* if there is a tree T and a decomposition $(A_t : t \in V(T))$ such that

1. $V(G) = \bigcup(A_t : t \in V(T))$,

2. $A_t \bigcap A_{t'} = \emptyset$ for all $t, t' \in V(T)$ with $t \neq t'$,

3. for each edge $e \in E(G)$ with endvertices a and b there is an element $t \in A_t$ or there are $t, t' \in V(T)$ which are adjacent in T such that $a \in A_t$ and $b \in A_{t'}$,

4. $|A_t| \leq n$.

The pair $(T, (A_t : t \in V(T)))$ is also denoted as strong tree decomposition of G of width n.

The minimal n, where G is n–bounded tree partite, is called the *strong tree width of G*. Both notions of width are close related via the following proposition.

(13) Fact: (a) If G is n–bounded tree partite, then G has tree width $\leq 2n - 1$.

(b) If G has tree width n, then there is a G' which has a minor isomorphic to G such that G' is $(n + 1)$–bounded tree partite.

We shall also need the following notion of mixed graphs.

Definition: A *mixed graph* is a relational structure $(A, V, E, D, P_1, \ldots, P_l, R_1, R_2, R_3)$, where A is a nonempty set (set of vertices and edges), $V, E, D, P_1, \ldots, P_l$ are unary predicates and R_1, R_2, R_3 are binary predicates, with the intended meaning, that

V	designates the set of vertices,
E	designates the set of edges,
D	designates the set of directed edges,
$P_1, \ldots, P_l$	are special sets of vertices and edges,
$R_1(a, b)$	holds if and only if a is a vertex which is incident with the edge b
$R_2(a, b)$	holds if and only if a is the tail of the directed edge b
$R_3(a, b)$	holds if and only if a is the head of the directed edge b

For mixed graphs the following notion of tree decomposition will be used during the paper.

Definition:
A *tree decomposition of a mixed graph* $G = (A, V, E, D, P_1, \ldots, P_1, R_1, R_2, R_3)$ is a pair (T, S), where T is a tree and S is a family of sets indexed by the vertices of T, such that:

1. $\bigcup_{X_t \in S} X_t = A$

2. For all c such that $E(c)$ holds there exists a unique $X_t \in S$ such that $c \in X_t$; and if a, b satisfies $R_1(a, c)$ and $R_1(b, c)$ respectively then $a, b \in X_t$.

3. For all c such that $D(c)$ holds there exists a unique $X_t \in S$ such that $c \in X_t$; and if a, b satisfies $R_2(a, c)$ and $R_3(b, c)$ respectively then $a, b \in X_t$.

4. For all $a \in A$ the subgraph of T induced by $\{t : a \in X_t\}$ is connected

The width of such a decomposition is $max_{X_t \in S} |\{a : a \in X_t, V(a)\}| - 1$.

Sometimes we shall allow also evaluations of the edges and vertices by non–negative integers or rational numbers. In this case one has to relate corresponding functions to the structures. When dealing with algorithmic problems of graphs it is natural to connect with a graph G a measure, its *size* $|G|$, for a graph without evaluation this is the number of vertices. In cases with integer valued evaluations we will add all values in these to get the size of the problem instance, and for rational valued evaluations we add the numerators and denominators and use a common denominator. For a graph G we denote its number of vertices by n.

3   Tilings

Tilings are an elegant and very useful tool to verify undecidability or complexity results in mathematical logic and theoretical computer science. Tiling problems, which are considered here can be described as in [Em83] as problems asking whether it is possible to cover a region in the square grid in the Cartesian plane (i.e., a subset of $Z \times Z$), which region may be bounded (e.g. a n by n square or an n by m rectangle) or unbounded (the whole plane, a half plane or quadrant), using "tiles" from a given finite set, such that specific boundary and adjacency conditions are satisfied. The tiles are represented by unit squares with coloured edges, with colours drawn from a fixed finite alphabet; rotations or reflections of tiles are not allowed. The adjacency conditions stipulate that tiles placed at adjacent squares (horizontally or vertically) have matching colours on their common edge. The boundary of the region to be covered have colours matching with a given colouring along this boundary, or that on specific places specific tiles have to be placed.

The following variants of the problem are here of interest.

Square tiling:

Instance: a finite set D of tiles; an N by N square with colourings on the border.

Question: Does there exist a tiling of the entire square, extending the colouring along the
border?

Unconstrained square tiling:

Instance: a finite set D of tiles; a natural number N.

Question: Does there exist a tiling of the N by N square?

Origin constrained tiling:

Instance: a finite set D of tiles; a tile T_0 in D.

Question: Is it possible to tile the entire plane with T_0 appearing at the origin?

Unconstrained tiling:

Instance: a finite set D of tiles

Question: Is it possible to tile the entire plane?

Tiling problems recommend themselves because of their simplicity as a tool which one should try first in every attempt to establish the undecidability of a theory or the "high" complexity of a problem.

(14) Theorem ([GaJoPa77]): Square tiling and Unconstrained square tiling are NP–complete

(15) Theorem ([Wa61]): Origin constrained tiling is undecidable.

(16) Theorem ([Ber66]): Unconstrained tiling is undecidable.

The unconstrained cases can be reduced via a sophisticated construction to the corresponding origin constrained cases. These and the square tiling are proved to be undecidable or NP–hard via a natural coding of Turing machines. During his research on the mechanization of mathematical arguments H. Wang was the first to use tilings in his result on the undecidability of the prefix–class $\forall\exists\forall$ of elementary logic (see [Wa75]). Afterwards several variants of this problem were considered in many applications to undecidability and complexity (see e.g. [Boe85], [MulSchu85], [Se75], [Se76] or the references of these articles).

Of special interest for our applications are generalizations of tiling problems which use not only unit squares as tiles and square grids as a ground to cover. In the following a generalization will be presented, where these conditions are generalized. This generalization was considered in [Se86] and it can be viewed as a weak version of the origin restricted tiling problems with forbidden patterns which were introduced in [MulSchu81], [MulSchu85]. We need some notations first.

A *rooted* graph (G, a) is a graph G together with a designated vertex a, denoted as *root*. The radius of a rooted graph is the maximal distance of an arbitrary vertex in it from the root. As a *stencil S* denote a rooted graph of radius 1 whose vertices are coloured by finitely many colours. Let D be a finite set of stencils and let Σ be the set of colours used for the stencils of D. A Σ–colouring F of an arbitrary graph G is said to be a D–colouring if for every vertex a of G there is an $S \in D$ with

$$(G|N(a), F|N(a), a) \cong S.$$

Where $|$ denotes in case $G|N(a)$ "subgraph induced by" and in case $F|N(a)$ "restricted to" respectively.

Now consider the following problem.

Stencil colouring:

Instance: a finite set D of stencils; a finite graph G.

Question: Does there exist a D–colouring of G?

(17) Lemma: Stencil colouring is NP–hard.

To prove this Lemma one can easily see that the problem "Chromatic number of $G \leq k$" can be reduced to the stencil colouring for each fixed k, since for each fixed k there is a set D_k of stencils such that

> G has a D_k–colouring if and only if the chromatic
> number of $G \leq k$.

Hence this problem can be viewed as a natural generalization of the chromatic number problem for fixed k (other generalizations can be found in [BrCor87], [Ru86]). But while the latter problem has polynomial solutions for a lot of classes of "very regular" graphs, there are classes of very regular graphs for which the first problem remains NP–hard.

(18) Lemma: If the sets of stencils are restricted to sets of rooted coloured trees of radius ≤ 1 and if the graphs G are choosen from the set of $n \times n$ grids (with n arbitrary and not fixed), then this restricted Stencil colouring remains NP–hard.

The proof can be done by reducing the square tiling problem to this restricted stencil colouring problem. The basic idea is to define a very special set of stencils D_0 with the help of which it is possible to create a special pattern on the $(n \times n)$–grid which "enables the orientation of the tiles". The stencils coding a given set of tiles are then built by giving the stencils of D_0 "additional" colours.

Thio ohowo that Stencil colouring can bo viowod ao a joint gonoraliaation of colouring and tiling problems. Most applications of tiling problems are negative in a sense that they are used to prove high complexity of given problems. A landmark for positive applications are the investigations in [MulSchu81], [MulSchu85] on generalized tiling languages with the help of which D.E. Muller and P.E. Schupp could show that the monadic second order theory of any context–free graph is decidable (see also Section 10). Here will be given a positive application of stencil colourings, which is based on the next lemma.

(19)Lemma: Let D be a fixed set of rooted coloured trees of radius ≤ 1. There is a linear time solution for the problem "G is D–colourable" if G varies over trees only.

The proof of this lemma is an easy application of dynamic programming. One chooses a root of the tree and starting in the leafes one runs once through all the vertices a checking locally all possible stencils fitting for a which are "consistent" with the possible stencils on the predecessors.

4   The message of the tile

Trying to prove that a given algorithmic problem P for a class K of graphs is NP–hard via coding of a tiling problem usually requires to find for each set D of tiles and each natural number n a graph $G \in K$ such that

(a) in G can be defined a $n \times n$–grid whose vertices code the positions of the tiles

and

(b) for each vertex of this grid there is found a configuration in G coding a possible tile lying at this position.

Moreover this should be possible in a uniform way, "describable" by problem P.

This heuristic argument suggests two ways to avoid the possibility to code a tiling problem and hence high complexity; at least till somebody proves that tiling problems have lowdegree polynomial solutions.

These two ways are:

(A) restrict the size of the embeddable or definable grids

(B) restrict the local structure of the graphs such that a coding of the tiles is not possible (e.g. by considering very regular graphs or graphs which are generated by simple operations).

This article concentrates mainly on the consequences of (A). To make it more clear what it means to follow (A) one has to define what the embeddability or definability of a grid in an graph means.

Investigating the proofs of the NP–hardness of many algorithmic problems suggests to avoid the embeddability or definability of large grids by avoiding thes grids as minors.

This first approximation to (A) is general enough to lead to interesting results. The structure theoretic background is Theorem 5 of N. Robertson and P. D. Seymour which shows that it is sufficient to concentrate the attention to classes of bounded tree width.

In the light of the tiling–paradigm this result is an essential step to explain the gap between high complexity and large grids on the one side and low complexity and structures close related to trees on the other side.

The next step on the way following this paradigm is to find a sufficiently comprehensive class of problems, which become solvable in polynomial (or even linear) time if they are restricted to classes of graphs not containing arbitrary large grids as minors. This will be done in the next section.

5 Monadic Second Order Graph Properties

There are several interesting approaches to find a characterization of linear or polynomial time properties for classes of graphs of bounded tree width. From these one has to mention the following approaches:

- on series parallel graphs in [TaNiSa82],

- on bandwidth constraind problems in [MoSu84],

- on partial k–trees in [Ar85], [ArPr89], [ArCorPr87a],

- on regular, homeomorphism–based or state–transition definable problems in [BeLawWo85], [BeLawWo87], [MahPe87],

- on recurrence–systems and k–terminal graphs in [Wim88], [WimHe85],

- on algebraic or logical defined local graph properties in [Bo87], [Sch86], [Sch87], [Sch88], [Sch89], [SchSe89], [Se85], [Se86],

- on graph properties definable in monadic second order logic or its extensions in [Cou87], [ArLaSe88], [ArLaSe89], [Bor88], [BorParTo88]

- on hyperedge replacement and graph grammars in [Lau87], [Lau89], [Lau90],

- on (k, μ)–decomposable graphs in [Ho89], [HoRe90],

- on nonconstructive ideas, well–quasi–orderings, graph minors, self–reductions in [RoSey86], [RoSey86b], [FeLan86], [FeLan87], [FeLan88], [FeLan89], [BrFeLan87], [Bo88].

Essentially all approaches use the fact that for a graph of bounded tree width one can easily find "the underlying tree structure" of the graph and use it to solve the algorithmic problem via dynamic programming.

Borrowing ideas from the investigation of the decidability of monadic second order theories here is presented a logical approach (see [Cou87], [ArLaSe88]). It is not only very general and elegant, but it shows also that, most algorithmic problems for graphs of bounded tree width can be reduced via interpretability (see Section 6) to a corresponding problem for trees which can then be solved via tree automata (see Section 7) and hence dynamic programming in linear time. The method is completely automatic and has not to be changed for each different problem.

First we shall introduce pure monadic second order properties, then we shall demonstrate the method of interpretability and tree automata, compare the results with results on decidability and illustrate how these ideas can be extended to more general graph properties

Let K be a class of structures of a fixed signature, especially the signatures of mixed graphes given in the introduction. The *monadic second order language* for K results from the usual first order language by adding variables $X, Y, Z, X_1, X_2, \ldots$ for sets of individuals and the membership symbol $\in$ and allowing quantification also over set variables. To complete the definition of monadic second order logic one has to define the satisfaction relation $G \models \Phi$ for a given structure G of the corresponding signature with domain A and a given formula Φ whose free individual variables have a given interpretation as elements of A and whose free set variables have a given interpretation as subsets of A. This is done as usual (see e.g. [Sh67]) by induction on the structure of Φ where the only new case $\exists X \Phi(X)$ with set variable X is handled as follows:

$G \models \exists X \Phi(X)$ iff   there is a subset B of A such that

$\qquad\qquad$ $G \models \Phi[B]$, where B is chosen as interpretation
$\qquad\qquad$ of X

Universal quantification and the usual set operators and relations can now be defined in the standard way.

More information on monadic second order logic one can find in [Gu85].

Definition: A property P is a monadic second order property (or MS property) over a class K of structures of a given signature if there is a monadic second order formula Φ such that for all G of K: P holds for G iff $G \models \Phi$.

In the following we shall also write $P(G)$ instead of "P holds for G". Many properties of graphs are monadic second order properties for the class of all graphs. Some of those which are NP–hard for the class of all graphs are gathered in the following theorem.

(20) Theorem ([Cou87], [ArLaSe89]): The following properties are MS properties over the class of all mixed graphs:

Domatic number for fixed k [GT4], graph K–colorability for fixed K [GT4] achromatic number for fixed K [GT5], monochromatic triangle [GT6], partition into triangles [GT11], partition into isomorphic subgraphs for fixed connected H [GT12], partition into Hamiltonian subgraphs [GT13], partition into forests for fixed K [GT14], partition into cliques for fixed K [GT15], partition into perfect matchings for fixed K [GT16], covering by cliques for fixed K [GT17], covering by complete bipartite subgraphs for fixed K [GT18], induced subgraph with property P (for monadic second order properties P and fixed K) [GT21], induced connected subgraph with property P (for monadic second order properties P and fixed K) [GT22], induced path for fixed K [GT23], Cubic subgraph [GT32], Hamiltonian completion for fixed K [GT34], Hamiltonian circuit [GT37], directed Hamiltonian circuit [GT38], Hamiltonian path (and directed Hamiltonian path) [GT39], subgraph isomorphism for fixed H [GT48], graph contractability for fixed H [GT51], graph homomorphism for fixed H [GT52], path with forbidden pairs for fixed n [GT54], kernel [GT57], degree constrained spanning tree for fixed K[ND1], disjoint connecting paths for fixed K[ND40], chordal graph completion for fixed K, chromatic index for fixed K, spanning tree partity problem, distance d chromatic number for fixed d and k, thickness $\leq K$ for fixed K, membership for each class C of graphs which is closed under minor taking

The notations in (20) are taken from [GaJo78]. The proof of (20) is easy. One just has to translate the corresponding definition of the problem into the monadic second order language. As an example is handled the case of the domatic number.

The domatic number of a graph G is at least K if $V(G)$ can be partitioned into $m \geq K$ disjoint sets $V_1, V_2, \ldots, V_m$ such that each V_i is a dominating set for G. V_i is a dominating set for G if for all $a \in V(G)|V_i$ there is $a \in V_i$ for which a and b are adjacent in G. Obviously the latter can be expressed by a formula $\varphi(V_i)$:

$$\forall x (V(x) \land \neg x \in V_i \longrightarrow \exists y \exists z V(y) \land E(z) \land y \in V_i \land R_1(x,z) \land R_1(y,z))$$

But the property of being a dominating set is monotone i.e. if A is a dominating set for G and A is contained in B, then also B is a dominating set for G. Hence the considered property is equivalent to:

"$V(G)$ can be partitioned into K disjoint sets $V_1, V_2, \ldots, V_K$ such that each V_i is a dominating set for G."

But this can be formalized in monadic second order logic by a formula Ψ:

$$\exists X_1 \ldots \exists X_K \forall x ((\bigwedge_{i=1}^{K} x \in X_i \longrightarrow \bigwedge_{\substack{j=1 \\ j \neq i}}^{K} x \notin X_j) \land (V(x) \longrightarrow \bigvee_{i=1}^{K} x \in X_i) \land \bigwedge_{i=1}^{K} \varphi(X_i))$$

Now it is easy to see $G \models \Psi$ if and only of the domatic number of G is at least K. Most other properties are also very easy. To formalize planarity or membership in minor closed classes one has to use Kuratowski's theorem or one has to know respectively that each such class has a finite number of minimal forbidden minors by [RoSey88] (see [ArLaSe89]).

6   Interpretability

The method of interpretability is widely used in mathematical logic, especially to deduce decidability or undecidability results (see e. g. [Ra65], [Ra77]) Here is used a variant of the semantic interpretability, which was introduced in [Ra65] as a useful tool to transfer decidability and undecidability results from one theory to another (see also [Ra77]). A short desription of this method will be given here.

Let K and K' be model classes in languages L and L' respectively. Here is assumed that L and L' are monadic second order languages, but the concept works also in other cases. An *interpretation* I assigns to every relational symbol R of L of arity s_R a formula $\Psi_R(x_1, \ldots, x_{s_R})$ of L' an additionally it constitutes of a formula $\varphi(x)$ of L'. The interpretation of the basic symbols of the language L is inductively extended to all formulas χ of L. The interpreted formula is then denoted by χ^I and is built according to the following rules:

   1. $(x = y)^I := (x = y)$;

2. $(R(x_1,\ldots,x_{s_R}))^I := \Psi_R(x_1,\ldots,x_{s_R})$ for each relational symbol R of L';

3. $(\neg\chi)^I := \neg(\chi^I)$;

4. $(\chi_1 \wedge \chi_2)^I := \chi_1^I \wedge \chi_2^I$;

5. $(\exists x\chi)^I := \exists x(\phi(x) \wedge \chi^I)$;

6. $(x \in X)^I := (x \in X)$; and

7. $(\exists X\chi)^I := \exists X\chi^I$.

In case that L and L' are not monadic second order languages one has to omit simply the clauses 6. and 7.

If $G = (A,\ldots)$ is a structure of K' with domain A, then we obtain – using I – a structure G^I for L. The domain of G^I is simply the set

$$\{a : a \in A \text{ and } G \models \varphi[a]\}.$$

Each symbol R of L is interpreted by the relation

$$\{(a_1,\ldots,a_{s_R}): \quad a_1 \in A,\ldots,a_{s_R} \in A \text{ and } G \models \varphi[a_1] \wedge \ldots \wedge \varphi[a_{s_R}]$$
$$\wedge \Psi_R[a_1,\ldots,a_{s_R}]\}.$$

The key result for interpretability is the following lemma.

(21) Lemma ([Ra65]): For each sentence χ of L and for each structure G of K' it holds:

$G \models \chi^I$ if and only if $G^I \models \chi$.

The lemma can be easily proven by induction on the structure of formulas. A *theory* of K in the language L, denoted as $Th_L(K)$, is the set

$$\{\chi : \chi \text{ is an } L \text{ sentence and } K \models \chi\}.$$

The theory $Th_L(K)$ is said to be *interpretable in* $Th_{L'}(K')$ if

1. for every structure $G \in K$ there is a structure $F \in K'$ so that $F^I \cong G$;

2. for every structure $F \in K'$, the structure F^I is isomorphic to a structure of K.

¿From Lemma 21 it follows the main property of interpretations with respect to decidability.

(22) Theorem ([Ra65]): Let K and K' be model classes and let L and L', respectively, be suitabel languages. If $Th_L(K)$ is interpretable in $Th_{L'}(K')$, then the decidability of $Th_{L'}(K')$ implies the decidability of $Th_L(K)$.

The interpretability can easily be generalized by handling the identity as a non–logical symbol; that is to say, it is interpreted by a congruence relation definable by a formula $\epsilon(x,y)$ of L. To handle algorithmic problems for finite graphs in [ArLaSe88] was introduced the following variant of interpretability.

K is said to be *linear (polynomial) time interpretable* into K' with respect to L, L' if there is an interpretation as above (possibly with interpretation of equality as congruence relation) and an algorithm A which constructs for each $G \in K$ a structure $A(G) \in K'$ in time linear (polynomial) in $|G|$ such that:

$$G \cong (A(G))^I.$$

A property P for structures from K is an *L–property* if there is a sentence χ of L such that for each $G \in K$ it holds:

$$P(G) \text{ if and only if } G \models \chi.$$

For an interpretation I, the sentence χ^I is a sentence of L' which defines an L'–property P^I by:

$$P^I(G) \text{ holds } \Longleftrightarrow G \models \chi^I,$$

for all L' structures G. The following results are useful for applications of this notion.

(23) Fact ([Ra65],[ArLaSe89]): Interpretability and linear (polynomial) time interpretability is transitive.

(24) Theorem ([ArLaSe89]): Assume that P is an L–problem on a class K and let I be a linear (polynomial) time interpretation of K into K' in linear (polynomial) time, then P can also be solved in linear (polynomial) time over K.

Before the next results can be given we need some notation. For a finite set (an alphabet) $\Sigma \neq \emptyset$ let Σ^* be the set of all finite words of elements from Σ.

The *length* $l(u)$ of a word a is the number of its symbols from Σ. Assume here and in the following section that $\Delta := \{0, 1\}$. Let $A \subseteq \Delta^*$ be a subset of Δ^* which is closed under initial segment relation, i.e.

$$\text{if } uv \in A, \text{ then } u \in A.$$

With each such A we connect in a canonical way a *binary tree*:

$$(A, s_0, s_1), \quad \text{with } s_0 := \{(u, u0) : u0 \in A\}$$
$$\text{and } s_1 := \{(u, u1) : u1 \in yA\}.$$

s_0, denotes left successor or left son relation, while s_1 denotes right successor or right son relation. The empty word, denoted by λ is the *root*. Such structures will be denoted as *binary trees*.

(25) Theorem([ArLaSe89]): For each natural number m every class K of graphs whose tree width is bounded by m is polynomial time interpretable into the class of binary trees. If the graphs from K are given together with a tree decomposition the the corresponding transformation A can be performed in linear time.

Using (24) this result enables us to reduce mondadic second order problems for graphs of universally bounded tree width to related monadic second order problems for binary trees and to solve them in polynomial time, if the corresponding problems for trees can be solved in polynomial time. How this can be done will be shown in the next section.

The related result concerning decidability is the following theorem.

(26) Theorem ([Se87a]): Let m be a fixed natural number. The (weak) monadic second order theory of the class of all graphs possibly (infinite) of tree width $\leq m$ is interpretable in the (weak) monadic second order theory of the class of all trees. Hence both theories are decidable.

7   Tree automata

Using the method of interpretability from the previous section it is easy to reduce monadic second order properties of graphs of bounded tree width to monadic second order properties of binary trees or trees. Here will be presented now a canonical method to solve such problems for trees. Before we need the following notions.

For a finite set $\Sigma \neq \emptyset$, a Σ-*tree* is a structure (a, s_0, s_1, τ) such that (A, s_0, s_1) is a binary tree and τ is a function $\tau : a \longrightarrow \Sigma$

To prove the decidability of the weak monadic second order theory of two successor functions and to characterize definability in this theory in [Do66] and [ThaWr68] was introduced a concept of tree automaton and this automata were related to monadic second order formulas.

Definition: A *tree automaton* $M = (S, \Sigma, \delta, s_0, F)$ is a quintuple where:

S is a finite set of states;
Σ is a finite set, the alphabet, disjoint from S;
δ is a transition function, $\delta : S \times S \times \Sigma \longrightarrow S$;
s_0 is the initial state, $s_0 \in S$;
F is the set of accepting states, $F \subset S$.

For a given Σ such an automaton will be denoted also as Σ-automaton. A *run* of M on a Σ-tree $T = (A, s_0, s_1, \tau)$ is a function $r : A \longrightarrow S$ defined by $r(a) := \delta(s_0, s_0, \tau(a))$ if a is a leaf and:

$$r(b) := \delta(s_1, s_2, \tau(a)) \text{ if } \quad s_1 = \begin{cases} r(b0) \text{ if } b0 \in A \\ s_0 \quad \text{ if } b0 \notin A \end{cases}$$

$$\text{and}$$

$$s_2 = \begin{cases} r(b1) \text{ if } b1 \in A \\ s_0 \quad \text{ if } b1 \notin A. \end{cases}$$

M *accepts* T if there is a run r of M on T with $r(\lambda) \in F$. Definability of monadic second order formulas can be connected with evaluated trees. Let $\phi(X_1, \ldots, X_m)$ be such a formula. Assume that (A, s_0, s_1) is a binary tree and that $B_1, \ldots, B_m$ are subsets of A. Let χ_{B_i} be the *characteristic function* of each B_i:

$$\chi_{B_i}(a) := \begin{cases} 0 \text{ if } a \notin B_i \\ 1 \text{ if } a \in B_i. \end{cases}$$

Define $\Delta^m := \{a : a \in \Delta^* \text{ and } l(a) = m\}$. Now define $\tau_{B_1,\dots,B_m} : A \longrightarrow \Delta^m$ by:

$$\tau_{B_1,\dots,B_m}(a) := \chi_{B_1}(a)\chi_{B_2}(a)\dots\chi_{B_m}(a).$$

One of the key observations in this area is now the following theorem, which is proved by induction on the structure of formulas.

(27) Theorem([Do66], [ThaWr68]): For each monadic second order formula $\Phi(X_1,\dots X_1)$ one can effectively construct a Δ^m–automaton M such that for each binary tree T and arbitrary subsets $B_1,\dots,B_1$ from the domain of T the following two conditions are equivalent:

1. $T \models \Phi[B_1,\dots,B_1]$

2. M accepts $(T, \tau_{B_1,\dots,B_m})$.

But each tree automaton works obviously in linear time on the corresponding trees, i.e. for a given automaton one can decide in linear time whether it accepts a tree or not (linear in the number of the vertices of the tree).

(28) Theorem (implicetely in [Do66], [ThaWr68]): Let K be a class of binary trees. Then for each fixed MS–property P over K there is an algorithm dediding $P(G)$ for $G \in K$ in time linear in $|V(G)|$.

To prove the decidability of a theory one has to solve yet an additional problem for a tree automata – the acceptance problem. For a given Σ–automaton M define $T(M)$ to be the set of all Σ–trees which are accepted by M. The *emptyness problem* for a given tree automaton M is the problem to decide wheter $T(M)$ is empty? To solve it one needs the following notion.

The *depth* of a finite binary tree (A, S_0, s_1) is the length of the largest word in A.

(29) Theorem([Do66], [ThaWr68]): Let $M = (S, \Sigma, \delta, s_0, F)$ be a Σ–automaton. Then there is a Σ–tree accepted by M if and only if M accepts a Σ–tree of depth $< |S|$.

To prove this theorem one assumes that T is a tree in $T(M)$ of depth $\geq |S|$. Now one can construct a tree $T' \in T(M)$ of smaller depth, since each accepting run of M on T has a state repitition.

(30) Corollary ([Do66], [ThaWr68]): For each Σ–automaton M one can solve the emptyness problem.

8 Polynomial time solvable problems

Now we can draw some conclusions from the results of the previous sections. The following theorem which was proved by B. Courcelle on a different way is an immediate consequence of (25), (28) and (24)

(31) Theorem ([Cou87]): For each monadic second order property P, and for each class K of graphs of universally bounded tree width, the problem $P(G)$ for $G \in K$ can be decided in polynomial time. If G is given together with a tree decomposition, then this can be done in linear time.

Theorem 20 shows that there is a large number of monadic second order properties. But there are many others which are not and can nevertheless be handled in a similar way. In such problems there can be asked for instance questions about the cardinanlity of a set of edges or vertices with a specific property or about the maximum or minimum of a sum of an evaluation of the vertices or edges of the graph. To handle also such kinds of properties, extended monadic second order properties (shortly denoted as EMS properties) were introduced in [ArLaSe88].

As size $|G|$ of a graph G here normally $|V(G)| + |E(G)|$ is used, but in case that graphs with rational valued evaluations are considered all numerators and denomerators of all values of these evaluations are added, assuming that all evaluations have common denominators.

To define EMS properties one has to combine an open monadic second order formula containing free set variables $X_1, \ldots, X_l$ with a formula involving the sums of the evaluations $f_1, \ldots, f_m$ over sets for which $X_1, \ldots, X_l$ stand.

An *evaluation term* is an expression built from the arithmetic operators $+, -, \times$ and with atoms being rational numbers, $|X_i|_j$ for $1 \leq i \leq 1$ and $1 \leq j \leq m$, and Y_i for $1 \leq i \leq t$.

We denote it by, e.g., $F(|X_1|_1, \ldots, |X_l|_m, Y_1, \ldots, Y_t)$. In case that the evaluations are all into the set of natural numbers, one can also allow *mod p* for a natural number p in evaluation terms. Most EMS problems have only one evaluation which is identically 1. In such cases $|X_i|_1$ represents the cardinality of X_i. An evaluation relation is a propositional formula with atoms being of the form $F \leq 0$ or $F = 0$, where F is an evaluation term.

The intended interpretation of evaluation terms and relations should be obvious. Using them one can define the class of EMS problems as in [ArLaSe88].

Definition: A property P is an *extended monadic second order property*, or *EMS property* over class K of structures $(G, f_1^G, \ldots, f_m^G)$ if there are constants l, t a monadic second order formula Φ (which does not contain the evaluations) with free set variables $X_1, \ldots, X_l$ and an evaluation relation $\psi_P(|X_1|_1, \ldots, |X_l|_m, Y_1, \ldots, Y_t)$ such that for each structure $(G, f_1^G, \ldots, f_m^G) \in K$ it holds:

$P(G, f_1^G, \ldots, f_m^G, C_1 \ldots, C_t)$ if and only if there are subsets $A_1, \ldots, A_l$ of the domain A of G such that $G \models \Phi[A_1, \ldots, A_l]$ and $\psi_p[\sum_{a \in A_1} f_1^G(a), \ldots, \sum_{a \in A_l} f_m^G(a), C_1, \ldots, C_t]$. Above the C_i correspond to numbers given with the problem instance.

A problem is an *extended monadic second order extremum problem* if it can be stated as finding the maximum of an evaluation term
$F(\sum_{a \in A_l} f_1^G(a), \ldots, \sum_{a \in A_1} f_m^G(a), C_1, \ldots, C_t)$ under the constraints $\Phi[A_1, \ldots, A_l]$ and $\psi_p[\sum_{a \in A_l} f_1^G(a), \ldots, \sum_{a \in A_1} f_m^G(a), C_1, \ldots, C_t]$. for a monadic second order formula Φ and evaluation relation ψ. Since negation is allowed in evaluation terms, one can express minimization in terms of maximization.

For a *linear* EMS *extremum problem* one requires that the associated evaluation term F is linear in the quantities $|X_i|_j$, and that the evaluation relation is missing (i.e., identically true).

The following collection of EMS problems demonstrates the applicability of the EMS–concept.

(32) Theorem ([ArLaSe89]): The following problems are linear EMS extremum problems such that $|G|$ is linear in n, where n is the number of vertices and edges in G: vertex cover [GT1], dominating set [GT2], feedback vertex set [GT7], feedback arc set [GT8], partial feedback edge set for fixed maximum cycle length l [GT9], minimum maximal matching [GT10], partition into cliques [GT15], clique [GT19], independent set [GT20], induced subgraph with property P (for monadic second order property P) [GT21], induced connected subgraph with property P (for monadic second order property P) [GT22], induced path [GT23], balanced complete bipartite subgraph [GT24], bipartite subgraph [GT25], degree–bounde connected subgraph for fixed d [GT26], planar subgraph [GT27], transitive subgraph [GT29], uniconnected subgraph [GT30], minimum K–connected subgraph for fixed K [GT31], minimum equivalent digraph [GT33], Hamiltonian completion [GT34], multiple choice matching for fixed J [GT55], k–closure [GT58], path distinguishers [GT60], maximum leaf spanning tree [ND2], maximum length–bounded disjoint paths for fixed J [ND42], minimum edge (vertex) deletion for any monadic second order property P.

(33) Theorem ([ArLaSe89]): The following problems are linear EMS extremum problems over the class of mixed graphs: bounded diameter spanning tree for fixed d [ND4], multiple choice branching for fixed M [ND11], Steiner tree in graphs [ND12], maximum cut [ND16], longest circuit [ND28], longest path [ND29].

The last two theorems use the terminology of [GaJo78]. The proof of these results is simply done by a straightforward transformation of the problem definition into an extended second order expression. Some further EMS problems can be found in [ArLaSe89]. The techniques used to prove (31) can be applied in a modified form also to EMS properties.

First the concept of linear (polynomial) time interpretability can be extended in such a way that structures with evaluations of elements of their domain can be handled. For this type of interpretability, denoted as evaluation preserving linear (polynomial) time interpretability, (23), (24) and (25) were generalized in [ArLaSe89]. But also the tree automata concept can be generalized by adding counters which keep track of possible values for the sums of the evaluations over the X_i restricted to the subtree below a node.

Using these ideas the following result can be proved similar as the here presented idea of proof of Theorem 31.

(34) Theorem ([ArLaSe89]): For each class K of graphs of universally bounded tree width, every linear EMS extremum problem P can be solved in linear time with the uniform cost measure, for $G \in K$, if G is given with a tree decomposition. If the tree decomposition is not given the time complexity is $O(|V(G)| \cdot \log^2 |V(G)|)$.

9 Local properties

Historically local properties (see e.g. [Bo87], [Sch86], [Se85], [Se86]) were considered before the here considered EMS properties, which are "global" in character. Here we will investigate the relation between EMS properties from the point of view of the interpretability concept.

In the following we shall concentrate on a subclass of the existential logically verifiable properties (denoted as ELV–properties for short), which were introduced in [Se85] and generalized in [Sch87], [Sch89]. This class of monadic ELV–properties is of interest because it can be considered as intersection of the class of ELV–properties with the class of EMS–properties and moreover there is a method to transform each EMS–property for a class of graph of universally bounded tree width to a monadic ELV–property on a class of trees.

Before monadic ELV properties can be introduced the following definition is needed.

Definition: A formula $\Phi(X_1, \ldots, X_l, y)$ describes a *local property* over a class K of graphs if there is a natural number r such that: for each graph $G \in K$, all subsets $A_1, \ldots, A_l$ of its domain and each element a of its domain it holds

$G \models \Phi[A_1, \ldots, A_l, a]$ iff
$G|B \models \Phi[A_1 \cap B, \ldots, A_l \cap B, a]$, where
$B := N_r(G, a)$, and $N_r(G, a)$ denotes the set of all vertices of G which have distance $\leq r$ form a.

Definition: A property P of graphs is *monadic existential locally verifiable (monadic ELV for short)* over a class K of structures $(G, f_1^G, \ldots, f_m^G)$ if it is such an EMS property that the corresponding monadic second order formula $\Phi(X_1, \ldots, X_l)$ is of the form

$$\forall y \psi(X_1, \ldots, X_l, y)$$

and $\psi(X_1, \ldots, X_l, y)$ describes a local property over K.

The following lemma gathers some of the monadic ELV properties.

(35) Lemma: The following properties are monadic ELV properties:

1. "G is m–colourable" for fixed m,

2. "The domatic number of G is at least m" for fixed m,

3. "There is no monochromatic triangel in G",

4. "G has a distance d chromatic number not greater than m" for fixed m and d,

5. "$V(G)$ can be partitioned into at most m perfect matchings" for fixed m,

6. "G has disjoint paths between m fixed pairs of vertices" for fixed m,

7. "The chromatic index of G is at most m for fixed m.

These properties were already considered in [Sch87] (see also [Se86]). The proof of the lemma is almost trivial and proceeds simply by translating the definitions of the properties into a suitable formalization.

For ELV–properties in [Se86] was proved the following result.

(36) Lemma: Let n and m be fixed natural numbers and assume that P is an arbitrary monadic ELV property. Then there is an algorithm deciding $P(G)$ in time polynomial in $|V(G)|$ for all graphs of strong tree width $\leq n$ and maximal degree $\leq m$. Moreover the algorithm becomes linear if the graphs G are given together with their strong tree decomposition.

Assume that (T, X) is a strong tree decomposition of G of strong tree width $\leq n$. In case that P is a pure monadic second order problem, i.e. the problem definition does not contain an evaluation relation the main idea of the proof is to transform the problem $P(G)$ to a stencil colouring problem on T and then to apply Lemma 19. In the general case one gets beside the stencil colouring problem a corresponding evaluation problem which has to be solved in a strict combination with the stencil colouring problem via dynamic programming. Extensions of this result without the degree bound and for graphs of bounded tree width instead of the strong tree width can be found in [Sch87], [Sch89].

Now considering tree automata it is easy to transform an arbitrary tree automaton M into an equivalent monadic second order formula. This can be done in a canonical way just by describing the behaviour of the automaton on the trees. But this formula is of a very special form and it is easy to see that it describes a local property. Using this ideas one can prove the following result.

(37) Lemma: For each linear EMS extremum property P over a class K of graphs of universally bounded tree width there exists an algorithm A, transforming each structure $G \in K$ into a labeled binary tree $A(G)$, and there exists a monadic ELV property Q over the class of labeled binary tree produced by A such that:

$P(G)$ is equivalent to $Q(A(G))$ for each $G \in K$.

The algorithm, which has time complexity $O(|V(G)| \cdot \log^2 |V(G)|)$, becomes linear if for each G a corresponding tree decomposition is given.

On the other side the above idea illustrates how algorithmic problems on graphs of bounded tree width which can be reduced via interpretability or other methods to automata problems for trees are reducible (using the same methods) to (eventually extended) monadic second order problems. This demonstrates again the importance of the concept of extended monadic second order properties since many other approaches use automata–like constructions and could hence possibly be reduced to EMS problems for tree.

10 Monadic second order theories

In this section also infinite structures are allowed. Here, a graph will be a relational structure (A, R), where $A \neq \emptyset$ is the set of vertices and R is a binary, irreflexive and symmetric relation on R. This representation is possible since only simple graphs are considered here. One simply writes $R(a, b)$, if one wants to say that there is an edge between the two vertices a and b. In the corresponding monadic second order theory of such a graph one can "speak" only about sets of vertices. In case that one wants to speak also about sets of edges then one should use a representation of the considered graphs as mixed graphs (see also [Cou89]).

In the preceeding sections of this article was underlined the heuristic statement that "most" algorithmic problems for classes of graphs "containing large grids and not being very regular" are NP–hard and to the contrary if a class of graphs does not "contain large grids" the "most" problems have polynomial or even linear–time solutions. This heuristic argumentation becomes more serious if instead of single extended monadic second order problems large classes of problems or better whole monadic second order theories are considered, while NP–hardness is changed to undecidability and polynomial or linear–time solvability to decidability.

The large expressive power of the monadic second order logic, which was demonstrated also in the preceeding sections, makes results concerning decidable monadic second–order theories interesting.

Unfortunately there is a large collection of undecidable monadic second order theories (see e.g. [She75], [Gu79a], [Gu82], [GarSchm74], [KB79], [Se75], [Se78]). A strong connection between generalized tilings of arbitrary finitely generated graphs and the decidability of monadic second order theories of such graphs was discovered by D.E. Muller and P.E. Schupp. A *finitely generated graph* (see [MulSchu81] for details) is a connected labeled graph G with a distinguished vertex v_0, called the *origin* of G, a finite label alphabet, and a uniform upper bound m on the degrees of vertices. Cayley diagrams of finitely generated groups can be considered as finitely generated graphs. In [MulSchu81] is defined for each such regular graph G at first a basic tiling problem, which is close related to the here defined stencil colouring problem. The origin restricted tiling problem is to find a colouring of the given graph with a given finite set C_0 of patterns and the neighbourhood of any other vertex does not match with a pattern in a given set F of forbidden patterns. The triple (C, C_0, F) is denoted as as tiling constraint. The *origin restricted tiling problem* is the problem to decide whether for a given tiling constraint (C, C_0, F) and a finitely generated graph G such a tiling exists.

For a given graph G and a given tiling constraint (C, C_0, F) all possible colourings of G satisfying (C, C_0, F) are a basic tiling language. From basic tiling languages are derived general tiling languages by certain closure operations as projection, intersection, union and complementation. In such a way general tiling languages generalize tree automata which accept evaluated trees. D.E. Muller and P.E. Schupp proved the following theorem.

(38) Theorem ([MulSchu81]): Let G be a finitely generated graph. The monadic second order theory of G is decidable if and only if the emptyness problem for general tiling languages on G is decidable.

Using this result D.E. Muller and P.E. Schupp proved that the monadic second order theory of any context free graph G is decidable

The most powerful positive result concerning decidable monadic second order theories is M.O. Rabin's proof [Ra69] that S2S, the monadic theory of two successor functions, which is equivalent to the monadic theory of the (infinite) full binary tree is decidable.

To prove it M.O. Rabin developed a generalization of tree automata for infinite trees. Contemporary in [GuHar82] was found a simple proof of this result. The theory S2S is up to interpretability equivalent with the monadic second order theory of the class of all countable trees. The monadic second order theory of the theory of all tree does not differ from the monadic second order theory of all countable trees and is hence also decidable (see [She75], [St75], [Tou68]). As consequence of this result could be found many other decidable theories (see e.g. [Ra69], [Ra77], [Gu85], [KoRau76], [Se75], [Se76], [Se79]). This led in [Se79] to the following conjecture (which is formulated here in different terminology, see also [Se87]):

Let n be an arbitrary fixed natural number. Each class K of planar graphs, no graph of which has a minor isomorphic to the $n \times n$–grid, only contained structures which are "similar to trees"; more exactly $Th_2(K)$ is interpretable in $Th_2(T)$, for a certain class T of trees.

This conjecture was confirmed in [Se79] in the very trivial case that no graph from K has a minor isomorphic to the graph $(\{1, 2, \ldots, 6\}, \{\{1, 2\}, \{2, 3\}, \{3, 5\}, \{2, 4\},$
$\{3, 6\}, \{1, 4\}, \{4, 5\}, \{5, 6\}, \{6, 1\}\})$ and all considered graphs from K have degree ≤ 3.

An essential step in the proof of this conjecture was the following result.

(39) Theorem ([Se75], [Se87a]): Let K be a class of graphs such that for each planar graph H there is a planar graph $G \in K$ such that H is isomorphic to a minor of G. Then $Th_2(K)$ and $Th_{w2}(K)$ are undecidable.

This result was proved using a coding of tilings and applying Theorem 16.

The conjecture follows then from the following results.

(40) Theorem ([Se87a]): If K is a class of planar graphs such that $Th_2(K)$ or $Th_{w2}(K)$ are decidable, then there is an n such that each $G \in K$ has tree width $\leq n$.

(41) Corollary ([Se87a]): If K is a class of planar graphs such that $Th_2(K)$ or $Th_{w2}(K)$ are decidable, then there is a class T of trees such that $Th_2(K)$ is interpretable in $Th_2(T)$.

The corollary immediately follows from (40) and (26).

The proof of (40) is an application of (39) and (6).

Beside the following result it is open whether (40) and (41) hold for arbitrary classes of graphs.

(42) Theorem: Let K be a class of graphs with one of the following properties
– K is a class of graphs of universally bounded degree
– K is a class of graphs of universally bounded genus
– there is a graph H such that no graph of K has a minor isomorphic to H and all graphs in K are finite.
Moreover assume that $Th_2(K)$ or $Th_{w2}(K)$ are decidable. Then there is an n such that each $G \in K$ has tree width $\leq n$. Moreover there is a class T of trees such that $Th_2(K)$ is interpretable in $Th_2(T)$. In case of classes K of graphs of universally bounded degree this result was proved together with B. Courcelle. In the general case it can be proved similar to (40) and (41) using a result of [Cou91], that quantification on edges and sets of edges is eliminable in the considered theories.

If the expressive power of the monadic second order logic is extended by allowing quantifications not only over sets of vertices, but also over sets of edges, then Theorem 40 and its Corollary 41 can be generalized (for this extended logic) for all classes of graphs.

Instead of using this logic one can consider mixed graphs in the sense of Section 2. Then the following results can be proved.

(43) Theorem: Let K be a class of mixed graphs such that for each planar graph H there is a graph $G \in K$ such that H is isomorphic to a minor of G. Then $Th_2(K)$ and $Th_{w2}(K)$ are undecidable.

(44) Corollary: If K is a class of mixed graphs with decidable monadic or weak monadic theory, then there is an n such that each $G \in K$ has tree–width $\leq n$. Moreover there is a class T of trees such that $Th_2(K)$ is interpretable in $Th_2(T)$.

The above results indicates that the monadic theory of all trees is and will be unrivalled with respect to the question of decidability for classes of graphs of universally bounded degree or genus or classes of graph excluding a fixed minor or mixed graphs. Moreover these results underline the close correlation between arbitrary grids and high complexity on the one side and tree–structured objects and low complexity on the other, which was the main theme of this article.

11   Remarks and Problems

This article was mainly devoted to the research direction originating in possibility (A) from Section 4. It seems to be very interesting to follow possibility (B). Promising approaches for a beginning of this direction could be hierarchically defined classes of graphs from [LeWan87] and [LeWag86].

But also a combination of bounded tree width with a refined investigation of the structure of the considered graphs is of interest. Since one can get more information on the constants in the time bounds (see e.g. [HoRe90]). Especially promising is the approach on graph minors (see [RoSey86], [RoSey86b] and [FeLan86], [FeLan87], [FeLan88], [FeLan89]). Here N. Robertson and P.D. Seymour proved the following famous result.

(45) Theorem ([RoSey86]): Each property P of graphs which is closed under minor taking can be tested for each graph G in time $O(|V(G)|^3)$. In case that P holds only for graphs of universally bounded tree width, then there is an $O(|V(G)|^2)$ algorithm.

Here a property P is closed under minor taking if with $P(G)$ it holds also $P(G')$ for each graph G' which is isomorphic to a minor of G. An example of such a property is the embeddability of a graph into a fixed surface.

Unfortunately the proof of the existence of these algorithms is only existential (see [FeLan86], [FeLan88]). Since one has to find the set of minimal forbidden minors of the property.

But in case that only graphs of universally bounded tree width and monadic second order properties for such graphs are considered the minimal forbidden minors and hence the efficient algorithm can be found using methods which are close related to the ideas presented in this article (see [FeLan89], [FeAb89] [Bo88a], [ArCouPrSe90], [ArPr84].

(46) Theorem ([ArCouPrSe90], [ArPrSe90]): Each monadic second order property P of graphs which excludes a planar graph can be tested in linear time. If P is closed under minor taking then the minimal forbidden minors for P can be found effectively.

This result is also an improvement of (31). With respect to decidability it would be of interest to know whether for each class K of countable structures with a decidable monadic second order theory $Th_2(K)$ there is a class T of trees such that $Th_2(K)$ is interpretable into $Th_2(T)$. The here observed analogies between results on the complexity of algorithmic problems and the decidability of corresponding monadic second order theories were merely obtained by using similar methods in both cases. One could assume that there is a deeper connection between both areas. It would be interesting to prove this.

References

[Ar85] **S. Arnborg** 1985 *Efficient Algorithmus for Combinatorial Problems of Graphs with Bounded Decomposability* – A Survey BIT 25, 2–33

[Ar90] **S. Arnborg** 1990 *Graph decompositions and tree automata in model–based reasoning* preprint July 23, The Royal Institute of Technology

[ArCorPr87a] **S. Arnborg, D.G. Corneil, A. Proskurowski**, 1987a *Complexity of Finding Embeddings in a k-tree*, SIAM J. Alg- and Discr. Methods 8

[ArCouPrSe90] **S.Arnborg, B.Courcelle, A. Proskurowski and D. Seese** 1990, *An algebraic theory of graph reduction*, Preprint No. 90–02, LaBRI, U.E.R. de Mathematiques et d'Informatique, Universite de Bordeaux I

[ArLaSe88] **Arnborg, S.; Lagergren, J; Seese, D.**, 1988, *Problems Easy for Tree-Decomposable Graphs* (extended abstract), Proc. 15th ICALP, Springer–Verlag, Lecture Notes in Comp. Se. 317, 38–51

[ArLaSe89] **Arnborg, S.; Lagergren, J; Seese, D.**, 1989, *Problems for Easy Tree-decomposable Graphs, to appear, J. of Algorithms*

[ArPr84] **Arnborg, S; Proskurowski, A.**, 1984, *Characterization and Recognition of Partial k-trees*, Preprint TRITA–NA–8402, The Royal Institute of Technology Stockholm

[ArPr89] **S. Arnborg and A. Proskurowski**, 1989, *Linear Time Algorithm for NP-hard Problems on Graphs Embedded in k-trees* Discr. Appl. Math. 23, 11–24.

[ArPrCor86] **S. Arnborg, A. Proskurowski and D.G. Corneil**, 1986, *Forbidden minor characterization of partial 3-trees*, Discrete Math., to appear.

[ArPrSe90] **S. Arnborg, A. Proskurowski, D. Seese**, 1990, *Monadic second Order Logic, Tree Automata and Forbidden Minors*, to appear Proceedings of CSL '90, LNCS.

[BauSeTu80] **A. Baudisch, D. Seese, H.–P. Tuschik**, 1980, *Decidability and generalized quantifiers*, Berlin: Akad.–Verl.

[BauSeTu85] **A. Baudisch, D. Seese, H.–P. Tuschik**, *Decidability and Quantifier Elimination, in Model–Theoretic Logics*, 1985, ed. by J. Barwise and S. Feferman, Springer–Verlag, New York, 235–258

[Ber66] **R. Berger**, 1966, *The undecidability of the domino problem*, Mem. Amer. Math. Soc., 72 pp.

[BeLawWo85] **M.W. Bern, E.L. Lawler, A.L. Wong**, 1985, *Why certain subgraph computations require only linear time*, 26th Annual (IEEE Symposium on the Foundations of Computer Science).

[BeLawWo87] **M.W. Bern, E.L. Lawler, A.L. Wong**, 1987, *Linear time computation of optimal subgraphs of decomposable graphs*, J. of Algorithms 8, 216–235.

[Bo86] **H.L. Bodlaender**, 1986, *Classes of Graphs with Bounded Treewidth*, RUU-CS-86-22, University of Utrecht.

[Bo87] **H.L. Bodlaender**, 1987, *Dynamic Programming on Graphs with Bounded Tree-width*, MIT/LCS/TR-394, MIT.

[Bo87a] **H.L. Bodlaender**, 1987, *Polynomial algorithms for chromatic index and graph isomorphism on partial k-trees*, preprint June 1986.

[Bo88] **H.L. Bodlaender**, 1988, *Planar graphs with bounded treewidth*, Technical Report RUU-CS-88-14, March 1988, University of Utrecht.

[Bo88a] **H.L. Bodlaender** 1988, *Improved self-reduction algorithms for graphs with bounded tree-width*, Technical Report RUU-CS-88-29, September 1988, University of Utrecht.

[Bo88b] **H.L. Bodlaender**, 1988, *NC-Algorithmus for Graphs with Bounded Tree-width*, RUU-CS-88-4, University of Utrecht.

[Bo89] **H.L. Bodlaender** 1989, *On Linear Time Minor Tests and Depth First Search*, Technical Report RUU-CS-89-1, January 1989, Dept. of Computer Science, University of Utrecht.

[BoKl90] **H.L. Bodlaender, T. Kloks**, 1990, *Better Algorithms for Pathwidth and Treewidth of Graphs*, (extended abstract), preprint Dept. of CS, University of Utrecht.

[Boe83] **E. Börger**, 1983, *Undecidable versus difficult to decide* (An introduction into Computational Complexity of Logical Decision Problems), Forschungsbericht Nr. 155, Universität Dortmund.

[Boe83a] **E. Börger**, 1983, *Decision problems in predicate logic*, Bericht Nr. 153, Universität Dortmund.

[Boe85] **E. Börger**, 1985, *Berechenbarkeit, Komplexität, Logik: E. Einf. in Algorithmen, Sprachen und Kalküle unter bes. Berücksichtigung ihrer Komplexität*, Braunschweig

[Bor88] **R.B. Borie**, 1988, *Recursively constructed Graph Families: Membership and Linear Algorithms*

[BorParTo88] **R.B. Borie, R.G. Parker, C.A. Tovey**, 1988, *Automatic Generation of Linear Algorithms from Predicate Calculus Descriptions of Problems on Recursively Constructed Graph Families*, Preprint Juli 21, 1988

[BrFeLan87] **D.J. Brown, M.R. Fellows, M.A. Langston**, 1987, *Polynomial-Time Self Reducibility: Theoretical Motivations and Practical Results*, Preprint CS-87-171, August 1987, Washington State University.

[BrCor87] **D.J. Brown, D.G. Corneil**, 1987, *On generalized graph colourings*, preprint

[Cou87] **B. Courcelle**, 1987, *Recognizability and Second-Order Definability for Sets of Finite Graphs*, Preprint, Universite de Bordeaux, I-8634, Januar 1987.

[Cou88] **B. Courcelle**, 1988, *The decidability of the monadic second order theory of certain sets of finite and infinite graphs*, submitted to LICS '88, Locic in computer Science, Edinburg.

[Cou89] **B. Courcelle**, 1989, *The monadic second order logic of graphs VI: on several representations of graphs by relational structures*, Preprint LaBRRI No 89-99, Novembre 1989.

[Cou91] **B. Courcelle**, 1991, *The monadic second order logic of graphs VI: on several representations of graphs by relational structures*, Preprint Bordeaux-I University.

[Di90] **R. Diestel**, 1990, *Graph Decompositions*, Oxford University Press.

[Di89] **R. Diestel**, 1989, *Simplical tree-decompositions of infinite graphs I*, to appear J. Comb. Theory B.

[Do66] **J.E. Doner** 1966, *Decidability of the Weak Second Order Theory of two Successors*, Abstract 65T-468, Notices Amer. Math. Soc. 12, 819,ibid., 513.

[Em83] **P. van Emde Boas**, 1983, *Dominoes are forever*, preprint, Department of Mathematics, University of Amsterdam, Report 83–04

[FeAb89] **M.R. Fellows, K. Abrahamson**, 1989, *Cutset–Regularity Beats Well–Quasi–Ordering for Bounded Treewidth* (Extended Abstract) Nov. 1989, University of Idaho, Moskow, Washington State University, Pullman.

[FeLan86] **M.R. Fellows, M.A. Langston**, 1986, *The end of innocence in polynomial–time complexity*, preprint CS–86–162, Washington.

[FeLan87] **M.R. Fellows, M.A. Langston**, 1987, *Layout permutation problems and well–partially–ordered sets* (extended abstract), to appear 5th. MIT Conf. on Advanced Research in VLSI, Oct. 1987, FOCS 89.

[FeLan89a] **M.R. Fellows, M.A. Langston**, 1989, *On search, decision and efficiency of polynomial–time algorithms* (extended abstract), June 1988, 21st STOC, 501–512.

[FeLan88] **M.R. Fellows, M.A. Langston**, 1988, *Nonconstructive Tools for proving polynomial–time decidability*, to appear JACM 35(3), 727–739.

[FeLan89] **M. Fellows** 1989, *Applications of an Analogue of the Myhill-Nerode Theorem, In Obstruction Set Computation*, preprint April 28, 1989.

[GaJo78] **M.R. Garey, D.S. Johnson**, 1978, *Computers and intractability: A guide to the theory of NP–completeness*, San Francisco.

[GaJoPa77] **M.R. Garey, D.S. Johnson, C.H. Papadimitriou**, 1977, unpublished results.

[GarSchm74] **S. Garfunkel, J.H. Schmerl**, 1974, *The undecidability of theories of groupoids with an extra predicate*, Proc. Amer. Math. Soc. 42, 286–289.

[Gu77] **Y. Gurevich**, 1977, *Monadic theory of order and topology*, I. Israel Journal of Mathematics, vol 27, 299–319.

[Gu79] **Y. Gurevich**, 1979, *Modest theory of short chains*, I. Journal of Symbolic Locic, vol. 44, 481–490.

[Gu79a] **Y. Gurevich**, 1979, *Monadic theory of order and topology*, II. Israel Journal of Mathematics, vol. 34, 45–71.

[Gu82] **Y. Gurevich**, 1982, *Monadic theory of order and topology in ZFC*, Anals of Mathematical Logic, vol 23, 179–198.

[Gu83] **Y. Gurevich**, 1983, *Interpreting second–order logic in the monadic theory of order* J. of Symbolic, vol 48, 816–828.

[Gu84] **Y. Gurevich**, 1984, *Toward Logic tailored for computational complexity*, Preprint, The University of Michigan, Computing Research Laboratory, CRL–TR–2–84, January 1984.

[Gu85] **Y. Gurevich**, 1985, *Monadic Second–Order Theories*, Ch. XIII in Model–Theoretic Logics, Ed. Barwise and Feferman, Springer–Verlag, New York, 479–506.

[Gu85a] **Y. Gurevich**, 1985, *Logic and the Challenge of Computer Science*, Chapter 1 from Current Trends in Theoretical Computer Science, ed. by E. Börger, Computer Science Press.

[Gu86] **Y. Gurevich**, 1986, *On the Strength of the Interpretation Method*, Preprint, July 1986.

[GuHar82] **Y. Gurevich, L. Harrington**, 1982, *Tree Automata and Games*, Proc. ACM STDC, San Francisco, May 1982, 60–65.

[GuShe69] **Y. Gurevich, S. Shelah**, 1969, *On the strength of the interpretation method*, Preprint Juli 1986.

[Ha69] **F. Harary**, 1969, *Graph Theory*, Reading, Mass., Addison–Wesley.

[He87] **S.T. Hedetniemi**, 1987, *Open problems concerning the theory of algorithms on partial k-trees*, working paper.

[Ho89] **W. Hohberg**, 1989, *Zerlegung von Graphen – ein allgemeines, sequentielles und paralleles Lösungsverfahren für Graphenprobleme*, Dissertation, Darmstadt.

[HoRe90] **W. Hohberg, R. Reischuk**, 1990, *A Framework to Design Algorithms for Optimization Problems on Graphs*, Preprint April 1990, Universität Dortmund, FH Darmstadt.

[Im87] **N. Immerman**, 1987, *Languages that capture complexity classes*, SIAM J. Comut. Vol. 16, No. 4, August 1987, 760–778.

[Im87a] **N. Immerman**, 1987, *Expressibility as a complexity measure: results and directions*, Preprint, Yale University, Dept. of Computer Science, YALEU/DCS TR 538, April 1987.

[Im88] **N. Immerman**, 1988, *Descriptive and Computational Complexity*, Preprint, Lecture Notes for the AMS Short Course in Computational Complexity Theory, Jan. 5–6, 1988.

[Jo84] **D.S. Johnson**, 1984, *The NP-Completeness Column: An Ongoing Guide*, J. of Algorithms 6 (1984) 434–451.

[Jo87] **D.S. Johnson**, 1987, *The NP-Completeness Column: An Ongoing Guide*, J. of Algorithms 8 (1984) 434–451.

[Ki89] **N.Kinnersley** 1989, *Obstruction set isolation for layout permutation problems*, Ph.D. thesis, Washington State University.

[KB79] **Kleine Büning**, 1979, *Unentscheidbare Klassen von Gruppoiden mit monadischen Prädikaten*, abstract, Tagungsbericht 18/1979, Math Logik, 22.4. – 28.4.1979 Math. Forschungsinstitut Oberwohlfach, p.8

[KoRau76] **W. Korec, W. Rautenberg**, 1976, *Model-Interpretability into Trees and Applications*, Arch. math. Logic 17, 97–104

[La90] **J. Lagergren**, 1990, *Efficient Parallel Algorithms for Tree-Decomposition and Related Problems*, Preprint 20 April 1990, The Royal Institute of Technology, Stockholm

[La90a] **J. Lagergren**, 1990, *Finding minimal forbidden minors using a finite congruence*, preprint NADA, KTH, Stockholm.

[Lau87] **C. Lautemann**, 1987, *Efficient Algorithms on Graphs represented by Decomposition and Related Problems*, Preprint June 1987, Universität Bremen.

[Lau89] **C. Lautemann**, 1989, *The complexity of graph languages generated by hyperedge replacement*, Report No 4/89.

[Lau90] **C. Lautemann**, 1990, *Tree automata, free decomposition and hyperedge replacement*, submitted to Proceedings gragra workshop.

[Le84] **T. Lengauer**, 1984, *Hierarchical graph algorithms*, Preprint, Universität des Saarlandes, SFB 124, B2, August 1984.

[Le84a] **T. Lengauer**, 1984, *Hierarchical planarity testing*, Preprint, Universität des Saarlandes, SFB 124, B2, August 1984.

[Le87] **T. Lengauer**, 1987, *Efficient algorithms for finding minimum spanning forests of hierarchically defined graphs*, J. of Algorithms 8, pp. 264–284.

[LeWag86] **T. Lengauer, E. Wagner**, 1986, *The correlation between the complexities of the non–hierarchical and hierarchical versions of graph problems*, Preprint Universität Paderborn, Dez. 1986.

[LeWan87] **T. Lengauer, E. Wanke**, 1987, *Efficient solution of connectivity problems on hierarchically defined graphs*, preprint Universität Paderborn, Mai 1987.

[MahPe87] **S. Mahajan, J.G. Peters**, 1987, *Algorithms for Regular Properties in Recursive Graphs*, Preprint TR 87–7, August 1987, Simon Fraser University.

[MaTh88a] **J. Matoušek, R. Thomas**, 1988, *On the complexity of finding iso– and other morphism for partial k–trees, preprint May 1988.*

[MaTh88] **J. Matoušek, R. Thomas** Algorithms finding tree – decompositions of graphs, *preprint 1988.*

[MoSu84] **B. Monien, J.H. Sudborough**, *1984,* Bandwidth Constrained NP–complete Problems, *Preprint, June 1984.*

[MulSchu81] **D.E. Muller, P.E. Schupp**, *1981,* Pushdown automata, graphs, ends, second–order Logic, and reachability problems, *STOC (Milwaukee 1981), 46–54.*

[MulSchu85] **D.E. Muller, P.E. Schupp**, *1985,* The theory of ends, pushdown automata and second–order Logic, *Vol. 37, No. 1, May 1985, pp. 51–75.*

[Ra65] **M.O. Rabin** *1965,* A simple Method of Undecidablility proofs and some applications, *in Log. Meth. Phil. Sci. Proc. Jerusalem, 58–68.*

[Ra69] **M.O. Rabin** *1969,* Desidability of second order and automata on infinite trees, *Trans. Am. Math. Soc. 141, 1–35.*

[Ra77] **M.O. Rabin** *1977,* Decidable Theories, *in Handbook of Mathematical Logic, ed. by J. Barwise, North–Holland, pp. 595–629.*

[RoSey83] **N. Robertson and P.D. Seymour** *1983,* Graph Minors I. Excluding a forest, *J. Combin. Theory Ser.B. 35, 39–61.*

[RoSey84] **N. Robertson and P.D. Seymour**, *1984*, Graph minors. III. Planar tree–width., *J. of Combinatorial Theory. Ser. B. 41, 1986, 92–114*

[RoSey86] **N. Robertson and P.D. Seymour** *1986* Graph Minors II. Algorithmic Aspects of Tree Width *Journal of Algorithms 7, 309–322.*

[RoSey86a] **N. Robertson and P.D. Seymour** *1986*, Graph Minors V. Excluding a planar graph *J. Combinatorial Theory, Ser. B, 41, 92–114.*

[RoSey86b] **N. Robertson and P.D. Seymour** *1986*, Graph Minors XIII. The Disjoint Path Problem *Preprint.*

[RoSey88] **N. Robertson and P.D. Seymour** *1988*, Graph Minors XV. Wagners conjecture *Preprint.*

[Rob Sey 89] **N. Robertson and P.D. Seymour** *1989*, Personal Communication, *Toronto, Eugene.*

[RoSey89a] **N. Robertson and P.D. Seymour**, *1989*, Graph Minors XVI, *well–quasi–ordering on a surface*, Preprint April 1989.

[RoSey89] **N. Robertson and P.D. Seymour**, 1989, *Excluding infinite minors*, Preprint, October 1989.

[RoSeyTh89a] **N. Robertson and P.D. Seymour, R. Thomas**, 1989, *Quickly excluding a planar graph*, Preprint August 1989, revised May 1990.

[RoSeyTh89] **N. Robertson, P.D. Seymour, R. Thomas**, 1989, *Excluding infinite trees*, preprint.

[Ru86] **V. Rutenburg**, 1986, *Complexity of generalized graph colouring*, in Lecture Notes in Computer Science 133, ed. by G. Goos and J. Hartmanis, Proceedings of MFCS'86, Springer, Berlin, pp. 573–581.

[SchSe89] **P. Scheffler, D. Seese**, 1989, *A combinatorial and logical approach to linera–time computability* (extended abstract), Lecture Notes in Computer Science, 378–380.

[Sch88] **P. Scheffler** 1986, *What graphs, have bounded tree width?*, to appear Rostock, Math Kolloq.

[Sch86] **P. Scheffler** 1986, *Dynamic programming algorithms for tree-descomposition problems*, Karl-Weierstrass-Institut für Mathematik, Preprint P-Math-28/86, Berlin.

[Sch87] **P. Scheffler** 1987, *Linear–time algorithms for NP–complete problems restricted to partial k–trees*, Preprint, Akademie der Wissenschaften der DDR, Karl–Weierstraa–Institut für Mathematik, Report R–MATH–03/87.

[Sch88] **P. Scheffler** 1988, *Linear–time algorithms for graphs of bounded tree–width. The disjoint-path–problem*, to appear in "Graphen–theorie und Anwendungen"–Proceedings 1988.

[Sch89] **P. Scheffler** 1989 *Die Baumwerte von Graphen als ein Maa für die Kompliziertheit algorithmischer Probleme*, Dissertation (A), AdW d. DDR, Berlin 1989.

[Schm80] **J.H. Schmerl** 1980 *Decidability and $\aleph_0$-Categoricity of Theories of Partially Ordered Sets*, The Journal of Symbolic Logic, Vol. 45, No. 3, Sept 1980, 585–611.

[Schm81] **J.H. Schmerl** 1981 *Decidability and Finite Axiomatizability of Theories $\aleph_0$-Categorical Partially Ordered Sets*, The Journal of Symbolic Logic, Vol. 46, No. 1, March 1981, 101–120.

[Schm81a] **J.H. Schmerl** 1981 *Arborescent Structures. II: Interpretability in the Theory of Trees*, Transactions of the American Mathematical Society, Vol. 266, No. 2, August 1981, 629–643.

[Se72] **D. Seese** 1972 *Entscheidbarkeits– und Definierbarkeitsfragen der Theorie "netzartiger" Graphen. T.1.* – Wiss. Z. der Humboldt–Universität zu Berlin, Math. –Naturwiss. R. 21, 513–517.

[Se75] **D. Seese** 1975 *Ein Unterscheidbarkeitskriterium.*–Wiss Z. der Humboldt–Universität zu Berlin, Math.–Naturwiss. R. 24, 772–780.

[Se75a] **D. Seese** 1975 *Zur Entscheidbarkeit der monadischen Theorie 2. Stufe baumartiger Graphen.*, Wiss. Z. der Humboldt–Universität zu Berlin, Math. –Naturwiss. R. 24, 768–772.

[Se76] **D. Seese** 1976 *Entscheidbarkeits– und Interpretierbarkeitsfragen monadischer Theorien zweiter Stufe gewisser Klassen von Graphen.*, Diss. A. – Berlin: Humboldt–Universität.

[Se78] **D. Seese** 1978 *Über Unentscheidbare Erweiterungen von SC*, Zeitschrift f. math. Logik und Grundlagen d. Math. Bd. 24, 63–71.

[Se79] **D. Seese** 1979 *Some graph theoretical operations and decidability*, Math. Nachr. 87, 15–21.

[Se85] **D. Seese** 1985, *Tree-partite graphs and the complexity of algorithms* (extended abstract), in FCT'85, ed. L. Budach, LNCS 199, Springer, Berlin, 412–421.

[Se86] **D. Seese** 1986, *Tree-partite graphs and the complexity of algorithms,*preprint P-Math 08/86, Karl-Weierstrass-Institute für Mathematik.

[Se87] **D. Seese** 1987, *Ordered Tree-Representations of Infinite Graphs*, preprint 1987.

[Se87a] **D. Seese** 1987, *The structure of the Models of Decidable Monadic Theories of Graph*, manuscript, submitted to annals of pure and applied logic.

[She75] **S. Shelah** 1975, *The monadic theory of order*, Annals of Mathematics, vol. 102, 379–419.

[Sh67] **J.R. Shoenfield** 1967, *Mathematical Logic*, Reading, Addison Wesley.

[St75] **J. Stupp**, 1975, *The Lattice-Model is Recursive in The Original Model*, Preprint, Institute of Mathematics, The Hebrew University, Jerusalem, January 1975

[TaNiSa82] **K. Takamizawa, T. Nishizeki, N. Saito**, 1982, *Linear-Time Computability of Combinatorial Problems on Series-Parallel Graphs*, J. ACM 29, 623–641.

[ThaWr68] **J.W. Thatcher and J.B. Wright** 1968, *Generalized Finite Automata Theory with an Application to a Decision Problem in Second-Order Logic*, Mathematical Systems Theory 2, 57-81.

[Th] **R. Thomas**, *Well–Quasi–Ordering Infinite Graphs with Forbidden Finite Planar Minor*, to appear Trans. AMS.

[Th88] **W. Thomas**, 1988, *Automata on infinite objects*, (preliminary version, April 1988) extended version of a chapter for the forthcoming "Handbook of Theoretical Computer Science".

[Thom] **C. Thomassen**, *Embeddings and Minors*, to appear in Handbook of Combinatorics, ed. by R.L. Graham, M. Grötschel, L. Lovasz, North Holland.

[Tou68] **J.J. Le Tourneau**, 1968, *Decision problems related to the concept of operation*, Ph. D. Thesis, Berkeley.

[Wag84] **K. Wagner**, 1984, *The complexity of problems concerning graphs with regularities*, Preprint N/84/52.

[Wa61] **H. Wang**, 1961, *Proving theorems by pattern recognition II*, Bell Systems Technical J. 40, pp. 1–41.

[Wa75] **H. Wang**, 1975, *Notes on a class of tiling problems*, Fundamenta Mathematica LXXXII (1975), pp. 295–305.

[Wim88] **T.V. Wimer** 1988, *Ph D Thesis URI-030*, Clemson.

[WimHe85] **T.V. Wimer, S.T. Hedetniemi** 1985, *K–terminal recursive families of graphs*, Technical Report 86–May–7.

Tree Automata and Languages
M. Nivat and A. Podelski (editors)
© 1992 Elsevier Science Publishers B.V. All rights reserved.

115

COMPUTING TREES
WITH
GRAPH REWRITING SYSTEMS WITH PRIORITIES

IGOR LITOVSKY AND YVES MÉTIVIER[1]

Abstract. This paper is a survey of classical graphs and networks algorithms using the tree structure encoded with graph rewriting systems with priorities.

0. Introduction

This paper gives illustrations of the graph rewriting systems with priorities introduced in [6] to encode and to give simple mathematical proofs of graph algorithms and network algorithms in relation with trees. These systems are a very general tool to encode and to prove algorithms on graphs or on distributed systems. The natural way to process a word is sequentially from left to right (or from right to left). There is no such natural way to process a graph, and its distributed nature suggests distributed computations on them. Consider the following distributed system. Given n sequential processors P_1, P_2, ...,P_n ($n \geq 1$), arranged in a connected graph such that each processor is endowed with a label (or state). Computations are done in a local way : the value of a label is modified, in function of those of the neighbours ones, by fixed rules called graph rewriting rules. These labels take values in a finite set. Local steps are considered as atomic and are iterated until stability is reached. In certain cases, priorities of rules simultaneously applicable on overlapping patterns are needed. In this way one can defined Priority Graph Rewriting Systems (PGRS for short). Let k be the maximum of the diameters of the rewriting rules. Two steps are said independent if the distance between the underlying graphs is greater than k. The distributed aspect comes from the fact that several independent steps can be performed simultaneously.

Thus with each vertex and each edge of a graph, we associate a label which represents the local state of the algorithm or of the system. A rewriting rule computes locally on these labels according to the underlying graph structure of the rule. By adding to this model the notion of priority, we can locally control the order in which the rewriting rules are applied.

---

[1] With the support of the european project EBRA No 3166-ASMICS. This paper has been done for the workshop "Trees, tree languages, tree grammars, tree automata" organised by M.Nivat and A.Podelski at Le Touquet.

One can view the behaviour of a PGRS on a given labelled graph G in one of the following ways :

1.Sequential behaviour : consider all the occurrences in graph G which can be correctly rewritten (with respect to the priorities), choose one of them and apply the corresponding rewriting rule. Repeat this process while possible.

2.Distributed behaviour : consider all the occurrences in graph G which can be correctly applied (with respect to the priorities), choose one or several independent occurrences among them and apply the corresponding rewriting rule(s). In this case two or more rewriting steps may be done at the same time. Repeat this process while possible.

Among properties making PGRS's different from cellular automata is that they have a decentralized control, they perform actions in an asynchronous way, and the underlying connected graph for the rewriting rule is whatever. The model used in this paper is closely related to the one used in [2], and to the one used in [4], and [11] where the basic computation step is to compute the next label of each vertex in the graph according to its current label and according to one or all its neighbors.

This paper is a survey of various problems solved with PGRS's and linked to the tree structure [1], [6], [7], [9], [10].

We give systems which compute spanning trees (in a sequential or in a distributed way). We present also a system which decides whether or not a connected graph is acyclic, in this case it elects a vertex. Then we present a rewriting system which recognizes oriented rooted-trees such that all leaves have the same depth; this system may be used to encode Wave and Echo algorithms in a network. We end by giving examples of attribute graph rewriting systems.

Our aim is not to obtain new algorithms but to show that PGRS's can be a useful and a nice tool to design and to prove algorithms on graphs or in a distributed context.

General results about PGRS's (expressive power, decidability, priority, PGRS's and words) may be found in [9].

1. Basic notions and notations

a. Graphs [3]

A simple graph is defined as a finite set V of vertices together with a set E of edges which is a set of pairs of vertices, $E \subseteq \{\{v, v'\} \mid v, v' \in V\}$. It will be denoted by $G(V, E)$. In the sequel we suppose that for each edge $e = \{v, v'\}$, $v \neq v'$, that is we consider graphs which contain no self-loops. Let $e = \{v, v'\}$ be an edge, then we say that e is incident with v, and v is said to be a neighbour of v'. We also say that two edges are adjacent if they have a common vertex. The degree of a vertex v, denoted by $d(v)$, is the number of edges incident with v. Any vertex of degree one is called a leaf. A path P from v_1 to v_i in $G(V, E)$ is a sequence $P = v_1, e_1, v_2, e_2, \ldots, e_{i-1}, v_i$ of alternating vertices and edges such that for $1 \leq j < i$, e_j is incident with v_j and

v_{j+1}, $i-1$ is the length of P. If $v_1 = v_i$ then P is said to be a cycle. If in a path each vertex only appears once, then the sequence is called a simple path. Two vertices v and w are connected if there is a path from v to w. A graph is a connected graph if any two vertices are connected. A tree is a connected graph containing no cycles. In a tree, any two vertices are connected by precisely one path. A tree in which one vertex, called the root, is distinguished is called a rooted-tree. The depth or level of a vertex in a rooted-tree is the length of the simple path from the root to that vertex.

Let $T(V, E)$ be a tree, let v and v' be two vertices and $P = (\{v_i, v_i'\})_{1 \leq i \leq l}$ the simple path from v to v', let $d(v, v') = l$. The diameter of T, denoted by $D(T)$, is :

$$D(T) = Max\{d(v, v') \mid v, v' \in V\}.$$

A spanning-tree of a connected graph $G(V, E)$ is a subgraph which is a tree and which contains all the vertices of G.

b. Graph rewriting systems [6]

Let $G(V, E)$ and $G'(V', E')$ be two graphs, G' is a subgraph of G if $V' \subseteq V$ and $E' \subseteq E$. A neighbour of the subgraph G' in G is a neighbour of a vertex of V' in G.

Two graphs $G_1(V_1, E_1)$ and $G_2(V_2, E_2)$ are isomorphic if there is a pair $\phi = (\phi_V, \phi_E)$ of one-to-one correspondance between V_1 and V_2, and between E_1 and E_2 such that :

$$\forall \ \{v, v'\} \in E_1 \qquad \phi_E(\{v, v'\}) = \{\phi_V(v), \phi_V(v')\}.$$

Let $C^{(V)}$ and $C^{(E)}$ be two finite sets. A labelled graph over $(C^{(V)}, C^{(E)})$, denoted by $G(V, E, \lambda)$, is a graph $G(V, E)$ equipped with a labelling function $\lambda = (\lambda^{(V)}, \lambda^{(E)})$ where $\lambda^{(V)}$ (resp. $\lambda^{(E)}$) is a mapping from V (resp. E) into $C^{(V)}$ (resp. $C^{(E)}$). The graph $G(V, E)$ is called the underlying graph, and the mapping λ is a labelling of it. A c-labelled vertex is a vertex v such that $\lambda^{(V)}(v) = c$.

Let $c \in C^{(V)}$ (resp. $C^{(E)}$), $|G|_c$ is the number of c-labelled vertices (resp. edges) in $G(V, E, \lambda)$. Let $G(V, E, \lambda)$ be a labelled graph, if $G'(V', E')$ is a subgraph of $G(V, E)$ the restriction of λ to G' is denoted by $\lambda_{|G'}$, we have $\lambda_{|G'} = (\lambda^{(V)}_{|V'}, \lambda^{(E)}_{|E'})$. Let $G(V, E, \lambda)$ be a labelled graph, and c an element of $C^{(V)}$ then a c-connected component is a connected maximal subgraph of G such that any vertex is labelled c.

Two labelled graphs $G_1(V_1, E_1, \lambda_1)$ and $G_2(V_2, E_2, \lambda_2)$ are isomorphic if there is an isomorphism $\phi = (\phi_V, \phi_E)$ between $G_1(V_1, E_1)$ and $G_2(V_2, E_2)$ such that

$$\lambda_2^{(V)} \circ \phi_V = \lambda_1^{(V)}, \qquad \lambda_2^{(E)} \circ \phi_E = \lambda_1^{(E)}.$$

We will write $G_1(V_1, E_1, \lambda_1) \cong G_2(V_2, E_2, \lambda_2)$.

A graph rewriting rule R is an ordered pair $R = (G_R(V_R, E_R, \lambda_R), G'_R(V_R, E_R, \lambda'_R))$ of labelled graphs with the same underlying connected graph. Indeed, it is defined by two labellings of the same graph. We define the rewriting relation $\xrightarrow{R}$ over labelled graphs by : $G(V, E, \lambda) \xrightarrow{R} G'(V, E, \lambda')$ if there exists a subgraph $G_S(V_S, E_S)$ of $G(V, E)$ and an isomorphism ϕ from $G_S(V_S, E_S, \lambda_{|G_S})$ into $G_R(V_R, E_R, \lambda_R)$ such that ϕ is an isomorphism from $G_S(V_S, E_S, \lambda'_{|G_S})$ into $G'_R(V_R, E_R, \lambda'_R)$, and for each vertex $v \in V \setminus V_S$ (resp. for each edge $e \in E \setminus E_S$) one has $\lambda^{(V)}(v) = \lambda'^{(V)}(v)$ (resp. $\lambda^{(E)}(e) = \lambda'^{(E)}(e)$. The graph $G_S(V_S, E_S)$ is the support of the rewriting step. There is a support overlapping with another rule R' if there exists a subgraph $G'_S(V'_S, E'_S)$ with $V_S \cap V'_S \neq \emptyset$ which is a support for R'.

A graph rewriting system with priorities (PGRS) is a finite set $\mathcal{R}$ of graph rewriting rules equipped with a partial ordering relation $\succ$, this relation is called the priority. The rewriting relation $\xrightarrow{\mathcal{R}}$ is defined by :

$$G(V, E, \lambda) \qquad \xrightarrow{\mathcal{R}} \qquad G'(V, E, \lambda')$$

if there is a rule $R \in \mathcal{R}$ such that :

$$G(V, E, \lambda) \qquad \xrightarrow{R} \qquad G'(V, E, \lambda')$$

and there is no support overlapping with rules R' of $\mathcal{R}$ such that $R' \succ R$. The reflexive and transitive closure of $\xrightarrow{\mathcal{R}}$ is denoted by $\xrightarrow{*}{\mathcal{R}}$.

A graph $G(V, E, \lambda)$ is irreducible if there is no graph $G'(V, E, \lambda')$ such that :

$$G(V, E, \lambda) \qquad \xrightarrow{R} \qquad G'(V, E, \lambda').$$

Let $\mathcal{R}$ be a PGRS, let $G(V, E, \lambda)$, and $G'(V, E, \lambda')$ labelled graphs such that :

$$G(V, E, \lambda) \qquad \xrightarrow[R]{*} \qquad G'(V, E, \lambda').$$

Let v be an element of V, the history associated to v between G and G' is the word formed by the successive different labels of v.

We recall some definitions and two technical results we shall use later [8].

DEFINITION 1.1. *A relation* $\longrightarrow$ *over a set* X *is noetherian if there is no infinite sequence* $(x_i)_{i \geq 1}$ *of elements of* X *such that :*

$$x_1 \longrightarrow x_2 \longrightarrow x_3 \cdots \longrightarrow x_n \longrightarrow \cdots$$

We say that a relation $\longrightarrow$ *over a set* X *is compatible with an order* $>$ *if for any elements* x *and* y *of* X *we have :*

$$x \longrightarrow y \qquad \Rightarrow \qquad x > y.$$

Let $(x_i)_{0 \leq i \leq n}$ *be a sequence of elements of* X *such that :*

$$x_0 \longrightarrow x_1 \longrightarrow x_2 \longrightarrow \cdots \longrightarrow x_n$$

we say that n *is the length of this sequence, or the length of the rewriting chain* $(x_i)_{0 \leq i \leq n}$ *if* $\longrightarrow$ *is a rewriting relation.*

LEMMA 1.2. *Let p be an integer, and the ordering relation $>_p$ over $\mathbf{N}^p$ defined by :*

$$(x_1, \ldots, x_p) >_p (y_1, \ldots, y_p)$$

if there exists an integer j such that $x_1 = y_1, \ldots, x_{j-1} = y_{j-1}$, and $x_j > y_j$. Then $>_p$ is noetherian.

LEMMA 1.3. *If a relation $\longrightarrow$ over a set X is compatible with a noetherian order then it is noetherian.*

2. Distributed recognition of trees

This section is an illutration of the use of PGRS's as a tool for recognizing sets of graphs. As every rewriting system, a PGRS can be considered as a recognizer by specifying an initial labelling, and a set of irreducible graphs. Of course the initial labelling and the set of irreducible graphs have to be defined in a simple way.

We suppose we have a connected graph $G(V, E)$. We want to decide if $G(V, E)$ is without cycles and in this case we want to compute a root. We give a PGRS which solves our problem and we present its complete proof.

The set $C^{(V)}$ of vertex-labels is equal to $\{\mathbf{N}, \mathbf{I}, \mathbf{F}\}$; the set $C^{(E)}$ of edge-labels is the emptyset.

Initially every vertex is labelled $\mathbf{N}$.

The rewriting system $\mathcal{R}_1$ has five rules :

$$R_1 : \overset{\mathbf{N}}{\circ} - \overset{\mathbf{N}}{\circ} - \overset{\mathbf{N}}{\circ} \quad \longrightarrow \quad \overset{\mathbf{N}}{\circ} - \overset{\mathbf{I}}{\circ} - \overset{\mathbf{N}}{\circ}$$

$$R_2 : \overset{\mathbf{I}}{\circ} - \overset{\mathbf{N}}{\circ} - \overset{\mathbf{N}}{\circ} \quad \longrightarrow \quad \overset{\mathbf{I}}{\circ} - \overset{\mathbf{I}}{\circ} - \overset{\mathbf{N}}{\circ}$$

$$R_3 : \overset{\mathbf{I}}{\circ} - \overset{\mathbf{N}}{\circ} - \overset{\mathbf{I}}{\circ} \quad \longrightarrow \quad \overset{\mathbf{I}}{\circ} - \overset{\mathbf{I}}{\circ} - \overset{\mathbf{I}}{\circ}$$

$$R_4 : \overset{\mathbf{N}}{\circ} - \overset{\mathbf{N}}{\circ} \quad \longrightarrow \quad \overset{\mathbf{N}}{\circ} - \overset{\mathbf{F}}{\circ}$$

$$R_5 : \overset{\mathbf{I}}{\circ} - \overset{\mathbf{N}}{\circ} \quad \longrightarrow \quad \overset{\mathbf{N}}{\circ} - \overset{\mathbf{F}}{\circ}$$

The priority (i.e. the partial order between rules) is defined by :

$$\forall i \quad 1 \leq i \leq 3 \quad \forall j \quad 4 \leq j \leq 5 \quad R_i \succ R_j.$$

NOTATION. *In this paper we will use the following notations. Let x and y (and z) be the two (or three) rewritten vertices with $\mathcal{R}_1$; we assume that vertices are associated from left to right. For example if we apply R_1, it means :*

$$\underset{x}{\overset{\mathbf{N}}{\circ}} - \underset{y}{\overset{\mathbf{N}}{\circ}} - \underset{z}{\overset{\mathbf{N}}{\circ}} \quad \longrightarrow \quad \underset{x}{\overset{\mathbf{N}}{\circ}} - \underset{y}{\overset{\mathbf{I}}{\circ}} - \underset{z}{\overset{\mathbf{N}}{\circ}},$$

or if we apply R_4 :

$$\underset{x}{\overset{\mathbf{N}}{\circ}} - \underset{y}{\overset{\mathbf{N}}{\circ}} \quad \longrightarrow \quad \underset{x}{\overset{\mathbf{N}}{\circ}} - \underset{y}{\overset{\mathbf{F}}{\circ}}.$$

As an immediate consequence of the priority rules, we have :

FACT 2.1. *If we apply the rule R_4 or the rule R_5 then the vertex x is the only neighbour of the vertex y labelled* **N** *or* **I**.

To prove the main theorem we need a lemma giving properties of labelled vertices.

LEMMA 2.2. *Let $G(V, E, \lambda)$ be a connected labelled graph such that every vertex is labelled* **N**. *Let $G'(V, E, \lambda')$ be a graph such that :*

$$G(V, E, \lambda) \quad \xrightarrow[\mathcal{R}_1]{\quad * \quad} \quad G'(V, E, \lambda').$$

Then the graph $G'(V, E, \lambda')$ satisfies :
    (P_1) *a vertex labelled* **F** *has at most one neighbour labelled* **N** *or* **I**,
    (P_2) *a vertex labelled* **F** *does not belong to a cycle,*
    (P_3) *a F-connected component has exactly one neighbour labelled* **I** *or* **N**,
    (P_4) *a vertex labelled* **I** *has at least two neighbours labelled* **I** *or* **N**.

PROOF: These properties are obviously true for the initial labelling of the graph. We only have to check that if $G_1(V, E, \lambda_1)$ satisfies this set of properties and

$$G_1(V, E, \lambda_1) \quad \xrightarrow[\mathcal{R}_1]{\quad\quad} \quad G_2(V, E, \lambda_2)$$

then $G_2(V, E, \lambda_2)$ also verifies it. We study the effect of each of the five rules.

(P_1) Obviously, if we apply R_1, R_2, or R_3 to the graph $G_1(V, E, \lambda_1)$, changing the label of the vertex y preserves the property (P_1). If we employ R_4 or R_5, by the Fact 2.1, after the rewriting step the vertex y is labelled **F** and it has exactly one neighbour labelled **N** or **I**. As one label **N** is removed, the property (P_1) remains true for the others vertices labelled **F**.

(P_2) When we apply R_4 or R_5, the vertex y is labelled **F**, and one neighbour of y is labelled **N** or **I**; by (P_1) all others vertices are labelled **F**. Thus by the inductive hypothesis only one neighbour of y, possibly, belongs to a cycle, and y cannot belong to a cycle.

(P_3) If $G_1(V, E, \lambda_1) \quad \xrightarrow[\mathcal{R}_1]{\quad\quad} \quad G_2(V, E, \lambda_2)$ using one of the following rules R_1, R_2, and R_3 we have nothing to do.
    If $G_1(V, E, \lambda_1) \quad \xrightarrow[\mathcal{R}_1]{\quad\quad} \quad G_2(V, E, \lambda_2)$ using R_4 or R_5, we study the **F**-connected component of the vertex y. Any **F**-connected component neighbour of the vertex y was a **F**-connected component neighbour of the vertex y labelled **N** before the rewriting step. By inductive hypothesis they have not others neighbours labelled **N** or **I**. Thus after the rewriting step the **F**-connected component of the vertex y has exactly one neighbour (namely x) with a label different from **F**.

(P_4) When we put a label **I** on the vertex y using one of the rules R_1, R_2, or R_3, by definition of the rule the vertex y has two neighbours labelled **N** or **I**. By inductive hypothesis and as we don't modify **F** labels, others **I**-labelled vertices verify the property. If we use R_4 or R_5 the Fact 2.1 ends the proof. $\square$

    Now we can state the main theorem.

THEOREM 2.3. *The rewriting system $\mathcal{R}_1$ defined above is noetherian. Let $G(V, E, \lambda)$ be a connected labelled graph such that every vertex is labelled* **N**. *Let $G'(V, E, \lambda')$ be an irreducible graph such that :*

$$G(V, E, \lambda) \quad \xrightarrow[\mathcal{R}_1]{*} \quad G'(V, E, \lambda').$$

Then the graph $G(V, E)$ is acyclic if and only if the graph $G'(V, E, \lambda')$ has one vertex labelled **N**, *all other vertices are labelled* **F**.

If $G(V, E)$ has cycles, a vertex is labelled **I** *if and only if it belongs to a cycle or if it is in a simple path between two cycles. If not, vertices are labelled* **F**.

PROOF:

a. Termination Proof.

It is easy to check that if :

$$G(V, E, \lambda) \quad \xrightarrow[\mathcal{R}_1]{} \quad G'(V, E, \lambda')$$

then $(|G|_\mathbf{N}, |G|_\mathbf{I}) >_2 (|G'|_\mathbf{N}, |G'|_\mathbf{I})$.

Thus from Lemma 1.2 and Lemma 1.3 the relation $\xrightarrow[\mathcal{R}_1]{}$ is noetherian.

b. Properties of the irreducible graph.

We suppose that $G(V, E)$ has at least two vertices, if not there is nothing to do. We consider $G'(V, E, \lambda')$ an irreducible graph obtained from $G(V, E, \lambda)$. If there is a vertex v labelled **I**, by (P_4), it has at least two neighbours labelled **N** or **I**; as the graph is irreducible we cannot apply the rule R_5, thus the vertex v has no neighbours labelled **N**. From this fact and as the vertex set is finite we deduce that a vertex labelled **I** belongs to a cycle or is in a simple path between two cycles.

Thus if $G(V, E)$ is acyclic, any vertex of $G'(V, E, \lambda')$ is labelled **N** or **F**, as the graph is irreducible we have not two neighbouring vertices labelled **N**, and by (P_3) as the graph is connected there is at most one vertex labelled **N**. Necessarily the last rule applied in the rewriting chain is R_4 or R_5, and the irreducible graph has at least one vertex labelled **N**; thus it has exactly one such vertex. From this we deduce the theorem. $\square$

Observe that the rewriting system proposed above characterizes acyclic graphs. In this case it may be considered as a solution of the election problem. Considering that the elected vertex is the vertex of the irreducible graph labelled **N**. Furthermore, this system has the local termination detection property for the election problem : the elected vertex knows the result when it is labelled **N** and all its neighbours are labelled **F**.

From this theorem and because of the priority rule, when we apply R_4 the only not **F**-labelled neighbour of x is y, we deduce :

REMARK 2.4. *Let $G(V, E, \lambda)$ be a connected graph such that every vertex is labelled* **N**. *The rule R_4 is used if and only if the graph is without cycle. In this case it is the last rewriting step of any rewriting chain computing an irreducible labelled graph.*

The next result gives the complexity of the rewriting system.

LEMMA 2.5. *Let $\mathcal{R}_1$ be the PGRS defined above, and let $G(V, E, \lambda)$ a connected labelled graph such that every vertex is labelled* **N**. *Let n be the cardinality of V. Then every irreducible graph is obtained after exactly $2n - 3$ rewriting steps if $G(V, E)$ is acyclic, otherwise it is obtained after exactly $n + n'$, where n' is the number of vertices of the irreducible graph labelled* **F** *(those which do not belong to a cycle and don't belong to a simple path between two cycles).*

PROOF: We consider a rewriting chain which computes an irreducible graph. If the graph is acyclic the rule R_4 is used exactly one time (Remark 2.4). Thus the rule R_5 is used exactly $(n - 2)$ times.

Each time we apply the rule R_5 a label **I** disappears and in the initial labelling there is no label **I**. From this we deduce that in the rewriting chain exactly $(n - 2)$ labels **I** are generated, and the length of the rewriting chain equals $2n - 3$.

In the same manner we get the result in the case where G has cycles. $\square$

3. An example of election

An elected vertex can be used to make decisions in a distributed system or to centralize some informations. The aim of an election algorithm is to choose exactly one element in a set. In this section the rewriting system is the mechanism by which the elected vertex is determined. The initiation and the annoucement (i.e. the mechanism by which vertices are notified of the outcome of the election) are not done here.

The more general hypothesis are the following, we suppose that initially every vertex and every edge has the same label. We want a noetherian PGRS such that when we get an irreducible graph there is a label which appears exactly once on a vertex; this vertex will be the elected vertex. Furthermore we require that for any vertex v there exists a rewriting chain such that the elected vertex is this one.

The election problem formulated here has not always a solution [9]. In this section we show that the rewriting system $\mathcal{R}_1$ may be used as an election algorithm in the case of connected graph without cycles. In fact it only remains to prove that any vertex may be elected by $\mathcal{R}_1$.

PROPOSITION 3.1. *If the connected graph is without cycles then any vertex may be elected by the rewriting system $\mathcal{R}_1$.*

To prove this proposition we need the two following technical lemmas.

LEMMA 3.2. *Let $G(V, E, \lambda)$ be a connected labelled graph without cycles such that every vertex is labelled* **N**. *Let $G'(V, E, \lambda')$ be a graph such that :*

$$G(V, E, \lambda) \quad \xrightarrow[\mathcal{R}_1]{*} \quad G'(V, E, \lambda').$$

Let v be a **I**-*labelled vertex. Then there exist at least two different simple paths $P = v e_1 v_1 e_2 \ldots v_n$ and $P' = v e'_1 v'_1 e'_2 \ldots v'_p$ such that :*

1) $v_1 \neq v'_1$,

2) $\forall i \quad 1 \leq i < n \qquad \lambda^{(V)}(v_i) = \mathbf{I}$,
3) $\forall i \quad 1 \leq i < p \qquad \lambda^{(V)}(v_i') = \mathbf{I}$,
4) $\lambda^{(V)}(v_n) = \mathbf{N}$,
5) $\lambda^{(V)}(v_p') = \mathbf{N}$.

PROOF: This property is true for the initial labelling of the graph; we check that if $G_1(V, E, \lambda_1)$ verifies this property and

$$G_1(V, E, \lambda_1) \quad \xrightarrow{\quad X \quad} \quad G_2(V, E, \lambda_2)$$

($X \in \mathcal{R}_1$) then $G_2(V, E, \lambda_2)$ also verifies it. We study the effect of each of the five rules.

If $X \in \{R_1, R_2, R_3\}$, the result follows from the fact that if we take two simple paths P and P', $P = ve_1v_1e_2 \ldots v_n$ and $P' = ve_1'v_1'e_2' \ldots v_p'$ such that $v_1 \neq v_1'$, as the graph has no cycles, the only simple path from v_n to v_p' is $v_n e_n \ldots e_2 v_1 e_1 v e_1' v_1' \ldots v_p'$.

If $X = R_4$, there is nothing to do because the set of $\mathbf{I}$-labelled vertices is empty.

If $X = R_5$, from the Fact 2.1, the only $\mathbf{I}$-labelled neighbour of y is x. The result follows remark done in the first case. $\square$

LEMMA 3.3. *Let $k \geq 2$. Let $G(V, E, \lambda)$ be the labelled graph defined by :*

$$V = \{r, r_1, \ldots, r_k\}, \quad E = \{\{r, r_i\} \mid 1 \leq i \leq k\},$$

$$\lambda^{(V)}(r) = \mathbf{I}, \quad \lambda^{(V)}(r_i) = \mathbf{N}, \quad \forall i \quad 1 \leq i \leq k.$$

Then there exists a rewriting chain such that the elected vertex is r.

PROOF: We apply R_5 to vertices r and r_1, the label of r becomes $\mathbf{N}$ and the label of r_1 becomes $\mathbf{F}$. The end of the rewriting chain is made using $k-2$ times the following sequence, we apply first R_1 and then we apply R_5. The last rewriting step is made employing the rule R_4 keeping the label of r equal to $\mathbf{N}$. $\square$

PROOF (OF THE PROPOSITION 3.1):

Let r be a distinguished vertex of the graph $G(V, E, \lambda)$. We are going to prove that there exists a rewriting chain such that the elected vertex is r.

Let $r_1, \ldots, r_k$ be the neighbours of r, and $T_1, \ldots, T_k$ the acyclic subgraphs of G defined by :

T_i is the connected component of the subgraph $G(V \setminus \{r\}, E \setminus \{\{r, r_1\}, \ldots, \{r, r_k\}\}$ including the vertex r_i.

First, we assume that $k = 1$. We do a rewriting step only when the label of the vertex r remains equal to $\mathbf{N}$. As the degree of r equals 1 and the rules R_4 and R_5 have the lower priority, it is correct. When this process stops, the vertex r is labelled $\mathbf{N}$. We cannot do a rewriting step thus the label of r_1 is different from $\mathbf{N}$. Suppose the vertex r_1 labelled $\mathbf{I}$. From Lemma 3.2 the $\mathbf{I}$-connected component defined by r_1 has at least two $\mathbf{N}$-labelled neighbours, and we get a contradiction with the fact that the process has stopped. Thus the vertex r is labelled $\mathbf{F}$, and the graph is irreducible. The elected vertex is r.

Assume that $k > 1$. The first rewriting rule we apply is R_1 with the vertices r_1, r and r_2 in such a way that the label of r becomes I. After, we rewrite only when the label of r remains equal to I. It is correct because the only rule which possibly changes the label of r is R_5 and it has the lower priority. We examine the labelled graph when this process stops.

Assume that there exists a vertex r_i labelled **F**. The vertex r is labelled **I** thus any vertex of T_i is labelled **F** by (P_3). If we consider the last rewriting step changing a label of a vertex of T_i, it creates a label **F**, it is necessarily made by the rule R_4 or the rule R_5 thus there exists in T_i a N-labelled vertex. This yields a contradiction.

Assume there exists a r_i labelled **I**. From the Lemma 3.2 we deduce there exists in T_i a N-labelled vertex, and there is a contradiction with the halting of the rewriting process.

Hence every vertex r_i is labelled **N**. This fact and the halting of the process imply that all other vertices of T_i are labelled **F**. Lemma 3.3 ends the proof. $\square$

4. Sequential computation of a spanning-tree

Algorithms often use graph traversal, the most successful and commonly used is the depth-first search. This systematic method of visiting the vertices of a graph may be described in the following way. Suppose we are currently visiting the vertex v, then we try to visit a vertex neighbour of v which has not yet been visisted. If no such vertex exists then we go back to the vertex visited just before v. This process is repeated until every vertex has been visted. The set of edges traversed during the visite forms a spanning-tree.

We give a graph rewriting system which encodes this informal strategy. The vertex we are currently visiting is labelled **A**, a traversed edge is labelled **t**. If a visited vertex has no neighbours which has not yet been visited it is labelled **F**.

The set $C^{(V)}$ of vertex-labels is equal to $\{N, A, M, F\}$; the set $C^{(E)}$ of edge-labels is $\{f, t\}$.

Initially exactly one vertex called the Root is labelled **A**, other ones are labelled **N**, and every edge is labelled **f**.

The rewriting system $\mathcal{R}_2$ has two rules :

$$R_1 : \overset{A}{\circ} \overset{f}{\underline{\qquad}} \overset{N}{\circ} \quad \longrightarrow \quad \overset{M}{\circ} \overset{t}{\underline{\qquad}} \overset{A}{\circ}$$

$$R_2 : \overset{M}{\circ} \overset{t}{\underline{\qquad}} \overset{A}{\circ} \quad \longrightarrow \quad \overset{A}{\circ} \overset{t}{\underline{\qquad}} \overset{F}{\circ}$$

The priority (i.e. the partial order between rules) is defined by :

$$R_1 \succ R_2.$$

LEMMA 4.1. *Let $G(V, E, \lambda)$ be a connected labelled graph such that exactly one vertex called the Root is labelled **A**, other ones are labelled **N**, and every edge is* *labelled* **f**. *Let* $G'(V, E, \lambda')$ *be a graph such that :*

$$G(V, E, \lambda) \quad \xrightarrow[\mathcal{R}_2]{*} \quad G'(V, E, \lambda').$$

Then the graph $G'(V, E, \lambda')$ *satisfies :*

(I_1) *if an edge is incident with an* **N***-labelled vertex then it is labelled with* **f**,

(I_2) *the set of vertices not labelled with* **N** *is connected by* **t***-edges,*

(I_3) *the graph defined by the set of* **t***-labelled edges and incident vertices is without cycles,*

(I_4) *there is exactly one* **A***-labelled vertex,*

(I_5) *the set of* **M***-labelled vertices is a simple path from the Root to the* **A***-labelled vertex, and every edge of this path is labelled with* **t**,

(I_6) *a* **F***-labelled vertex has no neighbours labelled with* **N**.

PROOF: These properties are obviously true for the initial labelling of the graph. We only have to check that if $G_1(V, E, \lambda_1)$ satisfies this set of properties and

$$G_1(V, E, \lambda_1) \quad \xrightarrow[\mathcal{R}_1]{} \quad G_2(V, E, \lambda_2)$$

then $G_2(V, E, \lambda_2)$ also verifies it. We study the effect of each of the two rules.

(I_1) If we apply R_2 there is nothing to do; if we apply R_1 then we label an edge **t** and incident vertices are not labelled **N**.

(I_2) Obvious.

(I_3) Applying R_2 changes nothing; if the rewriting step is done with R_1 then by (I_1) after the rewriting step there is exactly one **t**-labelled edge incident with y thus we don't create a cycle.

(I_4) Each rewriting rule keeps the unique label **A**.

(I_5) If the rewriting step is done with R_2 the property remains true, if it is done with R_1, the property is a consequence of the inductive hypothesis and (I_3).

(I_6) It is a consequence of the priority rule: if a rewriting step labels a vertex y with **F**, the priority rule implies that y has no **N**-labelled neighbours .

$\square$

PROPOSITION 4.2. *The rewriting system* $\mathcal{R}_2$ *is noetherian. Let* $G(V, E, \lambda)$ *be a connected labelled graph such that exactly one vertex called Root is labelled* **A**, *other ones are labelled* **N**, *and every edge is labelled* **f**. *Let* $G'(V, E, \lambda')$ *be an irreducible graph such that :*

$$G(V, E, \lambda) \quad \xrightarrow[\mathcal{R}_2]{*} \quad G'(V, E, \lambda').$$

Then the graph whose edge-set is the set of **t***-labelled edges of* $G'(V, E, \lambda')$ *and whose vertex-set is the set of incident vertices is a spanning-tree of* $G(V, E)$. *Furthermore, Root is labelled* **A** *and any other vertex is labelled* **F**.

PROOF: The system $\mathcal{R}_2$ is noetherian because executing a rewriting step implies that $(|G|_{\mathbf{N}}, |G|_{\mathbf{M}})$ is decreasing for $>_2$.

Assume now that the graph is irreducible, from (I_5), we deduce that there is no M-labelled vertex; from (I_4) and (I_5) Root is labelled **A**. Other vertices are labelled **F** or **N**. By (I_1), neighbours of **A** are necessary labelled **F**. From (I_6) we deduce that there is no vertices labelled **N**. This fact, (I_3), and (I_2) imply the result. $\square$

LEMMA 4.3. *Let $G(V, E, \lambda)$ be a connected labelled graph such that exactly one vertex called Root is labelled* **A**, *other ones are labelled* **N**, *and every edge is labelled* **f**. *Then any irreducible graph is obtained with exactly $2n - 2$ rewriting steps.*

PROOF: The cardinality of the vertex-set is n then any spanning-tree of $G(V, E)$ has $n - 1$ edges; thus the rule R_1 is applied exactly $n - 1$ times. The irreducible has exactly one **A**-labelled vertex, other ones are labelled **F**, hence the rule R_2 is used exactly $n - 1$ times. $\square$

5. Distributed computation of a spanning-tree

Among difficulties related to distributed systems, there is one which does not exist in sequential computations, it is the local termination detection property. A distributed system is said to have the local termination detection property if a vertex can decide from its label and those of its neighbourhood whether or not the computed graph is irreducible i.e. the distributed computation is finished.

In the previous algorithm, only one vertex may add a vertex in the computed spanning-tree, and Root knows the termination when it is labelled **A** and every neighbour is labelled **F**.

In this section we give two graph rewriting systems such that any vertex belonging to the spanning-tree may add a vertex which is not yet in. The first one is very simple, nevertheless it has not the local termination detection property. The second one is a few more sophisticated, and yields to a local termination detection property.

a. The first one

The set $C^{(V)}$ of vertex-labels is equal to $\{N, A\}$; the set $C^{(E)}$ of edge-labels is $\{f, t\}$.

Initially exactly one vertex is labelled **A**, the other ones are labelled **N** and every edge is labelled **f**.

The corresponding rewriting system $\mathcal{R}_3$ has one rule :

$$R_1 : \overset{\textbf{A}}{\underset{}{\circ} } \overset{\textbf{f}}{\rule{1.5cm}{0.4pt}} \overset{\textbf{N}}{\circ} \quad \longrightarrow \quad \overset{\textbf{A}}{\circ} \overset{\textbf{t}}{\rule{1.5cm}{0.4pt}} \overset{\textbf{A}}{\circ}$$

PROPOSITION 5.1. *The rewriting system $\mathcal{R}_3$ defined above is noetherian.*

Let $G(V, E, \lambda)$ be a connected labelled graph such that exactly one vertex is labelled **A**, *the other ones are labelled* **N**, *and every edge is labelled* **f**. *Let $G'(V, E, \lambda')$ be an irreducible graph such that :*

$$G(V, E, \lambda) \quad \xrightarrow[\mathcal{R}_3]{*} \quad G'(V, E, \lambda').$$

*Let E_t the set of t-labelled edges, and V_t the set of incident vertices. Then every vertex of $G'(V, E, \lambda')$ is labelled **A** (i.e. $V_t = V$) and the graph $T(V_t, E_t)$ is a spanning-tree of $G(V, E)$.*

PROOF: Each rewriting step removes a label **f**, thus an irreducible graph is obtained with at most the cardinality of E rewriting steps.

At each step, the computed labelled graph verifies the following simple properties.

(I_1) Every edge incident with an **N**-labelled vertex is labelled **f**,

(I_2) a **A**-labelled vertex has a **t**-labelled incident edge,

(I_3) the subgraph of $G(V, E)$ induced by the **t**-labelled edges is a tree.

From these properties and the fact that $G'(V, E, \lambda')$ is irreducible, we deduce the desired result. $\square$

Let n be the cardinality of V; as $T(V, E_t)$ is a spanning-tree the cardinality of E_t is equal to $n - 1$, hence :

COROLLARY 5.2. *The irreducible graph is obtained with exactly $n - 1$ rewriting steps.*

b. The second one

If we want the local termination detection property, we have to encode in the rewriting system an acknowledgment. If every neighbour of a vertex v has been visited then v sends an acknowledgement to its father; when every son of a vertex have sent an acknowledgement then the father sends an acknowledgement to its father, and this recursively. The solution given here uses the relation *"to be father of"*, this relation is encoded by the labels $\{0, 1, 2\}$ with the order : *"0 is the father of 1"* , *"1 is the father of 2"* , and *"2 is the father of 0"*. We need this relation between two vertices labelled **M** to be sure that the last rewriting step is done with Root. If we don't take this relation the rewriting system computes a spanning-tree, but we have not the local detection of the halting with Root. In the following rewriting system we use the modulo 3 adding table on these operations.

The set $C^{(V)}$ of vertex-labels is equal to

$$\{\mathbf{N}, (\mathbf{A}, 0), (\mathbf{A}, 1), (\mathbf{A}, 2), (\mathbf{M}, 0), (\mathbf{M}, 1), (\mathbf{M}, 2), \mathbf{R}, \mathbf{F}\},$$

the set $C^{(E)}$ of edge-labels is

$$\{\mathbf{f}, \mathbf{t}\}.$$

Initially Root is labelled $(\mathbf{A}, 0)$, any other vertex is labelled **N**, and every edge is labelled **f**.

We present the rewriting system, denoted $\mathcal{R}_4$, organized as a set of clusters; each one encodes a part of the algorithm. We assume that $k = 0, 1, 2$.

The first cluster, describes what happens when a node is visited : it is labelled **A**, and the edge traversed is labelled **t**. As soon as a visited node has a son, it is labelled **M**.

$$R_1 : \quad \overset{(\mathbf{A},k)}{\circ} \overset{\mathbf{f}}{\textemdash} \overset{\mathbf{N}}{\circ} \quad \longrightarrow \quad \overset{(\mathbf{M},k)}{\circ} \overset{\mathbf{t}}{\textemdash} \overset{(\mathbf{A},k+1)}{\circ}$$

$$R_2 : \quad \overset{(\mathbf{M},k)}{\circ} \overset{\mathbf{f}}{\textemdash} \overset{\mathbf{N}}{\circ} \quad \longrightarrow \quad \overset{(\mathbf{M},k)}{\circ} \overset{\mathbf{t}}{\textemdash} \overset{(\mathbf{A},k+1)}{\circ}$$

The second cluster encodes the acknowledgement; when a vertex is ready to send its acknowledgement it is labelled **R**, and when it has sent its acknowledgement it is labelled **F**.

$$R_3 \; : \; \overset{(\mathbf{M},k)}{\circ} \overset{t}{\underline{\quad\quad}} \overset{(\mathbf{A},k+1)}{\circ} \qquad \longrightarrow \qquad \overset{(\mathbf{M},k)}{\circ} \overset{t}{\underline{\quad\quad}} \overset{\mathbf{R}}{\circ}$$

$$R_4 \; : \; \overset{\mathbf{R}}{\circ} \overset{t}{\underline{\quad\quad}} \overset{(\mathbf{M},k)}{\circ} \overset{t}{\underline{\quad\quad}} \overset{\mathbf{R}}{\circ} \qquad \longrightarrow \qquad \overset{\mathbf{F}}{\circ} \overset{t}{\underline{\quad\quad}} \overset{(\mathbf{M},k)}{\circ} \overset{t}{\underline{\quad\quad}} \overset{\mathbf{R}}{\circ}$$

$$R_5 \; : \; \overset{\mathbf{R}}{\circ} \overset{t}{\underline{\quad\quad}} \overset{(\mathbf{M},k)}{\circ} \overset{t}{\underline{\quad\quad}} \overset{(\mathbf{M},k+1)}{\circ} \qquad \longrightarrow \qquad \overset{\mathbf{F}}{\circ} \overset{t}{\underline{\quad\quad}} \overset{(\mathbf{M},k)}{\circ} \overset{t}{\underline{\quad\quad}} \overset{(\mathbf{M},k+1)}{\circ}$$

$$R_6 \; : \; \overset{(\mathbf{M},k)}{\circ} \overset{t}{\underline{\quad\quad}} \overset{\mathbf{R}}{\circ} \qquad \longrightarrow \qquad \overset{\mathbf{R}}{\circ} \overset{t}{\underline{\quad\quad}} \overset{\mathbf{F}}{\circ}$$

The priority (i.e. the partial order between rules) is defined by :

$$\forall i \quad 1 \leq i \leq 2 \quad R_i \succ R_6 \quad and \quad \forall j \quad 3 \leq j \leq 5 \quad R_i \succ R_j \quad and \quad R_j \succ R_6.$$

LEMMA 5.3. *Let $G(V,E,\lambda)$ be a connected labelled graph such that exactly one vertex called Root is labelled $(\mathbf{A},0)$, other ones are labelled $\mathbf{N}$, and every edge is labelled $\mathbf{f}$. Let $G'(V,E,\lambda')$ be a graph such that :*

$$G(V,E,\lambda) \quad \xrightarrow[\mathcal{R}_4]{\quad * \quad} \quad G'(V,E,\lambda').$$

Then the graph $G'(V,E,\lambda')$ satisfies :

(I_1) *if an edge is incident with an $\mathbf{N}$-labelled vertex then it is labelled $\mathbf{f}$,*

(I_2) *the set of vertices not labelled $\mathbf{N}$ is connected with $\mathbf{t}$-labelled edges,*

(I_3) *the graph induced by the set of $\mathbf{t}$-labelled edges has no cycles,*

(I_4) *if $\overset{(L_1,k)}{\underset{x}{\circ}} \overset{t}{\underline{\quad\quad}} \overset{(L_2,j)}{\underset{y}{\circ}}$, where L_1 and L_2 are equal to $\mathbf{M}$ or to $\mathbf{A}$, is in the labelled graph then $k = j + 1$*

or $j = k+1$, and the vertex x has been visited before the vertex y (or equivalently x is the father of y)
if and only if $j = k+1$,

(I_5) *if $\overset{(\mathbf{M},k)}{\underset{x}{\circ}} \overset{t}{\underline{\quad\quad}} \overset{R}{\underset{y}{\circ}}$, is in the labelled graph then the vertex x has been visited before the vertex y (or equivalently x is the father of y),*

(I_6) *if $\overset{(\mathbf{M},k)}{\underset{x}{\circ}} \overset{t}{\underline{\quad\quad}} \overset{F}{\underset{y}{\circ}}$, is in the labelled graph then the vertex x has been visited before the vertex y (or equivalently x is the father of y),*

(I_7) *every $\mathbf{A}$-labelled vertex is a leaf (w.r. to the $\mathbf{t}$-labelled edges); all other leaves (w.r. to the $\mathbf{t}$-labelled edges) are labelled $\mathbf{R}$ or $\mathbf{F}$,*

(I_8) *a $\mathbf{N}$-labelled vertex has only $\mathbf{N}$- , $\mathbf{A}$- , or $\mathbf{M}$-labelled neighbours,*

(I_9) *a $\mathbf{M}$-labelled vertex has at least a son labelled $\mathbf{A}$, $\mathbf{M}$, or $\mathbf{R}$; if it has a son labelled $\mathbf{F}$ then it has no $\mathbf{N}$-labelled neighbours,*

(I_{10}) *a vertex labelled* **R** *has at most one neighbour, linked by a* **t**-*labelled edge, with a label different from*
  F, *this neighbour is labelled* **M**,

(I_{11}) *a maximal* **F**-*connected component has exactly one neighbour, linked by a* **t**-*labelled edge, with a label different*
  from **F**, *this neighbour is labelled* **M** *or* **R**,

(I_{12}) *the father of a* **A**-*labelled vertex is labelled* **M**,

(I_{13}) *if* $\overset{\mathbf{R}}{\underset{x}{\circ}} \,\xrightarrow{\;\;t\;\;}\, \overset{\mathbf{F}}{\underset{y}{\circ}}$, *then* x *has been visited before* y.

PROOF: These properties are obviously true for the initial labelling of the graph. We only have to check that if $G_1(V, E, \lambda_1)$ satisfies this set of properties and

$$G_1(V, E, \lambda_1) \quad \xrightarrow[\mathcal{R}_4]{} \quad G_2(V, E, \lambda_2)$$

then $G_2(V, E, \lambda_2)$ also verifies it. We study the effect of each of the rules.

The proofs of (I_1), (I_2), and (I_3) are similar to those of Lemma 4.1.

(I_4) The edge between x and y is labelled **t**, thus one of the two vertices has been visited before the ohter one, using R_1 or R_2; the result follows from the definition of the numerotation.

(I_5) Before to be labelled **R** the vertices was labelled **M** or **A**. The labelling with **R** has been done with R_3 or R_6, and the result is a consequence of (I_4).

(I_6) It is a direct consequence of (I_5) and the fact that before to be labelled **F** the vertex y was **R**.

(I_7) Before to be a **A**-labelled vertex, it was a **N** labelled vertex, hence it had no incident **t**-labelled edges. Furthermore, when we add a son it becomes **M**. Thus, a **A**-labelled vertex is a leaf.

(I_8) When a vertex y becomes a **R**-labelled vertex as R_1 and R_2 have a priority greater than R_3 or R_6, necessary the vertex y has no **N**-labelled neighbours, the end of the property follows from the fact that before to be labelled **F**, a vertex is labelled **R**.

(I_9) It is a direct consequence of the priority rule.

(I_{10}) If the vertex x is labelled **R** with R_3 then x has only one neighbour with a **t**-labelled edge, it is labelled **M**. If the vertex x is labelled **R** with R_6 then the result is a direct consequence of the relation R_4 and R_5 have a priority greater than R_6.

(I_{11}) It is a consequence of (I_{10}) and the fact that before to be labelled **F** a vertex is labelled **R**.

(I_{12}) It is true when **A** is created, and it remains true from (I_{11}) and (I_{10}).

(I_{13}) It is a consequence of R_6. $\square$

PROPOSITION 5.4. *Let $G(V, E, \lambda)$ be a connected labelled graph such that exactly one vertex called Root is labelled $(\mathbf{A}, 0)$, other ones are labelled $\mathbf{N}$, and every edge is labelled $\mathbf{f}$. Let $G'(V, E, \lambda')$ be an irreducible graph such that :*

$$G(V, E, \lambda) \quad \xrightarrow[\mathcal{R}_2]{\quad * \quad} \quad G'(V, E, \lambda').$$

Then the graph whose edge-set is the set of $\mathbf{t}$-labelled edges of $G'(V, E, \lambda')$ and whose vertex-set is the set of incident vertices is a spanning-tree of $G(V, E)$. Furthermore, Root is labelled $\mathbf{R}$ and any other vertex is labelled $\mathbf{F}$.

PROOF: From (I_9) we deduce that any irreducible graph has no $\mathbf{M}$-labelled vertices. From this fact, and (I_{12}) G' has no $\mathbf{A}$-labelled vertices. Thus there is no vertices labelled $\mathbf{N}$.

Thus the graph is labelled with only $\mathbf{F}$ and $\mathbf{R}$. The result follows from (I_{10}), (I_{11}) and (I_{13}). $\square$

We deduce from this the local termination detection.

COROLLARY 5.5. *Let $G(V, E, \lambda)$ be a connected labelled graph such that exactly one vertex called Root is labelled $(\mathbf{A}, 0)$, other ones are labelled $\mathbf{N}$, and every edge is labelled $\mathbf{f}$. A graph computed by the rewriting system is irreducible if and only if the vertex Root is labelled $\mathbf{R}$.*

LEMMA 5.6. *Let $G(V, E, \lambda)$ be a connected labelled graph such that exactly one vertex called Root is labelled $(\mathbf{A}, 0)$, other ones are labelled $\mathbf{N}$, and every edge is labelled $\mathbf{f}$. Let n be the cardinality of the vertex-set. Then any irreducible graph is obtained with exactly $2n - 2 + l$ rewriting steps, where l is the cardinality of the set of the leaves of the computed tree.*

PROOF: The spanning tree has $n - 1$ edges, thus we apply exactly $n - 1$ times R_1 or R_2. The irreducible graph has $n - 1$ vertices labelled $\mathbf{F}$, hence we apply R_4, R_5 or R_6 exactly $n - 1$ times. To achieve the calculus, we only have to note that R_3 is only applyed once for each leaf of the computed tree. $\square$

6. Echo graph rewriting system

An oriented graph $G(V, A)$ is defined by a finite set V of vertices and a set A of arcs; an arc is an ordered pair of vertices. In the sequel we assume that arcs are ordered pairs of different vertices. In this section we illustrate wave and echo algorithms, giving a graph rewriting system which recognizes rooted oriented trees such that all the leaves have the same depth.

From this system with simple modifications we can deduce a system which computes a spanning-tree with a minimal depth for any connected graph. The rewriting system works in the following way. The vertex Root broadcasts a wave to its neighbours; each neighbour rebroadcasts it along. The wave stops when it reachs an unvisited vertex. If the unvisited vertex is a leaf it echoes to its father an halting signal; if the unvisited vertex is an internal node it echoes to its father a continue signal. Then a vertex which receives an echo signal sends this signal to its father, and so on. If a vertex receives an halting echo signal and a continue echo signal then it detects that a leaf and an internal node have the same depth, and the tree is rejected. If the Root receives from all its sons a continue signal, it sends a new wave; if it receives from all its sons an halting signal then the tree is accepted. Otherwise the tree is rejected.

We assume that Root has both its label and an extra label **Root**, which identifies it in a unique way. The meaning of this extra label is quite simple : it enables Root to create a wave.

The set $C^{(V)}$ of vertex-labels is equal to

$$\{\mathbf{N}, \mathbf{D}, \mathbf{C}, \mathbf{H}, \mathbf{Q}, \mathbf{W}ave, \mathbf{W}ait, \mathbf{W}aitC, \mathbf{W}aitH, \mathbf{N}o\},$$

the set $C^{(E)}$ of edge-labels is the emptyset.

Initially Root is labelled **D**, and any other vertex is labelled **N**.

In the rewriting rules given below X denotes a label belonging to

$$\{\mathbf{D}, \mathbf{C}, \mathbf{H}, \mathbf{W}ave, \mathbf{W}ait, \mathbf{W}aitC, \mathbf{W}aitH\},$$

Y denotes a label of the same set plus $\mathbf{N}o$, and Z denotes any label.

We present the rewriting system, denoted $\mathcal{R}_5$, organized as a set of clusters; each one encodes a part of the algorithm. With each rule is associated an integer, the priority is the order induces by the associated integers.

The first cluster, describes what happens when a node is discovered i.e. when it is labelled **D**. If it has at least a son it echoes the signal **C** (i.e. Continue); it needs at least one more wave. If it is a leaf then it echoes an halting signal i.e. **H**.

$$R_1 \quad \overset{\mathbf{D}}{\circ}\!\rightarrow\!\!-\!\overset{\mathbf{N}}{\circ} \quad \longrightarrow \quad \overset{\mathbf{C}}{\circ}\!\rightarrow\!\!-\!\overset{\mathbf{N}}{\circ} \qquad\qquad 9$$

$$R_2 \quad \overset{\mathbf{D}}{\circ} \quad \longrightarrow \quad \overset{\mathbf{H}}{\circ} \qquad\qquad 8$$

The second cluster computes the Echo signal. A vertex is labelled **Q** i.e. Quiet, when it has been visited, it does not wait an echo signal, and it does not propagate a wave. When a vertex is labelled **W**ait, it is waiting an echo signal, and it has received nothings. If it has already received an echo signal **C** (resp. **H**) and it is still waiting at least another echo signal then it is labelled **W**aitC (resp. **W**aitH). The last label **N**o encodes a node which has received both an halting echo signal and a continue echo signal. It denotes that an internal node and a leaf with the same depth has been

132 *I. Litovsky and Y. Métivier*

discovered in the subtree induced by the No-labelled vertex.

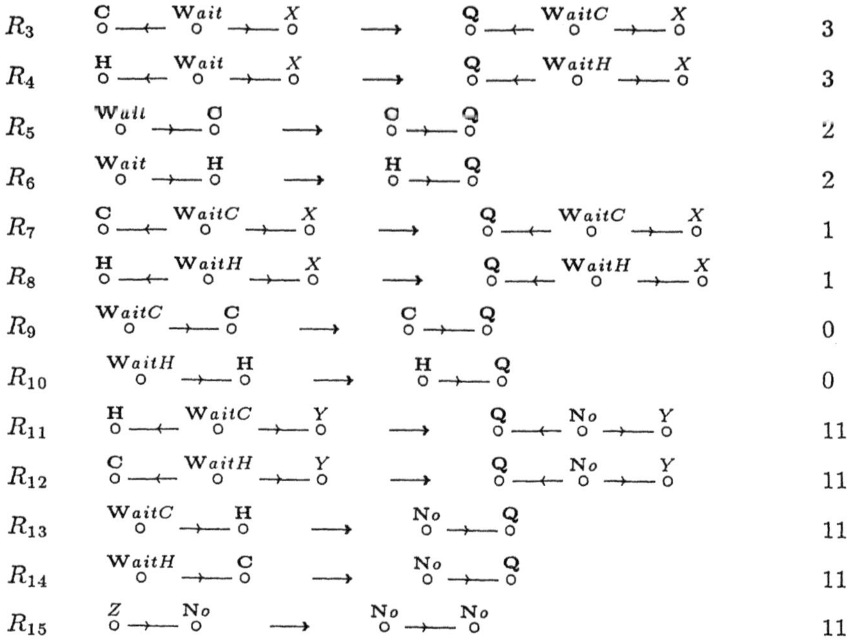

Rule		
R_3	C ←— Wait —→ X $\longrightarrow$ Q ←— WaitC —→ X	3
R_4	H ←— Wait —→ X $\longrightarrow$ Q ←— WaitH —→ X	3
R_5	Wait —→ C $\longrightarrow$ C —→ Q	2
R_6	Wait —→ H $\longrightarrow$ H —→ Q	2
R_7	C ←— WaitC —→ X $\longrightarrow$ Q ←— WaitC —→ X	1
R_8	H ←— WaitH —→ X $\longrightarrow$ Q ←— WaitH —→ X	1
R_9	WaitC —→ C $\longrightarrow$ C —→ Q	0
R_{10}	WaitH —→ H $\longrightarrow$ H —→ Q	0
R_{11}	H ←— WaitC —→ Y $\longrightarrow$ Q ←— No —→ Y	11
R_{12}	C ←— WaitH —→ Y $\longrightarrow$ Q ←— No —→ Y	11
R_{13}	WaitC —→ H $\longrightarrow$ No —→ Q	11
R_{14}	WaitH —→ C $\longrightarrow$ No —→ Q	11
R_{15}	Z —→ No $\longrightarrow$ No —→ No	11

The last cluster contains rules for the creation, and the progression of a wave. Root
creates a wave when it has received from all its sons a continue echo signal. When a
node has sent a wave to all its sons it waits the echo signals from. If the wave reachs
a new vertex then the wave stops and label the discovered vertex **D**.

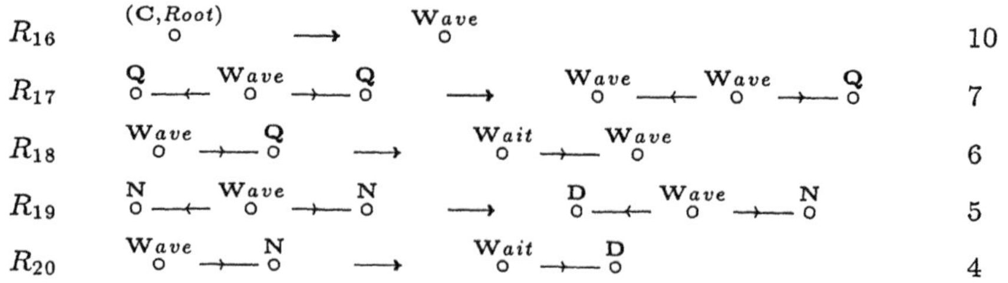

Rule		
R_{16}	(C, Root) $\longrightarrow$ Wave	10
R_{17}	Q ←— Wave —→ Q $\longrightarrow$ Wave ←— Wave —→ Q	7
R_{18}	Wave —→ Q $\longrightarrow$ Wait —→ Wave	6
R_{19}	N ←— Wave —→ N $\longrightarrow$ D ←— Wave —→ N	5
R_{20}	Wave —→ N $\longrightarrow$ Wait —→ D	4

In the next lemma we use the following. Let v be a vertex of $G(V, E)$, $T(v)$ denotes
the oriented subgraph of G whose root is v. By $ST(v)$, we denote the subgraph of
$T(v)$ induced by vertices not labelled **N**.

LEMMA 6.1. *Let $G(V, E, \lambda)$ be a connected labelled graph such that exactly one vertex called Root is labelled* **D**, *other ones are labelled* **N**. *Let $G'(V, E, \lambda')$ be a graph such that :*

$$G(V, E, \lambda) \quad \xrightarrow[\mathcal{R}_5]{*} \quad G'(V, E, \lambda').$$

Then the graph $G'(V, E, \lambda')$ satisfies :

(I_1) $\forall$ $v \in V$ *the graph $ST(v)$ is connected.*

(I_2) *A* **W**ait-*labelled vertex has a son labelled by an element of :*
$\{\mathbf{W}ave, \mathbf{D}, \mathbf{H}, \mathbf{C}, \mathbf{W}ait, \mathbf{W}aitC, \mathbf{W}aitH, \mathbf{No}\}$.

(I_3) *A* **W**ave-*labelled vertex has a son labelled by an element of* $\{\mathbf{Q}, \mathbf{N}\}$.

(I_4) *A* **W**aitC-*labelled vertex has a son labelled by an element of :*
$\{\mathbf{W}ave, \mathbf{D}, \mathbf{H}, \mathbf{C}, \mathbf{W}ait, \mathbf{W}aitC, \mathbf{W}aitH, \mathbf{No}\}$.

(I_5) *A* **W**aitH-*labelled vertex has a son labelled by an element of :*
$\{\mathbf{W}ave, \mathbf{D}, \mathbf{H}, \mathbf{C}, \mathbf{W}ait, \mathbf{W}aitC, \mathbf{W}aitH, \mathbf{No}\}$.

(I_6) *A* **C**-*labelled vertex has only* **Q**-*labelled sons, or only* **N**-*labelled sons.*

(I_7) *A* **H**-*labelled vertex has only* **Q**-*labelled sons.*

(I_8) *Let v be a* **Q**-*labelled vertex, the sons of v are all* **Q**-*labelled vertices, or all* **N**-*labelled vertices.*

(I_9) *Let v be a* **C**-*labelled vertex; every leaf of $ST(v)$ has at least a* **N**-*labelled son.*

(I_{10}) *Let v be a* **H**-*labelled vertex; every leaf of $ST(v)$ has no sons.*

(I_{11}) *The history of a* **W**ave-*labelled vertex is $h_1 \mathbf{CQW}ave$;*

The history of a **W**ait-*labelled vertex is $h_2 \mathbf{CQW}ave\mathbf{W}ait$;*

The history of a **W**aitC-*labelled vertex is $h_3 \mathbf{CQW}ave\mathbf{W}ait\mathbf{W}aitC$;*

The history of a **W**aitH-*labelled vertex is $h_4 \mathbf{CQW}ave\mathbf{W}ait\mathbf{W}aitH$;*

the history of a **Q**-*labelled son of a* **W**aitC *labelled vertex is $h_5 \mathbf{CQ}$;*

the history of a **Q**-*labelled son of a* **W**aitH *labelled vertex is $h_6 \mathbf{HQ}$;*

where h_1, h_2, h_3, h_4, h_5, h_6 are histories.

We can summarize these properties saying that :

the history of every vertex is a prefix of $\mathbf{NDCQ}(\mathbf{W}ave\mathbf{W}aitC\mathbf{CQ})^*(\mathbf{W}ave\mathbf{W}ait(\mathbf{H}+\mathbf{No}))$.

(I_{12}) *Let v be a vertex; assume that :*

$$h(v) = h_1 \mathbf{QW}ave\mathbf{W}ait\mathbf{W}aitC\mathbf{C},$$

or

$$h(v) = h_1 \mathbf{QW}ave\mathbf{W}ait\mathbf{W}aitH\mathbf{H}.$$

Let $T_1 = ST(v)$ corresponding to the history $h_1 \mathbf{Q}$, and $T_2 = ST(v)$ corresponding to the history $h(v)$.

Then T_2 is obtained from T_1 adding all the sons of the leaves of T_1.

(I_{13}) *Let v be a* **C**-*labelled vertex or a* **H**-*labelled vertex, then all the leaves of $ST(v)$ have the same depth.*

(I_{14}) *If a node v is labelled* **No** *then $ST(v)$ has a leaf and an interior node with the same depth.*

PROOF: Studying carefully the effect of each rule on these properties, we get the desired result. $\square$

PROPOSITION 6.2. *The rewriting system $\mathcal{R}_5$ is noetherian. Let $G(V, E, \lambda)$ be an oriented rooted directed tree such that the root is labelled* **D** *and any other vertex is labelled* **N**. *Then the root of the irreducible graph obtained from $G(V, E, \lambda)$ with $\mathcal{R}_5$ is labelled* **H** *if and only if all leaves have the same depth. Otherwise the root is labelled* **No**.

PROOF: In the proof of the termination, we use the following notations. Let v be a vertex, $d(Root, v)$ is the length of the simple path from the vertex Root to the vertex v. We define the integer $P(G, \mathbf{C})$ (resp. $P(G, \mathbf{H})$) as the sum of the length of the simple paths from Root to the **C**-labelled vertices (resp. **H**-labelled vertices), more precisely :

$$P(G, \mathbf{C}) = \sum_{v \in \phi_V^{-1}(\{\mathbf{C}\})} d(Root, v),$$

and

$$P(G, \mathbf{H}) = \sum_{v \in \phi_V^{-1}(\{\mathbf{H}\})} d(Root, v).$$

Let $\mathcal{B}$ be the set of labels different from **No**. By $|G|_{\mathcal{B}}$ we denote the cardinality of the set of vertices labelled with elements of $\mathcal{B}$. The termination argument is the strict decreasing of the tuple

$$(|G|_{\mathbf{N}}, |G|_{\mathbf{D}}, P(G, \mathbf{C}), P(G, \mathbf{H}), |G|_{(\mathbf{C}, Root)}, |G|_{\mathbf{Q}}, |G|_{\mathcal{B}}).$$

With properties of Lemma 6.1 we deduce that when the graph is irreducible then Root is labelled with one of the two following labels **H** or **No**. The result follows from (I_{13}) and (I_{14}). $\square$

7. Attribute graph rewriting system

In this section we illustrate attribute graph rewriting system with three computations associated to the tree structure. In the first one we associate semantic rules to the system $\mathcal{R}_1$ getting in this way the value of the diameter of a graph without cycles. In the second example we associate semantic rules to the system $\mathcal{R}_4$ solving the identity problem. In the third example we organize with $\mathcal{R}_4$ an election.

DEFINITION 7.1. *An attribute graph rewriting system with priorities is given by a graph rewriting system with priority $G(V, E, \lambda)$ together with semantic rules defined by*
(A, Att, Dom, Sem) where :
1) A is a finite set of symbols called attribute names or simply attributes,
2) Att is a function from $C^{(V)} \cup C^{(E)}$ into $\wp(A)$,
3) Dom is a mapping from A into domains which associates with each attribute a its domain of values $Dom(a)$,

4) for each rewriting rule R, $Sem(R)$ is the set of semantic rules associated to R, each semantic rule has the following form

$$a_0(x_0) = f(a_1(x_1), \ldots, a_k(x_k))$$

with for any i $x_i \in V \cup E$, $a_i \in Att(\lambda(x_i))$ and f is a mapping.

a. Calculus of the diameter

This problem can be solved by using the following attributes; with labels **N** and **I** we associate integer attributes $b(\mathbf{N})$, $b(\mathbf{I})$, and $d(\mathbf{N})$, $d(\mathbf{I})$.

The attribute $b(\mathbf{X})$, $(\mathbf{X} \in \{\mathbf{N}, \mathbf{I}\})$ stands for the maximum-length of a simple path from the vertex labelled $\mathbf{X}$ to a leaf of **F**-connected components neighbour of this vertex.

The attribute $d(\mathbf{X})$ stands for the diameter of the rooted tree which has the vertex labelled $\mathbf{X}$ as root and which has as vertex-set the vertices of the **F**-connected components connected to this vertex.

We add to the PGRS $\mathcal{R}_1$ the following semantic rules (we use subscripts to distinguish between the distinct occurrences of the same label), the attribute rewriting system will be denoted by $\mathcal{R}_{1_A}$:

$$R_1 : \quad \mathbf{N}_1 \!-\!\!-\! \mathbf{N}_2 \!-\!\!-\! \mathbf{N}_3 \quad \longrightarrow \quad \mathbf{N}_4 \!-\!\!-\! \mathbf{I} \!-\!\!-\! \mathbf{N}_5$$

$\{b(\mathbf{N}_4) = b(\mathbf{N}_1), \quad b(\mathbf{N}_5) = b(\mathbf{N}_3), \quad b(\mathbf{I}) = b(\mathbf{N}_2), \quad d(\mathbf{N}_4) = d(\mathbf{N}_1),$
$d(\mathbf{N}_5) = d(\mathbf{N}_3), \quad d(\mathbf{I}) = d(\mathbf{N}_2)\};$

$$R_2 : \quad \mathbf{I}_1 \!-\!\!-\! \mathbf{N}_1 \!-\!\!-\! \mathbf{N}_2 \quad \longrightarrow \quad \mathbf{I}_2 \!-\!\!-\! \mathbf{I}_3 \!-\!\!-\! \mathbf{N}_3$$

$\{b(\mathbf{I}_2) = b(\mathbf{I}_1), \quad b(\mathbf{I}_3) = b(\mathbf{N}_1), \quad b(\mathbf{N}_3) = b(\mathbf{N}_2), \quad d(\mathbf{I}_2) = d(\mathbf{I}_1),$
$d(\mathbf{I}_3) = d(\mathbf{N}_1), \quad d(\mathbf{N}_3) = d(\mathbf{N}_2)\};$

$$R_3 : \quad \mathbf{I}_1 \!-\!\!-\! \mathbf{N} \!-\!\!-\! \mathbf{I}_2 \quad \longrightarrow \quad \mathbf{I}_3 \!-\!\!-\! \mathbf{I}_4 \!-\!\!-\! \mathbf{I}_5$$

$\{b(\mathbf{I}_3) = b(\mathbf{I}_1), \quad b(\mathbf{I}_4) = b(\mathbf{N}), \quad b(\mathbf{I}_5) = b(\mathbf{I}_2), \quad d(\mathbf{I}_3) = d(\mathbf{I}_1), \quad d(\mathbf{I}_4) = d(\mathbf{N}),$
$d(\mathbf{I}_5) = d(\mathbf{I}_2)\};$

$$R_4 : \quad \mathbf{N}_1 \!-\!\!-\! \mathbf{N}_2 \quad \longrightarrow \quad \mathbf{N}_3 \!-\!\!-\! \mathbf{F}$$

$\{b(\mathbf{N}_3) = Max(b(\mathbf{N}_1), b(\mathbf{N}_2) + 1),$
$d(\mathbf{N}_3) = Max(d(\mathbf{N}_1), d(\mathbf{N}_2), b(\mathbf{N}_1) + b(\mathbf{N}_2) + 1)\};$

$$R_5 : \quad \mathbf{I} \!-\!\!-\! \mathbf{N}_1 \quad \longrightarrow \quad \mathbf{N}_2 \!-\!\!-\! \mathbf{F}$$

$\{b(\mathbf{N}_2) = Max(b(\mathbf{I}), b(\mathbf{N}_1) + 1), \quad d(\mathbf{N}_2) = Max(d(\mathbf{I}), d(\mathbf{N}_1), b(\mathbf{I}) + b(\mathbf{N}_1) + 1)\}.$

LEMMA 7.2. *Let $G(V, E, \lambda)$ be a labelled connected graph such that any vertex is labelled $\mathbf{N}$, with the following attribute values $b(\mathbf{N}) = 0$, and $d(\mathbf{N}) = 0$. Let $G'(V, E, \lambda')$ be a graph such that :*

$$G(V, E, \lambda) \quad \xrightarrow[\mathcal{R}_{1_A}]{*} \quad G'(V, E, \lambda').$$

Let x be a vertex labelled c ($c = \mathbf{N}$, or $c = \mathbf{I}$), the graph $G'(V, E, \lambda')$ satisfies :

(V_1) $b(c)$ is equal to the maximum-length simple path from x to a leaf of $\mathbf{F}$-connected components connected to x,

(V_2) $d(c)$ is equal to the diameter of the rooted tree which has x as root and whose vertex set is the $\mathbf{F}$-connected components connected to x.

PROOF: The properties (V_1) and (V_2) are obviously true for the initial labelling of the graph. We only have to check that if $G_1(V, E, \lambda_1)$ satisfies this set of properties and

$$G_1(V, E, \lambda_1) \quad \xrightarrow[\mathcal{R}_{1_A}]{} \quad G_2(V, E, \lambda_2)$$

then $G_2(V, E, \lambda_2)$ also verifies it. We study the effect of each of the five rules.

If $G_1(V, E, \lambda_1) \quad \xrightarrow[X]{} \quad G_2(V, E, \lambda_2)$, with $X \in \{R_1, R_2, R_3\}$ then the set of $\mathbf{F}$-labelled vertices and the values of attributes are kept unchanged. Thus properties (V_1) and (V_2) remain true.

Let T_1 (resp. T_2) be the rooted-tree which is a subgraph of $G_1(V, E, \lambda_1)$ whose root is the vertex x (resp. y) and whose vertex set is the set of $\mathbf{F}$-labelled vertices connected to x (resp. to y). Using the rule R_4 or the rule R_5 we add the tree T_2 to the rooted-tree T_1. Let T_3 be the tree obtained in this way. Its root is the vertex x. The diameter and the maximal-length of a simple path from x to a leaf of T_3 may be computed from the diameters of T_1 and T_2 and the maximal-length simple path from x (resp y) to a leaf of T_1 (resp. T_2) by the formula encoded in the semantic rules. Thus V_1 and V_2 remains true, after a rewriting step by R_4 or R_5. $\square$

If the graph $G(V, E)$ is acyclic and $G'(V, E, \lambda')$ is irreducible we deduce immediatly from Theorem 2.2 :

COROLLARY 7.3. *Let $G(V, E, \lambda)$ be a connected labelled graph such that every vertex is labelled $\mathbf{N}$. Let $G'(V, E, \lambda')$ an irreducible graph such that :*

$$G(V, E, \lambda) \quad \xrightarrow[\mathcal{R}_{1_A}]{*} \quad G'(V, E, \lambda').$$

If the graph $G(V, E)$ is acyclic then the only vertex labelled $\mathbf{N}$ in the irreducible graph $G'(V, E, \lambda')$ is such that $d(\mathbf{N})$ is equal to the diameter of the graph $G(V, E)$.

b. The identity problem

The aim of the following attribute graph rewriting system is to solve the problem of giving to each vertex an identifier, distinct from any other's one. For every vertex, we compute its dewey identifier. For short, we denote $i((\mathbf{A}, k))$ (resp. $i((\mathbf{M}, k))$) by $i(\mathbf{A})$ (resp. $i(\mathbf{M})$). With any label L different from $\mathbf{N}$, we associate an integer attribute denoted by $i(L)$, and with each label $(\mathbf{A}, k)$, or $(\mathbf{M}, k)$ ($0 \leq k \leq 2$) we associate an integer attribute a, which denotes the first free value to be assigned to a son of the labelled vertex. Initially the vertex Root is labelled $(\mathbf{A}, 0)$, $i(\mathbf{A}) = 1$, $a(\mathbf{A}) = 1$, and any other vertex is labelled $\mathbf{N}$.

The concatenation operation is denoted by $\bullet$. The attribute rewriting system $\mathcal{R}_{4_A}$ has six generic rules :

$$R_1 \; : \quad \overset{(\mathbf{A},k)_1}{\circ} \; \overset{f}{\underline{\quad\quad}} \; \overset{\mathbf{N}}{\circ} \quad \longrightarrow \quad \overset{(\mathbf{M},k)}{\circ} \; \overset{t}{\underline{\quad\quad}} \; \overset{(\mathbf{A},k+1)_2}{\circ}$$

$$\{i(\mathbf{M}) = i(\mathbf{A}_1), i(\mathbf{A}_2) = i(\mathbf{A}_1) \bullet a(\mathbf{A}_1), a(\mathbf{M}) = a(\mathbf{A}_1) + 1, a(\mathbf{A}_2) = 1\}$$

$$R_2 \; : \quad \overset{(\mathbf{M},k)_1}{\circ} \; \overset{f}{\underline{\quad\quad}} \; \overset{\mathbf{N}}{\circ} \quad \longrightarrow \quad \overset{(\mathbf{M},k)_2}{\circ} \; \overset{t}{\underline{\quad\quad}} \; \overset{(\mathbf{A},k+1)}{\circ}$$

$$\{i(\mathbf{M}_2) = i(\mathbf{M}_1), i(\mathbf{A}) = i(\mathbf{M}_1) \bullet a(\mathbf{M}_1), a(\mathbf{M}_2) = a(\mathbf{M}_1) + 1, a(\mathbf{A}) = 1\}$$

$$R_3 \; : \quad \overset{(\mathbf{M},k)_1}{\circ} \; \overset{t}{\underline{\quad\quad}} \; \overset{(\mathbf{A},k+1)}{\circ} \quad \longrightarrow \quad \overset{(\mathbf{M},k)_2}{\circ} \; \overset{t}{\underline{\quad\quad}} \; \overset{\mathbf{R}}{\circ}$$

$$\{i(\mathbf{M}_2) = i(\mathbf{M}_1), i(\mathbf{R}) = i(\mathbf{A}), a(\mathbf{M}_2) = a(\mathbf{M}_1)\}$$

$$R_4 \; : \quad \overset{\mathbf{R}_1}{\circ} \; \overset{t}{\underline{\quad}} \; \overset{(\mathbf{M},k)_1}{\circ} \; \overset{t}{\underline{\quad}} \; \overset{\mathbf{R}_2}{\circ} \quad \longrightarrow \quad \overset{\mathbf{F}}{\circ} \; \overset{t}{\underline{\quad}} \; \overset{(\mathbf{M},k)_2}{\circ} \; \overset{t}{\underline{\quad}} \; \overset{\mathbf{R}_3}{\circ}$$

$$\{i(\mathbf{M}_2) = i(\mathbf{M}_1), i(\mathbf{F}) = i(\mathbf{R}_1), i(\mathbf{R}_3) = i(\mathbf{R}_2), a(\mathbf{M}_2) = a(\mathbf{M}_1)\}$$

$$R_5 \; : \quad \overset{\mathbf{R}}{\circ} \; \overset{t}{\underline{\quad}} \; \overset{(\mathbf{M},k)_1}{\circ} \; \overset{t}{\underline{\quad}} \; \overset{(\mathbf{M},k+1)_2}{\circ} \quad \longrightarrow \quad \overset{\mathbf{F}}{\circ} \; \overset{t}{\underline{\quad}} \; \overset{(\mathbf{M},k)_3}{\circ} \; \overset{t}{\underline{\quad}} \; \overset{(\mathbf{M},k+1)_4}{\circ}$$

$$\{i(\mathbf{M}_3) = i(\mathbf{M}_1), i(\mathbf{M}_4) = i(\mathbf{M}_2), i(\mathbf{F}) = i(\mathbf{R}), a(\mathbf{M}_3) = a(\mathbf{M}_1), a(\mathbf{M}_4) = a(\mathbf{M}_2)\}$$

$$R_6 \; : \quad \overset{(\mathbf{M},k)}{\circ} \; \overset{t}{\underline{\quad}} \; \overset{\mathbf{R}_1}{\circ} \quad \longrightarrow \quad \overset{\mathbf{R}_2}{\circ} \; \overset{t}{\underline{\quad}} \; \overset{\mathbf{F}}{\circ}$$

$$\{i(\mathbf{R}_2) = i(\mathbf{M}), i(\mathbf{F}) = i(\mathbf{R}_1), i(\mathbf{F}) = i(\mathbf{R}), a(\mathbf{M}_3) = a(\mathbf{M}_1), a(\mathbf{M}_4) = a(\mathbf{M}_2)\}$$

Using previous notations, and an induction with Lemma 5.3 we get:

PROPOSITION 7.4. *Let $G(V, E, \lambda)$ be a connected graph with a distinguished vertex, called Root, labelled $\mathbf{A}$ (with $i(\mathbf{A}) = 1$ and $a(\mathbf{A}) = 1$) and any other vertex labelled $\mathbf{N}$. Then any irreducible graph obtained from $G(V, E, \lambda)$ with $\mathcal{R}_{4_A}$ satisfies the following property :*

let x and y be two any distinct vertices of G', and let $\mathbf{X}$ and $\mathbf{Y}$ be their associated labels then $i(\mathbf{X})$ and $i(\mathbf{Y})$ are distinct.

c. Organizing the election of a leader

We suppose here that the graph is such that each vertex has its proper identifier. We assume that the set of identifiers is ordering and that the function Max gives the greatest among two identifiers. The problem consists in choosing the vertex with the greatest identifier to be the leader, in such a way that at the end of the computation the vertex which has initiated the calculus knows which vertex has been elected. To select the leader we organize a graph traversal and at each step we keep the vertex which has the greater identifier among those seen. Hence we assume that with any label $\mathbf{N}$, $\mathbf{A}$, $\mathbf{M}$, $\mathbf{R}$, $\mathbf{F}$ is associated an attribute i which corresponds to the identifier of the labelled vertex. With the labels $\mathbf{M}$ and $\mathbf{R}$ we associate an attribute m whose value will be the leader of the current corresponding subtree. Initially the vertex Root (which initiates the calculus) is labelled $(\mathbf{A}, 0)$, and any other vertex is labelled $\mathbf{N}$. We assume that the value of the corresponding attribute for the identifier has been given, and that two different vertices have different values. In the semantic rules we have to keep the identifiers of the vertices, thus we have actions of the form $i(x) = i(y)$. The system is denoted by $\mathcal{R}_{4'_A}$.

$$R_1 \; : \; \overset{(\mathbf{A},k)_1}{\circ} \xrightarrow{\;\; f \;\;} \overset{\mathbf{N}}{\circ} \quad \longrightarrow \quad \overset{(\mathbf{M},k)}{\circ} \xrightarrow{\;\; t \;\;} \overset{(\mathbf{A},k+1)_2}{\circ}$$

$$\{i(\mathbf{M}) = i(\mathbf{A}_1), i(\mathbf{A}_2) = i(\mathbf{N}), m(\mathbf{M}) = i(\mathbf{A}_1)\}$$

$$R_2 \; : \; \overset{(\mathbf{M},k)_1}{\circ} \xrightarrow{\;\; f \;\;} \overset{\mathbf{N}}{\circ} \quad \longrightarrow \quad \overset{(\mathbf{M},k)_2}{\circ} \xrightarrow{\;\; t \;\;} \overset{(\mathbf{A},k+1)}{\circ}$$

$$\{i(\mathbf{M}_2) = i(\mathbf{M}_2), i(\mathbf{A}) = i(\mathbf{N}), m(\mathbf{M}_2) = m(\mathbf{M}_1)\}$$

$$R_3 \; : \; \overset{(\mathbf{M},k)_1}{\circ} \xrightarrow{\;\; t \;\;} \overset{(\mathbf{A},k+1)}{\circ} \quad \longrightarrow \quad \overset{(\mathbf{M},k)_2}{\circ} \xrightarrow{\;\; t \;\;} \overset{\mathbf{R}}{\circ}$$

$$\{i(\mathbf{M}_2) = i(\mathbf{M}_1), i(\mathbf{R}) = i(\mathbf{A}), m(\mathbf{R}) = i(\mathbf{A}), m(\mathbf{M}_2) = m(\mathbf{M}_1), \}$$

$$R_4 \; : \; \overset{\mathbf{R}_1}{\circ} \xrightarrow{\;\; t \;\;} \overset{(\mathbf{M},k)_1}{\circ} \xrightarrow{\;\; t \;\;} \overset{\mathbf{R}_2}{\circ} \quad \longrightarrow \quad \overset{\mathbf{F}}{\circ} \xrightarrow{\;\; t \;\;} \overset{(\mathbf{M},k)_2}{\circ} \xrightarrow{\;\; t \;\;} \overset{\mathbf{R}_3}{\circ}$$

$$\{i(\mathbf{M}_2) = i(\mathbf{M}_1), i(\mathbf{R}_3) = i(\mathbf{R}_2), i(\mathbf{F}) = i(\mathbf{R}_1), m(\mathbf{M}_2) = Max(m(\mathbf{M}_1), m(\mathbf{R}_1))\}$$

$$R_5 \; : \; \overset{\mathbf{R}}{\circ} \xrightarrow{\;\; t \;\;} \overset{(\mathbf{M},k)_1}{\circ} \xrightarrow{\;\; t \;\;} \overset{(\mathbf{M},k+1)_2}{\circ} \quad \longrightarrow \quad \overset{\mathbf{F}}{\circ} \xrightarrow{\;\; t \;\;} \overset{(\mathbf{M},k)_3}{\circ} \xrightarrow{\;\; t \;\;} \overset{(\mathbf{M},k+1)_4}{\circ}$$

$$\{i(\mathbf{M}_3) = i(\mathbf{M}_1), i(\mathbf{M}_4) = i(\mathbf{M}_2), i(\mathbf{F}) = i(\mathbf{R}), m(\mathbf{M}_3) =$$
$$Max(m(\mathbf{M}_1), m(\mathbf{R})), m(\mathbf{M}_4) = m(\mathbf{M}_2)\}$$

$$R_6 \; : \; \overset{(\mathbf{M},k)}{\circ} \xrightarrow{\;\; t \;\;} \overset{\mathbf{R}_1}{\circ} \quad \longrightarrow \quad \overset{\mathbf{R}_2}{\circ} \xrightarrow{\;\; t \;\;} \overset{\mathbf{F}}{\circ}$$

$$\{i(\mathbf{R}_2) = i(\mathbf{M}), i(\mathbf{F}) = i(\mathbf{R}_1), m(\mathbf{R}_2) = Max(m(\mathbf{M}), m(\mathbf{R}_1)), \}$$

By a simple inductive proof, using Lemma 5.3 we get :

PROPOSITION 7.5. *Let $G(V, E, \lambda)$ be a connected graph with a distinguished vertex, called Root, labelled $\mathbf{A}$ and any other vertex labelled $\mathbf{N}$. With each occurrence of label $\mathbf{X}$ in the graph $G(V, E, \lambda)$ is associated a value denoted by $i(\mathbf{X})$ which is the proper identifier of the corresponding vertex. Then any irreducible graph obtained from $G(V, E, \lambda)$ with $\mathcal{R}_{4'_A}$ satisfies the following properties :*

Root is labelled $\mathbf{R}$, all other vertices are labelled $\mathbf{F}$, and $m(\mathbf{R})$ contains the maximum value of the identifiers of the vertices of G.

REFERENCES

1. M.W.Alford, J.P.Ansart, G.Hommel, L.Lamport, B.Liskov, G.P.Mullery, F.B.Schneider, *Distributed Systems*, Lecture Notes in Comput. Sci. **190** (1985).
2. D.Angluin, *Local and global properties in networks of processors*, Proceedings of the twelfth annual ACM Symposium on Theory of Computing (1980), 82–93.
3. C.Berge, "Graphes et Hypergraphes," Dunod, Paris, 1970.
4. J.Berstel, *Résolution, par un réseau d'automates, du problème des arborescences dans un graphe*, C.R. Acad. Sc. Paris, Série A, t.264 (1967), 388–390.
5. M.Billaud, P.Lafon, Y.Métivier and E.Sopena, *Graph Rewriting Systems with Priorities*, Lecture Notes in Comput. Sci. **411** (1989), 94–106.
6. M.Billaud, P.Lafon, Y.Métivier and E.Sopena, *Graph Rewriting Systems with Priorities : Definitions and Applications*, Internal report, University Bordeaux I **8909** (1989).
7. B.Courcelle, *Recognizable sets of unrooted trees*, "Definability and recognizability of sets of trees," M.Nivat and A.Podelski, Elsevier Science Publishers B.V., Amsterdam, 1991.
8. G.Huet, *Confluent reductions: Abstract properties and applications to term rewriting systems*, J. Assoc. Comput. Mach. **27** (4) (1980), 797–821.
9. I.Litovsky and Y.Métivier, *Computing with Graph Rewriting Systems with Priorities*, Fourth International Workshop on Graph Grammars and their Applications to Computer Science, Bremen. (to appear in LNCS) Internal report, University Bordeaux I (1990).
10. M.Raynal, "Algorithmes distribués et protocoles," Eyrolles, Paris, 1985.
11. P.Rosenstiehl, J.R.Fiksel, and A.Holliger, *Intelligent graphs*, "Graph theory and Computing," Ronald C. Read, Academic Press, New York, 1972, pp. 219–265.

Keywords. Attribute, Diameter, Echo, Election, Graph, Identity Problem, Network, Rewriting System, Semantic Action,Spanning-tree, Tree, Wave.

Igor Litovsky, Yves Métivier LaBRI URA CNRS n° 1304 ENSERB 33405 TALENCE CEDEX FRANCE

Tree Automata and Languages
M. Nivat and A. Podelski (editors)
© 1992 Elsevier Science Publishers B.V. All rights reserved.

141

Recognizable Sets of Unrooted Trees

Bruno Courcelle

Laboratoire d'Informatique[1], Université Bordeaux-I, 351 Cours de la Libération, 33405 Talence, France

Abstract

We give an algebraic presentation of the set of graph theoretic, unrooted, undirected trees. We define finite-state automata that accept the recognizable sets of such trees.

INTRODUCTION

Many methods and tools of formal language theory can be fruitfully applied to the definition and study of graphs and sets of graphs. Grammars, systems of equations, congruences all have been generalized for this purpose from words and trees to graphs.

The notion of a recognizable set of finite graphs has been introduced by Courcelle [6]. It is based on the notion of a congruence on an appropriate algebra of finite graphs having finitely many classes. A challenging open question is the definition of finite-state graph automata that could describe recognizable sets of graphs and that could be implemented so as to yield efficient recognition algorithms.

Since the recognizable sets of finite graphs form an uncountable set (Courcelle [6]), there is no hope of representing all of them by finite-state automata (because each automaton would yield a recognition algorithm, and there are only countably many algorithms). On the other hand, every

---

[1] Laboratoire associé au CNRS ; email : courcell@geocub. greco-prog.fr.
This work has been supported by the "Programme de Recherches Coordonnées : Mathématiques et Informatique".

recognizable set of graphs *of tree-width at most some fixed integer k* can be also generated by a context-free graph-grammar of a certain type (a *hyperedge replacement* grammar to be precise). Hence, there are countably many recognizable sets of graphs of bounded tree-width, and the definition of a class of graph automata representing them is a reasonable goal. (See Courcelle [5,7] for more detail on graph-grammars and tree-width.)

Trees form a special class of graphs, and tree automata have been studied for a long time (Doner [8], Gecseg and Steinby [9]). However, the corresponding trees are actually terms over finite signatures. They are not trees in the usual sense of graph theory. Encoded as directed graphs, terms are ordered; they are of bounded degree, and they have a distinguished node called the root.

In the present paper, we treat the case of unbounded, unordered, undirected, unrooted finite trees, which we call simply *trees* in the remainder of this paper. This case is a necessary step towards a characterization of recognizable sets of finite graphs of bounded tree-width in terms of automata because trees are precisely the connected graphs (without loops and multiple edges) of tree-width at most 1.

We now describe the content of the present paper. For discussing recognizable sets of trees, we need an algebraic structure. A set of trees is then recognizable iff it is a union of classes of an equivalence relation that has finitely many classes and is a congruence with respect to the algebraic structure.

We define such a structure, together with a complete equational axiomatization of it. A similar axiomatization is done in Courcelle [4] for several classes of rooted trees, and in Bauderon and Courcelle [1] for graphs and hypergraphs. Having an appropriate algebraic structure, we obtain in a standard way several equivalent definitions of a recognizable set of trees, together with a related notion of automaton. These automata are actually finite-state automata on terms that satisfy certain equational conditions. They do not process trees, but rather terms denoting trees.

We show then how these automata can be implemented so as to process trees as opposed to terms. This implementation is based on certain graph relabelling systems introduced by Métivier and others (see [2,10]).

This paper is essentially an application of methods presented in more detail in Courcelle [4]. Hence, some proofs will be omitted.

1. TREES

A *tree* T is a finite connected undirected graph without multiple edges and cycles. The set of trees is denoted by $\mathbb{T}$. A *rooted tree* is a pair R=(T,r) consisting of a tree T and a distinguished node r called the *root*. The set of rooted trees is denoted by $\mathbb{R}$. The set of nodes of a rooted or unrooted tree T is denoted by **N**(T).

We now define a few operations on trees and rooted trees. The profiles of these operations will be given in terms of two sorts, **t** and **r**, denoting respectively the "types" "tree" and "rooted tree".

The first operation is $// : \mathbf{r} \times \mathbf{r} \longrightarrow \mathbf{r}$, called the *parallel composition*. For S and T in $\mathbb{R}$, we let S//T be the rooted tree obtained by fusing the roots of S and T (or rather, of two disjoint isomorphic copies of S and T).

The second operation is $\mathbf{ext} : \mathbf{r} \longrightarrow \mathbf{r}$ called the *extension*. For T in $\mathbb{R}$, we let **ext**(T) be the rooted tree obtained from T by the addition of a new node that becomes the root of **ext**(T), linked by a new edge to the former root of T.

We denote by $\mathbb{1}$ the rooted tree reduced to a root without any edge. Finally we let $\mathbf{fg} : \mathbf{r} \longrightarrow \mathbf{t}$ be the mapping that "forgets" the root of a rooted tree R. Formally, $\mathbf{fg}(R) = T$ where $R = (T,r) \in \mathbb{R}$.

Hence, we have defined an {**r**, **t**}-sorted signature $F := \{ //, \mathbf{ext}, \mathbb{1}, \mathbf{fg} \}$. We obtain a many-sorted F-magma (i.e., F-algebra) $\mathbb{TREE}$ having $\mathbb{R}$ as domain of sort $\mathbf{r}$, $\mathbb{T}$ as domain of sort $\mathbf{t}$, and the operations of F defined above. Hence, $\mathbb{TREE} = <\mathbb{R}, \mathbb{T}, //, \mathbf{ext}, \mathbb{1}, \mathbf{fg}>$. We shall denote by $\mathbb{M}(F)$ the initial F-magma. Its two domains are $\mathbb{M}(F)_{\mathbf{r}}$ and $\mathbb{M}(F)_{\mathbf{t}}$, the sets of finite well-formed terms over F of sort respectively $\mathbf{r}$ and $\mathbf{t}$. Every term t has a *value* in $\mathbb{TREE}$ that we denote by $\mathbf{val}(t)$. This value is a tree if t is of sort $\mathbf{t}$ and a rooted tree if t is of sort $\mathbf{r}$. The mapping $\mathbf{val}$ is the unique homomorphism from the initial F-magma $\mathbb{M}(F)$ into $\mathbb{TREE}$.

We let $\mathcal{E}$ be the following set of four equational axioms :

$$(\mathcal{E}_1) \qquad x // y = y // x$$
$$(\mathcal{E}_2) \qquad (x // y) // z = x // (y // z)$$
$$(\mathcal{E}_3) \qquad x // \mathbb{1} = x$$
$$(\mathcal{E}_4) \qquad \mathbf{fg}(x // \mathbf{ext}(y)) = \mathbf{fg}(\mathbf{ext}(x) // y),$$

where x,y,z are variables of sort $\mathbf{r}$, intended to denote rooted trees. It is clear from the definitions that these axioms are valid in $\mathbb{TREE}$. Moreover, they *define* $\mathbb{TREE}$, since we have :

Proposition 1 : $\mathbb{TREE}$ *is (isomorphic to) the initial* $(F,\mathcal{E})$-*magma*.

By an $(F,\mathcal{E})$-*magma*, we mean any F-magma satisfying $\mathcal{E}$. The initial $(F,\mathcal{E})$-magma is well-known to be isomorphic to $\mathbb{M}(F) / \overset{*}{\underset{\mathcal{E}}{\longleftrightarrow}}$, i.e., to the quotient of the initial F-magma $\mathbb{M}(F)$ by the congruence relation generated by $\mathcal{E}$.

Proof sketch :

Since $\mathbb{TREE}$ satisfies $\mathcal{E}$, we have $\mathbf{val}(t) = \mathbf{val}(t')$ if $t \overset{*}{\underset{\mathcal{E}}{\longleftrightarrow}} t'$. Let us conversely assume that $\mathbf{val}(t) = \mathbf{val}(t')$ and prove that $t \overset{*}{\underset{\mathcal{E}}{\longleftrightarrow}} t'$.

If the common sort of t and t' is $\mathbf{r}$, then t and t' $\in \mathbb{M}(\{//,\mathbf{ext},\mathbb{1}\})$. One can establish that $t \overset{*}{\underset{\mathcal{E}}{\longleftrightarrow}} t'$ by an induction on the structure of t.

Otherwise, the common sort of t and t' is **t** and t = **fg**(w) , t' = **fg**(w') for some w,w' in $\mathbb{M}(\{//, \textbf{ext}, \mathbb{1}\})$. Let W = **val**(w) and W' = **val**(w'). We have W = (T,r) and W'=(T,r') where T is the unrooted tree defined by both t and t'.

First case : r = r' . Then $w \xleftrightarrow[\mathcal{E}]{*} w'$ by the previous case and the result follows.

Second case : r and r' are at distance 1 in T, i.e., are linked by an edge. The graph T minus the edge linking r and r' has two connected components. Let T_1 be that containing r, and T_2 that containing r'.

We let t_1 and t_2 be terms in $\mathbb{M}(F)_\textbf{r}$ that define the rooted trees (T_1,r) and (T_2,r') respectively. Then we have:

$$W = \textbf{val}(w) \qquad \text{by definition of W and w,}$$
$$= \textbf{val}(t_1//\textbf{ext}(t_2)) \qquad \text{by definition of } T_1, T_2, t_1, \text{ and } t_2.$$

Similarly
$$W' = \textbf{val}(w') = \textbf{val}\,(\textbf{ext}(t_1)//t_2).$$

Hence, by the first part of the proof :

$$w \xleftrightarrow[\mathcal{E}]{*} t_1//\textbf{ext}(t_2)$$

$$w' \xleftrightarrow[\mathcal{E}]{*} \textbf{ext}(t_1)//t_2$$

$$t = \textbf{fg}(w) \xleftrightarrow[\mathcal{E}]{*} \textbf{fg}(t_1//\textbf{ext}(t_2))$$

$$\xleftrightarrow[\mathcal{E}]{} \textbf{fg}(\textbf{ext}(t_1)//t_2)$$

$$\xleftrightarrow[\mathcal{E}]{*} \textbf{fg}(w') = t'.$$

This completes the proof of the second case.

Third case : r and r' are at distance $n+1$ in T. Hence, we can find a path $r \text{---} r_1 \text{---} r_2 \text{---} ... \text{---} r_n \text{---} r'$ in T, and for each $i=1,...,n$, a term w_i in $\mathbb{M}(\{//, \textbf{ext}, \mathbb{1}\})_{\textbf{r}}$ such that $\textbf{val}(w_i) = (T, r_i)$. By the second case, we have :

$$t \xleftrightarrow[\mathcal{E}]{*} \textbf{fg}(w_1),$$

$$\textbf{fg}(w_i) \xleftrightarrow[\mathcal{E}]{*} \textbf{fg}(w_{i+1}) \qquad \text{for each } i=1,...,n-1,$$

$$\textbf{fg}(w_n) \xleftrightarrow[\mathcal{E}]{*} t' .$$

Hence, we get $t \xleftrightarrow[\mathcal{E}]{*} t'$ as desired. $\square$

2. RECOGNIZABLE SETS OF TREES

Having an algebraic presentation of the set of (unrooted) trees, we now apply to it the general definition of a recognizable set presented and investigated in Courcelle [4,6].

Let $L \subseteq \mathbb{T}$. The following properties are equivalent and characterize the *recognizability* of L :

      (1) L is saturated with respect to a congruence relation on $\mathbb{TREE}$ having finitely many classes. (In the definition of a congruence relation we require that any two congruent objects are of the same sort, either $\textbf{r}$ or $\textbf{t}$; saying that L is saturated means that it is a union of equivalence classes.)

      (2) $L = h^{-1}(L')$ where h is a homomorphism : $\mathbb{TREE} \rightarrow \mathbb{P}$ for some finite F-magma $\mathbb{P}$ and some subset L' of $\mathbb{P}_{\textbf{t}}$. (We denote by $\mathbb{P}_{\textbf{t}}$ the domain of $\mathbb{P}$ of sort $\textbf{t}$.)

(3) The equivalence relation $\sim$ on $\mathbb{R}$ such that :

$T \sim T'$ iff
for every $c \in \mathbb{LIN}(F,\{x\})_t$: $c_{\mathbb{TREE}}(T) \in L \Longleftrightarrow c_{\mathbb{TREE}}(T') \in L$

has finitely many classes. (We denote by $\mathbb{LIN}(F,\{x\})_t$ the set of terms of sort **t**, written with the symbols of F and having exactly one occurrence of the variable x, that we assume to be of sort **r**. We denote by $c_{\mathbb{TREE}} : \mathbb{R} \longrightarrow \mathbb{T}$ the derived operation of $\mathbb{TREE}$ associated with a term c in $\mathbb{LIN}(F,\{x\})_t$.)

(4) The set $\mathbf{val}^{-1}(L) \subseteq \mathbb{M}(F)_t$ is recognizable as a set of terms, i.e., in the classical sense of Gecseg and Steinby [9].

A few comments are in order. As in [4], we call the triple $(h,\mathbb{P},L')$ of condition (2) an *automaton* and we say that L is the set it *recognizes* . We shall give below an equivalent but more concrete notion of automaton.

In condition (3), the equivalence $\sim$ is analogous to the syntactical congruence, the finiteness of which characterizes the regular languages. If we let in addition, for T and T' in $\mathbb{T}$:

$$T \sim T' \text{ iff } T \in L \Longleftrightarrow T' \in L,$$

then $\sim$ is a congruence relation on $\mathbb{TREE}$, and is actually the coarsest of those for which L is saturated.

From condition (4), we get the existence of a *tree-automaton* (see Gecseg and Steinby [9]) recognizing $\mathbf{val}^{-1}(L)$. Such an automaton $\mathcal{A}$ (the set of terms it recognizes will be denoted by $\mathbf{L}(\mathcal{A})$) can be used in the following (nondeterministic) algorithm :

Input : A tree T .

Method : 1 - Construct some term t in $\mathbb{M}(F)_t$ such that $\mathbf{val}(t) = T$.

    2 - Decide whether $t \in \mathbf{L}(\mathcal{A})$.

Output : The answer " $T \in L$ " if $t \in \mathbf{L}(\mathcal{A})$ and the answer " $T \notin L$ " otherwise.

It is clear that this algorithm decides correctly (for all computations) whether $T \in L$. Note its nondeterminism: in step 1, we can take *any* term t defining T. Nevertheless, the final result is uniquely defined because $\mathbf{L}(\mathcal{A})$, the set recognized by $\mathcal{A}$, is saturated w.r.t. the congruence $\xleftrightarrow[\mathcal{E}]{*}$.

By Proposition 1, the following condition is nothing but a reformulation of (4) :

    (4') $L = \mathbf{val}(K)$ where K is a recognizable subset of $\mathbb{M}(F)_t$ that is $\xleftrightarrow[\mathcal{E}]{*}$-saturated.

We now give the concrete notion of a $\mathbb{T}$-automaton that we have anounced above. A $\mathbb{T}$-*automaton* is a tuple $\mathcal{A} = \langle Q, //_{\mathcal{A}}, \mathbf{ext}_{\mathcal{A}}, q_0, Q_F \rangle$ where Q is a set called the set of *states*, $q_0 \in Q$ is the *initial state*, $Q_F \subseteq Q$ is the set of *final states*, and $//_{\mathcal{A}}$ and $\mathbf{ext}_{\mathcal{A}}$ are two *transition functions* : $Q \times Q \to Q$ and : $Q \to Q$, respectively, such that :

    (1) $(Q, //_{\mathcal{A}}, q_0)$ is a commutative monoid;
    (2) for every q and q' in Q we have:

$$q //_{\mathcal{A}} \mathbf{ext}_{\mathcal{A}}(q') \in Q_F \quad \text{iff} \quad q' //_{\mathcal{A}} \mathbf{ext}_{\mathcal{A}}(q) \in Q_F.$$

We say that $\mathcal{A}$ is *finite* iff its set of states is finite. The behaviour of $\mathcal{A}$ can be described by a mapping $\mathbf{h}_{\mathcal{A}}: \mathbb{M}(F)_r \to Q$ defined as follows :

$$\mathbf{h}_{\mathcal{A}}(\mathbb{1}) = q_0$$
$$\mathbf{h}_{\mathcal{A}}(t_1 //t_2) = \mathbf{h}_{\mathcal{A}}(t_1) //_{\mathcal{A}} \mathbf{h}_{\mathcal{A}}(t_2)$$
$$\mathbf{h}_{\mathcal{A}}(\mathbf{ext}(t_1)) = \mathbf{ext}_{\mathcal{A}}(\mathbf{h}_{\mathcal{A}}(t_1)).$$

Note that by Proposition 1 and Condition (1) of the definition of a $\mathbb{T}$-automaton, the mapping $\mathbf{h}_\mathcal{A}$ is well-defined on $\mathbb{R}$, i.e., for every rooted tree T, one can let $\mathbf{h}_\mathcal{A}(T) = \mathbf{h}_\mathcal{A}(t)$, where t is any term in $\mathbb{M}(F)_\mathbf{r}$ defining T.

The *set of terms recognized by* $\mathcal{A}$ is defined as:

$$\mathbf{T}(\mathcal{A}) = \{t \in \mathbb{M}(F)_\mathbf{t} \ / \ t = \mathbf{fg}(r), \ \mathbf{h}_\mathcal{A}(r) \in Q_F ,$$
$$\text{for some r in } \mathbb{M}(F)_\mathbf{r} \}$$

and the *set of trees recognized by* $\mathcal{A}$ is defined as:

$$\mathbf{L}(\mathcal{A}) = \mathbf{val}(\mathbf{T}(\mathcal{A}))$$
$$= \{T \in \mathbb{T} \ / \ T = \mathbf{val}(t), \text{ for some t in } \mathbf{T}(\mathcal{A})\}.$$

Note that $\mathbf{T}(\mathcal{A})$ is a recognizable set of terms in the usual sense.

Proposition 2 - *A set of trees* $L \subseteq \mathbb{T}$ *is recognizable iff* $L = \mathbf{L}(\mathcal{A})$ *for some finite-state* $\mathbb{T}$-*automaton* $\mathcal{A}$.

Proof sketch - Let $\mathcal{A}$ be a $\mathbb{T}$-automaton. We shall prove that:

$$\mathbf{val}^{-1}(\mathbf{L}(\mathcal{A})) = \mathbf{T}(\mathcal{A}).$$

From this and Condition (4), it will follow that $\mathbf{L}(\mathcal{A})$ is recognizable, since $\mathbf{T}(\mathcal{A})$ is a recognizable set of terms in the sense of [9]. The inclusion $\supseteq$ holds by the definitions. For the other direction, let $t \in \mathbf{T}(\mathcal{A})$, and $t' \in \mathbb{M}(F)_\mathbf{t}$ be such that $\mathbf{val}(t) = \mathbf{val}(t')$. Then $t \underset{\mathcal{E}}{\overset{*}{\longleftrightarrow}} t'$ by Proposition 1. By Conditions (1) and (2) in the definition of a $\mathbb{T}$-automaton, we obtain that $t' \in \mathbf{T}(\mathcal{A})$. (The proof is as follows: One need only prove that t belongs to $\mathbf{T}(\mathcal{A})$ iff t' belongs to $\mathbf{T}(\mathcal{A})$ for a one-step rewriting of t into t'. By the definition of the signature F, two such terms are necessarily of the form t $= \mathbf{fg}(r)$ and t' $= \mathbf{fg}(r')$ with r, r' in $\mathbb{M}(\{//,\mathbf{ext},\mathbb{1}\})_\mathbf{r}$. If the rewriting of t into t' uses one of the rules $\mathcal{E}_1$ to $\mathcal{E}_3$, then $\mathbf{h}_\mathcal{A}(r) = \mathbf{h}_\mathcal{A}(r')$ by Condition (1), and t belongs to $\mathbf{T}(\mathcal{A})$ iff t' belongs to $\mathbf{T}(\mathcal{A})$. If it uses rule $\mathcal{E}_4$, then t belongs to $\mathbf{T}(\mathcal{A})$ iff t' belongs to $\mathbf{T}(\mathcal{A})$ by Condition (2).)

Let us assume conversely that $L \subseteq \mathbb{T}$ is recognizable. We let $\sim$ be a finite congruence relation on $\mathbb{M}(F)$ such that L is saturated. We denote by $\sim_r$ the restriction of $\sim$ to $\mathbb{M}(F)_r$ and by $\sim_t$ its restriction to $\mathbb{M}(F)_t$. We let $Q = \mathbb{M}(F)_r / \sim_r$, we let $//_Q$ be the quotient of $//$ by $\sim_r$, and we let q_0 be the equivalence class of $\mathbb{1}$. In order to obtain a finite-state $\mathbb{T}$-automaton, we need only define Q_F . We let $Q_F := \{[t] \ / \ \mathbf{val}(\mathbf{fg}(t)) \in L\}$ (where [t] denotes the equivalence class of t). It is then easy to verify that one obtains actually a $\mathbb{T}$-automaton in this way, and that it recognizes L.$\square$

3. A DISTRIBUTED IMPLEMENTATION OF $\mathbb{T}$- AUTOMATA

The $\mathbb{T}$-automata we have defined process terms from $\mathbb{M}(F)_r$ and not trees. That is, in order to use such an automaton Q to decide whether a given tree T belongs to the set $\mathbf{L}(Q)$, first one must find a term t in $\mathbb{M}(F)_r$ such that $\mathbf{val}(\mathbf{fg}(t)) = T$ (*any* such term will do), and then one lets Q work on t to decide whether $\mathbf{fg}(t)$ belongs to $\mathbf{T}(Q)$, i.e., whether T belongs to $\mathbf{L}(Q)$.

We now wish to avoid the construction of t, and to let Q work "directly" on T. $\mathbb{T}$-automata will be implemented by means of *distributed graph relabelling systems*, able to process several parts of the given tree simultaneously. These distributed systems have been introduced by Métivier and others in [2,10]. They do not add or delete edges and vertices of the graph to which they apply. They modify only the labellings of vertices and the orientations of edges.

We need a few technical definitions. Let T be a tree, and r be a node of T. Then (T,r) is a rooted tree. Let:

$$\mathbf{E}(T,r) = \{(x,y) \ / \ x \text{ and } y \text{ are the ends of some edge of T,}$$
$$\text{and } y \text{ is on the unique cycle-free path from x to r}\}.$$

This binary relation on $\mathbf{N}(T)$ defines an *orientation* of T: We shall say that an edge of T with ends x and y is *directed from* x *to* y if (x,y) belongs to $\mathbf{E}(T,r)$, and we shall say that the pair $(T, \mathbf{E}(T,r))$ is a *directed tree* .

Note that r can be determined in a unique way from the relation $\mathbf{E}(T,r)$. Hence, there is a canonical bijection between rooted trees and directed trees. In the following, we shall identify the two notions and say that a directed tree has a root. Note that not every binary relation on the set of nodes of a tree is an orientation.

Let T be a tree. Let A be a set of vertex labels. An *annotation* of T is a pair (f,E) consisting of a mapping f associating a label in A with every node of T, and of a binary relation E on $\mathbf{N}(T)$ such that :

$$\text{if } (x,y) \in E \text{ then } x \text{ and } y \text{ form an edge of T;}$$
$$\text{if } (x,y) \in E \text{ then } (y,x) \notin E.$$

Hence, E specifies a direction for some edges of T. Letting $G = (T,f,E)$, we shall distinguish between the *directed edges* of G, namely the edges of T having a direction defined by E, and the *undirected* ones, namely the other edges of T. We let $\mathbf{ANN}(T,A)$ denote the set of triples of the form (T,f,E) where (f,E) is an annotation of T, and we let $\mathbf{ANN}(\mathbb{T},A)$ denote $\cup\{\mathbf{ANN}(T,A) \ / \ T \in \mathbb{T}\}$.

With a finite $\mathbb{T}$-automaton $\mathcal{Q} = \langle Q, //_{\mathcal{Q}}, \mathbf{ext}_{\mathcal{Q}}, q_0, Q_F \rangle$, we shall associate a binary relation $\xrightarrow[\mathcal{Q}]{}$ on $\mathbf{ANN}(\mathbb{T},Q)$ defined as follows :

$$G = (T,f,E) \xrightarrow[\mathcal{Q}]{} G' = (T', g, E')$$

iff:

$$G \xrightarrow[(x,y)]{} G'$$

for some nodes x and y of T the following conditions hold :

(1) $T' = T$;

(2) G has an undirected edge $\{x,y\}$ such that no other undirected edge of G is incident with x;

(3) $E' = E \cup \{(x,y)\}$;

(4) $g(u) = f(u)$ for all nodes $u \neq y$, and $g(y) = \mathbf{ext}_{\mathcal{Q}}(f(x))//_{\mathcal{Q}}f(y)$.

Given T we shall consider the annotation $(f_0, \emptyset)$ of T where $f_0(x)$ = q_0 (the initial state of $\mathcal{Q}$) for all x, and we shall transform $(T, f_0, \emptyset)$ by successive applications of $\xrightarrow[\mathcal{Q}]{}$. We shall say that G is a *normal form of* $(T, f_0, \emptyset)$ if $(T, f_0, \emptyset) \xrightarrow[\mathcal{Q}]{*} G$ and $G \xrightarrow[\mathcal{Q}]{} G'$ for no G' in **ANN**$(\mathbb{T}, \mathcal{Q})$. With this notation, we can state :

Proposition 3 : *For every* T *in* $\mathbb{T}$:

(1) every sequence of $\xrightarrow[\mathcal{Q}]{}$*-rewritings starting at* $(T, f_0, \emptyset)$ *reaches some normal form* (T,f,E) *within* n *steps, where* n *is the number of edges of* T;

(2) if (T,f,E) *is such a normal form, then* (T,E) *is a directed tree, and* T *belongs to* **L**$(\mathcal{Q})$ *iff* $f(x) \in Q_F$, *where* x *is the root of* T.

Proof :   We let **P**(G) be the conjunction of the following two properties of $G = (T, f, E) \in$ **ANN**$(\mathbb{T}, \mathcal{Q})$:

P1(G) : The directed graph obtained from G by the deletion of all the undirected edges (while keeping all vertices) is the disjoint union of a set **F**(G) of directed trees, and every node of T that is an end of some undirected edge of G is the root of some tree in **F**(G).

P2(G) : For each tree U in **F**(G), we have $\mathbf{h}_{\mathcal{Q}}(U) = f(u)$ where u is the root of U.

Claim : *For every rewriting sequence of length* m : $(T, f_0, \emptyset) \xrightarrow[\mathcal{Q}]{m} (T, f, E)$, *the cardinality of* E *is* m, **P**(T,f,E) *holds and* (T,f,E) *is a normal form iff all edges of* (T,f,E) *are directed.*

Proof : By induction on m.

Case m=0. **P**$(T, f_0, \emptyset)$ clearly holds. (All trees in **F**$(T, f_0, \emptyset)$ are equal to $\mathbb{1}$.) $(T, f_0, \emptyset)$ is a normal form iff T has no edge, i.e., is reduced to a single vertex, hence, iff all its edges are directed.

Case m>0. Let $(T,f_0,\emptyset) \xrightarrow{\underset{Q}{m-1}} (T,g,E') \xrightarrow{\underset{Q}{}} (T,f,E)$. By the induction hypothesis, we can assume that $\mathbf{P}(T,g,E')$ holds and that m-1 = $\mathbf{card}(E')$.

We first check $\mathbf{P1}(G)$ by assuming that $\mathbf{P1}(G')$ holds. Let x,y be such that $G' = (T,g,E') \xrightarrow{\overline{(x,y)}} (T,f,E) = G$. The set $\mathbf{F}(G')$ contains two directed trees U and U' with respective roots x and y, and $\mathbf{F}(G) = \mathbf{F}(G')\cup\{U''\}-\{U,U'\}$ where U" = $\mathbf{ext}(U)//U'$. Let u be a node of T belonging to an undirected edge of G. It cannot be equal to x, because otherwise the $\xrightarrow{\overline{(x,y)}}$-rewriting step is not applicable. If u = y then u is the root of U" belonging to $\mathbf{F}(G)$ and otherwise, it is the root of a directed tree of $\mathbf{F}(G')$ that is also in $\mathbf{F}(G)$.

We now check $\mathbf{P2}(G)$. We only have to consider U". Its root is y and we have :

$$
\begin{aligned}
f(y) &= \mathbf{ext}_Q(g(x)) \; //_Q \; g(y) \\
&= \mathbf{ext}_Q(\mathbf{h}_Q(U)) \; //_Q \; \mathbf{h}_Q(U') \\
&\qquad\qquad \text{(by the induction hypothesis)} \\
&= \mathbf{h}_Q(\mathbf{ext}(U) \; // \; U') \\
&\qquad\qquad \text{(by the definition of } \mathbf{h}_Q) \\
&= \mathbf{h}_Q(U").
\end{aligned}
$$

The other trees in $\mathbf{F}(G)$ being in $\mathbf{F}(G')$, they satisfy the desired condition since $\mathbf{P2}(G')$ holds. Hence, $\mathbf{P2}(G)$ holds.

It is clear that $\mathbf{card}(E) = \mathbf{card}(E') + 1 = m$, and that (T,f,E) is in normal form iff there is no undirected edge in (T,f,E). This is so because T is a tree: if it has undirected edges, then there is a vertex incident with a single undirected edge, and (T,f,E) is not in normal form. This concludes the proof of the claim.□

We now continue the proof of the proposition.

Every rewriting sequence $(T,f_0,\emptyset) \xrightarrow{m} G$ has a number of steps bounded by n, the number of edges of T. If m=n, then G=(T,f,E) is a normal form, (T,E) is a directed tree, namely the unique element of $\mathbf{F}(G)$;

its root is the unique vertex x with no outgoing edge, $f(x) = \mathbf{h}_{\mathcal{Q}}(T,E)$ by the claim, and $T \in \mathbf{L}(\mathcal{Q})$ iff $f(x) \in Q_F$.

If $m < n$, then G is not a normal form, and by the claim, $G \xrightarrow[\mathcal{Q}]{} G'$ for some G'. One can continue to rewrite until one reaches a normal form. The number of necessary steps is as required, as shown by the beginning of the proof. $\square$

Let us make two observations. Two different computations starting from $(T,f_0,\emptyset)$ may reach different normal forms, say (T,f,E) and (T,g,E'). However, the resulting answers as to whether T belongs to $\mathbf{L}(\mathcal{Q})$ are the same: if R and R' denote the rooted trees (T,E) and (T,E'), with roots r and r', we have $f(r) = \mathbf{h}_{\mathcal{Q}}(R)$, $g(r') = \mathbf{h}_{\mathcal{Q}}(R')$ and $f(r) \in Q_F$ iff $g(r') \in Q_F$ by the proof of Proposition 2.

Here is the second observation. The relabelling relation $\xrightarrow[\mathcal{Q}]{}$ has the following confluence property :

$$\text{if } G \xrightarrow[(x,y)]{} G_1 \text{ and } G \xrightarrow[(u,v)]{} G_2 \text{ with } \{x,y\} \cap \{u,v\} = \emptyset$$
$$\text{then we have } G_1 \xrightarrow[(u,v)]{} G_3 \text{ and } G_2 \xrightarrow[(x,y)]{} G_3 \text{ for some } G_3.$$

This means that in an actual implementation, the rewritings of G by $\xrightarrow[(x,y)]{}$ and $\xrightarrow[(u,v)]{}$ can be executed *simultaneously*, yielding G_3.

4 - CONCLUDING REMARKS

It has been proved by Büchi [3] that a set of words is regular iff it is definable in monadic second-order logic. (Each word can be represented by a logical structure; a language L is definable in a logical language $\mathfrak{L}$ if there exists a closed formula φ in $\mathfrak{L}$ such that L is the set of all words where φ holds true, each word being considered as a logical structure of the appropriate type.)

The equivalence stated in Büchi's theorem can be proved as follows: One constructs a monadic second-order formula describing the behaviour of a given finite-state automaton. For the other direction, one uses the fact that the family of regular languages is closed under Boolean operations, and homomorphisms. This technique has been extended by Doner to recognizable sets of terms [8]. In both cases, the rôle of finite-state automata is important to establish that recognizability implies definability.

What about recognizable sets of (unrooted, unordered) trees? A similar case, namely that of rooted, unordered trees has been considered in Courcelle [6]. It has been proved there that monadic second-order logic is insufficient to characterize the recognizable sets of rooted, unordered trees. However, an extension of this language, called *counting* monadic second-order logic, where special atomic formulas are introduced in order to test whether the cardinality of a set is a multiple of some fixed integer, is introduced. It is proved, essentially by the technique of Büchi and Doner, that a set of rooted, unordered trees is recognizable iff it is definable in counting monadic second-order logic. This proof extends easily to the unrooted trees considered in this paper.

We now say a few words of graphs. Every set of finite graphs definable in counting monadic second-order logic is recognizable, but not conversely. (See Courcelle [5-7].) We pose the following two questions :

(4.1) Question : *Is it true that every recognizable set of graphs of bounded tree-width is definable in counting monadic second-order logic?*

We conjecture that the answer is positive. The case of sets of graphs of tree-width at most 2 is established in Courcelle [7].(The tree-width of a graph is an integer measuring how different it is from a tree; trees are of tree-width 1 and series-parallel graphs are of tree-width 2; see [5-7].) We are much less confident concerning the following question :

(4.2) Question : *Is it true that every recognizable set of graphs of bounded tree-width is recognized by a finite-state graph automaton ?*

Note that we have not defined any general notion of finite-state graph automaton. Hence, answering this question necessitates defining such a notion. From the cases of words and trees, one might try first to answer Question (4.2). The answer to Question (4.1) might follow easily if the behaviour of the introduced automata were expressible in counting monadic second-order logic. Unfortunately, Question (4.2) seems even more difficult than Question (4.1): The proof given in [7] for graphs of tree-width at most 2 does not use graph automata, and does not suggest any definition.

Acknowledgements: I thank A. Arnold, Y. Métivier, and A. Podelski for many helpful comments on a first draft of this paper. I thank K. Callaway for helping me with the English style.

References

[1]   BAUDERON M., COURCELLE B., Graph expressions and graph rewritings, Math. Systems Theory 20 (1987) 83-127.

[2]   BILLAUD M., LAFON P., METIVIER Y., SOPENA E., Graph rewriting systems with priorities, L.N.C.S. 411 (1989) 94-106.

[3]   BÜCHI J., Weak second order logic and finite automata, Z. Math. Logik Grundlag. Math. 5 (1960), 66-92 (reprinted in Büchi's collected works, Springer, 1990).

[4]   COURCELLE B.,   On recognizable sets and tree automata, in "Resolution of Equations in Algebraic Structures, Volume 1", H. Aït-Kaci and M. Nivat, eds, Academic Press, New-York, 1989, 93-126.

[5]   COURCELLE B., Graph rewriting : an algebraic and logic approach, in "Handbook of Theoretical Computer Science, Volume B", J. van Leeuwen ed., Elsevier, Amsterdam, 1990, 193-242.

[6] COURCELLE B., The monadic second-order logic of graphs I: Recognizable sets of finite graphs, Information and Computation 85 (1990) 12-75.

[7] COURCELLE B., The monadic second-order logic of graphs V: On closing the gap between definability and recognizability, Theoret. Comput. Sci., in press, 1989.

[8] DONER J., Tree acceptors and some of their applications, J. Comput. System Sci. 4, (1970) 406-451.

[9] GECSEG F., STEINBY M., Tree Automata, Akad. Kiado, Budapest, 1984.

[10] LITOVSKY I., METIVIER Y., Computing trees with graph rewriting systems with priorities, this volume.

Tree Automata and Languages
M. Nivat and A. Podelski (editors)
© 1992 Elsevier Science Publishers B.V. All rights reserved.

Fixed point characterization of weak monadic logic definable sets of trees

André Arnold[a] and Damian Niwiński[b]

[a]Université Bordeaux I, LaBRI

[b]Warsaw University, Institute of Mathematics

Abstract

In this paper, we show that the sets of trees which can be defined by formulas of the weak monadic second-order logic can also be characterized in the fixed point calculus.

INTRODUCTION

The importance of the monadic second order theory of the full binary tree (S2S) for Computer Science follows from two reasons. First, this theory is decidable as shown by Rabin ([12], see also [14]). Second, one can express in it many interesting properties of sequential and parallel programs. As a matter of fact, all the logics of programs that are known to be decidable, such as propositional dynamic logic and its extensions, propositional algorithmic logic, propositional μ-calculus, and several variants of temporal logic (cf. [5]), turn out to be reducible to S2S. A subsystem of this theory resulting from the restriction of the range of quantifiers to finite sets, called the weak monadic second order theory, WS2S, is also of interest. While this restriction is essentially weaker ([13]), several logics of programs can be reduced to WS2S, as, for instance, PDL.

Following the ideas of Büchi, Rabin proved his difficult results via a correspondence between the formulas of S2S and a kind of automata on infinite trees, presently called *Rabin automata*. A restricted variant of these automata, known as *Büchi automata*, correspond to existential formulas of S2S ([13]). An automata-theoretic characterization of WS2S was provided much later by Muller, Saoudi, and Schupp [6], in terms of alternating automata.

An alternative characterization of different levels of complexity of properties of infinite trees comes from the μ-calculus which can be viewed as a generalization of inductive definitions. Fixed points are known to fit very well with the semantics of recursive programs but they also turn out to be useful in the context of computational properties of infinite trees. Niwiński [9] characterizes Rabin automata, and, hence, S2S, in terms of fixed points equations in the powerset algebra of labelled infinite trees, where the basic operations are the binary set-union and, for each label f, the operation

$$\langle L_1, \ldots, L_k \rangle \mapsto \{ f(t_1, \ldots, t_k) \mid t_i \in L_i \},$$

where the t_i's range over trees. What is essential is the use of both least and greatest fixed point operators (respectively denoted by μ and ν) and the hierarchy resulting from alternating μ and ν turns out to be infinite [8]. The Büchi automata correspond to the $\nu\mu$-level of this hierarchy, as proved by Niwiński [8] and also, independently and in a different setting, by Takahashi [15]. Arnold and Niwiński [2] have proved that this correspondence continues to hold if we extend operations of the powerset algebra of trees by the binary set-intersection.

It is actually in this extension of the powerset algebra of trees that we now give a characterization of the weak monadic second order logic. The formulas of WS2S turn out to correspond exactly to the fixed point definitions where both least and gratest fixed point operators may occur but where no essential alternation between μ and ν is allowed. The examples of Niwiński [8] show that the use of intersection is essential for this result.

The paper is organized as follows. Sections 1, 2, 3, and 4 give the preliminaries concerning trees and automata, general fixed points and the μ-calculus considered as a logical language. Section 5 presents a useful property that allows us to characterize a set of trees definable by a fixed point term by the structure of its elements; this fact was previously established by Arnold [1]. Section 6 shows how the (restricted) fixed point definitions can be expressed by formulas of (weak) monadic second order logic. Section 7 shows how, conversely, the sentences of WS2S can be characterized by alternation-free fixed point expressions. In the last Section we place this result among other results establishing relationship between various ways of defining sets of trees.

Notations

The set of natural numbers $0, 1, \ldots$, or, equivalently, the first infinite ordinal, is denoted by ω. For a set X, $\wp(X)$ is the *powerset* of X, i.e. the set of all subsets of X; X^n is the cartesian product of n copies of X; $\mathrm{Card}(X)$ is the *cardinality* of X. The symbol $\leq$ is used with several meanings: the standard ordering of natural numbers (or ordinals), the initial segment orderings of words, as well as an abstract partial ordering in an abstract poset. Once the meaning of $\leq$ is understood from the context, the symbols $\geq$, $<$, have their usual sense.

Throughout the paper, we often use vector notation: a tuple $\langle a_1, \ldots, a_n \rangle$ is abbreviated $\vec{a}$. It is also applied in more complex terms, e.g., in writing $\vec{t}[\vec{s}/\vec{x}]$ instead of $\langle t_1[s_1/x_1, \ldots, s_m/x_m], \ldots, t_n[s_1/x_1, \ldots, s_m/x_m] \rangle$, or $\vec{a} \leq \vec{b}$ to mean "$a_1 \leq b_1$ and $\ldots$ and $a_n \leq b_n$", or $\vec{X} \subseteq \vec{Y}$ to mean "$X_1 \subseteq Y_1$ and $\ldots$ and $X_n \subseteq Y_n$".

1   TREES AND AUTOMATA

For a set X, X^* denotes the free monoid generated by X, i.e., the set of all finite words that can be written with X as an alphabet, including the empty word λ. A word $w \in X^*$ can be uniquely decomposed in $w = x_1 \cdots x_k$, where $x_1, \ldots, x_k \in X$; the number k is called the *length* of w, and denoted by $|w|$ (in particular $|\lambda| = 0$). The *concatenation* of two words $w = x_1 \cdots x_k$ and $v = y_1 \cdots y_m$ is the word $x_1 \cdots x_k y_1 \cdots y_m$ denoted by wv. The similar notation will be applied for sets of words: if $L, K \subseteq X^*$, $LK = \{uv \mid u \in L, v \in K\}$.

Note that $\lambda w = w\lambda = w$ for any $w \in X^*$.

The *initial segment* relation $\leq$ is defined by $u \leq w$ if and only if there exists v such that $uv = w$. Note that λ is the least element (with respect to $\leq$) of X^*.

Any nonempty subset T of X^* closed under initial segment is called a *tree*. The elements of T are usually called *nodes*. The word λ, which belongs to every tree, is called the *root*. The $\leq$-maximal elements are called *leaves*. If $u \in T$, $x \in X$, and $ux \in T$ then ux is an *immediate successor* of u in T. An infinite sequence $w_0, w_1, \ldots, w_m, \ldots$ such that $w_0 = \lambda$ and, for each m, w_{m+1} is an immediate successor of w_m is called a *path* in T.

If S is an arbitrary set and T is a tree, then a mapping $t : T \to S$ is called an *S-valued tree* or shortly an *S-tree*; in this context T is called the *domain* of t and denoted by $T = \mathrm{dom}(t)$. We say "root of t", "path in t", etc, referring to the corresponding objects in $\mathrm{dom}(t)$.

For an S-tree $t : \mathrm{dom}(t) \to S$, and a node $v \in \mathrm{dom}(t)$, the *subtree* of t induced by v is the S-tree denoted by $t \cdot v$ and defined by

- $\mathrm{dom}(t \cdot v) = \{w \mid vw \in \mathrm{dom}(t)\}$,

- $t \cdot v(w) = t(vw)$, for $w \in \mathrm{dom}(t \cdot v)$.

Now suppose $A \subseteq \mathrm{dom}(t)$ is an *antichain* with respect to $\leq$ (i.e., any two elements of A are incomparable), and let f be a function which associates an S-tree $f(w)$ with each $w \in A$. Then the *substitution* $t[f]$ is the S-tree defined by

- $\mathrm{dom}(t[f]) = \{w \in \mathrm{dom}(t) \mid \forall w' \leq w, w' \notin A\} \cup \bigcup_{w \in A} w\, \mathrm{dom}(f(w))$,

- $t[f](u) = \begin{cases} f(w)(v) & \text{if } u = wv, w \in A \\ t(u) & \text{otherwise.} \end{cases}$

In the case A is finite, say $A = \{w_1, \ldots, w_k\}$, we shall often express f explicitly, writing for example $t[w_1 \leftarrow t_1, \ldots, w_k \leftarrow t_k]$.

We also introduce a concept of *limit*. Suppose $t_0, t_1, \ldots$ is a sequence of S-trees such that $\mathrm{dom}(t_0) \subseteq \mathrm{dom}(t_1) \subseteq \cdots$, and, for each $w \in \bigcup_{n < \omega} \mathrm{dom}(t_n)$, there is some $m(w)$ such that $\forall m \geq m(w)$, $t_m(w) = t_{m(w)}(w)$. Then we define the S-tree t by

- $\mathrm{dom}(t) = \bigcup_{n < \omega} \mathrm{dom}(t_n)$,

- $t(w) = t_{m(w)}(w)$, for $w \in \mathrm{dom}(t)$.

By assumption, t is well-defined. We call it the limit of the sequence t_n and denote it by $\lim t_n$.

Now let Σ be a finite signature, i.e., a finite set the elements of which are considered as function symbols, each $f \in \Sigma$ given with a finite arity $\rho(f) \geq 0$. A *syntactic tree* over Σ is a Σ-tree $t : \mathrm{dom}(t) \to \Sigma$, where $\mathrm{dom}(t)$ is included in ω^* (recall that ω is the set of natural numbers) and the following condition is satisfied: if $t(w)$ is a symbol of arity k, then w has exactly k immediate successors in $\mathrm{dom}(t)$ which are $w1, \ldots, wk$. Note that a node of a syntactic tree is a leaf if and only if it is labelled by a constant (i.e., 0-ary) symbol. The collection of all syntactic trees over Σ is denoted by T_Σ. It is obvious that finite syntactic trees can be identified with closed terms, therefore, syntactic trees may be viewed as the infinitary extension of terms.

1.1 Algebra of Σ-trees

An algebra $\mathsf{A} = \langle A, \{f^{\mathsf{A}} \mid f \in \Sigma\} \rangle$ over a signature Σ consists of a set A which is the *universe* of the algebra, and for each $f \in \Sigma$, an *interpretation* of f in A, which is a mapping $f^{\mathsf{A}} : A^{\rho(f)} \to A$.

The set of syntactic trees over a signature Σ is made into an algebra $\mathcal{T}_\Sigma = \langle T_\Sigma, \{f^{\mathcal{T}_\Sigma} \mid f \in \Sigma\} \rangle$, defining $f^{\mathcal{T}_\Sigma}(t_1, \ldots, t_{\rho(f)})$, where $t_i \in T_\Sigma$, to be the unique Σ-tree t satisfying the following conditions:

- $\mathrm{dom}(t) = \{\lambda\} \cup 1\mathrm{dom}(t_1) \cup \cdots \cup \rho(f)\mathrm{dom}(t_{\rho(f)})$,

- $t(\lambda) = f$,

- $t(iw) = t_i(w)$, for $i \in \{1, \ldots, \rho(f)\}$ and $w \in \mathrm{dom}(t_i)$.

Clearly, the algebra $\mathcal{T}_\Sigma$ is an extension of the free algebra over Σ generated by the empty set (algebra of closed terms).

1.2 Büchi automata on trees

Here we slightly modify the classical definition of a Büchi automaton [13] to take into account the fact that a tree may have finite branches.

A Büchi automaton on Σ-trees is a tuple

$$\mathcal{A} = \langle Q, q_0, Tr, F, G \rangle,$$

where Q is a finite set of *states*, $q_0 \in Q$ is an *initial* state, $F \subseteq Q$ is a set of *repeated* states, $G \subseteq Q$ is a set of *final* states, and Tr is a set of *transitions*. A transition is a tuple $\langle y, f, x_1, \ldots, x_k \rangle$ where $f \in \Sigma$, $k = \rho(f)$, and $y, x_1, \ldots, x_k \in Q$. Such a transition can also be represented by the equality $y = f(x_1, \ldots, x_k)$.

A *run* of the automaton $\mathcal{A}$ on a tree $t \in T_\Sigma$ is a Q-tree $r : \mathrm{dom}(t) \to Q$ such that for each node $w \in \mathrm{dom}(t)$, $\langle r(w), f, r(w1), \ldots, r(wk) \rangle \in Tr$, where $f = t(w)$ and $k = \rho(f)$. A q-run is a run r such that $r(\lambda) = q$. A run r is *accepting* if, for each leaf w the state $r(w)$ is final and, on each path $P = v_0, v_1, \ldots$ of t, the automaton enters a repeated state infinitely often, i.e., the set $\{v_i \mid r(v_i) \in F\}$ is infinite. A tree is accepted by an automaton starting in state q if there exists an accepting q-run on it. The set of all Σ-trees accepted by an automaton $\mathcal{A}$ starting in state q is denoted by $L(\mathcal{A}, q)$. The set of all Σ-trees accepted by the automaton is $L(\mathcal{A}) = L(\mathcal{A}, q_0)$.

2   FIXED POINTS

In this section we discuss some basic properties of least and greatest fixed points operators in complete lattices.

Suppose $\langle L, \leq \rangle$ is a *complete lattice*, i.e., L is partially ordered by $\leq$ and each subset A of L has a least upper bound in L, denoted by $\bigcup A$. Notice that, consequently, each $A \subseteq L$ has also a greatest lower bound $\bigcap A$, which is the least upper bound of the set of lower bounds of A. In particular L itself has extremal elements $\bigcup L$ and $\bigcap L$, also denoted by $\top$ and $\bot$. A function $f : L \to L$ is said to be *monotonic* if $\forall a, b \in L$, $a \leq b \Rightarrow f(a) \leq f(b)$. The following basic observation is known as Knaster–Tarski theorem [4, 16]:

Theorem 2.1 *Any monotonic mapping $f : L \to L$ has a least fixed point*

$$\mu x.f(x) = \bigcap\{a \in L \mid f(a) \le a\}$$

and a greatest fixed point

$$\nu x.f(x) = \bigcup\{a \in L \mid a \le f(a)\}.$$

Proof   We shall show the second part of the result. Let r be $\bigcup\{a \in L \mid a \le f(a)\}$. If b is a fixed point of f then $b \le r$. It remains to verify that $r = f(r)$. Inequalities $a \le f(a)$ and $a \le r$ imply, by monotonicity of f, $a \le f(r)$. Thus $r \le f(r)$. Further, $f(r) \le f(f(r))$, and then $f(r) \le r$.   □

A useful representation of extremal fixed points is provided by the notion of transfinite iterations.

Let $f : L \to L$ be a monotonic function. Define inductively the transfinite sequences $f^\xi(\bot)$ and $f^\xi(\top)$, where ξ is an ordinal number, by

- $f^0(\bot) = \bot$,

- $f^{\xi+1}(\bot) = f(f^\xi(\bot))$,

- $f^\xi(\bot) = \bigcup\{f^\eta(\bot) \mid \eta < \xi\}$, if ξ is a limit ordinal,

- $f^0(\top) = \top$,

- $f^{\xi+1}(\top) = f(f^\xi(\top))$,

- $f^\xi(\top) = \bigcap\{f^\eta(\top) \mid \eta < \xi\}$, if ξ is a limit ordinal.

Theorem 2.2 *There exist ordinals α and α' such that*

$$\mu x.f(x) = f^\alpha(\bot)$$

and

$$\nu x.f(x) = f^{\alpha'}(\top).$$

Proof   Let us consider the case of the least fixed point, the other being symmetrical.

We first show that the sequence $f^\xi(\bot)$ is increasing, i.e., $f^\eta(\bot) \le f^\xi(\bot)$, whenever $\eta < \xi$. Let us prove this result by induction on ξ.

- If $\xi = 0$, there is nothing to prove.

- If ξ is a limit ordinal, we have $\eta < \xi \Rightarrow f^\eta(\bot) \le f^\xi(\bot)$ by definition of f^ξ.

- If $\xi = \chi + 1$, then $\eta < \xi \Rightarrow \eta \le \chi$ and we have three cases to consider.

    - If $\chi = 0$, $\eta \le \chi \Rightarrow \xi = 0$, and $\bot = f^\eta(\bot) \le f^\xi(\bot)$.

- If $\chi = \chi' + 1$, $f^\eta(\bot) \le f^\chi(\bot) = f(f^{\chi'}(\bot)) \le f(f^\chi(\bot)) = f^\xi(\bot)$ (by induction hypothesis and by monotonicity of f).
- Finally, if χ is a limit ordinal,
    * if $\eta < \chi$, then $\eta < \eta + 1 < \chi$, and $f^\eta(\bot) \le f^{\eta+1}(\bot), f^\eta(\bot) \le f^\chi(\bot)$, hence, $f^\eta(\bot) \le f^{\chi+1}(\bot)$,
    * if $\eta = \chi$, $\zeta < \chi \Rightarrow \zeta + 1 < \chi$, and $f^\zeta(\bot) \le f^\chi(\bot)$, hence, $f^\zeta(\bot) \le f^{\zeta+1}(\bot) \le f^{\chi+1}(\bot)$, and since this is true for all $\zeta < \chi$, we have $f^\chi(\bot) \le f^{\chi+1}(\bot)$, by definition of $f^\chi(\bot)$.

The transfinite sequence $f^\eta(\bot)$ must therefore stabilize, i.e., for some ordinal η, $f^{\eta+1}(\bot) = f^\eta(\bot)$ (it is enough to choose η greater than the cardinality of L but a less ordinal could suffice). So $f^\eta(\bot)$ is a fixed point of f. If a is any fixed point of f, an easy inductive argument shows that $f^\eta(\bot) \le a$ for any η. Hence, $f^\eta(\bot)$ is the least fixed point of f. $\qquad\qquad\square$

The reader may have noticed the apparent symmetry between the properties of the least and the greatest fixed points. An obvious justification of this comes from the fact that in the dual lattice $\langle L, \le^* \rangle$ of $\langle L, \le \rangle$, where $\le^* = \ge$, the monotonicity of mappings is preserved but the least fixed point of f in $\langle L, \le \rangle$ is the greatest fixed point in $\langle L, \le^* \rangle$, and vice versa. However, in a concrete model such as the powerset algebra of trees we shall consider in the sequel, the properties of μ and ν can be quite different.

If $\langle K, \le_K \rangle$ and $\langle M, \le_M \rangle$ are complete lattices (we shall usually omit the subscript of the symbol $\le$), a mapping $h : L \times K \to M$ is said to be *monotonic* if it is monotonic with respect to the product ordering on $L \times K$. In particular a mapping $g : L^n \to L$ is monotonic if, for any $a_1, \ldots, a_n, b_1, \ldots, b_n$, $\vec{a} \le \vec{b}$ implies $g(\vec{a}) \le g(\vec{b})$.

Suppose h is a monotonic mapping from $L \times K$ to L and we fix a value of K, say a, as a second argument of h. Then we obtain a monotonic mapping $h(x, a) : L \to L$. We can further consider the mapping from K to L which sends a on the least fixed point of $h(x, a)$; we shall denote it by $\mu x.h(x, y)$; the mapping $\nu x.h(x, y)$ is defined similarly. It is easy to see that these two mappings are still monotonic.

The monotonic mappings $f : L^n \to L$ will be called *operations* over L. In this context μ and ν can be viewed as operators that produce new operations such as $\mu x_1.f(x_1, \ldots x_n)$, $\nu x_3.\mu x_2.\nu x_1.f(x_1, \ldots, x_n)$.

The proofs of the following properties can be found in [7].

Proposition 2.3 *If $g(x, y)$ is an operation then*

$$\mu x.\mu y.g(x, y) = \mu x.g(x, x)$$

and

$$\mu x.\nu y.g(x, y) \le \nu y.\mu x.g(x, y).$$

2.1   The closures

We shall now introduce one of the central concepts of this paper.

Let $\mathcal{F}$ be a family of operations over a complete lattice $\langle L, \leq \rangle$. The μ-closure of $\mathcal{F}$ is the set of operations, denoted by $\mu(\mathcal{F})$, which is the closure the union of $\mathcal{F}$ and of the set of *projections* under composition and μ-operators. More formally $\mu(\mathcal{F})$ is the least set $\mathcal{G}$ of operations such that

1. $\mathcal{F} \subseteq \mathcal{G}$,

2. for each $n \geq 0$, for each $i \leq n$, the mapping $\pi_n^i : \langle a_1, \ldots, a_n \rangle \mapsto a_i$ is in $\mathcal{G}$,

3. if $f(x_1, \ldots, x_n), g_1(\vec{y}), \ldots, g_n(\vec{y})$ are in $\mathcal{G}$, then $f(g_1(\vec{y}), \ldots, g_n(\vec{y}))$ is in $\mathcal{G}$,

4. if $f(x_1, \ldots, x_n)$ is in $\mathcal{G}$, then, for $i = 1, \ldots, n$, $\mu x_i . f(x_1, \ldots, x_n)$ is in $\mathcal{G}$.

The ν-*closure* of $\mathcal{F}$, $\nu(\mathcal{F})$, is defined analogously, replacing μ by ν. Finally, the *compositional closure* of $\mathcal{F}$, $\mathrm{Comp}(\mathcal{F})$, is defined by the conditions 1,2,and 3 only. It follows from the definitions :

Proposition 2.4

$$\mu(\mathcal{F}) = \mu(\mu(\mathcal{F})) = \mu(\mathrm{Comp}(\mathcal{F})) = \mathrm{Comp}(\mu(\mathcal{F})),$$

$$\nu(\mathcal{F}) = \nu(\nu(\mathcal{F})) = \nu(\mathrm{Comp}(\mathcal{F})) = \mathrm{Comp}(\nu(\mathcal{F})).$$

2.2   Vectorial fixed points

It is sometimes useful to consider fixed points of systems of equations rather than of single equations. We shall see how this case may be reduced to the process of solving, possibly several times, one-dimensional fixed point equations.

Let $h_1(y_1, \ldots y_k, z_1, \ldots, z_m), \ldots, h_k(y_1, \ldots y_k, z_1, \ldots, z_m)$ be monotonic operations on L. The product mapping $\vec{h}(\vec{y}, \vec{z}) : L^{k+m} \to L^k$ is monotonic with respect to the product ordering. We call it a *vectorial operation* on L. According to our convention, $\mu \vec{y} . \vec{h}(\vec{y}, \vec{z})$ denotes the mapping from L^m to L^k that sends a vector $\vec{a}$ to the least solution of the system of equations

$$
\begin{aligned}
y_1 &= h_1(\vec{y}, \vec{a}), \\
&\vdots \\
y_k &= h_k(\vec{y}, \vec{a}).
\end{aligned}
$$

We shall show that the i-th component of that mapping, namely $\pi_m^i(\mu \vec{y} . \vec{h}(\vec{y}, \vec{z}))$, may be obtained from the operations $h_1, \ldots, h_k$, using composition and the μ-operator.

Lemma 2.5 *Let $\langle L, \leq_L \rangle$ and $\langle K, \leq_K \rangle$ be two complete lattices, and let $F_1 : L \times K \to L$, $F_2 : L \times K \to K$, and $\vec{F} = \langle F_1, F_2 \rangle : L \times K \to L \times K$ be monotonic mappings. The two components of the least fixed point $\mu xy . \vec{F}(x, y)$ of $\vec{F}$ satisfy the following equalities:*

$$
\begin{aligned}
\pi_2^1(\mu xy . \vec{F}(x, y)) &= \mu x . F_1(x, \mu y . F_2(x, y)) \\
\pi_2^2(\mu xy . \vec{F}(x, y)) &= \mu y . F_2(x, \mu x . F_1(x, y)).
\end{aligned}
$$

Proof Let us denote by a and a' the left-hand side and right-hand side of the first equation, and b and b' those of the second equation.

We have $a = F_1(a, b)$ and $b = F_2(a, b)$, since $\langle a, b \rangle$ is a fixed point of $\vec{F}$. Hence, $\mu x.F_1(x, b) \leq a$ and $\mu y.F_2(a, y) \leq b$. By monotonicity of F_1 and F_2, $F_2(\mu x.F_1(x, b), b) \leq F_2(a, b) = b$ and $F_1(a, \mu y.F_2(a, y)) \leq F_1(a, b) = a$. It follows that $b' \leq b$ and $a' \leq a$.

On the other hand, $a' = F_1(a', \mu y.F_2(a', y))$ and $b' = F_2(\mu x.F_1(x, b'), b')$. Let $b'' = \mu y.F_2(a', y))$ and $a'' = \mu x.F_1(x, b')$. We have $a' = F_1(a', b'')$, $b' = F_2(a'', b')$, $b'' = F_2(a', b'')$, $a'' = F_1(a'', b')$. It follows that $\langle a', b'' \rangle$ and $\langle a'', b' \rangle$ are fixed points of $\vec{F}$, hence, $a \leq a'$ and $b \leq b'$.   □

Clearly, a similar result can be proved for greatest fixed points.

This result can be extended in the following way:

Proposition 2.6 Let $h_i : L^{k+m} \to L$, for $1 \leq i \leq k$, be monotonic mappings. Each component of $\mu \vec{x}.\vec{h}(\vec{x}, \vec{y})$ belongs to $\mu(\{h_i \mid 1 \leq i \leq k\})$.

Proof Let us prove this result by induction on k. If $k = 1$, there is nothing to prove. Otherwise, let us write $F_1 = h_1$, $F_2 = \langle h_2, \ldots, h_k \rangle$, and $\vec{z} = \langle x_2, \ldots, x_k \rangle$ and let us apply Lemma 2.5 and the induction hypothesis. The first component of $\mu \vec{x}.\vec{h}(\vec{x}, \vec{y})$ is equal to $\mu x_1.h_1(x_1, \mu \vec{z}.F_2(x_1, \vec{z}, \vec{y}))$; since each component of $\mu \vec{z}.F_2(x_1, \vec{z}, \vec{y})$ is in $\mu(\{h_2, \ldots, h_k\})$, this first component is in $\mu(\{h_1\} \cup \mu(\{h_2, \ldots, h_k\})) \subseteq \mu(\{h_1, \ldots, h_k\})$. The other components are components of $\mu \vec{z}.F_2(\mu x_1.h_1(x\vec{z}, \vec{y}), \vec{z}, \vec{y})$ which are elements of $\mu(\mathrm{Comp}(\mu(\{h_1\}) \cup \{h_2, \ldots, h_k\})) \subseteq \mu(\{h_1, \ldots, h_k\})$.   □

The converse of this proposition is also true and we get

Proposition 2.7 Let $\mathcal{F}$ be a class of operations over a complete lattice, and let f be a m-ary operation over this lattice. Then $f \in \mu(\mathcal{F})$ if and only if there exist a vectorial mapping $\vec{h} : L^{k+m} \to L^k$ in $\mathrm{Comp}(\mathcal{F})^k$ and some i, $1 \leq i \leq k$ such that $f(\vec{y}) = \pi_k^i(\mu \vec{x}.\vec{h}(\vec{x}, \vec{y}))$.

Proof The "if" part of the equivalence was proved in the previous proposition. Let us prove the "only if" part by induction on the construction of f.

- if $f(y_1, \ldots, y_m)$ is a projection π_m^i, we have just to consider the equation $x = \pi_{m+1}^{i+1}(x, y_1, \ldots, y_m)$;

- if $f(y_1, \ldots, y_n)$ is in $\mathcal{F}$, we consider the equation

$$x = f(\pi_{m+1}^2(x, y_1, \ldots, y_m), \ldots, \pi_{m+1}^{m+1}(x, y_1, \ldots, y_m))$$

  which is equivalent to

$$x = f(y_1, \ldots, y_m);$$

- if $f(y_1, \ldots, y_m) = g(h_1(y_1, \ldots, y_m), \ldots, h_n(y_1, \ldots, y_m))$ then by induction hypothesis, there are equations

  − $\vec{x} = \vec{G}(\vec{x}, \vec{z})$ such that g is the j_0-th component of its least fixed point,

- $\vec{xi} = \vec{H}_i(\vec{xi}, \vec{y})$ such that h_i is the j_i-th component of its least fixed point, for $1 \leq i \leq n$

with $G = \{G_1, G_2, \ldots\}$, $H_i = \{(H_i)_1, (H_i)_2, \ldots\}$ included in $\mathrm{Comp}(\mathcal{F})$; then f is the j_0-th component of the least fixed point of the system

$$\begin{aligned}
\vec{x} &= \vec{G}(\vec{x}, x1_{j_1}, \ldots, xn_{j_n}), \\
\vec{x1} &= \vec{H}_1(\vec{x1}, \vec{y}), \\
&\ \vdots \\
\vec{xn} &= \vec{H}_n(\vec{xn}, \vec{y}).
\end{aligned}$$

which can be rewritten as $\vec{X} = F(\vec{X}, \vec{y})$ with $F \in \mathrm{Comp}(\{G, H_1, \ldots H_n\}) \subseteq \mathrm{Comp}(\mathcal{F})$.

- if $f(y_1, \ldots, y_m) = \mu z.g(z, y_1, \ldots, y_m)$ such that $g(z, \vec{y})$ is the j-th component of the least fixed point of $\vec{x} = \vec{G}(\vec{x}, z, \vec{y})$. Then we consider the system

$$\begin{aligned}
\vec{x} &= \vec{G}(\vec{x}, z, \vec{y}) \\
z &= x_j
\end{aligned}$$

where $x_j = \pi_k^j(\vec{x}, z, \vec{y})$. By Lemma 2.5, the last component of its least fixed point is precisely

$$\mu z.\pi_k^j(\mu \vec{x}.\vec{G}(\vec{x}, z, \vec{y}), z, \vec{y})$$

which is equal to $\mu z.g(z, \vec{y})$.

$\square$

The fact stated in this proposition is sometimes called the *Bekič–Scott Principle* [10].

2.3   Duality

Now suppose that $\langle L, \leq \rangle$ is a complete *boolean lattice*, i.e., L is a distributive lattice and every element a has a complement $\neg a$ in L, which is a unique element such that $\bigcup\{a, \neg a\} = \top$ and $\bigcap\{a, \neg a\} = \bot$. Let f be an operation on L; its *dual*, $\tilde{f}$, is defined by $\tilde{f}(x) = \neg f(\neg x)$. It is easy to show that $x \leq y$ if and only if $\neg y \leq \neg x$, (in fact: $\neg x \vee x = \top$ and $x \leq y$ imply $\neg x \vee y = \top$ and then $\neg y = \neg y \wedge \top = \neg y \wedge (\neg x \vee y) = (\neg y \wedge \neg x) \vee (\neg y \wedge y) = (\neg y \wedge \neg x) \vee \bot = \neg y \wedge \neg x$, hence, $\neg y \leq \neg x$) and then, $\tilde{f}$ is an operation, since it is monotonic. It can also be easily shown that a is the least fixed point of $\tilde{f}$ if and only if $\neg a$ is the greatest fixed point of f. Hence

$$\begin{aligned}
\mu x.f(x) &= \neg \nu x.\tilde{f}(x), \\
\nu x.f(x) &= \neg \mu x.\tilde{f}(x).
\end{aligned}$$

More generally, if f is an n-ary operation on L, its dual is $\tilde{f}$ defined by $\tilde{f}(x_1, \ldots, x_n) = \neg f(\neg x_1, \ldots, \neg x_n)$. Note that $f = \tilde{g}$ if and only if $\tilde{f} = g$, and that if f is a constant, $\tilde{f} = \neg f$. For a class $\mathcal{F}$ of operations, we denote by $\tilde{\mathcal{F}}$ the class $\{\tilde{f} \mid f \in \mathcal{F}\}$.

Proposition 2.8 *Let $\mathcal{F}$ be a class of operations on a complete boolean lattice $\langle L, \leq \rangle$. Then*

$$\widetilde{\mu(\mathcal{F})} = \nu(\tilde{\mathcal{F}}).$$

If, moreover, $\tilde{\mathcal{F}} \subseteq \mathrm{Comp}(\mathcal{F})$, then

$$\widetilde{\mu(\mathcal{F})} = \nu(\mathcal{F}).$$

Proof For the first equality, it is enough to prove $\widetilde{\mu(\mathcal{F})} \subseteq \nu(\tilde{\mathcal{F}})$, the reverse inclusion will follow from a symmetric argument. The proof is by induction on the construction of $f \in \mu(\mathcal{F})$.

- if f is π_n^i, then $\tilde{f} = \pi_n^i \in \nu(\tilde{\mathcal{F}})$,

- if $f \in \mathcal{F}$, $\tilde{f} \in \tilde{\mathcal{F}} \subseteq \nu(\tilde{\mathcal{F}})$,

- if $f(x_1, \ldots, x_n) = g(h_1(x_1, \ldots, x_n), \ldots, h_k(x_1, \ldots, x_n))$, then

$$\tilde{f}(x_1, \ldots, x_n) = \neg g(h_1(\neg x_1, \ldots, \neg x_n), \ldots, h_k(\neg x_1, \ldots, \neg x_n)),$$

which is equal to $\tilde{g}(\tilde{h_1}(x_1, \ldots, x_n), \ldots, \tilde{h_k}(x_1, \ldots, x_n))$, and is in $\nu(\tilde{\mathcal{F}})$, since so are $\tilde{g}, \tilde{h}_1, \ldots, \tilde{h}_k$,

- if $f(\vec{y}) = \mu x.g(x, \vec{y})$ then $\tilde{f}(\vec{y}) = \neg \mu x.g(x, \neg \vec{y})$. Let $h_{\vec{z}}(x) = g(x, \vec{z})$, hence, $\tilde{h}_{\vec{z}}(x) = \neg g(\neg x, \vec{z}) = \tilde{g}(x, \neg \vec{z})$. We have

$$\neg \mu x.g(x, \neg \vec{y}) = \neg \mu x.h_{\neg \vec{y}}(x) = \nu x.\widetilde{h_{\neg \vec{y}}}(x) = \nu x.\tilde{g}(x, \vec{y}).$$

Since $\tilde{g}$ is in $\nu(\tilde{\mathcal{F}})$, so is $\tilde{f}$.

If $\tilde{\mathcal{F}} \subseteq \mathrm{Comp}(\mathcal{F})$, then $\mathcal{F} \subseteq \widetilde{\mathrm{Comp}(\mathcal{F})} = \mathrm{Comp}(\tilde{\mathcal{F}})$ and $\mathrm{Comp}(\mathcal{F}) \subseteq \mathrm{Comp}(\tilde{\mathcal{F}}) \subseteq \mathrm{Comp}(\mathcal{F})$, hence $\nu(\tilde{\mathcal{F}}) = \nu(\mathrm{Comp}(\tilde{\mathcal{F}})) = \nu(\mathrm{Comp}(\mathcal{F})) = \nu(\mathcal{F})$. $\square$

3 THE EXAMPLE OF REGULAR ω-LANGUAGES

As an illustration of the notions introduced in the previous section, we give some results about regular languages, which can be also considered as the instanciation to words of our main result, proved in Section 7.

Let A be a fixed finite alphabet, and let us denote by A^ω the set of infinite words over A, i.e., the mappings of the form $u : \omega \to A$. Infinite words over A can be viewed as instances of Σ-trees, where $\Sigma = A$ is considered as a set of unary function symbols, and ω is identified with 1^* ($= \mathrm{dom}(u)$, for each $u \in A^\omega$). The definition of Büchi automaton can be therefore applied to infinite words without changes. A set of infinite words is said to be a *regular ω-language* if and only if it is accepted by a Büchi automaton [3].

The set $\wp(A^\omega)$, ordered by inclusion, is a complete lattice. Let us consider the following operations over $\wp(A^\omega)$:

- $\cup$ and $\cap$, the set union and intersection,

- for every element a of A, the unary operation $a\cdot$ defined by $a \cdot (L) = \{au \mid u \in L\}$, where L is a subset of $\wp(A^\omega)$.

We denote by $\mathcal{F}$ the set of operations $\{a\cdot \mid a \in A\} \cup \{\cup\}$ and by $\mathcal{G}$ the set $\{\mathcal{F}\} \cup \{\cap\}$. Let us remark that the dual $\tilde{a}$ of $a\cdot$ is defined by

$$\tilde{a}(L) = a \cdot (L) \cup \bigcup_{b \neq a} b \cdot (A^\omega),$$

since $\tilde{a}(L) = A^\omega - a \cdot (A^\omega - L)$. It follows that

$$\tilde{\mathcal{F}} \subseteq \tilde{\mathcal{G}} \subseteq \mathrm{Comp}(\mathcal{G}).$$

Let us consider also, for any subset L of A^*, the operation $L\cdot$ defined by $L \cdot (L') = \{uv \mid u \in L, v \in L'\}$, where L' is a subset of $\wp(A^\omega)$.

The following fact can be easily derived from the definition of a regular language.

Proposition 3.1 *If L is a regular language, then $L\cdot$ is in $\mu(\mathcal{F})$.*

Let us define, for a subset L of A^+, the subset L^ω of A^ω by

$$L^\omega = \{u_1 u_2 \cdots u_n \cdots \mid u_i \in L\}.$$

It is also easy to show (see [11] for instance) the following result.

Proposition 3.2 *If $L \subseteq A^+$, then L^ω is the greatest fixed point of the equation*

$$X = L \cdot (X).$$

Proof   Readily, $L^\omega = L \cdot (L^\omega)$. Suppose $M \subseteq A^\omega$ is any set such that $M = L \cdot (M)$, and let $u \in M$. Then $u \in L \cdot (M)$ and there exist some u_1 in L ($u_1 \neq \lambda$), and v_1 in M such that $u = u_1 v_1$. For the same reason, v_1 can be decomposed in $u_2 v_2$ with $u_2 \in L$ and $v_2 \in M$, and so on. Thus, $u = u_1 u_2 \cdots u_n \cdots$ with $u_i \in L$. $\qquad\square$

It follows that any regular ω-language, which is a finite union of sets in the form KL^ω, or, with our notations, $K \cdot (L^\omega)$, with K and L regular subsets of A^+, is a 0-ary operation in $\nu(\mu(\mathcal{F}))$.

If we allow us to use intersection, i.e., $\mathcal{G}$ instead of $\mathcal{F}$, we can improve this result and prove the following.

Theorem 3.3 *Every regular ω-language is a 0-ary operation in $\mathrm{Comp}(\mu(\mathcal{G}) \cup \nu(\mathcal{G}))$.*

An example   Before proving this theorem, let us consider a simple example.

Let A be the alphabet $\{a, b\}$ and let $L = (a^*b)^\omega$.

Let $f(Y)$ be the least fixed point of the equation

$$X = a \cdot (X) \cup b \cdot (Y).$$

For every subset L of A^ω, $f(L)$ is $a^*b \cdot (L) = a^*bL$, and is an operation in $\mu(\mathcal{F}) \subseteq \mu(\mathcal{G})$. Let $g(Y)$ be the greatest fixed point of the equation

$$X = a \cdot (X \cap Y) \cup b \cdot (X \cap Y).$$

Obviously g is in $\nu(\mathcal{G})$, and, since A^ω is also in $\nu(\mathcal{G})$ (because it is the greatest fixed point of $X = X$), $g(f(A^\omega))$ is a 0-ary operation in $\mathrm{Comp}(\mu(\mathcal{G}) \cup \nu(\mathcal{G}))$.

It is not difficult to prove that $g(f(A^\omega))$ is exactly $(a^*b)^\omega$. It is easy to show that, if L is a subset of A^ω, then

$$g(L) = \{a_1 a_2 \cdots a_n \cdots \in A^\omega \mid \forall i > 1, a_i a_{i+1} \cdots \in L\}.$$

It follows that $g(f(A^\omega)) = g(a^*bA^\omega)$ is the set of all infinite words containing an infinite number of b's.

The proof   Obviously, the family of subsets of A^ω that are 0-ary operations in $\mathrm{Comp}(\mu(\mathcal{G}) \cup \nu(\mathcal{G}))$ is closed under union and intersection. It is also closed under complementation because the set of duals of operations in $\mathrm{Comp}(\mu(\mathcal{G}) \cup \nu(\mathcal{G}))$ is

$$\mathrm{Comp}(\widetilde{\mu(\mathcal{G})} \cup \widetilde{\nu(\mathcal{G})})$$

and since $\widetilde{\mathcal{G}} \subseteq \mathrm{Comp}(\mathcal{G})$, by Proposition 2.8, this set is equal to $\mathrm{Comp}(\mu(\mathcal{G}) \cup \nu(\mathcal{G}))$.

By McNaughton's theorem, every regular ω-language is a boolean combination of *deterministic* ones. Hence, it suffices to prove that any deterministic regular ω-language is a 0-ary operation in $\mathrm{Comp}(\mu(\mathcal{G}) \cup \nu(\mathcal{G}))$.

Let $L \subseteq A^\omega$ be a deterministic regular ω-language recognized by the deterministic Büchi automaton $\mathcal{A} = \langle Q, q_0, \delta, F \rangle$. Since this automaton is deterministic and can be completed, the relation δ can be cosidered as a mapping $\delta : Q \times A \to Q$ and this mapping is extended, as usual, into a mapping from $Q \times A^*$ into Q, denoted also by δ.

Let us denote by $L_{q,q'}$ the set of all nonempty finite words u such that $\delta(q, u) = q'$. The associated unary operations $L_{q,q'}\cdot$ are all in $\mu(\mathcal{F})$. To prove this fact we have just to consider the following system of equations:

$$X_{q,q'} = e_{q,q'}$$

for all q and q' in Q, where $e_{q,q'}$ is

$$\bigcup_{a \in A : \delta(q,a)=q'} a \cdot (X) \cup \bigcup_{a \in A, q'' \in Q : \delta(q,a)=q''} a \cdot (X_{q'',q'}).$$

The components of the least solution of this system are all in $\mu(\mathcal{F})$ (because of Proposition 2.7) and it is easy to see that the component corresponding to the variable $X_{q,q'}$ is $L_{q,q'}$.

Now let us consider $L_q = \bigcup_{q' \in F} L_{q,q'}$. The associated mapping $L_q \cdot$ is also in $\mu(\mathcal{F})$. Finally, let us consider the following set of equations

$$Z_q = \bigcup_{a \in A, q' \in Q : \delta(q,a) = q'} a \cdot (Z_{q'} \cap Y_{q'})$$

for all q in Q, and let $h_q(Y_{q_0}, Y_{q_1}, \ldots)$ be the components of its greatest vectorial fixed point. We claim that

$$L = h_{q_0}(L_{q_0} \cdot (A^\omega), L_{q_1} \cdot (A^\omega), \cdots)$$

hence, L is a 0-ary operation in $\mathrm{Comp}(\mu(\mathcal{G}) \cup \nu(\mathcal{G}))$.

To prove our claim, we consider an infinite word $u = a_1 a_2 \cdots$ and its unique computation $q_0' = q_0, q_1', q_2', \ldots$ such that $\delta(q_j', a_j) = q_{j+1}'$ This word is in $h_{q_0}(Y_{q_0}, Y_{q_1}, \ldots)$ if and only if for all $j > 0$, the word $a_j a_{j+1} \ldots$ is in $Y_{q_j'}$. Moreover, an infinite word u is in $L_{q,q'} \cdot (A^\omega)$ if and only if its unique computation starting in the state q contains the state q'. It follows that a word u is in $h_{q_0}(L_{q_0} \cdot (A^\omega), L_{q_1} \cdot (A^\omega), \cdots)$ if and only if its unique computation starting in q_0 contains an infinite number of states belonging to F. $\qquad\square$

4 FIXED POINT CALCULUS

In this section we introduce a formal language for fixed point definable operations. Indeed it will be a formalization of the notation that has been already used in Section 2.

4.1 Poweralgebras

Let Σ be a finite signature and let $\mathsf{A} = \langle A, \{f^\mathsf{A} \mid f \in \Sigma\}\rangle$ be a Σ-algebra. This Σ-algebra is a Σ-*poweralgebra* if A is a complete boolean lattice such that every interpretation f^A is monotonic, i.e., is an operation on this lattice.

As an example, with every Σ-algebra A, we can associate a Σ-poweralgebra $\wp\mathsf{A} = \langle \wp(A), \{f^{\wp\mathsf{A}} \mid f \in \Sigma\}\rangle$, where $\wp(A)$ is the boolean complete lattice of the subsets of A, and, for each $f \in \Sigma$, $f^{\wp\mathsf{A}}$ is the operation on the lattice $\wp(A)$ defined by

$$f^{\wp\mathsf{A}}(L_1, \ldots, L_{\rho(f)}) = \{f^\mathsf{A}(a_1, \ldots, a_{\rho(f)}) \mid a_i \in L_i, i = 1, \ldots, \rho(f)\}.$$

Such a poweralgebra is called a *powerset algebra*.

Our most important example of powerset algebra will be the powerset algebra of Σ-trees, $\wp\mathcal{T}_\Sigma$. We will also consider the following Σ-poweralgebras associated with the Σ-trees. For a tree t, we define the poweralgebra $\wp t = \langle \wp(\mathrm{dom}(t)), \{f^{\wp t} \mid f \in \Sigma\}\rangle$, where $\wp(\mathrm{dom}(t))$ is the lattice of subsets of $\mathrm{dom}(t)$, and $f^{\wp t}$ is defined by

$$f^{\wp t}(L_1, \ldots, L_{\rho(f)}) = \{v \in \mathrm{dom}(t) \mid t(v) = f, \forall i \in \{1, \ldots, \rho(f)\}, vi \in L_i\}.$$

An important relationship between these poweralgebras will be investigated in the next section.

4.2   The fixpoint-terms

The fixpoint-terms (called μ-terms in [8]) are the syntactic descriptions of operations in a poweralgebra, built from the interpretations of symbols, boolean operations, and least and greatest fixed points operators.

Let us consider a set $V = \{v_0, v_1, \dots\}$ of variables and let us extend the signature Σ by two (infix) binary symbols $\vee$ and $\wedge$. The set fpT_Σ of Σ-fixpoint-terms τ is defined inductively, together with the sets $FV(\tau)$ and $BV(\tau)$ of free and bound variables of τ by:

- $V \subseteq fpT_\Sigma$, $FV(v) = \{v\}$, $BV(v) = \emptyset$,

- if $f \in \Sigma$, and $\tau_1, \dots, \tau_{\rho(f)} \in fpT_\Sigma$, then $\tau = f(\tau_1, \dots, \tau_{\rho(f)}) \in fpT_\Sigma$, $FV(\tau) = \bigcup_{i=1}^{\rho(k)} FV(\tau_i)$, $BV(\tau) = \bigcup_{i=1}^{\rho(k)} BV(\tau_i)$,

- if τ_1 and τ_2 are in fpT_Σ, then $\tau_1 \vee \tau_2$ and $\tau_1 \wedge \tau_2$ are in μT_Σ, $FV(\tau_1 \vee \tau_2) = FV(\tau_1 \wedge \tau_2) = FV(\tau_1) \cup FV(\tau_2)$, $BV(\tau_1 \vee \tau_2) = BV(\tau_1 \wedge \tau_2) = BV(\tau_1) \cup FV(\tau_2)$,

- if $\tau \in fpT_\Sigma$, and if $v \in FV(\tau)$, then $\tau' = \mu v.\tau$ and $\tau'' = \nu v.\tau$ are in fpT_Σ, $FV(\tau') = FV(\tau'') = FV(\tau) - \{v\}$, $BV(\tau') = BV(\tau'') = BV(\tau) \cup \{v\}$.

The definition of the composition of terms needs the usual manipulations on variables (renaming, or α-conversion) in order to get the hypothesis given below. We do not present these manipulations here.

Let $\vec{y} = \langle y_1, \dots, y_n \rangle$ be a vector of variables and let $\tau_0, \tau_1, \dots, \tau_n$ be terms such that

- $FV(\tau_0) \supseteq \{y_1, \dots, y_n\}$,

- the sets $BV(\tau_i)$, $i = 0, \dots, n$ are pairwise disjoint,

- the sets $\bigcup_{i=0}^{n} FV(\tau_i)$ and $\bigcup_{i=0}^{n} BV(\tau_i)$ are disjoint.

Under these conditions the term $\tau = \tau_0[y_1 \leftarrow \tau_1, \dots, y_n \leftarrow \tau_n]$ can be defined by induction on the construction of τ_0 and $FV(\tau) = \bigcup_{i=1}^{n} FV(\tau_i) \cup (FV(\tau_0) - \{y_1, \dots, y_n\})$, $BV(\tau) = \bigcup_{i=0}^{n} BV(\tau_i)$.

4.3   Semantics

Let $\mathsf{A} = \langle A, \{f^\mathsf{A} \mid f \in \Sigma\} \rangle$ be a Σ-poweralgebra. With each Σ-fixpoint-term τ, and with each vector $\vec{x} = \langle x_{i_1}, \dots, x_{i_n} \rangle$ of distinct variables such that $\forall y \in FV(\tau), \exists j : y = x_{i_j}$, we associate an operation $\tau^\mathsf{A}[\vec{x}] : A^n \to A$ defined by induction on the construction of τ:

- if $\tau = x_{i_j}$, then $\tau^\mathsf{A}[\vec{x}] = \pi_n^j$,

- if $\tau = f(\tau_1, \dots, \tau_n)$, then $\tau^\mathsf{A}[\vec{x}] = f^\mathsf{A}(\tau_1^\mathsf{A}[\vec{x}] \dots, \tau_n^\mathsf{A}[\vec{x}])$,

- $(\tau_1 \vee \tau_2)^\mathsf{A}[\vec{x}] = \tau_1^\mathsf{A}[\vec{x}] \cup \tau_2^\mathsf{A}[\vec{x}]$, $(\tau_1 \wedge \tau_2)^\mathsf{A}[\vec{x}] = \tau_1^\mathsf{A}[\vec{x}] \cap \tau_2^\mathsf{A}[\vec{x}]$,

- $(\mu v.\tau)^\mathsf{A}[\vec{x}] = \mu v.\tau^\mathsf{A}[v, \vec{x}]$, $(\nu v.\tau)^\mathsf{A}[\vec{x}] = \nu v.\tau^\mathsf{A}[v, \vec{x}]$.

In the previous definition, the operation $f_1 \cup f_2$ (resp. $f_1 \cap f_2$) is the mapping which sends $\vec{a}$ on the least upper bound (resp. greatest lower bound) of $f_1(\vec{a})$ and $f_2(\vec{a})$.

These definitions are obviously made to allow us to prove

Lemma 4.1 *If* $\tau = \tau_0[y_1 \leftarrow \tau_1, \ldots, y_n \leftarrow \tau_n]$, *then* $\tau^{\mathsf{A}}[\vec{x}] = \tau_0^{\mathsf{A}}[\vec{y}](\tau_1^{\mathsf{A}}[\vec{x}] \ldots, \tau_n^{\mathsf{A}}[\vec{x}])$.

If F is a set of Σ-fixpoint-terms, we can define the sets of terms $\mu(F)$, $\nu(F)$, $\mathrm{Comp}(F)$ in a way similar to the corresponding notions for operations. We also define F^{A} as the set of all operations which can be associated with an element of F (recall that an infinity of operations can be associated with a term). We get, obviously:

Proposition 4.2 *If* F *is a set of* Σ-*fixpoint-terms and* A *is a* Σ-*poweralgebra, then*

$$\begin{aligned}
\mu(F^{\mathsf{A}}) &= (\mu(F))^{\mathsf{A}}, \\
\nu(F^{\mathsf{A}}) &= (\nu(F))^{\mathsf{A}}, \\
\mathrm{Comp}(F^{\mathsf{A}}) &= (\mathrm{Comp}(F))^{\mathsf{A}}.
\end{aligned}$$

We can also formulate syntactic versions of the propositions of Section 2 concerning vectorial operations. When examining the proofs of these propositions, we can easily see that the involved transformations can be done uniformly for all Σ-poweralgebras, since they do not depend on a particular interpretation. In particular, we have:

Proposition 4.3 *Let* F *be a set of* Σ-*fixpoint-terms closed under renaming of variables. Let* θ *be a* Σ-*fixpoint-term in* $\mu(F)$. *Then there exists* Σ-*fixpoint-terms* $\tau_1, \ldots \tau_n$ *in* $\mathrm{Comp}(F)$, *and a vector* $\vec{y} = \langle y_1, \ldots, y_n \rangle$ *such that*

- $FV(\theta) \cap \{y_1, \ldots, y_n\} = \emptyset$,

- $\forall i, FV(\tau_i) \subseteq \{y_1, \ldots, y_n\} \cup FV(\theta)$

which satisfies: for any vector $\vec{x}$ *containing all the variables of* $FV(\theta)$, *for any* Σ-*poweralgebra* A

$$\theta^{\mathsf{A}}[\vec{x}] = \pi_n^1(\mu\vec{y}.\langle \tau_1^{\mathsf{A}}[\vec{y}, \vec{x}], \ldots, \tau_n^{\mathsf{A}}[\vec{y}, \vec{x}]\rangle).$$

Conversely, if $\tau_1, \ldots \tau_n$ *are in* $\mathrm{Comp}(F)$ *and if* $\vec{y} = \langle y_1, \ldots, y_n \rangle$ *and* $\vec{x} = \langle x_1, \ldots, x_k \rangle$ *are two vectors which contain all the free variables of each* τ_i, *then for any* $i \leq n$, *there exists a term* θ_i *in* $\mu(F)$ *such that all its free variables are in* $\vec{x}$, *and, for any* Σ-*poweralgebra* A,

$$\theta_i^{\mathsf{A}}[\vec{x}] = \pi_n^i(\mu\vec{y}.\langle \tau_1^{\mathsf{A}}[\vec{y}, \vec{x}], \ldots, \tau_n^{\mathsf{A}}[\vec{y}, \vec{x}]\rangle).$$

5   INTERNALIZATION

We show in this section that a set of trees which is defined as the interpretation, in the powerset algebra $\wp\mathcal{T}_\Sigma$, of a closed Σ-fixpoint-term τ can be also characterized by the structure of its elements, using the interpretations of τ in the poweralgebras $\wp t$. This follows from the observation that each $\wp t$ is a kind of homomorphic image of $\wp\mathcal{T}_\Sigma$.

Let Σ be a signature, and let A and B be two Σ-poweralgebras. A mapping $h : A \to B$ is a *homomorphism* of Σ-poweralgebras if

- it is a homomorphism of Σ-algebras, i.e., for any $f \in \Sigma$, for any $a_1, \ldots, a_{\rho(f)}$,

$$h(f^{\mathsf{A}}(a_1, \ldots, a_{\rho(f)})) = f^{\mathsf{B}}(h(a_1), \ldots, h(a_{\rho(f)})),$$

- it preserves the least upper bounds and the greatest lower bounds, i.e., for any subset X of A,

$$h(\bigcup X) = \bigcup \{ h(a) \mid a \in X \},$$
$$h(\bigcap X) = \bigcap \{ h(a) \mid a \in X \}.$$

The following property is a consequence of this definition.

Proposition 5.1 *If $h : A \to B$ is a homomorphism of Σ-poweralgebras, then for every Σ-fixpoint-term τ, for any two operations $\tau^{\mathsf{A}}[\vec{x}]$ and $\tau^{\mathsf{B}}[\vec{x}]$ of the same arity n, for any $a_1, \ldots, a_n \in A$, we have*

$$h(\tau^{\mathsf{A}}[\vec{x}](a_1, \ldots, a_n)) = \tau^{\mathsf{B}}[\vec{x}](h(a_1), \ldots, h(a_n)).$$

Proof   Here again, the proof is by induction on the construction of τ. Most of the cases are straightforward. We consider only the case $\tau' = \mu v.\tau$. Let us fix $a_1, \ldots, a_n \in A$, and let us define the two mappings

$$
\begin{array}{rcl}
f &:& A \ \to \ A \\
  & & a \ \mapsto \ \tau^{\mathsf{A}}[v, \vec{x}](a, a_1, \ldots, a_n)
\end{array}
\quad \text{and} \quad
\begin{array}{rcl}
g &:& B \ \to \ B \\
  & & b \ \mapsto \ \tau^{\mathsf{B}}[v, \vec{x}](b, h(a_1), \ldots, h(a_n))
\end{array}
$$

We have to prove that the image under h of the least fixed point of f is the least fixed point of g, knowing that $g(h(a)) = h(f(a))$ (this is the induction hypothesis).

Because of Theorem 2.2, it suffices to show that for every ordinal ξ, $h(f^{\xi}(\bot_{\mathsf{A}})) = g^{\xi}(\bot_{\mathsf{B}})$.

- if $\xi = 0$, then $f^{\xi}(\bot_{\mathsf{A}}) = \bot_{\mathsf{A}}$, $g^{\xi}(\bot_{\mathsf{B}}) = \bot_{\mathsf{B}}$, and $h(\bot_{\mathsf{A}}) = \bot_{\mathsf{B}}$, since h preserves the least upper bound of the empty set,

- if $\xi = \eta + 1$, then

$$h(f^{\xi}(\bot_{\mathsf{A}})) = h(f(f^{\eta}(\bot_{\mathsf{A}}))) = g(h(f^{\eta}(\bot_{\mathsf{A}}))) = g(g^{\eta}(\bot_{\mathsf{B}})) = g^{\xi}(\bot_{\mathsf{B}}),$$

- if ξ is a limit ordinal,

$$h(f^{\xi}(\bot_{\mathsf{A}})) = h(\bigcup_{\eta < \xi} f^{\eta}(\bot_{\mathsf{A}})) = \bigcup_{\eta < \xi} h(f^{\eta}(\bot_{\mathsf{A}})) = \bigcup_{\eta < \xi} g^{\eta}(\bot_{\mathsf{B}}).$$

$\square$

Now, with each Σ-tree t, we associate the mapping $h_t : \wp\mathcal{T}_\Sigma \to \wp t$ defined by $h_t(L) = \{v \in \mathrm{dom}(t) \mid t \cdot v \in L\}$.

The following properties are straightforward consequences of the definitions.

Proposition 5.2 *The mapping h_t is a homomorphism of Σ-poweralgebras.*
For any subset L of T_Σ, we have

$$v \in h_t(L) \Leftrightarrow t \cdot v \in L.$$

In particular, we have:

Corollary 5.3 *For any closed Σ-fixpoint-term τ, for any Σ-tree t, for any $v \in \mathrm{dom}(t)$,*
$t \cdot v \in \tau^{\wp\mathcal{T}_\Sigma}$ *if and only if $v \in \tau^{\wp t}$.*
For any closed Σ-fixpoint-term τ, $\tau^{\wp\mathcal{T}_\Sigma} = \{t \in \mathcal{T}_\Sigma \mid \lambda \in \tau^{\wp t}\}$.

6 LOGICAL DEFINABILITY OF FIXED POINTS

In this section we shall see how the fixed point definable operations on the powerset algebra of trees can be represented in monadic second order logic. We shall also see that the operations that can be defined without alternation of μ and ν can be expressed in the weak monadic second order logic.

We fix a signature Σ and let m be the maximum of the arities of symbols in Σ. A tree $t \in \mathcal{T}_\Sigma$ will be identified to the logical structure $t = \langle \mathrm{dom}(t), \{\sigma_i \mid i = 1, \ldots, m\}, \{P_f \mid f \in \Sigma\}\rangle$, where each σ_i is a binary relation on $\mathrm{dom}(t)$ defined by $\sigma_i(v, w) \Leftrightarrow w = vi$, and each P_f is a unary relation defined by $P_f(v) \Leftrightarrow t(v) = f$.

Properties of nodes and sets of nodes of a tree may be expressed by formulas of monadic second order logic over the language $\mathcal{L}_\Sigma$ whose nonlogical symbols are the σ_i and the P_f.

To build formulas, it is convenient to use three types of variables: *individual variables* $x_0, x_1, \ldots$, ranging over elements of $\mathrm{dom}(t)$, *set variables* $X_0, X_1, \ldots$, ranging over (arbitrary) subsets of $\mathrm{dom}(t)$, and *finite-set variables* $X_0^F, X_1^F, \ldots$, ranging over finite subsets of $\mathrm{dom}(t)$. Atomic formulas are $x_i = x_j$, $P_f(x_i)$, $\sigma(x_i, x_j)$, $X_i(x_j)$. The other formulas are built using propositional connectives $\vee, \wedge, \neg, \Rightarrow, \Leftrightarrow$, and quantifiers $\forall, \exists$ ranging over all types of variables. Let $M(\mathcal{L}_\Sigma)$ be the collection of these formulas. We shall also consider two important subclasses of $M(\mathcal{L}_\Sigma)$. $WM(\mathcal{L}_\Sigma)$ consists of formulas where the quantifiers range only over individual and finite-set variables (but they may contain free set variables). These formulas are called formulas of weak monadic second order logic. $FO(\mathcal{L}_\Sigma)$ are first-order formulas, i.e., where only individual variables are quantified.

The semantics of a formula is defined by the *satisfaction relation* $\models$. If $\varphi(\xi_1, \ldots, \xi_n)$ is a formula with free variables (of either type) among $\xi_1, \ldots, \xi_n$, if t is in $\mathcal{T}_\Sigma$, and if $\alpha_1, \ldots, \alpha_n$ are elements, subsets, finite subsets of $\mathrm{dom}(t)$, according to the type of the corresponding variables, we write $t \models \varphi[\xi_1 : \alpha_1, \ldots, \xi_n : \alpha_n]$, or, shortly, $t \models \varphi[\alpha_1, \ldots, \alpha_n]$ to mean that φ is satisfied in the model associated with t when α_i is the interpretation of ξ_i. Remember that $t \models X(x)[X : V, x : v]$ if and only if $v \in V$.

A *sentence* is a formula without free variables. If φ is a sentence, $\mathrm{Mod}(\varphi) = \{t \mid t \models \varphi\}$ is the set of models of φ.

The first observation is that all fixed point definable operations can be described by monadic second order formulas, where the exact meaning of this notion is given by the following proposition:

Proposition 6.1 *For any Σ-fixpoint-term τ with $FV(\tau) = \{x_1, \ldots, x_n\}$, there is a monadic second order formula $\varphi_\tau(x, X_1, \ldots, X_n) \in M(\mathcal{L}_\Sigma)$, such that for any vector $\vec{y} = \langle y_1, \ldots, y_k \rangle$ containing the free variables of τ, for any Σ-tree t and any subsets $K_1, \ldots, K_k$ of $\mathrm{dom}(t)$,*

$$\tau^{\wp t}[\vec{y}](K_1, \ldots, K_k) = \{w \in \mathrm{dom}(t) \mid t \models \varphi_\tau[x : w, Y_1 : K_1, \ldots, Y_k : K_k]\}.$$

Proof   The proof is by induction on the construction of τ.

- If τ is the variable x_i, then $\varphi_\tau = X_i(x)$.

- If $\tau(x, X_1, \ldots, X_n) = f(\tau_1, \ldots, \tau_{\rho(f)})$, then

$$\varphi_\tau = P_f(x) \wedge \bigwedge_{i=1}^{\rho(f)} (\forall x'(\sigma_i(x, x') \Rightarrow \varphi_{\tau_i}(x', X_1, \ldots, X_n))).$$

- $\varphi_{\tau_1 \vee \tau_2} = \varphi_{\tau_1} \vee \varphi_{\tau_2}, \; \varphi_{\tau_1 \wedge \tau_2} = \varphi_{\tau_1} \wedge \varphi_{\tau_2}.$

- If $\tau = \mu v.\tau'(v, x_1, \ldots, x_n)$, then

$$
\varphi_\tau(x, X_1, \ldots, X_n) = \\
\exists Y \quad \{\forall y(Y(y) \Leftrightarrow \varphi_{\tau'}(y, Y, X_1, \ldots, X_n)) \\
\wedge \forall Y'(\forall y(Y'(y) \Leftrightarrow \varphi_{\tau'}(y, Y', X_1, \ldots, X_n)) \Rightarrow \forall y(Y(y) \Rightarrow Y'(y))) \\
\wedge Y(x)\}.
$$

Let t be a tree and let L_i be subsets of $\mathrm{dom}(t)$. We have, by hypothesis induction, $t \models \varphi_{\tau'}[y : w, Y : K, X_i : L_i]$ if and only if $w \in \tau'^{\wp t}[y, \vec{x}](K, L_1, \ldots, L_n)$. Hence $t \models \forall y Y(y) \Leftrightarrow \varphi_{\tau'}[Y : K, X_i : L_i]$ if and only if $K = \tau'^{\wp t}[y, \vec{x}](K, L_1, \ldots, L_n)$. Also $t \models \forall y(Y(y) \Rightarrow Y'(y))[Y : K, Y' : K']$ if and only if $K \subseteq K'$. It follows that $t \models \forall y(Y(y) \Leftrightarrow \varphi_{\tau'}) \wedge \forall Y'(\forall y(Y'(y) \Leftrightarrow \varphi_{\tau'}[Y \leftarrow Y']) \Rightarrow \forall y(Y(y) \Rightarrow Y'(y)))[Y : K, X_i : L_i]$ if and only if K is the least solution of the equation $K = \tau'^{\wp t}[y, \vec{x}](K, L_1, \ldots, L_n)$. Hence, $t \models \varphi_\tau[y : w, X_i : L_i]$ if and only if w is in this least fixed point.

- If $\tau = \nu v.\tau'(v, x_1, \ldots, x_n)$, then

$$
\varphi_\tau(x, X_1, \ldots, X_n) = \\
\exists Y \quad \{\forall y(Y(y) \Leftrightarrow \varphi_{\tau'}(y, Y, X_1, \ldots, X_n)) \\
\wedge \forall Y'(\forall y(Y'(y) \Leftrightarrow \varphi_{\tau'}(y, Y', X_1, \ldots, X_n)) \Rightarrow \forall y(Y'(y) \Rightarrow Y(y))) \\
\wedge Y(x)\}.
$$

We have just modified the previous formula to express that Y is a greatest fixed point instead of a least fixed point. $\square$

Considering the principle of internalization (Corollary 5.3), we get

Corollary 6.2 *For a closed Σ-fixpoint-term τ, there is a sentence φ_τ such that*

$$\tau^{\wp T_\Sigma} = Mod(\varphi_\tau).$$

Proof Let $\varphi'_\tau(x)$ be the formula provided by the previous proposition. We have $t \models \varphi'_\tau[x : w]$ if and only if $w \in \tau^{\wp t}$. By Corollary 5.3, we get $t \in \tau^{\wp T_\Sigma}$ if and only if $t \models \varphi'_\tau[x : \lambda]$. Let $\mathrm{root}(x)$ be the formula $\neg\exists y(\bigvee_{i=1}^M \sigma_i(x, y))$. This formula is such that $t \models \mathrm{root}[x : v]$ if and only if $v = \lambda$. Therefore $t \models \varphi'_\tau[x : \lambda]$ if and only if $t \models \exists x(\mathrm{root}(x) \wedge \varphi'_\tau(x))$. $\qquad\square$

We now exhibit a class of Σ-fixpoint-terms τ such that the formulas φ_τ associated with (in the sense of Proposition 6.1) can be choosen in the weak monadic second order logic. This class is $\mathrm{Comp}(\mu(\mathrm{Base}_\Sigma) \cup \nu(\mathrm{Base}_\Sigma))$, which we will denote shortly by $\mathrm{Comp}(\mu, \nu)_\Sigma$, where $\mathrm{Base}_\Sigma = \{f(x_1, \ldots, x_{\rho(f)}) \mid f \in \Sigma\} \cup \{x_1 \vee x_2, x_1 \wedge x_2\}$.

To prove this result we need a definition and a lemma which will show that certains operations can be finitely approximated.

Let E be a set. We consider the product $\wp(E)^n$ as a complete lattice ordered by the componentwise inclusion, and the notations used for the complete lattice $\wp(E)$, such as $\cup, \subseteq, \ldots$ are also used in this case.

An element $\vec{X} = \langle X_1, \ldots, X_n \rangle$ of $\wp(E)^n$ is said to be *finite* if every X_i is finite. A mapping $f : \wp(E)^n \to \wp(E)$ is *compact* if, for each $\vec{X} \in \wp(E)^n$, and for each $x \in f(\vec{X})$, there exists a finite $\vec{Y}$ included in $\vec{X}$ such that $x \in f(\vec{Y})$. A vectorial mapping $\vec{f} = \langle f_1, \ldots, f_m \rangle : \wp(E)^n \to \wp(E)^m$ is compact if each f_i is compact.

If f is compact and monotonic, if A is a finite subset of $f(X)$, then there exists a finite subset B of X such that $A \subseteq f(B)$. To see that, let B_x, for any $x \in A$, the finite subset of X such that $x \in f(B_x)$, and let $B = \bigcup_{x \in A} B_x$. Since f is monotonic $f(B_x) \subseteq f(B)$, hence, for any $x \in A$, $x \in f(B_x) \subseteq f(B)$.

For $A \in \wp(E)^n$, we denote by $\wp(E)^n_A$ the set $\{\vec{Y} \in \wp(E)^n \mid \vec{Y} \subseteq A\}$. Clearly, $\wp(E)^n_A$ is a complete lattice for componentwise inclusion. If $\vec{f}$ is a mapping from $\wp(E)^n$ to $\wp(E)^n$, we denote by $\vec{f}_A$ the mapping from $\wp(E)^n_A$ to $\wp(E)^n_A$ defined by $\vec{f}_A(\vec{X}) = \vec{f}(\vec{X}) \cap A$. If $\vec{f}$ is monotonic, then $\vec{f}_A$ is also monotonic.

Lemma 6.3 *If $\vec{f}$ is a monotonic and compact mapping from $\wp(E)^n$ to $\wp(E)^n$, then*

$$\mu\vec{x}.\vec{f}(\vec{x}) = \bigcup\{\mu\vec{x}.\vec{f}_A(\vec{x}) \mid A \in \wp(E)^n \text{ is finite}\}.$$

Proof "$\supseteq$". If $S = \vec{f}(S)$ then $S \cap A = \vec{f}(S) \cap A \supseteq \vec{f}(S \cap A) \cap A = \vec{f}_A(S \cap A)$, hence, $\mu\vec{x}.\vec{f}_A(\vec{x}) \subseteq S \cap A \subseteq S$.

"$\subseteq$". It is enough to prove that if $a \in \pi^i_n(\mu\vec{x}.\vec{f}(\vec{x}))$ then there exists a finite A such that $a \in \pi^i_n(\mu\vec{x}.\vec{f}_A(\vec{x}))$. Thus, let a be in $\pi^i_n(\mu\vec{x}.\vec{f}(\vec{x}))$. We construct a decreasing sequence of ordinals $\xi_0 > \xi_1 > \cdots$, and a sequence $B_0, B_1, \ldots$ of finite elements of $\wp(E)^n$ as follows. Let $B_0 = \langle \emptyset, \ldots, \emptyset, \{a\}, \emptyset, \ldots, \emptyset \rangle$ where $\{a\}$ occurs in i-th position, and let ξ_0 be the least ordinal such that $B_0 \subseteq \vec{f}^{\xi_0}(\vec{\emptyset})$. Since ξ_0 cannot be a limit ordinal, then $\xi_0 = \eta_0 + 1$ and

$B_0 \subseteq \vec{f}^{\xi_0}(\vec{\emptyset}) = \vec{f}(\vec{f}^{\eta_0}(\vec{\emptyset}))$. By compactness, there exists $B_1 \subseteq \vec{f}^{\eta_0}(\vec{\emptyset})$ such that $B_0 \subseteq \vec{f}(B_1)$. If $B_1 = \vec{\emptyset}$, then the construction is finished. Otherwise, let ξ_1 the least ordinal such that $B_1 \subseteq \vec{f}^{\xi_1}(\vec{\emptyset})$. Note that since $B_1 \subseteq \vec{f}^{\eta_0}(\vec{\emptyset})$, we have $\xi_1 \leq \eta_0 < \xi_0$. We continue in the same way, until we get $B_p = \vec{\emptyset}$. Since the sequence of ordinals $\xi_0, \xi_1, \ldots$, is strictly decreasing, it cannot be infinite, and thus such a B_p exists. Moreover, by definition, $B_j \subseteq \vec{f}(B_{j+1})$. Let $A = \bigcup_{i=0}^{p} B_i$. We claim that $a \in \pi_n^i(\mu\vec{x}.\vec{f}_A(\vec{x}))$. For that, we prove by downward induction that, for $j \leq p$,

$$B_j \subseteq \vec{f}_A^{p-j}(\vec{\emptyset}). \tag{1}$$

If $j = p$, we have $B_j = \vec{\emptyset} = \vec{f}_A^0(\vec{\emptyset})$, and (1) holds. Let us assume that (1) holds for some $j > 0$. Then $B_{j-1} = B_{j-1} \cap A \subseteq \vec{f}(B_j) \cap A = \vec{f}_A(B_j) \subseteq \vec{f}_A(\vec{f}_A^{p-j}(\vec{\emptyset})) = \vec{f}_A^{p-(j-1)}(\vec{\emptyset})$.

It follows that $B_0 \subseteq \vec{f}_A^p(\vec{\emptyset})$, which completes the proof. □

We are ready to prove the following.

Proposition 6.4 *Let τ be a Σ-fixpoint-term in $\mathrm{Comp}(\mu,\nu)_\Sigma$, and let $\vec{x} = \langle x_1, \ldots, x_n \rangle$ be a vector containing all the free variables of τ. Then there exists a formula*

$$\varphi_\tau(x, X_1, \ldots, X_n)$$

of $WM(\mathcal{L}_\Sigma)$, such that for any Σ-tree t and any subsets $K_1, \ldots, K_n$ of $\mathrm{dom}(t)$,

$$\tau^{\wp t}[\vec{x}](K_1, \ldots, K_n) = \{w \in \mathrm{dom}(t) \mid t \models \varphi_\tau[x : w, X_1 : K_1, \ldots, X_n : K_n]\}.$$

Proof We first show the claim for a term $\tau \in \mu(\Sigma \cup \{\vee, \wedge\})$. By proposition 4.3, there exist vectors of variables $\vec{y}$ and $\vec{x}$, Σ-fixpoint-terms $\tau_1, \ldots, \tau_k$ in $\mathrm{Comp}(\Sigma \cup \{\vee, \wedge\})$ such that for any Σ-tree t

$$\tau^{\wp t}[\vec{x}] = \pi_k^1(\mu\vec{y}.\vec{\tau}^{\wp t}[\vec{y}, \vec{x}]).$$

It is straightforward to show that with each τ_i is associated a *first-order* formula $\varphi_i(x, \vec{Y}, \vec{X})$ such that $w \in \tau_i^{\wp t}[\vec{y}, \vec{x}](\vec{L}, \vec{K})$ if and only if $t \models \varphi_i[x : w, \vec{X} : \vec{L}, \vec{Y} : \vec{K}]$. Next, it is also easy to show that, for any t and $\vec{K}$, the mapping $\vec{f}_{t,\vec{K}}$ defined by $\vec{f}_{t,\vec{K}}(\vec{L}) = \vec{\tau}^{\wp t}[\vec{y}, \vec{x}](\vec{L}, \vec{K})$ is monotonic and compact; this is easily done by induction on the construction of $\vec{\tau}$, observing that compactness is preserved by composition. Hence, by Lemma 6.3, we have $v \in \tau^{\wp t}[\vec{x}](\vec{K})$ if and only if there exist some finite element $\vec{A}$ of $\wp(\mathrm{dom}(t))^k$ such that $v \in \pi_k^1(\mu\vec{x}.(\vec{f}_{t,\vec{K}})_{\vec{A}}(\vec{x}))$. This fact can be expressed by a formula of the weak monadic second order logic, using the techniques of the proof of Proposition 6.1, and observing that all quantified variables can be restricted to range over finite sets, since all the fixed points of $(\vec{f}_{t,\vec{K}})_{\vec{A}}(\vec{x})$ are finite.

We now show the claim for a term $\tau \in \nu(\Sigma \cup \{\vee, \wedge\})$. Firstly, let us prove that for such a term, there exists a term θ in $\mu(\Sigma \cup \{\vee, \wedge, tt\})$ such that, for any t and $\vec{x}$, $\theta^{\wp t}[\vec{x}]$ is the dual of $\tau^{\wp t}[\vec{x}]$. Here, tt is a new constant such that $tt^{\wp t}[\vec{x}] = \mathrm{dom}(t)$. We prove that by induction on the construction of τ. In what follows, let θ_i be the term already associated with τ_i.

- If τ is a variable, then $\theta = \tau$.

- If $\tau = \tau_1 \vee \tau_2$, then $\theta = \theta_1 \wedge \theta_2$.

- If $\tau = \tau_1 \wedge \tau_2$, then $\theta = \theta_1 \vee \theta_2$.

- If $\tau = \nu x.\tau_1$, then $\theta = \mu x.\theta_1$.

- If $\tau = f(\tau_1, \ldots, \tau_n)$, then $\theta =$

$$\bigvee_{g \neq f} g(tt, \ldots, tt) \vee \bigvee_{i=1}^{n} f(tt, \ldots, tt, \theta_i, tt, \ldots, tt).$$

The proof that this construction is correct is straightforward. As to $\nu x.\tau_1$, we have already proved, in Proposition 2.8, that the dual of the greatest fixed point of an operation is the least fixed point of the dual operation.

Now, let ψ be the formula of weak monadic second order logic associated with the dual θ of τ, which has been previously proved to exist (the formula $x = x$ being associated with the constant tt). We have

$$t \models \psi[y : w, \vec{X} : \vec{K}] \Leftrightarrow w \in \theta^{\wp t}[\vec{x}](\vec{K}).$$

But θ is the dual of τ, hence,

$$w \in \tau^{\wp t}[\vec{x}](\vec{K}) \Leftrightarrow w \notin \theta^{\wp t}[\vec{x}](\overrightarrow{\mathrm{dom}}(t) - \vec{K}),$$

and thus,

$$w \in \tau^{\wp t}[\vec{x}](\vec{K}) \Leftrightarrow t \models \neg\psi[y : w, \vec{X} : \overrightarrow{\mathrm{dom}}(t) - \vec{K}].$$

Let φ be the formula obtained by substituting $\neg X_i(z)$ (where X_i is a component of $\vec{X}$) for each subformula $X_i(z)$ in ψ. We have

$$t \models \psi[y : w, \vec{X} : \overrightarrow{\mathrm{dom}}(t) - \vec{K}] \Leftrightarrow t \models \varphi[y : w, \vec{X} : \vec{K}],$$

hence,

$$t \models \neg\varphi[y : w, \vec{X} : \vec{K}] \Leftrightarrow w \in \tau^{\wp t}[\vec{x}](\vec{K}).$$

Finally, we have to examine the composition of terms. If $\varphi_0, \varphi_1, \ldots, \varphi_n$ are the formulas associated with the terms $\tau_0, \tau_1, \ldots, \tau_n$, the formula associated with the term $\tau_0[\tau_1/y_1, \ldots, \tau_n/y_n]$ is obtained by substituting φ_i for every subformula $Y_i(z)$ in φ_0. $\quad\square$

From Proposition 6.4 and Corollary 6.2 we deduce

Corollary 6.5 *For a closed Σ-fixpoint-term τ in* $\mathrm{Comp}(\mu, \nu)_\Sigma$*, there is a sentence φ_τ in* $WM(\mathcal{L}_\Sigma)$ *such that*

$$\tau^{\wp T_\Sigma} = Mod(\varphi_\tau).$$

7   FIXED POINT CHARACTERIZATION OF WEAK MONADIC SECOND ORDER LOGIC

In this section we show the converse of Corollary 6.5: any Σ-tree language definable by a sentence of the weak monadic second order logic can be defined by a closed Σ-fixpoint-term in $\mathrm{Comp}(\mu, \nu)_\Sigma$.

To prove our main result, we shall use one of the fundamental results of M. O. Rabin.

Theorem 7.1 (Rabin) *A set $L \subseteq T_\Sigma$ is definable by a Büchi automaton if and only if $L = Mod(\varphi)$ for some sentence φ of the form $\exists X_1 \cdots \exists X_n \psi$, where ψ is a formula of the weak monadic second order logic.*

In fact, Rabin proved this result for full binary trees, which corresponds, in our setting, to the case of a signature containing only binary symbols. But an analysis of the proof shows that the argument can be easily adapted to an arbitrary signature.

An immediate consequence of the above theorem is the following.

Theorem 7.2 (Rabin) *If ψ is a sentence of the weak monadic second order logic, then $Mod(\psi)$ and its complement in T_Σ are both definable by Büchi automata.*

In fact Rabin also proved the converse of this property.

Because of this theorem, the converse of Corollary 6.4 we are going to prove, can be stated as

Theorem 7.3 *Let $L \subseteq T_\Sigma$. If L and $T_\Sigma - L$ are both definable by Büchi automata, then there exists a closed Σ-fixpoint-term τ in $\mathrm{Comp}(\mu, \nu)_\Sigma$ such that $L = \tau^{\wp T_\Sigma}$.*

Proof We shall essentially follow the ideas used by Rabin [13] to prove the converse of Theorem 7.2. We begin this proof with some definitions.

Some definitions Let t be a tree. A *cut* is a finite nonempty subset A of $\mathrm{dom}(t)$ satisfying the following properties

- any two elements of A are incomparable for the initial segment relation $\leq$,

- for all $v \in \mathrm{dom}(t) - A$ such that v is on a path, $\exists w \in A : v < w$ or $\exists w \in A : v > w$.

Note that cuts are not exactly finite maximal antichains, for a node which does not belong to any path need not to be in a cut even if it is incomparable with any other element of this cut.

If A is a cut we denote by $[\lambda, A]$ the set of all nodes $v \in \mathrm{dom}(t)$ such that there exists a node $w \in \mathrm{dom}(t)$ which is either a leaf or an element of A with $v \leq w$. In particular $\{\lambda\}$ is a cut and $[\lambda, \{\lambda\}]$ is the set all nodes less than or equal to a leaf.

If A and A' are two cuts, we denote by $A < A'$ the relation defined by $\forall v \in A, \forall w \in A', w \not\leq v$. In particular, the set $\{\lambda\}$ is a cut, and if A is any cut not equal to $\{\lambda\}$, then $\{\lambda\} < A$. If A and A' are two cuts such that $A \leq A'$, then $[\lambda, A] \subset [\lambda, A']$.

If r is any mapping from $\mathrm{dom}(t)$ into any set, and if E is a subset of $\mathrm{dom}(t)$, we denote by $r \uparrow E$ the restriction of r to E.

Now let $\mathcal{A} = \langle Q, q_0, Tr, F, G \rangle$ be a Büchi automaton. We can assume, without loss of generality, that $\mathcal{A}$ is complete, i.e., for every symbol f in Σ, and every state q in Q, there exists at least one transition $\langle q, f, q_1, \ldots, q_{\rho(f)} \rangle$ in Tr, so that every Σ-tree has at least one run.

It is easy to see that the definition of an accepting run of $\mathcal{A}$ on a tree t can be restated as follows

Lemma 7.4 *The run r on t is accepting if and only if for any leaf w, $r(w) \in G$ and for any cut A, there exists a cut A' such that $A < A'$ and $r(A') \subseteq F$.*

Let us define, by induction, the sets K_q^n of Σ-trees, for $n \geq 0$ and $q \in Q$.

- $t \in K_q^0$ if and only if there exists a q-run r on t such that if v is a leaf then $r(v) \in G$,

- $t \in K_q^{n+1}$ if and only if there exists a q-run r on t such that for any leaf v, $r(v) \in G$, and for any cut A, there exists a cut A' and a q-run r' satisfying

  - $A < A'$,
  - $r \uparrow [\lambda, A] = r' \uparrow [\lambda, A]$,
  - $\forall w \in A'$, $r'(w) \in F$ and $t \cdot w \in K_{r'(w)}^n$.

Let us remark that since every leaf belongs to $[\lambda, A]$, the fact that $r(v) \in G$ for every leaf v and the condition $r \uparrow [\lambda, A] = r' \uparrow [\lambda, A]$ imply that $r'(v) \in G$ for every leaf v .

Lemma 7.5 *Let $\mathcal{A}$ be an automaton and the sets K_q^n associated with it. Then, for any $q \in Q$ and any $n \geq 0$, we have*
$$L(\mathcal{A}, q) \subseteq K_q^n.$$

Proof Let us prove this result by induction on n. Obviously, $L(\mathcal{A}, q) \subseteq K_q^0$. Let t be in $L(\mathcal{A}, q)$ and let us consider an accepting q-run on t. Let A be any cut. By Lemma 7.4, there exists a cut A' with $A < A'$ and $r(A') \subseteq F$. Thus, it remains to prove that each $t \cdot w$, for $w \in A'$, is in $K_{r(w)}^n$. By induction hypothesis, this is implied by $t \cdot w \in L(\mathcal{A}, r(w))$ which is obviously true since if r is an accepting q-run on t, the $r(w)$-run r' on $t \cdot w$ defined by $r'(w) = r(wv)$ is still accepting. $\square$

Now, let us assume that the complement of $L = L(\mathcal{A}, q_0)$ is also defined by a Büchi automaton $\mathcal{B} = \langle Q', q_0', Tr', F', G' \rangle$, and let $p = \mathrm{Card}(Q)$, $p' = \mathrm{Card}(Q')$ and $n = 2^{pp'} + 1$. We have

Lemma 7.6 $L = K_{q_0}^n$.

Proof We already know that $L \subseteq K_{q_0}^n$. Let us prove the converse inclusion: let us assume that there is a tree t in $K_{q_0}^n$ which is not in L and let us derive a contradiction.

Since t is not in L, there exists an accepting q_0'-run r' of $\mathcal{B}$ on t. Let us show by induction that for all $i \leq n$ there is a sequence $A_1 < A_1' \cdots < A_i < A_i'$ of cuts and a q_0-run r_i of $\mathcal{A}$ such that

- if v is a leaf, $r_i(v) \in G$ and $r'(v) \in G'$,

- $\forall j \leq i, r_i(A_j) \subseteq F$,

- $\forall j \leq i, r'(A_j') \subseteq F'$,

- $\forall w \in A_i, t \cdot w \in K_{r_i(w)}^{n-i}$.

Firstly, let us remark that Lemma 7.4 immediately implies that for any cut A_i there is a cut A_i' such that $A_i < A_i'$ and $r'(A_i') \subseteq F'$. Thus we have only to examine how to construct the cuts A_i.

Let $A_0 = \{\lambda\}$ and let r_0 be any q_0-run of $\mathcal{A}$ on t such that for any leaf v, $r(v) \in G$. Such a run exists since $t \in K_{q_0}^n$. Moreover, there is a cut A_1 and a run r_1 with $r_1(A_1) \subseteq F$, $\forall w \in A_1, t \cdot w \in K_{r_1(w)}^{n-1}$, and, since $[\lambda, A_0] = \{\lambda\}$, $r_1(\lambda) = q_0$.

Now, we assume that we have the cuts $A_1, A_1', \ldots A_i, A_i'$, and a q_0-run r_i satisfying the hypothesis induction. Let w be any element of A_i. Let us consider the cut A_w' of $t \cdot w$ defined by $A_w' = \{v \mid wv \in A_i'\}$, which is not equal to $\{\lambda\}$. Since $t \cdot w$ is in $K_{r_i(w)}^{n-i}$, there is a $r_i(w)$-run r_w of $\mathcal{A}$ on $t \cdot w$ and a cut A_w such that $A_w' < A_w$, $r_w(A_w) \subseteq F$, and $\forall v \in A_w$, $t \cdot wv \in K_{r_w(v)}^{n-i-1}$. Thus we define A_{i+1} and r_{i+1} as follows

$$A_{i+1} = \bigcup_{w \in A_i} A_w,$$

$$r_{i+1}(u) = \begin{cases} r_i(u) & \text{if } u \in [\lambda, A_i], \\ r_w(v) & \text{otherwise}, \end{cases}$$

where v and w are such that $u = wv$ and $w \in A_i$.

It is easy to check that $A_i' < A_{i+1}$, $r_i \uparrow [\lambda, A_i] = r_{i+1} \uparrow [\lambda, A_i]$, and for any $u \in A_{i+1}, u = wv$ with $w \in A_i$, $v \in A_w$, and thus, $r_{i+1}(u) = r_w(v) \in F$ and $t \cdot u \in K_{r_{i+1}(u)}^{n-i+1}$.

Now let us denote by r the run r_n. For each $i \leq n$, let E_i be the subset of $Q \times Q'$ equal to $\{\langle r(w), r'(w) \rangle \mid w \in A_i\}$. Since $\wp(Q \times Q')$ has $2^{pp'}$ elements, there exists i and j with $i < j$ such that $E_i = E_j$. For any pair $\langle q, q' \rangle \in E_i = E_j$ we choose an arbitrary node $w_{\langle q, q' \rangle}$ in A_i (see Figure 1) and we construct a new tree t' by simultaneously replacing in t any subtree $t \cdot v$ with v in A_j by $t \cdot w_{\langle r(v), r'(v) \rangle}$ (see Figure 2). We define two new runs r_1 and r_1' of $\mathcal{A}$ and $\mathcal{B}$ on t' by transforming the two runs r and r' in the same way. This new tree has now at least five cuts $A_i < A_i' < A_j < C_1' < C_1$ where C_1 is the cut $\{uv \mid u \in A_j, w_{\langle r(u), r'(u) \rangle} v \in A_j\}$ and C_1' is the cut $\{uv \mid u \in A_j, w_{\langle r(u), r'(u) \rangle} v \in A_i'\}$. By construction, we have $r_1(A_i) \subseteq F$, $r_1(A_j) \subseteq F$, $r_1(C_1) \subseteq F$, $r_1'(A_i') \subseteq F'$, $r_1'(C_1') \subseteq F'$, and $\{\langle r_1(u), r_1'(u) \rangle\} \subseteq E_i = E_j$. We can again apply the same transformation to this new tree by replacing the subtrees rooted in C_1. We obtain a tree with two more cuts C_2 and C_2' and two new runs r_2 and r_2' with $r_2 \uparrow [\lambda, C_1] = r_1 \uparrow [\lambda, C_1]$, $r_2' \uparrow [\lambda, C_1] = r_1' \uparrow [\lambda, C_1]$, $r_2(C_2) \subseteq F$, and $r_2'(C_2') \subseteq F'$. By iterating this process we get a tree which has both an accepting q_0-run of $\mathcal{A}$ and a q_0'-run of $\mathcal{B}$, a contradiction. $\square$

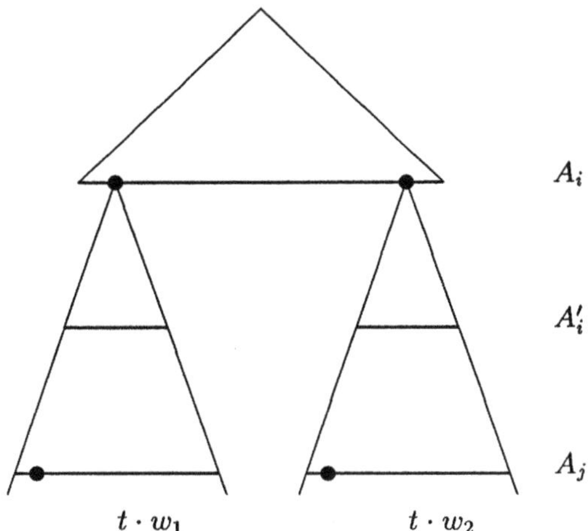

Figure 1: The tree t

Now, it remains to prove that the sets K_q^n are all interpretations of Σ-fixpoint-terms in $\mathrm{Comp}(\mu, \nu)_\Sigma$. Indeed, in order to avoid heavy notations we just show how these sets can be defined using least and greatest fixed points of systems of equations.

First we assume that the states of Q are numbered from 1 to $p = \mathrm{Card}(Q)$ in such a way that the states in F are numbered from 1 to k ($k \leq p$), the initial state q_0 is numbered p_0 and we identify states with their number. Thus $Q = \{1, \ldots, p\}$ and $F = \{1, \ldots, k\}$.

Let us consider the three following vectors of variables

$$\begin{aligned}
\vec{x} &= \langle x_1, \ldots, x_p \rangle, \\
\vec{y} &= \langle y_1, \ldots, y_k \rangle, \\
\vec{z} &= \langle z_1, \ldots, z_p \rangle.
\end{aligned}$$

We consider the vectorial operation $\vec{g}(\vec{x}, \vec{y}) = \langle g_1(\vec{x}, \vec{y}), \ldots, g_p(\vec{x}, \vec{y}) \rangle$ on $\wp T_\Sigma$ where, for each $i \in \{1, \ldots, p\}$,

$$g_i(\vec{x}, \vec{y}) = \bigcup_{\langle i, f, i_1, \ldots, i_{\rho(f)} \rangle \in Tr} f(u_{i_1}, \ldots, u_{i_{\rho(f)}})$$

and

$$u_i = \begin{cases} x_i & \text{if } i > k \\ x_i \cup y_i & \text{otherwise.} \end{cases}$$

Let $g_i^*(\vec{y})$ be the i-th component of the least fixed point of $\vec{g}$, i.e.,

$$g_i^*(\vec{y}) = \pi_p^i(\mu \vec{x}. \vec{g}(\vec{x}, \vec{y})).$$

For $\vec{L} \in \wp(T_\Sigma)^k$, let us denote by $G_i(\vec{L})$ the set of all Σ-trees t such that

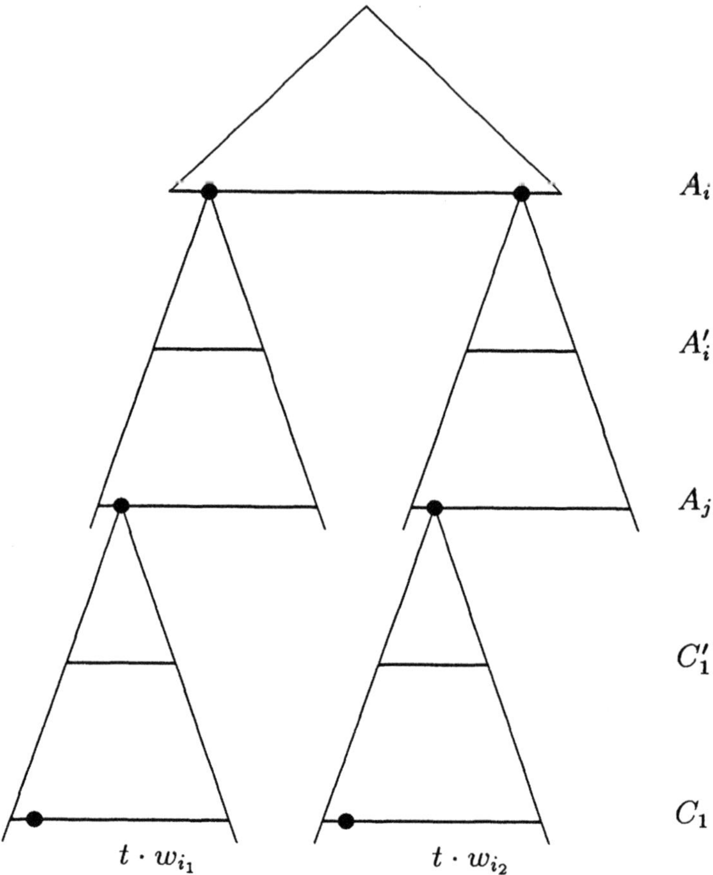

Figure 2: The tree t'

(i) there exist a i-run r on t and a cut $A \neq \{\lambda\}$ such that for all $w \in A$,

- $r(w) \leq k$ and
- $t \cdot w \in L_{r(w)}$.

Lemma 7.7 *For all i and $\vec{L}$, $G_i(\vec{L}) = g_i^*(\vec{L})$.*

Proof Let $\langle i, f, i_1, \ldots, i_{\rho(f)} \rangle$ be in Tr, and let $t_1, \ldots, t_{\rho(f)}$ such that $t_j \in G_{i_j}(\vec{L})$ if $i_j > k$ and $t_j \in G_{i_j}(\vec{L}) \cup L_{i_j}$ otherwise. Obviously, $f(t_1, \ldots, t_{\rho(f)})$ is in $G_i(\vec{L})$, hence, $g_i(\vec{G}(\vec{L}), \vec{L}) \subseteq G_i(\vec{L})$ and thus $g_i^*(\vec{L}) \subseteq G_i(\vec{L})$.

Conversely we can show by induction on the maximum length of elements in the cut A of a tree t in $G_i(\vec{L})$ that $t \in g_i^*(\vec{L})$. Such a t can be written as $f(t_1, \ldots, t_{\rho(f)})$ and let r be the i-run of $\mathcal{A}$ on t. It can be easily shown that if $r(j) \leq k$ then $t_j \in L_{r(j)} \cup G_{r(j)}(\vec{L})$ and if $r(j) > k$ then $t_j \in G_{r(j)}(\vec{L})$. Moreover, if $t_j \in G_{r(j)}(\vec{L})$ the induction hypothesis can be applied to t_j, thus, $t \in g_i(\vec{g}^*(\vec{L}), \vec{L}) = g_i^*(\vec{L})$. $\qquad\square$

Now, we consider the vectorial operation $\vec{h}(\vec{x}, \vec{z}) = \langle h_1(\vec{x}, \vec{y}), \ldots, h_p(\vec{x}, \vec{y}) \rangle$ on $\wp T_\Sigma$ where, for each $i \in \{1, \ldots, p\}$,

$$h_i(\vec{x}, \vec{z}) = \bigcup_{\langle i, f, i_1, \ldots, i_{\rho(f)} \rangle \in Tr} f(x_{i_1} \cap z_{i_1}, \ldots, x_{i_{\rho(f)}} \cap z_{i_{\rho(f)}}).$$

Let $h_i^\omega(\vec{z})$ be the i-th component of the greatest fixed point of $\vec{h}$, i.e.,

$$h_i^\omega(\vec{z}) = \pi_p^i(\nu \vec{x}.\vec{h}(\vec{x}, \vec{z})).$$

For $\vec{L} \in \wp(T_\Sigma)^p$, let us denote by $H_i(\vec{L})$ the set of all Σ-trees t such that

(ii) there exists a i-run r on t such that for any $w \in \mathrm{dom}(t), w \neq \lambda$, $t \cdot w \in L_{r(w)}$.

Lemma 7.8 *For all i and $\vec{L}$, $H_i(\vec{L}) = h_i^\omega(\vec{L})$.*

Proof Obviously $h_i^\omega(\vec{L}) \subseteq H_i(\vec{L}) = h_i(\vec{H}(\vec{L}), \vec{L})$. $\qquad\square$

It is then easy to show that for any i, $1 \leq i \leq p$, $K_i^{n+1} = h_i^\omega(\vec{g}^*(K_1^n, \ldots, K_k^n))$, and since $K_i^0 = T_\Sigma = \nu x.x$, we get that for every i and n, K_i^n is the interpretation of a Σ-fixpoint-term in $\mathrm{Comp}(\mu, \nu)_\Sigma$.

end of the proof.

8 CONCLUSION

The correspondence between weak monadic second order logic and fixed point definitions which we have presented in this paper should be considered as a part of a more general picture of relations between different modes of definability of properties of infinite trees: logic fixed points, and automata. Recall that definability in monadic second order logic ($M(\mathcal{L}_\Sigma)$, in the notations of this paper) corresponds to definability by Rabin automata ([12], see also [17]) while $WM(\mathcal{L}_\Sigma)$ has been characterized by Muller, Saoudi, and Schupp [6] in terms of alternating automata with a weak acceptance condition (that we shall call, for short, weak alternating automata). Recall that we have defined Base_Σ as being the set $\{f(x_1,\ldots,x_{\rho(f)}) \mid f \in \Sigma\} \cup \{x_1 \vee x_2, x_1 \wedge x_2\}$ and $\text{Base}_{\Sigma\setminus\wedge}$ as being the set $\{f(x_1,\ldots,x_{\rho(f)}) \mid f \in \Sigma\} \cup \{x_1 \vee x_2\}$. We also use the following abbreviations:

$$(\nu\mu)_\Sigma = \nu(\mu(\text{Base}_\Sigma))$$
$$(\nu\mu)_{\Sigma\setminus\wedge} = \nu(\mu(\text{Base}_{\Sigma\setminus\wedge}))$$

and we denote by $fpT_{\Sigma\setminus\wedge}$ the set of all Σ-fixpoint-terms where $\wedge$ does not occur. Finally, let $\exists WM(\mathcal{L}_\Sigma)$ be the set of existential sentences of $M(\mathcal{L}_\Sigma)$, i.e., the sentences φ in Theorem 7.1 above. Then we can summarize the known facts in the following table :

Logic	Fixed points	Automata
$M(\mathcal{L}_\Sigma)$	$fpT_\Sigma = fpT_{\Sigma\setminus\wedge}$ [9], [this paper]	Rabin automata [12]
$\exists WM(\mathcal{L}_\Sigma)$	$(\nu\mu)_\Sigma = (\nu\mu)_{\Sigma\setminus\wedge}$ [8, 15, 2]	Büchi automata [13]
$WM(\mathcal{L}_\Sigma)$	$\text{Comp}(\mu,\nu)_\Sigma$ [this paper]	weak alternating automata [6]

(This table should be understood in the following way: for each of its row, a set $L \subseteq T_\Sigma$ is definable by a formula φ of the logic of this row if and only if it is definable by a fixpoint-term of the set given in this row if and only if it is definable by an automaton of the form given in this row.

An intriguing open problem concerns the hierarchy of fixed point definitions induced by the number of alternations between μ and ν. For fixpoint-terms *without* $\wedge$, the hierarchy is known to be infinite and coincides with the hierarchy of Rabin indices [8]. For the fixpoint-terms containing $\wedge$, we only know, because of a counterexample given by Rabin [12] to show that the family of Büchi-definable sets is not closed under complementation, that $(\nu\mu)_\Sigma^{\wp T_\Sigma} \neq (\mu\nu)_\Sigma^{\wp T_\Sigma}$, as soon as Σ contains at least two symbols, one of them being at least binary.

We also know that the examples used in [8] to show the infiniteness of the hierarchy do no longer work in presence of intersection. Thus, we would like to know whether the hierarchy of sets definable by fixpoint-terms is infinite or not. As a first step in answering this question we would like to know more about the set $\text{Comp}(\mu\nu, \nu\mu)_\Sigma^{\wp T_\Sigma}$ and we set the following question:

Open problem Is $\text{Comp}(\mu\nu, \nu\mu)_\Sigma^{\wp T_\Sigma}$ equal or not to $fpT_\Sigma^{\wp T_\Sigma}$.

ACKNOWLEDGEMENTS The authors are pleased to thank B. Courcelle and G. Mirkowska for carefully reading this text and for their numerous comments and remarks.

References

[1] A. Arnold. Logical definability of fixed points. *Theor. Comput. Sci.*, 61:289–297, 1988.

[2] A. Arnold and D. Niwiński. Fixed point characterization of Büchi automata on infinite trees. *J. Inf. Process. Cybern. EIK*, 26:453–461, 1990.

[3] J. R. Büchi. On a decision method in restricted second order arithmetic. In E. Nagl, editor, *Logic, Methodology, and Philosophy of Science*, pages 1–11. Stanford Univ. Press, 1960.

[4] B. Knaster. Un théorème sur les fonctions des ensembles. *Ann. Soc. Polon. Math*, 6:133–134, 1928.

[5] D. Kozen and J. Tiuryn. Logics of programs. In J. Van Leeuwen, editor, *Handbook of Theoretical Computer Science (vol. B)*, pages 789–840. 1990.

[6] D. E. Muller, A. Saoudi, and P. Schupp. Alternating automata, the weak monadic theory of the tree, and its complexity. In L. Kott, editor, *Proc. 13th ICALP*, pages 275–283. Springer–Verlag, Lect. Notes Comput. Sci. 226, 1986.

[7] D. Niwiński. Equational μ-calculus. In *Computation Theory*, pages 169–176. Springer–Verlag, Lect. Notes Comput. Sci. 208, 1985.

[8] D. Niwiński. On fixed point clones. In L. Kott, editor, *Proc. 13th ICALP*, pages 464–473. Springer–Verlag, Lect. Notes Comput. Sci. 226, 1986.

[9] D. Niwiński. Fixed points vs. infinite generation. In *Proc. 3rd IEEE Symp. on Logic in Comput. Sci.*, pages 402–409, 1988.

[10] D. Park. On the semantics of fair parallelism. In *Abstract Software Specifications*, pages 504–526. Springer–Verlag, Lect. Notes Comput. Sci. 86, 1980.

[11] D. Park. Concurrency and automata on infinite sequences. In *5th GI Conf. on Theoret. Comput. Sci.*, pages 167–183, Karlsruhe, 1981. Lect. Notes Comput. Sci. 104.

[12] M. O. Rabin. Decidability of second-order theories and automata on infinite trees. *Trans. Amer. Soc.*, 141:1–35, 1969.

[13] M. O. Rabin. Weakly definable relations and special automata. In Y. Bar-Hillel, editor, *Mathematical Logic In Foundations of Set Theory*, pages 1–23. 1970.

[14] M. O. Rabin. Decidable theories. In J. Barwise, editor, *Handbook of Mathematical Logic*, pages 595–630. North-Holland, 1977.

[15] M. Takahashi. The greatest fixed points and rational omega-tree languages. *Theor. Comput. Sci.*, 44:259–274, 1986.

[16] A. Tarski. A lattice theoretical fixpoint theorem and its applications. *Pacific J. of Math.*, 5:285–309, 1955.

[17] W. Thomas. Automata on infinite objects. In J. Van Leeuwen, editor, *Handbook of Theoretical Computer Science (vol. B)*, pages 133–191. 1990.

Tree Automata and Languages
M. Nivat and A. Podelski (editors)
© 1992 Elsevier Science Publishers B.V. All rights reserved.

Automata on Infinite Trees and Rational Control

A. Saoudi (1) and P. Bonizzoni (2)

(1) L.I.P.N, Universite Paris XIII, Centre Scientifique et Polytechnique
Av J. B. Clément, 93430 Villetaneuse, France
E-mail: saoudi@LIPN.univ-paris13.fr or as@litp.ibp.fr

(2) Università degli Studi di Milano, Dipartimento di Scienze dell'Informazione
Via Moretto da Brescia 9, 20133 Milano, Italie
E-mail: bonizzon@imiucca.unimi.it

INTRODUCTION

The theory of automata is well known to be strongly connected to both logic and computer science. For example, D. Muller[12] uses automata on strings to characterize a class of circuits and then obtains an analysis method of these circuits. In another example, J. R. Büchi[3] uses automata to characterize a class of second order logic formulas and then proves the decidability of the satisfiability problem for these formulas. This theory has been extended to infinite trees ([13],[14],[15],[18],[19],[20], [21],[22],[24] ,[25],[26]). The idea of extending both Büchi and Muller automata to infinite trees is due to M. O. Rabin. In [20], Rabin gives a characterization of the weak monadic theory of the tree by proving that a set of infinite trees L is weakly definable if and only if L and its complement are recognizable by Büchi tree automata. He proves also that Muller tree automata are more powerful than Büchi tree automata. So, it is natural to look for the relationship between Büchi tree automata and Muller tree automata. In [19], M. Nivat et al. introduce the notion of controlled tree automata and show that the relationship betwen Büchi tree automata and Muller tree automata is the passage from deterministic Büchi control and deterministic Muller control.
A more general notion of controlled tree automata is then given in [3]; generalized controlled tree automata are used to characterize the Boolean closure of the family accepted by deterministic Muller tree automata.
The organization of this paper is as follows. In the first section, we give some basic

definitions. In the second section, we recall some definitions and results on automata on infinite words. In the third section, we define the notion of automata on infinite trees and prove that nondeterministic automata on infinite trees are more powerful than deterministic ones. In the last two sections, we characterize recognizable and rational infinite tree sets in terms of controlled tree automata. We relate family of sets accepted by deterministic Muller automata to generalized controlled tree automata.

1. BASIC DEFINITIONS

Before defining the classes of infinite tree automata, we will give some basic definitions.

Infinite words and ω-languages :

An infinite word over Σ is a mapping from $[\omega]$, the set of all natural numbers, to Σ. We denote the set of infinite words over Σ by Σ^ω. An ω-language is a set of infinite words. Let L be a language over Σ. We define :

$(i) \, L^\omega = \{u \in \Sigma^\omega : \, u = u_1 u_2 \ldots \text{ and } u_i \in L\}.$

$(ii) \, K.L^\omega = \{u : u = u_1 u_2, u_1 \in K, u_2 \in L^\omega, \text{ and } u \text{ is an infinite word }\}$

The limit of L is the set of infinite words u such that the set of initial segments of u meets infinitely many time the set L. An ω-language is called rational iff there exist two sequences $(A_i)_{0 \le i \le n}$ and $(B_i)_{0 \le i \le n}$ of rational (i.e regular) languages such that

$$L = \bigcup_{i=1}^{n} A_i B_i^\omega$$

The ω-language L defines what Gold and Cohen [5] call the ω-Kleene closure of $(A_i)_{0 \le i \le n}$ and $(B_i)_{0 \le i \le n}$. Let u be an infinite word, we denote by $Inf(u)$ the set of symbols that occur infinitely often in u.

Finite and infinite k-ary trees :

Let Σ be a finite alphabet, the set of k-ary finite trees over Σ, denoted by T_Σ, is defined inductively as follows :

(i) If $a \in \Sigma$ then $a \in T_\Sigma$.

(ii) If $t_1, t_2, \ldots, t_k \in T_\Sigma$ and $a \in \Sigma$ then $a(t_1, \ldots, t_k) \in T_\Sigma$.

Let t be a k-ary finite tree, called tree for short, then the domain of t, denoted by $dom(t)$, is defined inductively as follows :

(i) If $t \in \Sigma$ then $dom(t) = \{\lambda\}$, where λ denotes the empty word.

(ii) If $t = a(t_1, \ldots, t_k)$ then $dom(t) = \{\lambda\} \cup (\cup_{i=1}^{k} i.dom(t_i))$.

An infinite tree (i.e. ω-tree) over Σ is a mapping from $D = \{1, \ldots, k\}^*$ to Σ . We denote the set of infinite trees over Σ by T_Σ^ω .

Let t be an infinite tree, then we define an infinite branch of t starting at the root as an infinite word $(t(u_i))_{i \ge 0}$, where :

(i) $u_0 = \lambda$, and

(ii) For each i, there exists a $j_i \in \{1, \ldots, k\}$ such that $u_i = u_{i-1} j_i$.

A projection is a mapping from a set Σ to a set Δ . A projection determines a mapping from the set of trees on Σ to the set of trees on Δ.

2. AUTOMATA ON INFINITE WORDS

Before defining automata on infinite trees, we will define ω-automata (i.e. finite automata on infinite words) to make clear the passage from words to trees.

The idea of using automata for recognizing infinite sequences is due to the late Büchi [4]. Büchi used ω-automata to prove the decidability of the monadic second order theory of natural numbers with the successor relation, which is called S1S.

Definition 2.1 A Büchi automaton is a structure $M=< Q, \Sigma, q_0, \delta, F >$, where :
(i) Q is a finite set of states,
(ii) Σ is the input alphabet,
(iii) q_0 is the initial state,
(iv) $\delta : Q \times \Sigma \to 2^Q$ is the transition function, and
(v) $F \subseteq Q$ is the set of designated final states.
A computation of M over an infinite word u is a mapping C: $[\omega] \to Q$ satisfying the following conditions :
(i) $C(0) = q_0$, and
(ii) for each $i \in [\omega]$, $C(i + 1) \in \delta(C(i), u(i))$.
M accepts the infinite word u if and only if there exists a designated state which occurs infinitely often in this computation.
In 1963, D. Muller [12] introduced a new condition for accepting infinite sequences and used it, with finite state automata, to analyze asynchronous circuits.
Definition 2.2
A **Muller automaton** is a structure $M= < Q, \Sigma, q_0, \delta, F >$, where :
(i) Q, Σ, q_0, δ are defined as before, and
(ii) F is the family of designated sets of states.
A computation C is accepted in the sense of Muller if and only if the set of states occurring infinitely often in this computation belongs to F.
M is called deterministic if and only if for each state q and each input symbol a, we have $Card(\delta(q, a)) \leq 1$.
Proposition 2.1 For each ω-language L the following conditions are equivalent.
(i) L is accepted by a Büchi (resp. Muller) automaton, and
(ii) L is a projection of an ω-language accepted by a deterministic Büchi (resp. Muller) automaton.
Proof :
Let L be an ω-language accepted by a Büchi (resp. Muller) automaton
$M =< Q, \Sigma, q_0, \delta, F >$, we shall exhibit a deterministic Büchi (resp. Muller) automaton
$\bar{M} =< \bar{Q}, \bar{\Sigma}, \bar{q}_0, \bar{\delta}, \bar{F} >$ accepting K, and a projection Π such that $L = \Pi(K)$.
Construction :
(1) $\bar{Q} = Q$
(2) $\bar{q}_0 = q_0$

(3) $\bar{\Sigma} = \Sigma \times Q$

(4) For each $< a, q >\in \Sigma \times Q$ and $q_1 \in Q$

If $q_2 \in \delta(q_1, a)$ then $q_2 = \bar{\delta}(q_1, < a, q_2 >)$

(5) $\bar{F} = F$.

Now let Π be the projection such that, for each $< x, q >$, $\Pi(< x, q >) = x$. It is clear that $L = \Pi(K)$. To finish the proof one can use the fact that the family of ω-languages accepted by nondeterministic Büchi (resp. Muller) automata is closed under projection.

Theorem 2.1 (McNaughton[16])

For each ω-language L, the following conditions are equivalent :

(i) L is accepted by a deterministic Muller automaton, and

(ii) L is a finite Boolean combination of ω-languages accepted by deterministic Büchi automata, and

$(iii) L = \bigcup_{i=1}^{n} L_i K_i^{\omega}$, where L_i and K_i are regular languages.

The equivalence between (i) and (iii) was known as Muller's conjecture.

Theorem 2.2 (J. R. Büchi[4])

The family of ω-languages accepted by nondeterministic Büchi automata is a Boolean algebra.

Proposition 2.2

For each ω-language L, the following conditions are equivalent :

(i) L is accepted by a deterministic Büchi automaton, and

(ii) L is a limit of a rational (i.e regular) language.

The equivalence between (i) and (ii) is due to S. Eilenberg [7].

3. AUTOMATA ON INFINITE TREES

In this section, we will give the definitions of automata on infinite trees and example of automata on infinite trees.

Definition 3.1

A Büchi (resp. Muller) k-ary tree automaton is a structure $M =< Q, \Sigma, q_0, \delta, F >$, where Q, Σ, q_0, F are defined as before, and $\delta : Q \times \Sigma \rightarrow 2^{Q^k}$ is the transition function.

A **computation** of Büchi (resp. Muller) k-ary tree automaton M on the infinite tree t is the mapping from $[k]^*$ to Q, where $C(\lambda) = q_0$ and for each node u, $((C(u1), \ldots, C(uk)) \in \delta(C(u), t(u))$

A computation C is called Büchi (resp. Muller) accepted if for each branch, of this computation, satisfies the Büchi (resp. Muller) condition .

M is called deterministic if for each state q and each input symbol a, there is at most one transition in $\delta(q, a)$.

Proposition 3.1(A. Saoudi[22])

The family of sets accepted by deterministic Büchi (resp. Muller) tree automata is properly included in the family of those accepted by non deterministic Büchi (resp. Muller) tree automata.

Proof:

Let $K = a(b^\omega, b^\omega, \ldots, b^\omega, c^\omega) + a(c^\omega, b^\omega, \ldots, b^\omega)$,where x^ω is the unique infinite tree on x. One can easily construct a Büchi (resp. Muller) tree automaton accepting K. Now, assume that K is accepted by a deterministic Büchi (resp. Muller) tree automaton M . Let $C1$ (resp. $C2$) be the unique computation of M on the tree $a(b^\omega, \ldots, c^\omega)$ (resp. $a(c^\omega, \ldots, b^\omega)$). $C1$ and $C2$ are accepted. Let $C1 = q_0(A1, \ldots, A2)$ and $C2 = q_0(A3, \ldots, A4)$. Since M is deterministic, the tree $q_0(A1, \ldots, A4)$ is a computation of M on the tree $a(b^\omega, \ldots, b^\omega)$, which is accepted, this contradicts the fact that $a(b^\omega, \ldots, b^\omega)$ is not a member of K.

Proposition 3.2 (M.O. Rabin [20][21])

The family of sets accepted by Büchi (resp. Muller) automata is closed under union and intersection.

This can be easily proved using the classical construction.

Example 1 :

Let $k = 2$ and $M = < Q, \Sigma, q_0, \delta, F >$ be Büchi tree automaton, where $Q = \{q_0, q_1, q_2\}$, $\Sigma = \{a, b, c\}$, $F = \{q_1, q_2\}$, and

$\delta(q_0, a) = \{(q_1, q_2), (q_2, q_1)\}$
$\delta(q_1, b) = \{(q_1, q_1)\}$
$\delta(q_2, c) = \{(q_2, q_2)\}$

Then the set of infinite trees accepted by M is $L(M)$, where

$L(M) = \{a(b^\omega, c^\omega), a(c^\omega, b^\omega)\}$.

Proposition 3.3 (A. Saoudi [23])

For each set L of infinite trees, the following conditions are equivalent.

(i) L is accepted by a Büchi (resp. Muller) tree automaton, and

(ii) L is a projection of a set accepted by a deterministic Büchi (resp. Muller) tree automaton.

Proof :

Since the family of sets accepted by Büchi (resp. Muller) tree automata is closed under projection, it follows that condition (ii) implies condition (i).

Let $M = < Q, \Sigma, q_0, \delta, F >$ be a Büchi (resp. Muller) tree automaton accepting L. We now exhibit a deterministic Büchi (resp. Muller) tree automaton $M_1 = < Q_1, q_1, \Sigma_1, \delta_1, F_1 >$ and a projection Π, such that the set accepted by M is a projection by Π of the set accepted by M_1 .

Construction :

(i) $Q_1 = Q$

(ii) $\Sigma_1 = \Sigma \times Q \times [m]$, where m=Max(Card($\delta(q, a)$))

(iii) $q_1 = q_0$

(iv) For each $q \in Q$ and $a \in \Sigma$ do

If $\delta(q, a) = \{(q_1^1, \ldots, q_1^k), \ldots, (q_n^1, \ldots, q_n^k)\}$ then $\delta_1(q, < a, q, i >) = (q_i^1, \ldots, q_i^k)$

(v) $F_1 = F$

It is clear that $L^\omega(M_1) = p_1(L^\omega(M))$.

Proposition 3.4 (A. Saoudi [22])

The family of sets accepted by deterministic Büchi (resp. Muller) tree automata is not closed under union.

Proof :

Let $L_1 = a(b^\omega, \ldots, b^\omega, c^\omega)$ and $L_2 = a(c^\omega, \ldots, c^\omega, b^\omega)$, then one can easily construct a deterministic Di-automaton M_i (i=1,2) which recognizes L_i. We have seen that $L_1 \cup L_2$ cannot be accepted by any deterministic Muller tree automaton.

A characterization in terms of deterministic tree automata of the Boolean closure of the family of sets accepted by deterministic Muller tree automata is possible, using the following notion of M-automaton introduced by M. Nivat and A. Saoudi [19].

Definition 3.2

A M-automaton is a structure $M = < Q, \Sigma, q_0, \delta, H >$, where Q, Σ, q_0, δ are defined as for top-down tree automata and $H \subseteq 2^{2^Q}$ is a family of designated sets of sets of states. A computation C is accepted by a M-automaton if the family of sets occuring infinitely often on infinite branches of the computation belongs to H.

Proposition 3.5 (M. Nivat and A. Saoudi [19])

The Boolean closure of the family of sets accepted by deterministic Muller tree automata is equal to the family of sets of infinite trees accepted by deterministic M-automata.

In [19] it is proved that M-automata are equivalent to Muller automata. In the last section we shall see that M-automata can be related to the notion of controlled tree automata. In fact, generalized controlled tree automata, defined in [3], are equivalent to deterministic M-automata.

4. CONTROLLED AUTOMATA ON INFINITE TREES

In this section, we introduce controlled tree automata (i.e. tree automata with control language) and then compare them with the Büchi and Muller tree automata.

Definition 4.1

A **Controlled k-ary ω-tree automaton**, is a structure $M = < Q, \Sigma, q_0, \delta, K >$, where Q, Σ, q_0, δ are defined as for top-down tree automata and K is an ω-language on Q, called a control language.

A C-automaton (resp. $\bar{C}$-automaton) is a controlled tree automaton, where the control ω-language is accepted by a deterministic Büchi (resp. Muller)-automaton. A computation C of the controlled tree automaton M, is called accepted if the control language of M contains all infinites sequences lying on its branches.

Example 2 :

Let $k = 2$ and $M = < Q, \Sigma, q_0, \delta, K >$ be Büchi tree automaton, where $Q = \{q_0, q_1, q_2\}$,

$\Sigma = \{a, b, c\}$, $K = \{q_0 q_1^\omega, q_0 q_2^\omega\}$, and

$\delta(q_0, a) = \{(q_1, q_2), (q_2, q_1)\}$

$\delta(q_1, b) = \{(q_1, q_1)\}$

$\delta(q_2, c) = \{(q_2, q_2)\}$

Then the set of infinite trees accepted by M is $L(M)$, where

$L(M) = \{a(b^\omega, c^\omega), a(c^\omega, b^\omega)\}$.

Theorem 4.1(M. Nivat and A. Saoudi[18][19])

For each set K of infinite trees, the following conditions are equivalent :

(1) K is accepted by a nondeterministic (resp. deterministic) Büchi (resp. Muller) tree automaton,

(2) K is accepted by a nondeterministic (resp. deterministic) C-automaton (resp. $\bar{C}$-automaton).

Proof:

That part (1) implies (2) is obvious. Let $M_c = < Q, \Sigma, q_0, \delta, L >$ be a C-automaton (resp. $\bar{C}$-automaton) accepting K and let $A = < S, Q, s_0, \delta_c, F >$ be the deterministic Büchi (resp. Muller) automaton accepting L.

We now exhibit a Büchi (resp. Muller) tree automaton

$M_1 = < Q_1, \Sigma, q_1, \delta_1, F_1 >$ accepting K.

Construction:

(1) $Q_1 = S \times Q$

(2) $q_1 = (s_0, q_0)$

(3) For each $(s, q) \in Q_1$ $and\, a \in \Sigma\ do$

If $(q_1, \ldots, q_k) \in \delta(q, a)$ $Then$ $((\delta_c(s, q), q_1), \ldots, (\delta_c(s, q), q_k)) \in \delta_1((s, q), a)$

(4) $F_1 = \{A \subseteq Q_1 : p_1(A) \in F\}$.

One can easily prove that M_c and M_1 are equivalent (i.e they accept the same set). Note that if M_c is deterministic, M_1 is also deterministic.

5. GENERALIZED CONTROLLED TREE AUTOMATA

In this section, we generalize the notion of controlled tree automata by extending the control to a control set of ω-languages. We will use generalized controlled automata to characterize the family of sets accepted by deterministic generalized Muller tree automata.

Definition 5.1

A **generalized controlled k-ary ω-tree automaton**, is a structure $M = < Q, \Sigma, q_0, \delta, K_1, \ldots, K_n >$ where Q, Σ, q_0, δ are defined as for top-down tree automata and for every $i \in \{1, \ldots, n\}, K_i \subseteq Q^\omega$ is a control ω-language accepted by a deterministic Muller automaton.

Let C be a computation of the generalized controlled tree automaton. We denote by $s(C)$ the set of all infinite sequences lying on its branches.

Definition 5.2

An AND-automaton is a generalized controlled tree automaton, $M = < Q, \Sigma, q_0, \delta, K_1, \ldots, K_n >$. A computation C is called accepted iff $s(C)$ is contained in the language given by the union of all control languages and each control language L_i contains at least an infinite sequence which belongs to $s(C)$.

Definition 5.3

An OR-automaton is a generalized controlled tree automaton.

A computation C of OR-automaton, is called accepted iff $s(C)$ is contained in K_i for some $i \in \{1, \ldots, n\}$.

Example 3:

Let $k = 2$ and $M = <Q, \Sigma, q_0, \delta, K_1, K_2>$ be a deterministic AND-automaton, where:

$Q = \{q_0, q_1, q_2, q_3\}$,

$\Sigma = \{a, b, c\}$,

$K_1 = \{q_0 q_1 q_2^\omega\}, K_2 = \{q_0 q_1 q_3^\omega\}$

$\delta(q_0, a) = \{(q_1, q_1)\}$

$\delta(q_1, c) = \delta(q_3, c) = \{(q_3, q_3)\}$

$\delta(q_1, b) = \delta(q_2, b) = \{(q_2, q_2)\}$.

Then $L(M) = \{a(b^\omega, c^\omega), a(c^\omega, b^\omega)\}$ is the language of example 2. So we exhibit a deterministic tree automaton accepting language $L(M)$.

Theorem 5.1 (P. Bonizzoni)

For each set K of infinite trees, the following conditions are equivalent:

(1) K is accepted by an AND-automaton, and

(2) K is accepted by an M-automaton.

Proof:

Let M be an M-automaton, then there exists a Muller automaton equivalent to M. From theorem 4.1, one can construct a $\bar{C}$-automaton $M_c = <Q, \Sigma, q_0, \delta, K_1>$ which is equivalent to M. Note that M_c is an AND-automaton with only one control language. We now consider the part (1) implies (2). Let $M_a = <Q, \Sigma, q_0, \delta, K_1, \ldots, K_n>$ be an AND-automaton accepting K. We shall exhibit an M-automaton

$M = <Q', \Sigma, q_0', \delta', F'>$ accepting K.

Let K_i ($i \in \{1, \ldots, n\}$) be an ω-language accepted by the deterministic Muller automaton on infinite sequences A_i, where $A_i = <S_i, Q, s_{i0}, \delta_i, F_i>$.

Construction: (given for binary trees)

(1) $Q' = S_1 \times \ldots \times S_n \times Q$

(2) $q_0' = (s_{10}, \ldots, s_{n0}, q_0)$

(3) For each $(s_1, \ldots, s_n, q) \in Q'$ and $a \in \Sigma$ do

(δ is a total function)

if $(q_1, q_2) \in \delta(q, a)$ then

$((\delta_1(s_1, q), \ldots, \delta_n(s_n, q), q_1), (\delta_1(s_1, q), \ldots, \delta_n(s_n, q), q_2)) \in \delta'((s_1, \ldots, s_n, q), a)$

(4)$F' = \{H \subseteq 2^{Q'} : \forall D_j \in H, p_i(D_j) \in F_i$ for some $i \in \{1, \ldots, n\}$ and $\forall k \in \{1, \ldots, n\}$, there exists $D_j \in H, p_k(D_j) \in F_k\}$.

One can easily verify that M is equivalent to M_a.

Note that if M_a is deterministic then M is deterministic, while deterministic M-automata are not equivalent to deterministic AND-automata.

Proposition 5.1(P. Bonizzoni)

The family of sets accepted by deterministic OR-automata is closed under union and intersection.

Proof:

Let M be a deterministic OR-automaton $M = <Q, \Sigma, q_0, \delta, K_1, \ldots, K_n>$. It is easy to verify that $L(M) = \cup_{i=1}^{n} L(\bar{C}_i)$, where $\bar{C}_i = <Q, \Sigma, q_{i0}, \delta_i, K_i>$ is a deterministic $\bar{C}$−automaton.

Now let M_1, M_2 be deterministic OR-automata, then $L(M_1) \cup L(M_2) = \cup_{i=1}^{m} L(\bar{C}_i) = K$, where $\bar{C}_i = <Q, \Sigma, q_{i0}, \delta_i, K_i>$, and $\bar{C}_i$ is equivalent to a deterministic Muller automaton (see theorem 4.1).

Using the property of closure under intersection of deterministic Muller automata, as seen above for the union, the intersection of sets of trees accepted by deterministic OR-automata can be expressed as the union of sets accepted by Muller deterministic automata.

We now exhibit the deterministic OR-automaton $M = <Q', \Sigma, q', \delta', K_1', \ldots, K_m'>$ that accept $K = \cup_{i=1}^{m} L(\bar{C}_i)$. Let $K_i = L(A_i)$ with $A_i = <S, Q, s_i, \rho_i, F_i>$.

Construction:

(1) $Q' = q_0 \cup \{(q_1, \ldots, q_m) : q_i \in Q\}$

(2) For each $a \in \Sigma, q_0 \in Q'$:

$\delta(q_0, a) = \{((p_1(\rho_1(q_{10}, a)), \ldots, p_1(\rho_m(q_{m0}, a))), (p_2(\rho_1(q_{10}, a)) \ldots, p_2(\rho_m(q_{m0}, a))))\}$ (3) for each $(q_1, \ldots, q_m) \in Q^m$

$\delta((q_1, \ldots, q_m), a) = \{((p_1(\rho_1(q_1, a)), \ldots, p_1(\rho_m(q_m, a))), (p_2(\rho_1(q_1, a)), \ldots,$

$p_2(\rho_m(q_m, a)))\}$, where p_i is a projection.

(4) $K_i' = L(B_i)$ with $B_i = <S, Q', s_i, \rho_i', F_i>$ and for each $s \in S, (q_1, \ldots, q_m) \in Q'$:

$\rho_i'(s, (q_1, \ldots, q_m)) = \rho_i(s, p_i(q_1, \ldots, q_m))$. Now let $t \in K$ and C a computation of $\bar{C}_i$-automaton on t, $s(C) \subseteq K_i$ if and only if $s(C') \subseteq K_i'$, for C' computation of M on t.

Any deterministic Muller automaton is equivalent to a deterministic $\bar{C}$-automaton. But any $\bar{C}$-automaton is a deterministic OR-automaton with only one control language. Then it is immediate to prove the following proposition.

Proposition 5.2 (P. Bonizzoni)

The closure under union and intersection of sets accepted by deterministic tree Muller automata is equal to the family of sets accepted by deterministic OR-automata.

Proof: It follows from proposition 5.1.

Example 4:

Let $L(M_1) = a(c^\omega, b^\omega)$ and $L(M_2) = a(b^\omega, d^\omega)$, where M_1, M_2 are deterministic Muller automaton. Then $L(M_1) \cup L(M_2)$ is accepted by M, deterministic OR-automaton, while is not accepted by any deterministic Muller automaton.

Let k=2, $M = <Q, \Sigma, q, \delta, K_1, K_2>$ where:

$Q = \{q_0, q_1, q_2, q_3, q_4, q_5, q_6\}$,

$\Sigma = \{a, c, b, d\}$,

$K_1 = \{q_0 q_1 q_3^\omega, q_0 q_2 q_4^\omega\}, K_2 = \{q_0, q_1 q_5^\omega, q_0 q_2 q_6^\omega\}$ and

$\delta(q_0, a) = \{(q_1, q_2)\}$

$\delta(q_1, c/b) = \{(q_3, q_3)\}/\{(q_5, q_5)\} \quad \delta(q_2, b/d) = \{(q_4, q_4)\}/\{(q_6, q_6)\}$.

ACKNOWLEDGEMENT

The second author would like to thank G. Mauri for his suggestions.

REFERENCES

[1]**A. Arnold and M. Nivat**, Formal computations of nondeterministic recursive programm schemes, Math. Syst. Theory 13(1980)219-236.

[2]**A. Arnold and M. Nivat** , The metric space of infinite trees: Algebraic and topological properties, Fundamenta Informatica 4(1980)445-476.

[3]**P. Bonizzoni and G. Mauri**, On automata on infinite trees, to appear on Theoret. Comput. Sci.

[4] **J.R. Büchi**, On a decision method in restricted second order arithmetic, Proc. Cong. Logic Method and Philos. of Sci. 1960, Univ. Press, Stanford California .

[5] **R. S. Cohen and A. Y. Gold**, Theory of ω-languages I and II : A study of various models of ω-type generation and recognition, J. Comput. Syst. Sci. 15 (1977).

[6]**B. Courcelle**, Fundamental properties of infinite trees, Theoret. Comput. Sci. 25(1985)95-169.

[7] **S. Eilenberg**, Automata, Languages and Machines, vol. A, Academic Press (1974).

[8]**I. Guessarian**, On push-down tree automata, Math. Syst. Theory 16(1983) 237-263.

[9]**Y. Gurevich and L. Harrington**, Trees, Automata, and Games, Proc. 14th ACM Symp. on Theory of Computing (1982)237-263.

[10] **H.J.Hoogeboom and G. Rozenberg**, Infinitary languages: basic theory and applications to concurrent systems, Lect. Not. in Comp. Sci. 224, Springer, Berlin (1986) 266-342.

[11]**R. Hosseley and C. Rackoff**, The emptiness problem for automata on infinite trees, 13th IEEE on switching and automata theory(1972)121-124.

[12]**D. E. Muller**, Infinite sequences and finite machines, Proc. 4th IEEE on Switching Circuit Theory and Logical Design(1963)3-16.

[13]**D. E. Muller and P. E. Schupp**, Alternating automata on infinite objects, determinancy and Rabin's theorem, Actes de l'école de printemps d'Informatique Théorique, L.N.C.S no 109, (D. Perrin and M. Nivat, Eds.), Springer-Verlag, Berlin(1984)100-107.

[14]**D. Muller, A. Saoudi and P. Schupp**, Alternating automata, the weak monadic theory and its complexity, 13th Int. Coll. on Automata, Languages and Programming, L.N.C.S no. 226, Springer-Verlag, Berlin(1986).

[15]**D. Muller, A. Saoudi and P. Schupp**, Weak alternating automata give a simple explanation of why most temporal and dynamic logic are decidable in exponential time,Proc. of the 3rd Annual Symp. on Logic in Computer Science, July (1988).

[16]**McNaughton**, Testing and generating infinite sequences by a finite automaton, Inf. and Control no 9(1966)521-530.

[17] **M. Nivat**, Infinite words, infinite trees and infinite computations, Foundations of computer science III, part 2, Math. Center.Tract 109(1979).

[18]**M. Nivat et A. Saoudi**, Automata on infinite objects and their applications to logic and programming, Information and Computation, vol. 83, no. 1(1989)41-64.

[19]**M.Nivat et A. Saoudi**, Automata on infinite trees and rational infinite tree sets, Univ. Paris 7, Report 87-59 (1987).

[20]**M. O. Rabin**, Decidability of second order theories and automata on infinite trees, Trans. Amer. Math. Soc. 141(1969)1-35.

[21]**M. O. Rabin**, Weakly definable relations and special automata, Math. Logic and Foundation of set theory, (Y.Bar Hillel, Edit.), North-Holland, Amsterdam (1970)1-23.

[22] **A. Saoudi**, Infinite tree languages recognized by ω-automata, Inf. Proc. Letters 18(1984)15-19.

[23] **A. Saoudi**, Variétés d'automates descendants d'arbres infinis, Theoret. Comput. Sci. 43(1986)315-335.

[24]**M.P. Schutzenberger**, Sur les relations rationelles fonctionelles, Col. Automata, Languages, and Programming, (M. Nivat, Edit.), North-Holland, Amsterdam (1973)103-114.

[25]**W. Thomas**, A Hierarchy of sets of infinite trees ,"in Theoretical Computer Science", G.I Conference, (A.B. Cremers and H.P. Kriegel, Eds), L.N.C.S no 145, Springer-Verlag, Berlin (1982)335-342.

[26]**W. Thomas**, Automata on infinite objects, in "Handbook of Theoretical Computer Science", North-Holland(1990).

Tree Automata and Languages
M. Nivat and A. Podelski (editors)
1992 Elsevier Science Publishers B.V.

Recognizing Sets of Labelled Acyclic Graphs

P. Bonizzoni, G. Mauri, G. Pighizzini and N. Sabadini

Dipartimento di Scienze dell'Informazione, Università degli Studi di Milano
Via Moretto da Brescia, 9 – 20133 Milano – ITALY

Abstract
Labelled acyclic graphs with some restrictions on the labelling function can be used to describe concurrent processes. In this paper, we show how the restrictions on the labelling functions can be related with the assumptions made on the properties of the dependence and independence relations between actions. Furthermore, we compare two different recognizing devices for a particular class of labelled acyclic graphs, i.e. finite state automata on a free partially commutative monoid and finite state asynchronous automata, and give some results on the existence of minimal automata recognizing a given language.

1. INTRODUCTION

Most of the structures introduced as formal models of concurrent processes, e.g. pomsets [Pr], semitraces [Oc2] and traces [Ma1] can be represented by labelled acyclic graphs with particular conditions on the labelling function. In particular, dependence graphs related with traces have been studied from different points of view [Ma2], [AR].

In this paper, we give a classification of labelled acyclic graphs based on the above conditions, so as to define classes of graphs corresponding to the different models of concurrent processes. Furthermore, we show how these conditions are strictly related with the properties of the dependence and independence relations between actions which are assumed.

In Section 3, the construction of a dependence graph from a string of symbols, based on a given dependence structure for the alphabet, is shown, and an algebraic structure on dependence graphs is defined. Then, in Section 4, the particular case of concurrent alphabets $\langle\Sigma,\theta\rangle$ and traces is treated, and the well known result about the isomorphism between the free partially commutative monoid over $\langle\Sigma,\theta\rangle$ and the monoid of dependence graphs on the same alphabet is recalled.

For each class of graphs, we can now try to define a notion of "recognizable subset", based on a suitable recognition device. In Section 5 the recognition of traces is treated, with a comparison between finite state automata based on a free partially commutative monoid and finite state asynchronous automata, which assume concurrency as a primitive concept. The equivalence between these two recognizing devices has been proved by Zielonka [Zi1]. The above two kinds of automata have, beside the deeply differences in the structure, different algebraic properties, such as for example the existence of minimal automata recognizing a given language, as it is shown in Section 6.

Finally, in Section 7 an informal discussion is given about the problem of recognizing sets of infinite dependence graphs, which is an open problem.

2. ACYCLIC LABELLED GRAPHS AND DEPENDENCE STRUCTURES

In this section, we will give some basic definitions, starting from the one of acyclic labelled graph.

Def.2.1 – An *acyclic labelled graph* is a quadruple $L = (V,E,\Sigma,\lambda)$, where V is a set of *nodes*, $E \subseteq V^2$ is the set of *edges*, Σ is the *label alphabet* and $\lambda: V \to \Sigma$ is the *labelling function*. Furthermore, does not exist any sequence $v_1, v_2, ..., v_n$ of nodes such that $(v_i,v_{i+1}) \in E$ for $i=1,...,n-1$, and $v_1=v_n$. Equivalently, we can say that the reflexive and transitive closure of the relation E is antisymmetric, and hence it is an ordering relation $\leq$ on V.

As we said in the introduction, some of the formal models of concurrent processes, such as Mazurkiewicz's traces [Ma1] and Pratt's pomsets [Pr] can be described by labelled acyclic graphs. In both cases, however, the graphs have the property that they don't contain arcs which can be obtained by composition of other arcs (they are Hasse diagrams of ordering relations). Furthermore, graphs representing traces have strong restrictions on the labelling function. The following definition summarizes the restrictions we will impose on graphs.

Def.2.2 – An acyclic labelled graph $L = (V,E,\Sigma,\lambda)$ is:
- *reduced* if $\forall n>1(E \cap E^n = \emptyset)$, where E^n denotes the n–th power of E;
- *in line labelled* if $\forall x,y \in V(\lambda(x)=\lambda(y) \Rightarrow (x,y) \in E^+)$, where E^+ denotes the transitive closure of E;
- *coherent* if $\{(\lambda(x),\lambda(y)) \mid (x,y) \in E\} \cap \{(\lambda(x),\lambda(y)) \mid (x,y) \notin E\} = \emptyset$.

It should be clear that the reduction property excludes the presence of arcs which are composition of other arcs; the in line labelling means that nodes with the same labelling must belong to the same path in the graph; the coherence property means that if symbols a and b label nodes connected by an arc, then the same symbols cannot label two disconnected nodes.

In Def.2.1, Σ is assumed to be a generic set, without further structure. Now, we can consider the case where Σ represents a set of (names of) actions which can be executed in a given system. Some of these actions are independent, in the sense for example that they use completely different resources, and hence can be executed in a completely independent way, and some others depend from each other, in the sense that one can be executed only after the occurrence of another. This is captured by the following definition.

Def.2.3 – A *dependence structure* is a triple $S = \langle\Sigma,D,\theta\rangle$ where $\Sigma = \{\sigma_1,\sigma_2,...,\sigma_m\}$ is a finite alphabet and $D,\theta\subseteq\Sigma^2$ are binary relations called *dependence* and *independence* (or *concurrency*) relation, respectively.

The simpler, and most studied, case of dependence structures is that of concurrent alphabets, which lead to the definition of traces.

Def. 2.4 – A *concurrent alphabet* is a dependence structure $\langle\Sigma,D,\theta\rangle$ in which θ is symmetric and irreflexive and D is the complement of θ, that will be denoted by θ^c.

Hence, in a concurrent alphabet two symbols are necessarily dependent (concurrent) or independent, but not both. Since $D=\theta^c$, we can denote a concurrent alphabet simply by $\langle\Sigma,\theta\rangle$.

3. DEPENDENCE GRAPHS

In this section we will show how, for concurrent alphabets (and for more general dependence structures), it is possible to associate with each string $v\in\Sigma^*$ a labelled acyclic graph, called dependence graph, and will relate properties of the dependence structure with properties of the corresponding graphs.

The theory of dependence graphs for concurrent alphabets is strictly connected with the development of the theory of traces. In fact a trace corresponds in a natural way to a partial order on the occurrences of its symbols and hence to a directed node–labelled graph, that is defined as a dependence graph. Thus dependence graphs give a representation of traces that gets explicit ordering of symbols and allows to overcome some limitations caused by a representation of traces as an equivalent class of strings, as shown in the next section. A nice example of this is the definition of the concatenation for infinite traces, given by Diekert [Di] using a representation of traces as dependence graphs.

Since infinite graphs are treated in a very natural way, in the following we will consider finite and infinite graphs together. Let us recall that an infinite word over a finite alphabet Σ, is a function w: $N-\{0\} \rightarrow \Sigma$. We will denote by Σ^ω, the set of infinite words on Σ and pose $\Sigma^\infty = \Sigma^\omega \cup \Sigma^*$.

Def.3.1 – A *dependence graph* over a concurrent alphabet $\langle\Sigma,\theta\rangle$ is a labelled acyclic graph (V,E,Σ,λ) (with a finite or a countable number of nodes) in which two nodes are connected if and only if they are labelled with dependent symbols, i.e.:

    a) $\forall x,y \in V\ [(x,y) \in E \cup E^{-1} \text{ iff } (\lambda(x),\lambda(y)) \in \theta^c]$.

We will denote by $\mathbf{G}(\Sigma,\theta)$, the set of finite and infinite dependence graphs, while by $\mathbf{FG}(\Sigma,\theta)$ we will denote the set of finite dependence graphs.

Two dependence graphs are isomorphic if there exists a bijection between their nodes preserving labelling and edges connections. As usual two isomorphic graphs are identified; all properties of dependence graphs are given up to isomorphism.

Now we define a mapping G: $\Sigma^\infty \rightarrow \mathbf{G}(\Sigma,\theta)$ that associates a dependence graph with each string $u \in \Sigma^\infty$.

Def.3.2 – Let $u \in \Sigma^\infty$, the *graph G(u) associated to u* is $L_u = (V,E,\Sigma,\lambda)$ where:

    (i) $V = \{(a,i): a \in \Sigma, \text{ is i–th occurrence of a in u}\}$

    (ii) $E = \{((a,i), (b,k)): a\theta^c b, \text{ and i–th a occurs before k–th b in u}\}$

    (iii) $\lambda(a,i) = a$.

It is easy to verify that for each $u \in \Sigma^\infty$, G(u) is a dependence graph, that is an acyclic graph satisfying condition a). Conversely, for any dependence graph G over Σ, there exists a string $w \in \Sigma^\infty$, such that G is isomorphic to G(w).

Example 3.1 – Let $\langle\Sigma,\theta\rangle$ be a concurrent alphabet, with $\Sigma=\{a,b,c\}$ and $\theta=\{(a,b),(b,a)\}$ and w=acaabbc. The graph associated with w is presented by the following figure:

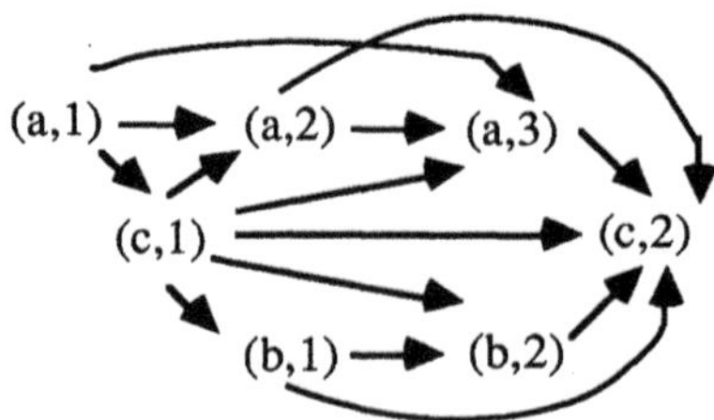

A reduced dependence graph can be associated with w, by deleting all edges (x,y) from E such there is some z with $(x,z)\in E$ and $(z,y)\in E$. Such a graph is called *Hasse diagram*. For example, the Hasse diagram associated with the graph of the previous example is:

$$
\begin{array}{ccccc}
(a,1) & & (a,2) \longrightarrow (a,3) & & \\
 & (c,1) & & & (c,2) \\
 & & (b,1) \longrightarrow (b,2) & &
\end{array}
$$

We have:

Th.3.1 – A reduced dependence graph on a concurrent alphabet is in line labelled and coherent.

A concatenation operation on dependence graphs can be naturally defined as follows:

Def.3.3 – Let $G_1=(V_1,E_1,\Sigma,\lambda_1)$ and $G_2=(V_2,E_2,\Sigma,\lambda_2)$ be two dependence graphs in $G(\Sigma,\theta)$ such that $V_1 \cap V_2 = \varnothing$.

The *concatenation* of G_1 and G_2, denoted by $G_1 \otimes G_2$, is the graph (V,E,Σ,λ), where

 i) $V = V_1 \cup V_2$,

 ii) $E = E_1 \cup E_2 \cup \{(x,y)\in V_1 \times V_2 \mid (\lambda_1(x),\lambda_2(y))\in \theta^c\}$,

 iii) for all $x\in V_1$, $\lambda(x) = \lambda_1(x)$, while for all $y\in V_2$, $\lambda(y) = \lambda_2(y)$.

It can be easily verified that concatenation is well defined and associative. A neutral element for the concatenation can be defined as the graph with empty sets of edges and labels, and will be denoted by 1_G.

Example 3.2 – Let $\langle\Sigma,\theta\rangle$ be the concurrent alphabet $\Sigma = \{a,b,c\}$ and $\theta = \{(a,b),(b,a)\}$, and $w_1 = acaabb$, $w_2 = aabcc$. The concatenation $G(w_1)\otimes G(w_2)$ is represented by the following Hasse diagram:

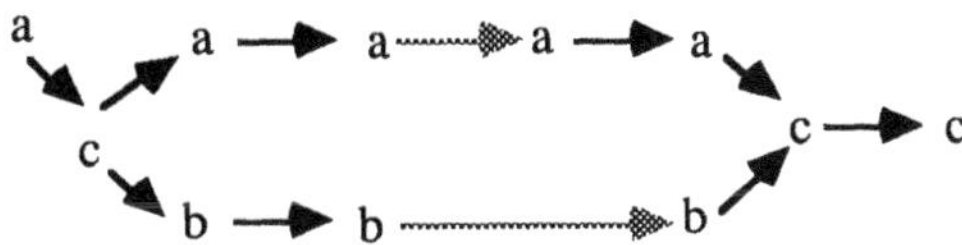

where the gray arrows represent arcs added for concatenating $G(w_1)$ and $G(w_2)$.

It is verified that, for a concurrent alphabet $\langle\Sigma,\theta\rangle$, two strings $u,v\in\Sigma^*$ have isomorphic dependence graphs if and only if they are equivalent in the sense that they define the same trace. This fact will be illustrated in the following section where we will introduce the standard notion of trace as equivalence class of strings in the free partially commutative monoids.

4. TRACES AND THEIR ALGEBRAIC STRUCTURE

Following an algebraic approach traces are defined by means of concurrent alphabets and free partially commutative monoids, as follows.

Def.4.1 – The *free partially commutative monoid* (fpcm, for short) $F(\Sigma,\theta)$ generated by a concurrent alphabet $\langle\Sigma,\theta\rangle$ is the quotient structure $F(\Sigma,\theta) = \Sigma^*/\equiv_\theta$, where $\equiv_\theta$ is the least congruence on Σ^* which extends the set of "commutativity laws": $\{\sigma_1\sigma_2{=}\sigma_2\sigma_1 \mid (\sigma_1,\sigma_2)\in\theta\}$.

Def.4.2 – A *trace* t on a concurrent alphabet $\langle\Sigma,\theta\rangle$ is an element $t\in F(\Sigma,\theta)$ i.e. a trace is a congruence class of words. $[w]_\theta$ will denote the class containing the string w. A *trace language* T is a subset of $F(\Sigma,\theta)$.

As a consequence of Def.4.1, the concatenation of the traces $[u]_\theta$ and $[v]_\theta$, containing the strings $u,v\in\Sigma^*$, denoted by $[u]_\theta{\cdot}[v]_\theta$, is the trace $[uv]_\theta$. The unit of $F(\Sigma,\theta)$, is the class $[\varepsilon]_\theta$, where ε is the empty string in Σ^*.

Given a language $L\subseteq\Sigma^*$ it is possible to associate with it a trace language as follows:

Def.4.3 – Given a language L on S and a concurrency relation $\theta\subseteq\Sigma\times\Sigma$, the set $[L]_\theta{=}\{[w]_\theta \mid w\in L\}$ is the *trace language generated by L* under θ.

Finite dependence graphs are an intuitive representation of finite traces. In fact finite words are equivalent under the congruence $\equiv_\theta$, defined above, iff they define the same dependence graph. More precisely, the relationship between traces and finite dependence graphs is expressed by the following theorem.

Th.4.1 [AR,Ma2] – Let $u,v\in\Sigma^*$; then $u\equiv_\theta v$ iff $G(u){=}G(v)$.

Hence, a dependence graph is really associated with a trace rather than with a string. This leads us to the following definition.

Def.4.4 – Let $t \in F(\Sigma, \theta)$. A dependence graph of t, denoted by G(t), is a graph isomorphic to
G(x), where $t = [x]_\theta$.

Connections between graphs and finite traces are more clearly expressed by the following
theorem, stating that the monoid of finite dependence graphs is isomorphic to the monoid
$F(\Sigma, \theta)$ of finite traces. In fact concatenation of dependence graphs naturally corresponds to
concatenation of traces.

Th.4.2 [Ma2] – Given a concurrent alphabet $\langle \Sigma, \theta \rangle$, the monoid $(\mathbf{FG}(\Sigma, \theta), \otimes, 1_G)$ of finite
dependence graphs over Σ is isomorphic with the monoid $(F(\Sigma, \theta), \cdot, [\varepsilon]_\theta)$ of traces over
Σ. The isomorphism is specified by $G(u) \rightarrow [u]_\theta$, for each string $u \in \Sigma^*$.

Dependence graphs can be defined in the same way for dependence structures more
general than concurrent languages. Relationships between properties of the relations D, θ and
properties of the labelling functions are as follows.

Th.4.3 – Let $S = \langle \Sigma, D, \theta \rangle$ be a dependence structure, and G(x) the reduced dependence
graph induced by the string $x \in \Sigma^*$. Then:
– if θ is irreflexive and $D \cup \theta = \Sigma^2$, then G(x) is in line labelled, and is called a
semitrace;
– if $(\theta - \Delta_\Sigma)^c - \Delta_\Sigma = D - \Delta_\Sigma$, where $\Delta_\Sigma = \{(\sigma, \sigma) \mid \sigma \in \Sigma\}$, then G(x) is coherent, and is
called a *coherent pomset.*

If we don't give any assumption on D and θ, then we obtain generic labelled acyclic
graphs, which corresponds to Pratt's *pomsets.*

5. RECOGNIZING DEVICES FOR TRACE LANGUAGES

At this point, we have defined four classes of reduced acyclic labelled graphs: pomsets,
coherent pomsets, semitraces and traces, and we are interested in recognizing subclasses of
these classes. Here, we will restrict ourselves to the recognition of sets of traces, for which
some results have been given. The extension to more general cases is an open problem.

We will consider finite state devices accepting trace languages. In particular, we will
analyze two models: automata over the free partially commutative monoid $F(\Sigma, \theta)$ ($F(\Sigma, \theta)$–
automata, for short) and asynchronous automata. While both models characterize the same
class of traces languages, they are deeply different from the point of view of their internal

structure. In fact, an $F(\Sigma,\theta)$–automaton recognizes a trace reading in a *sequential* way the letters of such a trace, i.e. considering a linear order compatible with the dependence graph, while an asynchronous automaton works directly on the dependence graph. Then, from the point of view of the concurrency theory, $F(\Sigma,\theta)$–automata represent only the *commutativity* among actions, while asynchronous automata are able to represent the *concurrency*.

We start giving the notion of automata over a free partially commutative monoid $F(\Sigma,\theta)$. Since $F(\Sigma,\theta)$ is a particular monoid, the general notion of automaton over a monoid M (M–automaton, for short) can be used.

Def.5.1 – Let M be a monoid with unit 1. A *M–automaton* A is a quadruple $A=\langle Q,q_0,\delta,F\rangle$, where:
- Q is a finite set of states;
- δ: Q×M $\rightarrow$ Q is a transition function such that:

$$\delta(q,1) = q \quad \text{for every } q \in Q$$
$$\delta(q,mm') = \delta(\delta(q,m),m') \quad \text{for every } m,m' \in M,\ q \in Q$$

- $q_0 \in Q$ is the initial state
- $F \subseteq Q$ is the set of final states.

Def.5.2 – The *language recognized* by a M–automaton A is the set:
$$L(A) = \{m \in M \mid \delta(q_0,m) \in F\}.$$

The class $\mathbf{Rec}(\Sigma,\theta)$ of *recognizable trace languages* over the fpcm $F(\Sigma,\theta)$ is defined as the class of languages accepted by $F(\Sigma,\theta)$–automata. On the other hand, the class $\mathbf{Rat}(\Sigma,\theta)$ of *rational trace languages* is defined as the least class of trace languages containing finite trace languages and closed with respect to the standard rational operations (union, concatenation and Kleene's closure). It is well–known that a free partially commutative monoid $F(\Sigma,\theta)$ is a Kleene monoid, i.e. $\mathbf{Rat}(\Sigma,\theta) = \mathbf{Rec}(\Sigma,\theta)$, if and only if θ is the empty relation [BBMS], while, in the general case, $\mathbf{Rec}(\Sigma,\theta) \subsetneq \mathbf{Rat}(\Sigma,\theta)$ [BMS1].

A different characterization of the class $\mathbf{Rec}(\Sigma,\theta)$ in terms of operations on sets was obtained by E. Ochmanski [Oc1].

As usual $F(\Sigma,\theta)$–automata will be represented by transition diagrams (where bold circles represent final states). We recall that a state $q \in Q$ is *reachable* in A if and only if there exists $m \in M$ such that $\delta(q_0,m)=q$; the automaton A is *reachable* iff every state in Q is reachable. From a unreachable automaton it is always possible to obtain a reachable automaton accepting the same trace language, removing all unreachable states.

Example 5.1 – Let $\langle\Sigma,\theta\rangle=\langle\{a,b,c\},\{(a,c),(c,a)\}\rangle$ be a concurrent alphabet and A the $F(\Sigma,\theta)$–automaton represented in the following figure:

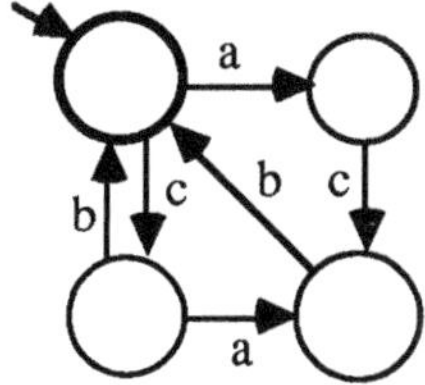

It is easy to see that A is a reachable automaton recognizing the trace language $T = [\{acb,cb\}^*]_\theta$. Observe that the previous transition diagram is also the diagram of a $F(\Sigma,\emptyset)$–automaton accepting the language $L = \{acb,cb,cab\}^*$. Then, transition diagrams do not give informations about the concurrency relation.

As a consequence of the definition, given a $F(\Sigma,\theta)$–automaton $A = \langle Q,q_0,\delta,F\rangle$, for every state $q\in Q$ and for every pair $(a,b)\in\theta$, it holds: $\delta(q,ab) = \delta(q,ba)$. So, $F(\Sigma,\theta)$–automata are devices representing the commutativity among independent actions. Whatever, as remarked in the previous example, from the transition diagram of the automaton (or, equivalently from the tuple $\langle Q,q_0,\delta,F\rangle$) it is impossible to deduce the concurrency relation. Then, the independence among actions is not represented by the automaton.

In order to obtain a device representing the concurrency among actions and working directly on a "concurrent representation" (as dependence graphs) W. Zielonka [Zi1] introduced *asynchronous automata*.

An asynchronous automaton appears as a set of control units which can act independently or synchronize each other. Every action, represented by a symbol, is processed by a subset of control units; two actions independent are processed by disjoint sets of control units. Thus, asynchronous automata can be seen as abstractions of distributed systems.

In the following, given a finite number of sets $A_1,...,A_k$, we will denote by $\prod_{i=1}^{k} A_i$ the cartesian product of $A_1,...,A_k$.

Def.5.3 – A *finite state asynchronous automaton* (FSAA) A with n processes is a tuple $A=\langle P_1,P_2,...,P_n,\Delta,F\rangle$ such that:

- $\forall i$, $1\leq i\leq n$, $P_i=\langle\Sigma_i,S_i,s_{i0}\rangle$ is the i–th process where: Σ_i is a finite non empty *local alphabet*, S_i is a finite set of *local states* s.t. $S_i\cap S_j=\emptyset$ for $j\neq i$; $s_{i0}\in S_i$ is the *initial state* of P_i;

- Δ is a set of *transition functions* containing exactly one transition function δ_σ for every action σ belonging to the *global alphabet* $\Sigma_A = \bigcup_{i=1}^{n} \Sigma_i$. The transition function δ_σ is a (possibly partial) function from $\prod_{i \in \mathrm{Dom}(\sigma)} S_i$ in itself; that is $\delta_\sigma : \prod_{i \in \mathrm{Dom}(\sigma)} S_i \rightarrow \prod_{i \in \mathrm{Dom}(\sigma)} S_i$, where $\mathrm{Dom}(\sigma) = \{ i \mid \sigma \in \Sigma_i \}$ is the *domain* of σ, that is the set of processes that can execute σ;

- $F \subseteq \prod_{i=1}^{n} S_i$ is the set of *final states*.

  Every $(s_1, \ldots, s_n) \in \prod_{i=1}^{n} S_i$ will be called *global state* of A.

We remark that given two actions $\sigma, \sigma' \in \Sigma_A$ whose domains $\mathrm{Dom}(\sigma)$ and $\mathrm{Dom}(\sigma')$ are disjoint, the functions δ_σ and $\delta_{\sigma'}$ act on disjoint set of local states and, consequently, the actions σ and σ' can be executed concurrently. Thus we can associate to each finite state asynchronous automaton A a *concurrency (independence) relation* denoted by θ_A and defined as:

$$\theta_A = \{ (\sigma, \sigma') \in \Sigma_A \times \Sigma_A \mid \mathrm{Dom}(\sigma) \cap \mathrm{Dom}(\sigma') = \emptyset \}.$$

For describing the "global behaviour" of a given asynchronous automaton A, we associate to every trace t a *t–reachability relation* $\overset{t}{\Rightarrow}$ on the set of global states, such that a pair (q,p) belongs to $\overset{t}{\Rightarrow}$ if and only if the automaton A starting from q and executing the trace t reaches the state p.

In order to formally define $\overset{t}{\Rightarrow}$, first we define for every $\sigma \in \Sigma_A$ the reachability relation $\overset{\sigma}{\Rightarrow}$ on the set of global states. Intuitively a pair (q,p) of global states belongs to $\overset{\sigma}{\Rightarrow}$ if the local states in q corresponding to processes executing σ are modified in p according to δ_σ and the remaining local states do not change. More formally:

Def.5.4 – Given a finite state asynchronous automaton $A = \langle P_1, \ldots, P_n, \Delta, F \rangle$, for every $\sigma \in \Sigma_A$ the *σ–reachability relation* $\overset{\sigma}{\Rightarrow} \subseteq (\prod_{i=1}^{n} S_i) \times (\prod_{i=1}^{n} S_i)$ is defined as follows:

$\forall (s_1, \ldots, s_n), (u_1, \ldots, u_n) \in \prod_{i=1}^{n} S_i$,

$\qquad (s_1, \ldots, s_n) \overset{\sigma}{\Rightarrow} (u_1, \ldots, u_n)$ iff $\forall i \notin \mathrm{Dom}(\sigma)$ $s_i = u_i$ and $(u_{i_1}, \ldots, u_{i_k}) = \delta_\sigma(s_{i_1}, \ldots, s_{i_k})$

where $\{ i_1, \ldots, i_k \} = \mathrm{Dom}(\sigma)$.

Now, given $\overset{\sigma}{\Rightarrow}$ the t–reachability relation $\overset{t}{\Rightarrow}$ can be inductively defined in the following way:

Def.5.5 – The ε–reachability relation $\overset{\varepsilon}{\Rightarrow}$ is the identity relation, that is

$$\overset{\epsilon}{\Rightarrow} = \{(q,p)\in (\prod_{i=1}^{n}S_i)\times(\prod_{i=1}^{n}S_i) \mid q=p\}$$

and, for every $x\in \Sigma_A^*$ and $\sigma\in \Sigma_A$, the $x\sigma$–reachability relation is the set:

$$\overset{x\sigma}{\Rightarrow} = \{(q,p)\in (\prod_{i=1}^{n}S_i)\times(\prod_{i=1}^{n}S_i) \mid \exists r\in \prod_{i=1}^{n}S_i \text{ s.t. } q \overset{x}{\Rightarrow} r \text{ and } r \overset{\sigma}{\Rightarrow} p\}.$$

Observe that $\backslash O(^{\sigma\sigma'};\Rightarrow) = \backslash O(^{\sigma'\sigma};\Rightarrow)$ for every $(\sigma,\sigma')\in \theta_A$, so we can define the t–reachability relation $\overset{t}{\Rightarrow}$, for every trace t, as:

$$\overset{t}{\Rightarrow} = \overset{x}{\Rightarrow}$$

where x is a string such that $[x]_{\theta_A}=t$.

Now, using the notion of t-reachability we can define the *trace language* $T\subseteq F(\Sigma_A,\theta_A)$ *recognized* by a given asynchronous automaton A as the set of traces t such that a final state is t–reachable in A from the initial state, that is:

$$T = \{t\in F(\Sigma_A,\theta_A) \mid (s_{10},..., s_{n0}) \overset{t}{\Rightarrow} (s_1,...,s_n) \text{ and } (s_1,...,s_n)\in F\}.$$

As pointed out by Zielonka [Zi1], a pictorial representation of Asynchronous Automata can be obtained using Petri Nets. The meaning of such a representation will be immediately clear from the next example.

Example 5.2 – Let A be the asynchronous automaton with two processes so defined: $\Sigma_1=\{a,b\}$, $\Sigma_2=\{b,c\}$ (then $\mathrm{Dom}(a)=\{1\}$, $\mathrm{Dom}(b)=\{1,2\}$ and $\mathrm{Dom}(c)=\{2\}$), $S_1=\{s_0,s_1\}$, $S_2=\{r_0,r_1\}$ (s_0 and r_0 initial local states), $F=\{(s_0,r_0)\}$, $\delta_a(s_0)=s_1$, $\delta_c(r_0)=r_1$, $\delta_b(s_0,r_1)=(s_0,r_0)$, $\delta_b(s_1,r_1)=(s_0,r_0)$. The concurrent alphabet associated with A is $<\{a,b,c\},\{(a,c),(c,a)\}>$.

The representation of A as Petri Net is given in the following figure:

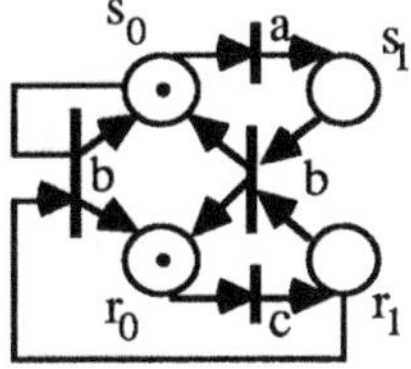

The language accepted by A is the set $T = [\{acb,cb\}^*]$.

In the next section we are interested in minimizing asynchronous automata. For formalizing the concept of minimum asynchronous automaton, we have to introduce the preliminary notions of reachable state, of reachable finite state asynchronous automaton and some important properties of this type of automata.

We start giving a preliminary example:

Example 5.3 – Consider the asynchronous automaton A on the concurrent alphabet $\langle\Sigma,\theta\rangle = \langle\{a,b,c\},\{(a,c),(c,a)\}\rangle$ represented by the following Petri Net (where $S_1=\{u_0,u_1,u_2,u_3,u_4\}$, $S_2=\{v_0,v_1,v_2,v_3\}$, $F=\{(u_0,v_0),\ (u_1,v_1),\ (u_2,v_2),\ (u_3,v_3)\}$):

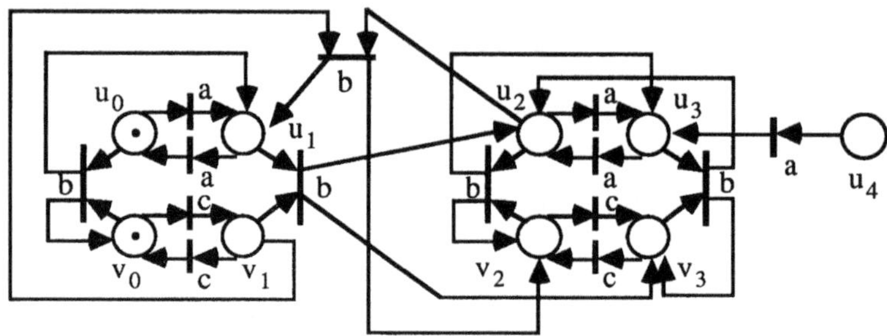

Observe that for any trace $t\in F(\Sigma,\theta)$, starting from the initial state (u_0,v_0), the automaton cannot go in a global state whose first component is u_4. Then, u_4 (and the transition $\delta_a(u_4)=u_3$) is not used and it can be removed obtaining another automaton recognizing the same trace language. Moreover, while the local states $v_1\in S_1$ and $u_2\in S_2$ are reachable (in fact $(u_0,v_0) \stackrel{a}{\Rightarrow} (u_1,v_0)$ and $(u_0,v_0) \stackrel{acb}{\Rightarrow} (u_2,v_3)$) the global state (v_1,u_2) is not reachable. So, the transition $\delta_b(v_1,u_2)=(u_1,v_2)$ is never used, and can be removed obtaining the following "smaller" automaton A_r:

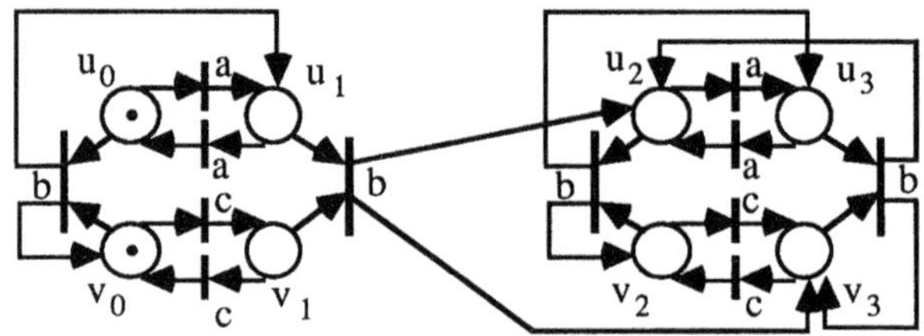

where all states and all transitions are used in the computation of same trace. An automaton with such a property is said to be *reachable*.

Now, we give the formal definition of reachable state and of reachable automaton and from this point on *we consider only reachable automata*.

Def.5.6 – Given an asynchronous automaton A and a set $\alpha=\{i_1,...,i_k\}\subseteq\{1,...,n\}$, a tuple of local states $(s_{i_1},...,s_{i_k})$, $s_{ij}\in S_{ij}$ $1\leq j\leq k$, is said to be *reachable* whether there exist a trace $t\in F(\Sigma_A,\theta_A)$ and a global state $(u_1,...,u_n)\in\prod_{i=1}^{n}S_i$ such that $(s_{10},...,s_{n0})\overset{t}{\Rightarrow}(u_1,...,u_n)$, and $u_{ij}=s_{ij}$, for $j=1,...,k$.

Def.5.7 – An asynchronous automaton A is reachable if and only if the following conditions hold:

(1) every local state $s\in S_i$, $1\leq i\leq n$, is reachable;

(2) for every $\sigma\in\Sigma$, with $Dom(\sigma)=\{i_1,...,i_k\}$ and $(s_{i_1},...,s_{i_k})\in\prod_{i\in Dom(\sigma)}S_i$, if $(s_{i_1},...,s_{i_k})$ is not reachable, then $\delta_\sigma(s_{i_1},...,s_{i_k})$ is not defined.

As shown in the Example 5.3, given a unreachable asynchronous automaton A, it is easy to obtain a reachable automaton A' recognizing the same trace language, by removing all unreachable local states and all transitions from unreachable tuples of local states.

If we impose that a given asynchronous automaton A executes the symbols sequentially, we can obtain a $F(\Sigma_A,\theta_A)$–automaton with the following characteristics:

Def.5.8 – Given a finite state asynchronous automaton A, the *sequential version* of A is the $F(\Sigma_A,\theta_A)$–automaton $\hat{A}=<Q,q_0,\delta,F>$, such that:

$-$ $Q\subseteq\prod_{i=1}^{n}S_i$ is the set of reachable global states of A;

$-$ $q_0=(s_{10},...,s_{n0})$;

$-$ $\forall\sigma\in\Sigma_A$, $(s_1,...,s_n),(u_1,...,u_n)\in Q$:
$$(u_1,...,u_n) = \delta((s_1,...,s_n),\sigma) \text{ iff } (s_1,...,s_n)\overset{\sigma}{\Rightarrow}(u_1,...,u_n)$$

$-$ F is the set of final reachable global states of A.

Example 5.4 – The sequential version of the automaton A_r of Example 5.3 is the $F(\Sigma,\theta)$–automaton represented by the following transition diagram:

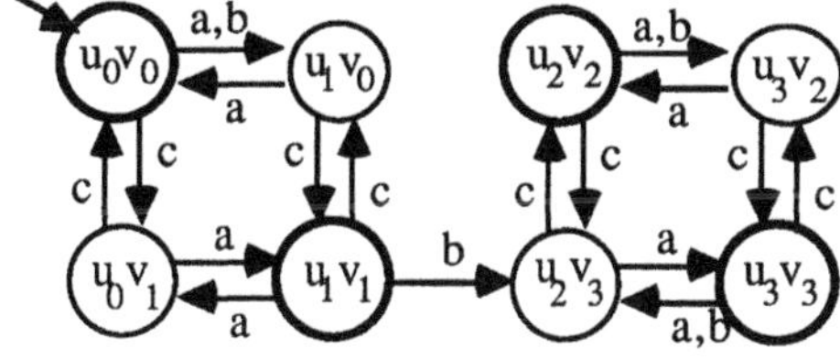

As shown in [Du], for every asynchronous automaton A accepting a language $T \subseteq F(\Sigma,\theta)$ there exists another asynchronous automaton A' such that every local alphabet Σ_i of A' is a maximal clique of the dependence relation and $\hat{A}$, $\hat{A}'$ are isomorphic. Thus in the following, we will consider, without loss of generality, only asynchronous automata whose local alphabets $\Sigma_1,\ldots,\Sigma_n$ are the maximal cliques of the dependence graph.

By definition, asynchronous automata are distributed devices that can execute independent actions in a concurrent way, while $F(\Sigma,\theta)$–automata use the interleaving in the execution of independent actions.

A natural problem is then the investigation in the relationships between the classes of behaviours (languages accepted) of $F(\Sigma,\theta)$–automata and of asynchronous automata.

From Definition 5.8, it is immediate to see that any asynchronous automaton A and its sequential version $\hat{A}$ recognize the same trace language. Then, the class of recognizable trace languages contains the class of languages accepted by asynchronous automata, and, in other words, the "concurrency" can be reduced to the "commutativity" among actions.

The converse problem has been investigated by W. Zielonka [Zi1], obtaining the following surprising result:

Th.5.1 – For every trace language $T \in \mathbf{Rec}(\Sigma,\theta)$ there exists a finite state asynchronous automaton A that recognizes T such that $<\Sigma_A,\theta_A>=<\Sigma,\theta>$.

In other words, Theorem 5.1 states that for every "commutative" system ($F(\Sigma,\theta)$–automaton A), there exists a concurrent system (asynchronous automaton A' with the *same* concurrent alphabet) with the same behaviour (language accepted). We remark that the concurrent alphabet associated to A' is $<\Sigma,\theta>$ and then *all* possible concurrency of the system is represented in the automaton A', i.e. independent actions of $<\Sigma,\theta>$ are executed by disjoint sets of resources.

The proof of Theorem 5.1 (see [Zi1,CM,CMZ]) is rather complicated and requires the use of many properties of ideal orders associated to traces. For concurrent alphabets with acyclic dependence graph are known more simple proofs ([BMS2,Me,Pe]).

An interesting variant of the models considered here and called *Asynchronous Cellular Automata* has been defined in [Zi2] and considered in [CMZ] and [Pi2]. This model is characterized, as Von Neumann's Cellular Automata [Vo], by a collection of elementary automata with local interconnections. However, while all components of Cellular Automata change state at the same time, in Asynchronous Cellular Automata only non–connected units can act concurrently.

In [Pi2], the author shows that Asynchronous Cellular Automata and Asynchronous Automata are polynomially related, by exhibiting polynomial time reductions between them. Then, also Asynchronous Cellular Automata are distributed devices characterizing the class of recognizable trace languages.

An extension of Asynchronous Automata to the probabilistic case has been recently introduced in [JPS]. In such a model the transition functions are substituted by transition matrices and the initial global state is chosen by means of a stochastic distribution on the set of global states. The generalization to the probabilistic case of Zielonka's main result, that is the equivalence between the class of behaviours of $F(\Sigma,\theta)$–Probabilistic Automata and of Probabilistic Asynchronous automata with concurrent alphabet $<\Sigma,\theta>$, has been proved in [JPS] for concurrent alphabets with acyclic dependence graph. The problem for generic concurrent alphabets is still open.

An equivalent definition of Asynchronous Automata is given in [Th]. In such a definition the automaton behaviour is defined in terms of runs on dependence graphs. Using this notion, it is proved that a trace language is recognizable by an asynchronous automaton if and only if the corresponding set of dependence graphs is definable in monadic second order logic.

6. MINIMAL ASYNCHRONOUS AUTOMATA

The proof of the Zielonka's Theorem ([Zi1],[CM]) describes an algorithm that, given a $F(\Sigma,\theta)$–automaton, produces an Asynchronous Automaton with concurrent alphabet $<\Sigma,\theta>$ that recognizes the same trace language. However, the automaton so constructed is very big, being its number of states super–exponential in the number of states of the given $F(\Sigma,\theta)$–automaton. So, it is useful to find a way of minimizing asynchronous automata. Such a problem has been investigated in [BPS] and in [Pi1]; the corresponding problem for Asynchronous Cellular Automata is considered in [Pi2].

In [BPS], the authors show that given a concurrent alphabet $<\Sigma,\theta>$, every recognizable language on $F(\Sigma,\theta)$ admits a minimum asynchronous automaton if and only if the dependence relation is transitive. Such a result is extended in [Pi1], showing that there are recognizable languages accepted by infinitely many minimal (non isomorphic) asynchronous automata. This section is devoted to present these results.

For monoid automata, the minimum automaton recognizing a language T is usually defined as the automaton with the minimum number of states. An equivalent Def. can be obtained in terms of universal algebras, defining the minimum automaton as the final object in the category of reachable automata accepting T. In a similar way, we can define the minimum asynchronous automaton recognizing T (if any) as the final object in the category of asynchronous automata accepting T. To consider such a category, we need of the definition of morphism between asynchronous automata.

Informally, a morphism between two automata A and A' is a family of maps from the set of states of A to the set of states of A', mapping the initial state of A in the initial state of A', preserving the transitions, mapping final states in final states and nonfinal states in nonfinal states. Intuitively, this map describes how the states of A can be grouped in order to obtain a "smaller" automaton A' recognizing the same trace language. For example consider the reachable automaton A_r of the Example 5.3 and the asynchronous automaton A', represented in the following figure (where the final states are (s_0,r_0) and (s_1,r_1)):

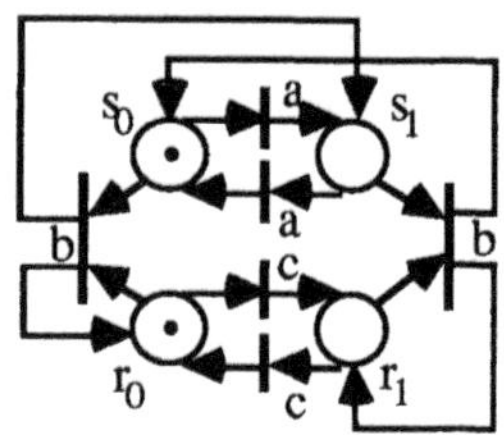

The pair of maps ϕ_1, ϕ_2 given by $\phi_1(u_0)=\phi_1(u_2)=s_0$, $\phi_1(u_1)=\phi_1(u_3)=s_1$, $\phi_2(v_0)=\phi_2(v_2)=r_0$, $\phi_2(v_1)=\phi_2(v_3)=r_1$, is a morphism from A_r to A'. Intuitively, the state s_0 (s_1,r_0, r_1, resp.) of A' is obtained identifying the states u_0 and u_2 (u_1 and u_3, v_0 and v_2, v_1 and v_3, resp.) of A_r.

Formally the definition of morphism is given as follows:

Def.6.1 – Given two finite state asynchronous automata $A=<P_1,\ldots,P_n,\Delta,F>$ and $A'=<P'_1,\ldots,P'_n,\Delta',F'>$ on $F(\Sigma,\theta)$, a *morphism* φ *between A and A'* (φ: A$\to$A') is a family of functions $<\varphi_i:S_i \to S'_i>_{i=1,\ldots,n}$ such that:

(1) for every i, $1\leq i\leq n$, $\varphi_i(s_{i0})=s'_{i0}$, that is φ preserves the initial states;

(2) for every $\sigma\in\Sigma$ with $Dom(\sigma)=\{i_1,\ldots,i_k\}$ and for every reachable tuple $(s_{i_1},\ldots,s_{i_k})\in S_{i_1}\times\ldots\times S_{i_k}$ and $\forall j$, $1\leq j\leq k$:

$$\varphi_{ij}(Proj [\delta_\sigma(s_{i_1},\ldots,s_{i_k})]) = Proj [\delta'_\sigma(\varphi_{i_1}(s_{i_1}),\ldots,\varphi_{i_k}(s_{i_k}))]$$

(where $Proj (x_1,\ldots,x_k)=x_j$), that is φ preserves the transitions;

(3) for every reachable global state $(s_1,\ldots,s_n)\in\prod_{i=1}^{n}S_i$, $(s_1,\ldots,s_n)\in F$ if and only if $(\varphi(s_1),\ldots,\varphi(s_n))\in F'$, that is φ preserves the set of final states.

It is easy to prove that for every pair A and A' of reachable finite state asynchronous automata, if there exists a morphism φ: A$\to$A', then this morphism is unique.

In the following we consider the category AA_T of the *reachable* finite state asynchronous automata that recognize a given trace language T with their morphisms.

We briefly recall that an object F in a category $\mathcal{C}$ is called *final object* if and only if for every object A in $\mathcal{C}$ there exists exactly one morphism $\varphi\colon A \to F$; the final object of a category $\mathcal{C}$, whether it exists, is unique up to isomorphism. Moreover, if $\mathcal{C}$ is a category of reachable objects, every morphism is surjective.

The following categorical definition of minimal automaton and of minimum automaton in a category $\mathcal{C}$ of reachable automata are from [EK].

Def. 6.2 – An automaton A in a category $\mathcal{C}$ is called *minimal* or *reduced* if and only if for every automaton A' in $\mathcal{C}$, every morphism $\varphi\colon A \to A'$ is an isomorphism. A is *minimum* if and only if A is the final object of $\mathcal{C}$.

From the above definition., the *minimum finite state asynchronous automaton* recognizing a trace language T is the final object in AA_T. Since the final object is unique, such an automaton, if any, is unique.

It is known that for every recognizable trace language T, there exists a unique (up to isomorphism) minimum $F(\Sigma,\theta)$–automaton accepting T [Ne]. In [BPS], the authors show that this result, in general, is not true for asynchronous automata. In fact, they prove that there exists a trace language that *does not admit* the minimum asynchronous automaton.

Example 6.1 – Let $T=[\{\{a,b,c\}\{a,c\}\}^*]_\theta$ be a language on the concurrent alphabet $<\Sigma,\theta> = <\{a,b,c\},\{(a,c),(c,a)\}>$. Observe that the dependence graph has the form:

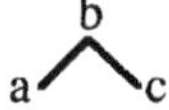

so the maximal cliques of the dependence relation, i.e. the local alphabets of asynchronous automata with concurrent alphabet $<\Sigma,\theta>$, are $\{a,b\}$ and $\{b,c\}$.

It is possible to verify that the following finite state asynchronous automata A_1 and A_2 (represented as Petri Nets) recognize the language T (set of final states $F_1=\{(s_0,r_0),(s_1,r_1)\}$ and $F_2=\{(s_2,r_2),(s_3,r_3)\}$):

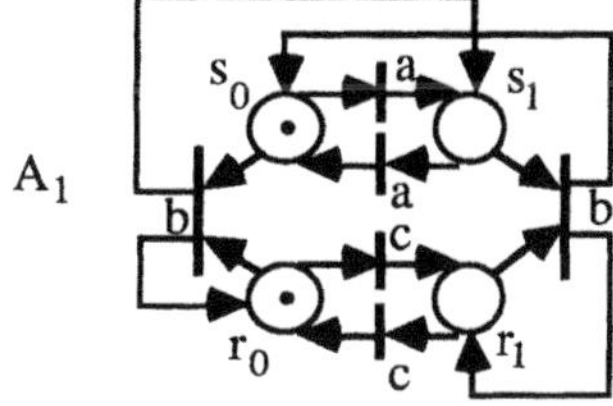

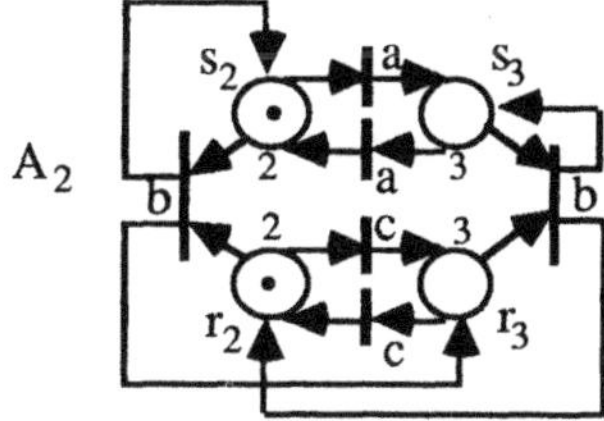

The sequential versions $\hat{A}_1$ and $\hat{A}_2$ of previous asynchronous automata can be represented by the following transition diagrams (where bold circles denote final states):

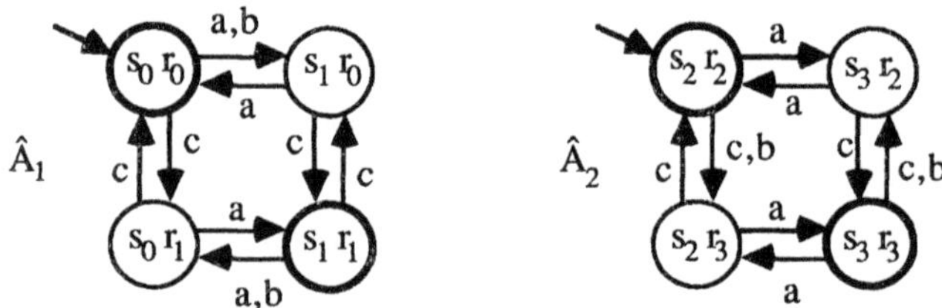

The minimum $F(\Sigma,\theta)$–automaton accepting T, is the $F(\Sigma,\theta)$–automaton A represented in the following picture:

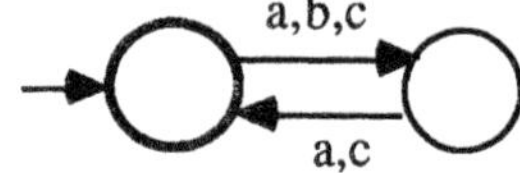

Since A recognizes the trace language $T=[\{\{a,b,c\}\{a,c\}\}^*]_\theta$, also $\hat{A}_1,\hat{A}_2$ and A_1,A_2 recognize T.

Using the definition of morphism, it is possible to verify that every morphism from A_1 (A_2, resp.) to another asynchronous automaton accepting T is an isomorphism. On the other hand, it is immediate to see that A_1 and A_2 are non isomorphic automata in AA_T.

Then, the automata A_1 and A_2 are two minimal non isomorphic objects in the category AA_T and the final object does not exist.

As a consequence of the previous example, for every concurrent alphabet $<\Sigma,\theta>$ with nontransitive dependence relation, there exists a recognizable trace language $T\subseteq F(\Sigma,\theta)$ that does not admit a minimum asynchronous automaton.

On the other hand, if the dependence relation of $<\Sigma,\theta>$ is transitive then the fpcm $F(\Sigma,\theta)$ is isomorphic to the cartesian product $\Sigma_1^*\times\ldots\times\Sigma_n^*$, where $\Sigma_1,\ldots,\Sigma_n$ are the maximal (disjoint) cliques of the dependence relation. In [BPS], the authors prove that for every recognizable trace language T defined over a such a monoid $F(\Sigma,\theta)$ there exists the *minimum* asynchronous automaton. The proof of that result is obtained giving a construction that extends the standard one for monoid automata, based on Nerode equivalence [Ne].

So, the following characterization holds:

Th.6.1 [BPS] – Let $<\Sigma,\theta>$ be a concurrent alphabet. Then the following sentences are equivalent:

(1) Every recognizable trace language $T \subseteq F(\Sigma,\theta)$ admits a unique (up to isomorphism) minimum finite state asynchronous automaton.

(2) The dependence relation θ^c is transitive.

Example 6.2 – Given the concurrent alphabet $\langle \Sigma,\theta \rangle$, where $\Sigma=\{a,b,c\}$, $\theta=\{(a,c),(c,a)\}$, for every integer $m \geq 1$, we consider the asynchronous automaton $A_m=\langle P_1,P_2,\Delta,F \rangle$ so defined:

- $\Sigma_1 = \{a,b\}$, $\Sigma_2 = \{b,c\}$;
- $S_1 = \{s_0,s_1\}$, $S_2 = \{r_1,u_1,r_2,u_2,\ldots,r_m,u_m\}$, initial local states s_0,s_1;
- $\delta_a(s_0)=s_1$, $\delta_a(s_1)=s_0$,

  for $i=1,\ldots,m$:

$$\delta_b(s_0,r_i)=(s_1,r_i), \qquad \delta_b(s_1,u_i) = \begin{cases} (s_0,u_{i+1}) & \text{if } i<m \\ (s_1,r_m) & \text{elsewhere} \end{cases}$$

$$\delta_c(r_i)=u_i, \quad \delta_c(u_i)=r_i;$$

- $F=\{(s_0,r_i) \mid i=1,\ldots,n\} \cup \{(s_1,u_i) \mid i=1,\ldots,m\}$.

The automaton A_m has a particular structure: the transition diagram associated to its sequential version $\hat{A}_m$ contains m strong connected components jointed one after the other by edges with label b. All such connected components, except the last one, are isomorphic graphs. For example, the graph of the automaton $\hat{A}_3$ is as in the following figure:

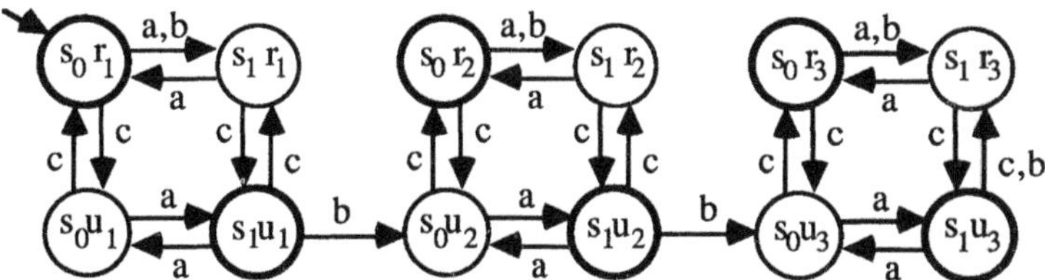

First, we observe that we can minimize the automaton $\hat{A}_m$ identifying all final states in a unique state and all nonfinal states in another state. In such a way, we obtain the minimum $F(\Sigma,\theta)$–automaton of Example 6.1. Then, for every $m \geq 1$, the asynchronous automaton A_m recognizes the trace language $T=[\{\{a,b,c\}\{a,c\}\}^*]_\theta$.

Now, we want to prove that A_m is a minimal asynchronous automaton. To achieve this goal, we will prove, using the definition of morphism, that every morphism φ from A_m to another asynchronous automaton A' is an isomorphism.

We recall that, as stated above, we consider only reachable automata and then only surjective morphisms. Suppose that φ is not an isomorphism; then there are two states $q,p \in S_i$, for some $i \in \{1,2\}$, whose images by φ coincide, i.e. $\varphi_i(q)=\varphi_i(p)$.

We consider all the possible cases:

$- \varphi_1(s_0)=\varphi_1(s_1)$:

this case is contradictory since, by Def.6.1(3), $(\varphi_1(s_0),\varphi_2(r_1)) \in F'$, $(\varphi_1(s_1),\varphi_2(r_1)) \notin F'$ and $(\varphi_1(s_0),\varphi_2(r_1))=(\varphi_1(s_1),\varphi_2(r_1))$.

$- \varphi_2(r_i)=\varphi_2(u_j)$, $i,j=1,\ldots,n$: contradictory for the same argument of the previous case.

$- \varphi_2(u_i)=\varphi_2(u_n)$: $i=1,\ldots,n-1$: it turns out that $(\varphi_1(s_1),\varphi_2(u_i))=(\varphi_1(s_1),\varphi_2(u_n))$ and then: $\delta_b'(\varphi_1(s_1),\varphi_2(u_i))=\delta_b'(\varphi_1(s_1),\varphi_2(u_n))$, and, being by Definition 6.1(2) $\delta_b'(\varphi_1(s_1),\varphi_2(u_i))=(\varphi_1(s_0),\varphi_2(u_{1+1}))$, $\delta_b'(\varphi_1(s_1),\varphi_2(u_n))=(\varphi_1(s_1),\varphi_2(r_3))$, it follows $(\varphi_1(s_0),\varphi_2(u_{1+1}))=(\varphi_1(s_1),\varphi_2(r_3))$ that is contradictory by the previous cases.

$- \varphi_2(u_i)=\varphi_2(u_j)$ or $\varphi_2(r_i)=\varphi_2(r_j)$: $i,j=1,\ldots,n-1$ and $i<j$: it is possible to verify that these equalities imply $\varphi_2(u_{i+1})=\varphi_2(u_{j+1})$, and then, for some $k>0$, $\varphi_2(u_{i+k})=\varphi_2(u_n)$, the previous case.

Then we can conclude that for every m the finite state asynchronous automaton A_m is minimal.

As a consequence of the previous example (obtained from the first example, presented in [Pi1], of a countable family of non isomorphic minimal finite state asynchronous automata recognizing a same trace languages), we obtain the following proposition:

Th.6.2 – Let $\langle\Sigma,\theta\rangle$ a concurrent alphabet with nontransitive dependence relation. Then, there exist trace languages $T \subseteq F(\Sigma,\theta)$ that admit infinitely many minimal finite state asynchronous automata.

Similar results have been obtained also for asynchronous automata with an infinite number of states. In particular in [Pi1] it is proved that for every concurrent alphabet with nontransitive dependence relation there exists a trace language accepted by infinitely many minimal finite state asynchronous automata and by infinitely many minimal *infinite* state asynchronous automata.

7. RECOGNIZING SETS OF INFINITE DEPENDENCE GRAPHS

As seen in Section 4, finite dependence graphs give a graphical representation of traces that makes explicit the ordering of symbol occurrences within traces. Analogously infinite dependence graphs, introduced in [Ma2], can represent infinite traces.

In order to extend the trace theory to the mathematical description of nonterminating concurrent systems behaviours, a theory of infinite traces is being developing [BMP,Di,Ga1,Ga2,GR,Kw].

While a finite trace is an equivalence class of finite sequential observations (words) of a concurrent system behaviour, an infinite trace is an equivalence class of infinite observations which can abstract an infinite behaviour of concurrent systems that are not supposed to eventually stop, such as distributed systems. A notion of infinite trace as an equivalent class of infinite words was independently introduced in [GR], [Kw].

In [GR] a notion of equivalence module concurrency relation is introduced. This notion leads to introduce commutations on infinite words. Thus a notion of infinite trace based (as in the finitary case) on equivalence relation among infinite words can be given.

Recent results in the theory of infinite traces ([Ga2], [Di]) show that a direction for developing a theory of infinite traces is strictly connected with graph representation of traces.

A general result showing that infinite dependence graphs can represent infinite traces, can be stated as a generalization of Th.4.1. In fact infinite words that have the same dependence graph are equivalent in the sense of the definition introduced by Gastin and Rozoy [GR].

The study of recognizable and rational languages of infinite traces is still at the beginning. Some research direction concerning recognizable languages are suggested in [Ga2]. In the theory of recognizable languages for infinite words, recognizability can be defined by means of morphisms over finite monoids. Following this approach in [Ga2] a definition of recognizable language of infinite traces is given by characterizing some properties of a syntactic congruence introduced for infinite traces. In [GPZ] a Kleene's like theorem is proved for languages of infinite traces showing that the recognizable languages of infinite traces are exactly the co-rational languages. The recognizability of infinitary trace languages by automata is an open problem; the definition of a notion of automaton with infinite behaviour and Büchi or Muller acceptance conditions for infinitary trace languages recognition seems not to be trivial. Furthermore, generalizing asynchronous automata to the infinitary case using similar acceptance conditions seems to be a difficult problem.

Note – This work has been supported by ESPRIT Basic Research Action N. 3166: Algebraic and Syntactic Methods in Computer Science (ASMICS).

REFERENCES

[AR]   IJ.J. Aalbersberg, G. Rozenberg,*Theory of traces*, Theoretical Computer Science **60**, pp.1-82, 1988

[BBMS]   A. Bertoni, M. Brambilla, G. Mauri, N. Sabadini, *An application of the theory of free partially commutative monoids: asymptotic densities of trace languages*, Proc.10th Symp. MFCS, Lecture Notes in Computer Science **118**, pp. 205-215, 1981

[BMP]   P. Bonizzoni, G. Mauri, G. Pighizzini, *About infinite traces*, Proc. Workshop "Free Partially Commutative Monoids", Tech. Rep. I9002, Technische Univerisität München, pp. 1-10, 1990

[BMS1]   A. Bertoni, G. Mauri, N. Sabadini, *A hierarchy of regular trace languages and some combinatorial applications*, Proc. II World Conf. on Math. at the Service of Men, Las Palmas, pp. 146-153, 1982

[BMS2]   A. Bertoni, G. Mauri, N. Sabadini, *Concurrency and commutativity,* Workshop on Petri Nets, Varenna, 1982

[BPS]   D. Bruschi, G. Pighizzini, N. Sabadini, *On the existence of the minimum asynchronous automaton and on decision problems for unambiguous regular trace languages*, Proc. 5th STACS, Lecture Notes in Computer Science **294**, pp. 334-346, 1988

[CM]   R. Cori, Y. Métivier, *Approximation of a trace, asynchronous automata, and the ordering of events in a distributed system,* Proc. 15th ICALP, Lecture Notes in Computer Science **317**, pp. 147-161, 1988

[CMZ]   R. Cori, Y. Métivier, W. Zielonka, *Asynchronous mappings and asynchronous cellular automata,* Tec. Rep. 89-97 - LaBRI, Univ. Bordeaux I, 1989

[Di]   V. Diekert, *On the concatenation of infinite traces*, Proc. STACS 1991, to appear in Lecture Notes in Computer Science, 1991

[Du]   C. Duboc, *Commutations dans les monoïdes libres: une cadre theorique pour l'étude du parallélisme*; Thèse de Doctorat, Université du Rouen, 1986

[EK]   H. Ehrig, K.D. Kiermeier, H.J. Kreowski, W. Kühnel, *Universal Theory of Automata,* Teubner Studienbücher, 1974

[Ga1]   P. Gastin, *Un modèle asynchrone pour les systèmes distribués*, Theoretical Computer Science **74**(2), 1990

[Ga2]   P. Gastin, *Recognizable and rational languages of finite and infinite traces*, Proc. STACS 91, to appear in Lecture Notes in Computer Science

[GPZ]   P. Gastin, A. Petit, W. Zielonka, *A Kleene theorem for infinite trace languages*, Tech. Rep. 90–93, LITP, Université Paris 6, France, 1990

[GR]   P. Gastin and B. Rozoy, *The Poset of infinitary traces*, Tech. Rep. 90-24, LITP, Université Paris 6, France, 1990

[JPS]   S. Jesi, G. Pighizzini, N. Sabadini, *Probabilistic asynchronous automata,* Proc. Workshop "Free Partially Commutative Monoids", Technische Univerisität München, pp. 99-114, 1990

[Kw]   M. Z. Kwiatkowska, *On infinitary trace languages*, Tech. Rep. 31, University of Leicester, England, 1989

[Ma1]   A. Mazurkiewicz, *Concurrent program schemes and their interpretations*, DAIMI Rep. PB-78, Aarhus University, 1977

[Ma2]   A. Mazurkiewicz, *Trace theory*, Advanced Course on Petri Nets, Lecture Notes in Computer Science **255**, pp. 279-324, 1986

[Me]   Y. Métivier, *An algorithm for computing asynchronous automata in the case of acyclic non-commutation graphs*, Proc. 14th ICALP, Lecture Notes in Computer Science **267**, pp. 226-236, 1987

[Ne]   A. Nerode, *Linear automaton transformations*, Proc. Amer. Math. Soc. **9**, pp. 541-544, 1958

[Oc1]   E. Ochmanski, *Regular behaviour of concurrent systems,* EATCS Bullettin **27**, pp. 56-67, 1985

[Oc2]   E. Ochmanski, *Semi-commutation for Place/Transition Systems,* EATCS Bullettin **38**, pp. 191-198, 1989

[Pe]   D. Perrin, *Partial commutations,* Proc 16th ICALP, Lecture Notes in Computer Science **372**, pp. 637-651, 1989

[Pi1]   G. Pighizzini, *Asynchronous automata: analysis of a category that admits infinitely many minimal objects,* Internal Report 47-89, Dip. Scienze dell'Informazione, Univ. Milano, 1989

[Pi2]   G. Pighizzini, *About asynchronous cellular automata,* Internal Report 76-90, Dip. Scienze dell'Informazione, Univ. Milano, 1990

[Pr]   V.R. Pratt, *Modelling concurrency with partial orders*, Int. J. of Parallel Programming **15**, 1986

[Th]   W. Thomas, *On logical definability of trace languages,* Proc. Workshop "Free Partially Commutative Monoids", Tech. Rep. TUM–I9002, Technische Univerisität München, pp. 172-182, 1990

[Vo]   J. Von Neumann, *Theory of Self-Reproducing Automata,* University of Illinois Press, 1966

[Zi1]   W. Zielonka, *Notes on finite asynchronous automata,* RAIRO Inf. Theor. **21**(2), pp. 99-135, 1987

[Zi2]   W. Zielonka, *Safe executions of recognizable trace languages,* Logic at Botik '89, Lecture Notes in Computer Science **363**, pp. 278-289, 1989

Tree Automata and Languages
M. Nivat and A. Podelski (editors)
© 1992 Elsevier Science Publishers B.V. All rights reserved.

Rational and Recognizable Infinite Tree Sets

A. Saoudi

L.I.P.N, Universite Paris XIII, Centre Scientifique et Polytechnique
Av J. B. Clément, 93430 Villetaneuse, France
E-mail: saoudi@LIPN.univ-paris13.fr

INTRODUCTION

The theory of automata has many applications such as in compiling, logic, and parallel processing. This theory has been extended to infinite words by defining various generating and recognizing devices (see Boasson and Nivat [2] and Büchi [3]). The above theory has been extended to infinite trees, first by Rabin [24,25] and then by Arnold et al.[1], Gurevich and Harrington [12], Muller and Schupp [9], Muller et al. [15,16,17], Nivat et al.[12], Hafer[14], Thomas [34,35], and Vardi et al. [36]. Rabin [24] gives a characterization of $S2S$ in terms of automata on infinite trees and then proves that $S2S$ is decidable. In [12], Gurevich and Harrington used a special game and reduce the complementation problem to a determinancy result. This allows them to simplify the proof of the decidability of $S2S$. Likwise, Muller and Schupp [13] introduce alternating automata on infinite trees and then give an alternative proof. It is also well known from Muller et al. [16] and Vardi et al. [36] that the complexity of decision procedures for most Temporal and Dynamic Logics depends on the complexity of tree automata.

Recently, Emerson and Jutla [8] have obtained essentially optimal decision procedures for various modal logics using an optimal algorithm for testing non-emptiness of tree automata. These give some ideas about applications of ω-tree automata theory to logic and parallel computations.

The first aim of this work is to define the notion of regular (i.e. rational) ω-tree sets and then generalize the Kleene theorem. The second aim is the simplification of Rabin's proof of the decidability of $S2S$. For this, we will try to simplify the proof of the closure under complementation of sets accepted by Rabin's tree automata. This can be done by extending Nerode's theorem to infinite trees. By using Nerode's theorem only a non-constructive proof of complementation lemma is attainable. The third aim is to study the difference between recognizable and rational (i.e. regular) ω-tree sets and the difference between recognizability by automata on ω-words and automata on ω-trees.

The organisation of this paper is as follows:
In the first section, we give some basic definitions. Some characterizations of tree sets accepted by finite state tree automata are given in the second section. In the third section, we give an extension of Kleene's theorem to automata on infinite trees, where Kleene's theorem is understood as the equality between regular (i.e. rational) languages and recognizable languages. In the fourth section, we prove that Nerode's theorem cannot extended to ω-tree automata. In the last section, we prove that deterministic top-down automata are essentially weaker for all the well known acceptance conditions and they define a proper subfamilies of recognizable ω-tree sets. Then we give an extension of Eilenberg's theorem to automata on infinite trees.

1. BASIC DEFINITIONS

Let Σ be a finite alphabet. Then the set of k-ary finite trees over Σ, called finite trees for short, denoted by T_Σ is defined inductively as follows :
(i) If $a \in \Sigma$ then $a \in T_\Sigma$.
(ii) If $t_1, t_2, ..., t_k \in T_\Sigma$ and $a \in \Sigma$ then $a(t_1, ..., t_k) \in T_\Sigma$.
Let t be a k-ary finite tree, called tree for short, then the domain of t, denoted by $dom(t)$, is defined inductively as follows :
(i) If $t \in \Sigma$ then $dom(t) = \{\lambda\}$, where λ denotes the empty word.
(ii) If $t = a(t_1, ..., t_k)$ then $dom(t) = \{\lambda\} \cup (\cup_{i=1}^{k} i.dom(t_i))$.
Note that a tree t can be viewed as a mapping from $dom(t)$ to Σ. Here trees are denoted as terms. An infinite tree (i.e. k-ary ω-tree) over Σ is a mapping from $D_k = \{1, ..., k\}^*$ to Σ. We will denote by a^ω the infinite tree over $\{a\}$. We denote by T_Σ^ω (resp. T_Σ) the set of infinite (resp. finite) k-ary trees over Σ. We define an infinite branch of the tree t, starting at the root, as an infinite word $(t(u_i))_{i\geq0}$, where :
(i) $u_0 = \lambda$, and
(ii) For each i, there exists a $j_i \in \{1, ..., k\}$ such that $u_{i+1} = u_i j_i$.
Let t_i (i=1,2) be a finite k-ary tree with F_i (i.e. Fr(t_i)) as a frontier (i.e. $F_i = \{u : u \in dom(t_i)$ and uj is not in $dom(t_i)$, for $1 \leq j \leq k\}$). We define the relation $\sqsubseteq$ between two frontiers as : $F_1 \sqsubseteq F_2$ if for each node $y \in F_2$ there exists a node $x \in F_1$ such that

$x \leq y$. t_1 is an *initial tree* of t_2 ($t_1 \leq t_2$) iff (i) $dom(t_1) \subseteq dom(t_2)$, and (ii)for each $u \in dom(t_1)$: $t_1(u) = t_2(u)$. Then t_1 is called a *proper initial tree* of t_2 (i.e $t_1 < t_2$) if t_1 is an initial tree of t_2 and $F_1 \sqsubset F_2$. Let L be a set of finite k-ary trees; we define the limit of L in the sense of Rabin as:

$Rlim(L) = \{t/$ there is an infinite sequence $(t_i)_{i \geq 0}$ with $t_i \in L$, $t_i < t_{i+1}$, and $t_i < t\}$.

Let $X = x_1, ..., x_n$ be a set of variables such that $X \cap \Sigma = \emptyset$ and let $(L_i)_{0 \leq i \leq n}$ be a sequence of sets of finite k-ary trees over $\Sigma \cup X$ such that $L_i \cap X = \emptyset$ and L_i contains finite k-ary tree such that all variables occur as values of frontier nodes. If $a \in \Sigma$ and L_i is a set of finite trees then $a(L_1, ..., L_k) = \{a(t_1, ..., t_k) : t_i \in L_i\}$. We define $L_0 < L_1, ..., L_n >$ as the set of trees that are obtained by taking t in L_0 and grafting elements of L_i at the nodes of t valued by x_i. $L_0 < L_1, ..., L_n >$ is formally defined as follows :

(i) If $t = x_i$ then $t < L_1, ..., L_n >= L_i$

(ii) If $t = a(t_1, ..., t_k)$ then $t < L_1, ..., L_n >= a(t_1 < L_1, ..., L_n >, ..., t_k < L_1, ..., L_n >)$

(iii) $L < L_1, ..., L_n >= \cup_{t \in L} t < L_1, ..., L_n >$.

Let $L_0 < L_1, ..., L_n >^p = (L_0 < L_1, ..., L_n >^{p-1}) < L_1, ..., L_n >$,

where $L_0 < L_1, ..., L_n >^0 = L_0$

$L_0 < L_1, ..., L_n >^\omega = \{t \in T_\Sigma^\omega : \exists (t_n)_{n \geq 0}$ such that $t = Rlim\{t_n : 0 \leq n \leq \omega\}, t_o \in L_o$

and $t_p \in t_{p-1} < L_1, ..., L_n >\}$.

A set L of infinite trees is said to be *rational*(i.e. regular) if there is a sequence $(L_i)_{0 \leq i \leq n}$ of rational finite tree sets such that L is equal to $L_0 < L_1, ..., L_n >^\omega$.

A *projection* is a mapping from a set Σ to a set Δ. A projection determines a mapping from the set of trees on Σ to the set of trees on Δ. For more details about properties of trees see Courcelle [4].

2. AUTOMATA AND GRAMMARS ON FINITE TREES

In this section we define top-down tree automata on finite trees and give some characterizations of tree sets accepted by top-down tree automata.

Definition 2.1

A *top-down k-ary tree automaton*, a top-down tree automaton for short, is structure $M =< Q, \Sigma, q_0, \delta, F >$, where :

(i) Q is a finite set of states,

(ii) Σ is the input alphabet,

(iii) q_0 is the initial state,

(iv) $\delta : Q \times \Sigma \to 2^{Q^k}$ is the transition function, and

(v) $F \subset Q$ is the set of terminal states .

Let t be a finite tree, then we denote by $dom^+(t) = dom(t) \cup Fr(t)\{1, ..., k\}$.

A *Computation* of the top-down tree automaton M on a tree t is a mapping from $dom^+(t)$ to Q satisfying the following conditions :

(i) $C(\lambda) = q_0$, and

(ii) for each $u \in dom(t)$: $(C(u1), ..., C(uk)) \in \delta(C(u), t(u))$.

Let C be a computation of the tree t then C is an accepting computation if and only if

$\forall u \in Fr(t)\{1, ..., k\} : C(u) \in F$.

A tree is accepted by a top-down tree automaton if it has an accepting computation.

A top-down tree automaton M is called deterministic if and only if δ is defined from $Q \times \Sigma$ to Q^k .

Definition 2.2

A *bottom-up k-ary tree automaton*, a bottom-up tree automaton for short, is a structure $M = < Q, \Sigma, \delta, Q_0, F >$, where Q, Σ, F are defined as for a top-down k-ary tree automaton, $\delta : Q^k \times \Sigma \to 2^Q$ is the transition function, and $Q_0 \subseteq Q$ is the set of initial states. A *computation of the bottom-up automaton* M on the tree t is a tree C over Q with $\forall u \in dom^+(t)\, C(u) \in Q_0$ and for each node $u \in dom(t)$, $C(u) \in \delta(< C(u1), ..., C(uk) >, t(u))$. M accepts t iff there is a computation C of M on t such that $C(\lambda)$ belongs to F.

M is said to be deterministic iff for each $< q_1, ..., q_k > \in Q^k$ and $f \in \Sigma$, $Card(\delta(< q_1, ..., q_k >, f)) \leq 1$.

Definition 2.3

A *regular k-ary tree grammar* is a structure $G = < V, \Sigma, v_0, R >$, where V is a finite set of nonterminals, Σ is an alphabet, v_0 is a start synmbol and R is a set of rewriting rules of the form $v \to t(v_1, ..., v_n)$ or $v \to t'$, where $v, v_i \in V$, $t' \in T_\Sigma$, and $t(v_1, ..., v_n)$ is a finite tree over $\Sigma \cup V$.

$t(v_1, ..., v_n)$ means that the leaves of t are labelled by letters in the set $\{v_1, ..., v_n\} \subseteq V$ and the interior nodes are labelled by letters from Σ.

Example 1:

Let G be the following grammar :

$v_0 \to a(b(v_1, v_1), v_2)$

$v_0 \to a(v_2, v_1)$

$v_1 \to b$

$v_2 \to c$.

$L(G) = \{a(b(b, b), c), a(c, b)\}$ is the language generated by G.

Theorem 2.1

For each set K of finite trees, the following conditions are equivalent :

(i) K is accepted by a top-down automaton,

(ii) K is accepted by a bottom-up automaton,

(iii) K is accepted by a deterministic bottom-up automaton,

(iv) K is rational (i.e. regular), and

(v) K is generated by a regular tree grammar.

This corresponds to the extension of Kleene's theorem, where Kleene's theorem is understood as the equality between the set of rational languages and the set of recognizable languages. For the definition of rational (i.e. regular) tree language see Gécseg and Steinby [9], Thatcher [33], and Thatcher and Wright[32]. In the next section we define the notion of regular sets of infinte trees and then gives an extension of Kleene's theorem.

Let us mention that Determinism is a real limitatition in the case of top-down tree automata. In other words, deterministic top-down tree automata define a proper subfamily of recognizable tree sets.

3. RATIONAL ω-TREE SETS AND KLEENE'S THEOREM

Let L be a set of infinite trees over Σ, L is called *rational* if and only if there is a sequence $(L_i)_{0 \leq i \leq n}$ of rational (i.e. regular) sets of finite trees such that
$L = L_0 < L_1, ..., L_n >^\omega$.
Now, we will introduce new types of infinite tree automata and then characterize both recognizable and rational ω-tree sets.

Definition 3.1

A *Büchi k-ary ω-tree automaton*, a Büchi tree automaton for short, is a structure $M =< Q, \Sigma, I, \delta, F >$, where Q is a set of finite states, Σ is a finite set of input symbols, I is the set of initial states, $\delta : Q \times \Sigma \to 2^{Q^k}$ is the transition function and F is a set of a designated states of Q.

A *computation* of M on the tree t is a tree C on Q with $C(\lambda) = q_0$ and for each node u $(C(u1), ..., C(uk)) \in \delta(C(u), t(u))$. The tree t is accepted by the Büchi tree automaton M, if for some computation of M on t, and for each branch of this computation, a state from F occurs infinitely often in this branch.

Example 2

Let $k = 2$ and $M =< Q, \Sigma, q_0, \delta, F >$ be Büchi tree automaton, where $Q = \{q_0, q_1, q_2\}$,
$\Sigma = \{a, b, c\}$, $F = \{q_1, q_2\}$, and
$\delta(q_0, a) = \{(q_1, q_2), (q_2, q_1)\}$
$\delta(q_1, b) = \{(q_1, q_1)\}$
$\delta(q_2, c) = \{(q_2, q_2)\}$
The set of infinite trees accepted by M is $L(M) = \{a(b^\omega, c^\omega), a(c^\omega, b^\omega)\}$.

Definition 3.2

A *Büchi k-ary ω-tree T-automaton*, a Büchi tree T-automaton for short, is a structure $M =< Q, \Sigma, I, \delta, T >$, where Q, Σ, δ, I are defined as before and T is a set of designated transitions.

An infinite tree t is accepted by a Büchi tree T-automaton M, if there is a computation of M on t such that each branch of this computation meets T infinitly many time.

Theorem 3.1

For each set L of infinite trees, the following conditions are equivalent :
(i) L is accepted by a Büchi tree automaton,
(ii) L is accepted by a Büchi tree T-automaton, and
(iii) L is rational .

Proof Sketch :

The equivalence between (i) and (iii) is due to Nivat et al. [19, 20]. The equivalence between (i) and (ii) can be done easily by coding computation with transitions.

Nivat et al. [19, 20] characterized rational ω-tree sets in terms of tree grammars. Rabin [25] proved that a set L os infinite trees is weakly definable iff L and $\bar{L}$ are recognizable by Büchi tree automata.

Arnold et al [1] gives another characterization of rational ω-tree sets in terms of $\nu\mu$-calculus.

4. RECOGNIZABLE ω-TREE SETS AND NERODE'S THEOREM

In this section we define several types of top-down tree automata on infinite trees. we give some characterization of recognizable infinite tree sets and show that Nerode's theorem cannot be extentded to infinite trees.

Definition 4.1

A *Muller k-ary ω-tree automaton*, a Muller tree automaton for short, is a structure $M =< Q, \Sigma, I, \delta, F >$, where Q, Σ, I, δ are defined as before, and $F \subseteq 2^Q$ is a collection of designated sets of states.

Definition 4.2

A *Muller k-ary ω-tree T-automaton*, a Muller tree T-automaton for short, is a structure $M =< Q, \Sigma, I, \delta, T >$, where Q, Σ, I, δ are defined as before and T is a family of designated sets of transitions.

The tree t is accepted by the Muller tree automaton (resp. T-automaton) M iff there is a computation of M on t such that for each infinite branch of this computation, the set of states (resp. transitions) occurring infinitely often in this branch belongs to F (resp. T).

Definition 4.3

A *generalized Muller k-ary ω-tree automaton*, a generalized Muller tree automaton, is a structure $M =< Q, \Sigma, I, \delta, F >$, where Q, Σ, I, δ are defined as before, and $F \subseteq 2^{2^Q}$ is a collection of designat ed sets.

Definition 4.4

A *generalized Muller k-ary ω-tree T-automaton*, a generalized Muller tree automaton for short, is a structure $M =< Q, \Sigma, I, \delta, T >$, where Q, Σ, I, δ are defined as before and T is a family of sets of sets of transitions.

The generalized (resp. T-automaton) Muller tree automaton M accepts t iff there is a computation of M on t such that the set of all sets of states(resp. transitions), occurring infinitely often on infinite branches belongs to F (resp. T).

Proposition 4.1

For each set of infinite trees K, the following conditions are equivalent :

(i) K is accepted by a Muller tree automaton,

(ii) K is accepted by a Muller tree T-automaton,

(iii) K is accepted by a genralized Muller tree automaton, and

(iv) K is accepted by a generalized Muller tree T-automaton.

The equivalence between (i) and (iii) is due to Nivat et al.[20, 21]. This equivalence is due to the fact that the generalized Muller condition is a boolean combination of the Muller condition. By using the same technique as for Büchi tree automata, one can easily prove that (i) is equivalent to (ii) and (iii) is equivalent to (iv).

Rabin [24] proved that a set of infinite trees is recognizable iff it is definable by a monadic second order formula. This gives another characterization of recognizable infinite tree sets. In [20], Nivat and Saoudi characterize recognizable ω-tree sets, where a recognizable set is a set accepted by a Muller tree automaton, in terms of tree grammars.

Next we show that each recognizable ω-tree set is union of some of the equivalence classes of an invariant relation of finite index. For this, we first recall some definitions and results.

Definition 4.5

A *bottom-up Muller k-ary ω-tree automaton* is a structure $M =< Q, \Sigma, \delta, Fin, F >$, where Q, Σ, F are defined as for a Muller tree automaton, $\delta : Q^k \times \Sigma \to 2^Q$ is the transition function, and $Fin \subseteq Q$ is the set of terminal states.

A *computation of the bottom-up automaton M* on the tree t is a tree C over Q with $C(\lambda) \in Fin$ and for each node u, $C(u) \in \delta(< C(u1), ..., C(uk) >, t(u))$. M accepts t iff there is a computation C of M on t such that for each infinite branch of this computation, the set of states occurring infinitely often in this branch belongs to F.

M is said to be deterministic (resp. total) iff for each $< q_1, ..., q_k >\in Q^k$ and $f \in \Sigma$, $Card(\delta(< q_1, ..., q_k >, f)) \leq 1$ (resp. $Card(\delta(< q_1, .., q_k >, f)) \geq 1$).

Theorem 4.1 (M. Nivat and A. Saoudi)

For each set L of infinite trees, the following conditions are equivalent:

(i) L is accepted by a deterministic bottom-up Muller tree automaton, and

(ii) L is recognizable.

Let R be an equivalence relation over T_Σ^ω then R is said to be k-invariant if $t_i R t'_i$, where $i = 1, ..., k$, then $f(t_1, ..., t_k) R f(t'_1, ..., t'_k) \ \forall f \in \Sigma$. The index of R is the number of its equivalence classes.

Theorem 4.2

Let L be a recognizable set of k-ary ω-trees. Then L is the union of some of the equivalence classes of a k-invariant equivalence relation of finite index.

Proof:

Let L be a recognizable set of k-ary ω-trees then one can exhibit a deterministic and total bottom-up Muller tree automaton $M =< Q, \Sigma, \delta, Fin, F >$ accepting L. Let R_M be the relation defined by :

$t R_M t'$ iff for each $q \in Q$ the following sets are equal :

$\bullet \{Inf(C) : C$ is a computation of M on t, and $C(\lambda) = q\}$, and

$\bullet \{Inf(C) : C$ is a computation of M on t', and $C(\lambda) = q\}$.

$Inf(C)$ denotes the set of sets of states occurring infinitely often on some infinite branch of C. R_M is an eqivalence relation. Since Q is finite, the index of R_M is also finite.

$L = \cup_{i=1}^n [t_i]$, where

$[t_i] = \{t : \exists q \in Fin \exists H \subseteq F \exists C$ such that $C(\lambda) = q$ and $Inf(C) = H\}$.

We shall prove that the converse of the last theorem is not true in general (i.e. $Card(\Sigma) \geq 2$). But, if Σ contains one symbole the converse of the last theorem is true. Let t be an infinite tree over $\{a, b\}$, then $t|_{1^*} = a^{n_1} b^{k_1} a^{n_2} b^{k_2} .....$, where $t|_{1^*}$ is the ω-word lying on the first branch (i.e. 1^*). Let $A(t)$ denotes the set $\{n_i : i \leq 1\}$.

Let R be the relation over T_Σ^ω such that $t R t'$ iff $A(t)$ and $A(t')$ are recursively enumerable. R is an equivalence relation of finite index which is k-invariant. Let L_0 be the set of k-ary infinite trees t over $\{a, b\}$ such that $A(t)$ is recursively enumerable. Since bottom-up Muller tree automaton are not able to count arbitrarily large, L_0 cannot be accepted by a bottom-up Muller tree automaton. This shows that Nerode's theorem cannot be extended to infinite trees.

5 . DETERMINISTIC AUTOMATA AND EILENBERG'S THEOREM

In this section, we will extend Eilenberg's theorem to infinite trees. It depends upon the equality between sets accepted by determistic Büchi tree automata on trees and the limit of finite tree sets accepted by deterministic top-down tree automata .

Theorem 5.1

For each set of infinite trees K, the following conditions are equivalent :

(i) K is accepted by a deterministic Büchi tree automaton,

(ii) K is the limit of a set of finite trees accepted by a deterministic top-down tree automaton.

The proof of theorem 5.1 is left as an exercise.

Theorem 5.2

For each set of infinite trees K, the following conditions are equivalent :

(i) K is accepted by a Büchi (resp. Muller) tree automaton, and

(ii) K is a projection of a set accepted by a deterministic Büchi (resp. Muller) tree automaton.

Note that the relationship between determinism and nodeterminism, in case of tree automata, can be expressed in terms of projections.

Proposition 5.1

The family of sets of infinite trees accepted by deterministic M-automata is closed under Boolean operations (i.e. union, intersection and complementation).

The proof of Proposition 5.1 use only classical constructions.

Theorem 5.3

The family of sets of infinite trees accepted by deterministic M-automata is equal to the Boolean closure of the family of sets of infinite trees accepted by deterministic Muller tree automata.

Theorem 5.4

The family of sets accepted by deterministic M-automata is properly included in the family of those accepted by M-automata.

Proof:

Let $T(f, a, b)$ be the set of finite trees t such that :

If $u \in Fr(t)$ then $t(u) = a$ or $t(u) = b$ else $t(u) = f$

Let $K = \{t \in T(f, a, b) : |t|_a \leq 2\}$, K is accepted by a bottom-up automaton, where $|t|_a$ is the number of a's occurring on t. Consider the substitution s such that :

(i) If $t = a$ then $s(t) = a^\omega$

(ii) If $t = b$ then $s(t) = b^\omega$

(iii) $s(a(t, t')) = a(s(t), s(t'))$

$s(K)$ is a set of infinite trees accepted by an M-automaton. Assume that $s(K)$ is accepted by a deterministic M-automaton M. Choose $t \in T(f, a, b)$ large enough with only b's on frontier, such that one state q' occurs at least three times on the frontier of t. Let t_1 be the tree obtained by changing one of the three b's into an a and t_2 be the tree obtained by changing one more b into an a. By definition $s(t_1) \in s(K)$ and $s(t_2) \in s(K)$. Now let C_i be the computation of M on t_i $(i = 1, 2)$ and F_i be the set of all subsets occurring infinitely often on branches of C_i. Since M is deterministic, F_1 is equal to F_2 and $s(t_1) \in s(K)$. This contradicts the fact that $s(t_1)$ is not in $s(K)$.

ACKNOWLEDGEMENT
We thank Prof. M. Nivat for many fruitful discussions.

REFERENCES

1. **A. Arnold and D. Niwinski.** *Fixed point characterization of Büchi automata on infinite trees*, RI 87-28, Univ. de Bordeaux (1987).

2. **L. Boasson and M. Nivat.** *Adherences of context-free languages*, J. Comput. Syst. Sci. 20(1980)285-309.

3. **J. R. Büchi.** *On a decision method in restricted second order arithmetic*, Proc. Cong. Logic in Methodology and Phil. of Sci., Standford University Press, Calif. (1960)1-11.

4. **B. Courcelle.** *Fundamentals of infinite trees*, Theoretical Computer Science no. 25(1983)95-169.

5. **B. Courcelle.** *Recognizable sets of unrooted trees*, this volume(1991).

6. **S. Eilenberg.** *Automata, Languages and Machines*, vol. A, Academic Press (1974).

7. **A. E. Emerson.** *Automata, tableaux, and temporal logics*, Proc. Workshop on Logics of Programs, Brooklyn (1985).

8. **A. E. Emerson, and C. Jutla.** *The complexity of tree automata and logics of programs*, Proc. 29th IEEE Symp. on Foundations of Comput. Sci. (1988)328-337.

9. **F. Gécseg and M. Steinby.** *Tree automata*, Academiai Kiado, Budapest (1984).

10. **I. Guessarian.** *On push-down tree automata*, Mathematical System Theory 16(1983) 237-263.

11. **I. Guessarian.** *Trees and algebraic semantics*, this volume (1991).

12. **Y. Gurevich and L. Harrington** . *Trees, Automata, and Games*, Proc. 14th ACM Symp. on Theory of Computing (1982)237-263.

13. **D.E. Muller and P. E Schupp.** *Alternating automata on infinite objects, determinancy, and Rabin's theorem*, in "Automata on infinite words" (M. Nivat and D. Perrin, eds) L.N.C.S no.192(1985)100-107.

14. **T. Hafer.** *On the Boolean closure of Büchi tree automaton definable sets of ω-trees*, Aachner Inform. Ber. Nr. 87-16, RWTH Aachen(1987).

15. **D. Muller, A. Saoudi and P. Schupp.** *Alternating automata, the weak monadic theory and its complexity*, 13th Int. Coll. on Automata, Languages and Programming, L.N.C.S no. 226, Springer-Verlag, Berlin(1986).

16. **D. E. Muller, A. Saoudi, and P. E. Schupp.** *Weak alternating automaton give a simple explanation of why most Temporal and Dynamic Logic are decidable in exponential time*, Proc. of the third IEEE Symposium on Logic in Computer Science(1988).

17. **D. E. Muller, A. Saoudi, and P. E. Schupp**. *Alternating Automaton, the weak monadic theory and its complexity*, to appear in Theoretical Computer Science(1991).

18. **A. Mostowski**. *Determinancy of sinking automata and various Rabin's pair indices*, Inf. Proc. Letters 15(1982)153-183.

19. **M. Nivat and A. Saoudi**. *Automata on infinite trees and Kleene closure of regular tree sets*, Bulletin of the E.A.T.C.S no. 36(1988)131-136.

20. **M. Nivat and A. Saoudi**. *Rational, Recognizable and Computable languages*, Univ. Paris VII, L.I.T.P. publication no. 85-75.

21. **M. Nivat et A. Saoudi**. *Automata on infinite objects and their applications to logic and programming*, Information and Computation, vol. 83, no. 1(1989)41-64.

22. **D. Niwinski**. *A note on indices of Rabin's pairs automata*, manuscript, The University of Warsaw (1986).

23. **R. Parikh**. *Propositional Game Logic*, Proc. 25th IEEE Symp. on Foundations of Comput. Sci. (1983)195-200.

24. **M. O. Rabin**. *Decidability of second order theories and automata on infinite trees*, Trans. Amer. Math. Soc. 141(1969)1-35.

25. **M. O. Rabin**. *Weakly definable relations and special automata, Math. Logic and Foundation of set theory*, (Y. Bar Hillel Edit.), Amsterdam North Holland (1970)1-23.

26. **S. Safra** . *On the complexity of ω-automata*, Proc. 29th Symp. on Foundations of Computer Sci. (1988)319-327.

27. **A. Saoudi**. *Infinite tree languages recognized by ω-automata*, Inf. Proc. Letters 18(1984)15-19.

28. **A. Saoudi**. *Generalized automata on infinite trees and Muller-McNaughton theorem*, to appear in Theoretical Computer Science (1991).

29. **A. Saoudi**. *Forêts infinitaires reconnaissables*, Thèse de 3ème Cycle, Univ. Paris VII, L.I.T.P. publication no. 82-19(1982).

30. **A. Saoudi**. *Contribution à la théorie des automates: Automates d'arbres infinis et applications à la logique et à la programmation*, Thèse d'Etat, Univ. Paris VII(1987).

31. **A. Saoudi**. *Variétés d'automates d'arbres infinis*, Theoretical Computer Science 44(1986)1-21.

32. **J. W. Thatcher and J. B. Wright**. *Generalized finite automata theory with applications to a decision problem of second order logic*, Math. Syst. Theory(1968)57-81.

33. **J. W. Thatcher**. *"Tree automata : An informal survey"*, in Currents in the theory of computing (A. V. Aho, edit.), Prentice Hall(1973).

34. **W. Thomas**. *A Hierarchy of sets of infinite trees*, G.I Conference, L.N.C.S no 145, Springer-Verlag, Berlin(1982)335-342.

35. **W. Thomas**. *Automata on infinite objects*, in "Handbook of Theoretical Computer Science", North-Holland(1990).

36. **M. Y. Vardi and P. Wolper**. *Reasoning about fair concurrent program*, Proc. 18th Symp. on Theory of Computing, Berkeley (1986).

37. **M. Y. Vardi**. *Verification of concurrent programs: The automata-theoretic framework*, Proc. of the IEEE Symposium on Logic in Computer Science (1987).

Tree Automata and Languages
M. Nivat and A. Podelski (editors)
© 1992 Elsevier Science Publishers B.V. All rights reserved.

235

ALGEBRAIC SPECIFICATION OF ACTION TREES AND RECURSIVE PROCESSES

M.Große-Rhode , C.Dimitrovici

TU Berlin, FB 20 (Informatik), Sekr. FR 6-1
Franklinstr. 28/29, D-1000 Berlin 10

Abstract
Projection algebras and -specifications turned out to be a convenient framework for the formal specification of combined data type and process algebras with mathematically well defined semantics. Projection spaces can be considered as an algebraic version of pseudo-ultrametric spaces and allow a generalization of the metrical approach to program semantics. This includes action trees or general tree spaces as semantic domains for concurrent process specifications. It turns out that the theory of projection specifications is compatible with that of usual algebraic specifications, which means that we have a suitable formalization of combined data type and process specifications.

0. INTRODUCTION

Besides the syntactical structures represented as trees there are also a lot of semantical domains which have an obvious and intuitive representation as tree spaces. One example with which we are dealing here is that of non deteministic concurrent processes.

During the last years there have been several approaches to formal specification of distributed, concurrent systems and parallel processes. The problem that always appears is the specification of infinite objects as description of virtually infinite processes or solutions of recursive equation systems. This means that a topological structure of the specified domains should be considered. Besides the approach with topologies induced by partial orders, ideal completion and the Knaster-Tarski fixed point theorem, e.g. worked out by Scott, Möller (see [29], [30], [31]), Tarlecki/Wirsing (see [33]) and the ADJ group (see [1], [2]), another approach has been introduced by the group in Amsterdam (deBakker/Zucker, Bergstra/Klop, Kranakis and others): On the set of processes a metric is defined by projections, which can be interpreted as finite approximation of a process. Completion of the metric process space then generates the infinite processes as unique limits of their finite approximations and yields a complete metric space as convenient model for a process space. Compatibility of the operations then means uniform continuity, which ensures the existence of unique extensions of the operations to the completed space. The whole model for the concurrent system becomes a continuous algebra.

In the papers of Bergstra/Klop (see [5], [6]) and others such models are constructed basing on the axioms for parallel processes given for the specification language CCS by R.Milner (see [27], [28]). Their topologies are investigated and there is already detailed material about communication protocolls, verification etc..

During the last year an algebraic version of the metrical approach using projection spaces has been developed (see [15], [16], [17], [18]). One of the main new ideas was to specify the projections instead of the distance function. Then it is possible to define an initial algebra semantics for the specification, that already carries the metrical structure. In the former approach it was necessary to define the metric outside the specification, after a model was constructed. Furthermore the specification of concurrent systems showed that it is necessary to integrate process and data type specifications. Using trivial (discrete) projections, data types that are used in the process specification can be specified together with the processes in one specification. So we have a unique well defined semantics for the combined data type- and process specification.

All definitions given in terms of the metric can be translated to statements about the projections, e.g. uniform continuity of the operations becomes a first order formula with the signature given in the specification. Thus projection specifications have the following advantages:

- specification of combined data type- and process algebras with arbitrary signature;
- well defined semantics : complete initial algebra semantics for projection specifications and free functor semantics for parameterized projection specifications;
- unique approach to correctness proofs.

The first part of this paper is a categorical investigation of projection spaces. An equivalence between the category of projection spaces with projection compatible functions and the category of (special) pseudo-ultrametric spaces with non expansive functions can be used to show the existence of some categorical objects. Furthermore completion and separation of projection spaces are introduced, together with the standard construction known from the preceeding papers about projection spaces, that are needed for the definition of the semantics of a projection specification.

The remaining parts of this paper are concerned with projection specifications and their applications.

In part two simple projection specifications are introduced. A projection specification has a projection operation symbol for each sort, that induces a metric on each base set of an algebra. A discrete projection (p(n,x)=x) leaves the base set invariant, whereas the completion of a sort with non discrete projection may introduce new infinite elements. Thus the data sorts can be characterized by discrete projections, the process sorts by non discrete projections. In order to construct the semantics of a projection specification the results from the first chapter of this paper are carried over from projection spaces to projection algebras, which are families of projection spaces together with projection compatible operations. In correspondance with usual algebraic specifications the semantics of a projection specification PS is defined as the isomorphism class of initial complete separated projection algebras satisfying PS. Completeness ensures that all infinite processes, which can be approximated by processes denoted by terms of the signature of the specification are contained in the semantics. The semantics of a projection specification is in general not finitely generated by the operations. But the projections of elements are always denoted by a term under the usual evaluation. Thus separatedness of the semantics ensures, that each process, thus the whole algebra, is completely determined by the generated part of the semantics. The relation between the usual initial algebra semantics of a projection specification considered as algebraic specification and its projection semantics can be expressed by a free functor, which is also needed in the third chapter of this paper. Furthermore correctness of a projection specification means, that the (initial) quotient term algebra of the specification is a dense subalgebra of the projection semantics algebra.

Part three introduces parameterized projection specifications. It is developed along chapters seven and eight of the book [13], where parameterized algebraic specifications are introduced. This book contains the fundamentals of algebraic specifications, which are used throughout this paper. The basic results of [13] are used without explicit references. The combination of the free functor induced by the underlying parameterized algebraic specification, and the projection semantics functor defined in chapter two of this paper yields a free functor between the categories of complete separated projection parameter algebras and complete separated projection target algebras, which will be called projection free functor and serves as semantics of the parameterized projection specification. This close relation between the semantics of a parameterized algebraic specification and the semantics of a parameterized projection specification allows to carry over the results concerning semantics (compositionality) and correctness w.r.t. a model. Internal correctnes of a parameterized projection specification concernes the relation between free functor and projection free functor. As demonstration for the theory some examples are given, especially parameterized projection specifications of Bergstra/Klop´s process algebras.

A special case of (parameterized) projection specifications are recursive (parameterized) projection specifications, where new processes on a given projection algebra are defined by a system of recursive equations. These recursive projection specifications, which are of special interest for practical applications, are discussed in the fourth chapter of this paper. The semantics of a recursive projection specification with contracting equation system can be constructed using the Banach fixed point theorem stated for projection spaces.

A proof of projection compatibility of the recursive processes, which is needed for general extensions of projection specifications, becomes unnecessary for recursive projection specifications, because contracting recursive projection specifications are always continuous enrichments.

A summary of the results concerning projection specifications is given in the conclusion (part five).

Concerning algebraic specifications this paper is based on [13]. Most of its results are quoted without explicit reference. We would like to thank H.Ehrig for his constant support.

1. PROJECTION SPACES

Projection spaces were introduced as an algebraic version of ultrametric spaces to render possible a purely algebraic describtion of combined data type and process algebras. (See [15]). To define the (initial complete algebra) semantics of a projection specification a completion of projection spaces is needed. Investigation of the completion of a projection space by standard construction (a special sequence space) showed that the separation axiom for projection spaces (resp. Approximation Induction Principle **AIP** : two elements are equal if and only if all their projections are equal) is not needed as a projection space axiom : standard constructions are always separated. Thus, in contrast with the preceeding papers, we only require one projection space axiom in this paper. The notion of non-separated projection spaces (i.e. projection spaces that do not satisfy the separation axiom) also makes the category of projection spaces much more convenient.

Another aspect of projection spaces is the possibility to describe approximations (e.g. projections in inner product spaces), topologies induced by depth-n-tests (Hennessy-Milner-Logics , see [21]) etc.. This aspect is not pointed out in this paper, but the categorical framework is developed, which is also interesting in its own right.

In the first part of this section projection spaces are defined and the topological (resp. metrical) notions are given in terms of projections. The notions *separation* and *finiteness* are introduced, which are central for the construction of the completion of a projection space (by standard construction) and the semantics of a projection specification.

The morphisms corresonding to projection spaces have to be compatible with the corresponding pseudo-ultrametric structure and the algebraic structure induced by the projections. Uniformly continuous functions, i.e.

functions $f:(A,p)\to(B,q)$ satisfying $\forall n\in\mathbb{N}_1\ \exists m\in\mathbb{N}_1\ q(n,\_)\circ f=q(n,\_)\circ f\circ p(m,\_)$, respect the pseudo-ultrametric structure, but (according to the occurence of the $\exists$-quantor) are not fully compatible with the algebraic structure; e.g. products and initial algebras in the corresponding category (see 1.8 and 2.4) do not exist. Therefore the class of morphisms is restricted to the functions $f:(A,p)\to(B,q)$ satisfying the *equation* $q(n,\_)\circ f=q(n,\_)\circ f\circ p(n,\_)$.

With regard to the application to specifications of process algebras projection compatible functions also seem to be appropriate; in all the examples discussed yet the operations were projection compatible.

At the end of this part a fixed-point theorem for projection spaces is given that is needed for the definition of the semantics of recursive process specifications. Since the comparison with ultrametric spaces is easier to formulate in the categorical framework it will be given later.

<u>1.1 Definition</u> (Projection Space) :

(1) A *projection space* (A,p) is a set A together with a function $p:\mathbb{N}_1\times A\to A$,

satisfying $\forall n,m\in\mathbb{N}_1\ \forall a\in A\ \ p(n,p(m,a))=p(\min(n,m),a)$;

(2) a projection space (A,p) is called *separated*, if for all $a,b\in A$

$\forall n\in\mathbb{N}_1\ p(n,a)=p(n,b)\ \to\ a=b$;

(3) an element a of a projection space (A,p) is called *finite*, if there exists an $n\in\mathbb{N}_1$ with $p(n,a)=a$; (A,p) is called *finite* if each $a\in A$ is finite, and $A_{fin}:=\{a\in A : a\text{ is finite}\}$.

(4) Given projection spaces (A,p) and (B,q) a function $f:(A,p)\to(B,q)$ is called *projection compatible*, if $\forall n\in\mathbb{N}_1\ q(n,\_)\circ f=q(n,\_)\circ f\circ p(n,\_)$

The following remark states that each separated projection space is a quotient space w.r.t. the congruence induced by the separation axiom. Furthermore the standard pseudo-ultrametric is introduced, which allows to interpret projection spaces as pseudo-ultrametric spaces. Recall that a pseudo-ultrametric d on a set A is a function $d:A\times A\to\mathbb{R}$ such that for all $a,b,c\in A$

(i) $d(a,b)\geq 0$, $d(a,a)=0$

(ii) $d(a,b)=d(b,a)$

(iii) $d(a,c)\leq\max\{d(a,b),d(b,c)\}$.

<u>1.2 Remark and Definition</u> (Separation and Standard Pseudo-Ultrametric) :

Given a projection space (A,p), then

(1) $a\equiv b$ if $\forall n\in\mathbb{N}_1\ p(n,a)=p(n,b)$ $(a,b\in A)$ defines a congruence relation on (A,p) and $(A/\!\equiv,p_\equiv)$, $p_\equiv(n,[a]):=[p(n,a)]$, is a separated projection space.

(2) $d_p:A\times A\to\mathbb{R}$, defined by

$d_p(a,b)=2^{-\min\{n\in\mathbb{N}_1 : p(n,a)\neq p(n,b)\}}$ if $\{n\in\mathbb{N}_1 : p(n,a)\neq p(n,b)\}\neq\varnothing$, and

$d_p(a,b)=0$ if $\forall n\in\mathbb{N}_1\ p(n,a)=p(n,b)$

is a pseudo-ultrametric on A; d_p is called *standard pseudo-ultrametric* induced by p and (A,d_p) the corresponding pseudo-ultrametric space; d_p is an ultrametric if and only if (A,p) is separated.

(3) $(A_{fin},p|A_{fin})$ is a separated projection space.

<u>Proof</u> : (1) Obviously $a\equiv b$ implies $p(n,a)=p(n,b)$, i.e. $\equiv$ is a congruence, whence the assertion follows.

(2) It is obvious that $d_p(a,b)\geq 0$ and $d_p(a,b)=d_p(b,a)$.

So we need to prove that $d_p(a,c)\leq \max\{d_p(a,b),d_p(b,c)\}$ for all $a,b,c\in A$.

If $a=b$ or $b=c$ the assertion holds, so let $i(a,b):=\min\{k\in \mathbb{N}_1:p(k,a)\neq p(k,b)\}=m$ and $i(b,c)=n$, then $p(k,a)=p(k,b)$ for $k=1,...,m-1$ and $p(k,b)=p(k,c)$ for $k=1,...,n-1$

hence $p(k,a)=p(k,c)$ for $k=1,...\min\{m-1,n-1\}$

thus $i(a,c)\geq\min\{m,n\}$

thus $d_p(a,c)\leq 2^{-\min\{i(a,b),i(b,c)\}}=\max\{d_p(a,b),d_p(b,c)\}$.

(3) If $a=p(n,a)$ and $b=p(m,b)$ then $a=p(k,a)$ and $b=p(k,b)$ for $k=\max(n,m)$. Thus $\forall n\in \mathbb{N}_1\ p(n,a)=p(n,b)$ includes the equation $a=p(k,a)=p(k,b)=b$. $\square$

Next Cauchy sequences and convergence in a projection space are defined. The definitions are equivalent to convergence in the corresponding pseudo-ultrametric space (see 1.4). Projective sequences are special Cauchy sequences that are uniquely determined by classes of equivalent Cauchy sequences and therefore form a complete system of representatives (see 1.5).

<u>1.3 Definition</u> (Sequences,Convergence and Completeness in Projection Spaces):

Given a projection space (A,p), a sequence $(a_n)_{n\geq 1}$ with $a_n\in A$, and $a\in A$

(1) $(a_n)_{n\geq 1}$ is *Cauchy* in (A,p), if $\forall n\in \mathbb{N}_1 \exists m\in \mathbb{N}_1 \forall k\in \mathbb{N}_1\ \ p(n,a_m)=p(n,a_{m+k})$

(2) $(a_n)_{n\geq 1}$ is *projective*, if $\forall n\in \mathbb{N}_1\ \ p(n,a_{n+1})=a_n$

(3) $(a_n)_{n\geq 1}$ is *convergent* with limit a in (A,p), written $\lim a_n=a$, if

    $\forall n\in \mathbb{N}_1 \exists m\in \mathbb{N}_1 \forall k\in \mathbb{N}_1\ \ p(n,a_{m+k})=p(n,a)$.

(4) (A,p) is a *complete* projection space, if each Cauchy sequence in (A,p) converges in (A,p).

<u>1.4 Fact</u> :

Given a projection space (A,p) with standard metric d_p and a sequence $(a_n)_{n\geq 1}$

$a_n\in A$

(1) $(a_n)_{n\geq 1}$ is Cauchy in (A,p) if and only if it is Cauchy in (A,d_p) ;

(2) $(a_n)_{n\geq 1}$ is convergent with limit a in (A,p) if and only if it is convergent with limit a in (A,d_p) ;

(3) (A,p) is a complete projection space if and only if (A,d_p) is a complete pseudo-ultrametric space .

<u>Proof</u> :
(1) The following formulations are equivalent :

$(a_n)_{n\geq 1}$ is Cauchy in (A,d_p)

$\Leftrightarrow \forall \varepsilon > 0 \, \exists n \in \mathbb{N}_1 \, \forall k \in \mathbb{N}_1 \; d_p(a_n, a_{n+k}) < \varepsilon$

$\Leftrightarrow \forall m \in \mathbb{N}_1 \, \exists n \in \mathbb{N}_1 \, \forall k \in \mathbb{N}_1 \; i(a_n, a_{n+k}) > m$

$\Leftrightarrow \forall m \in \mathbb{N}_1 \, \exists n \in \mathbb{N}_1 \, \forall k \in \mathbb{N}_1 \; p(m, a_n) = p(m, a_{n+k})$;

(2) the following formulations are equivalent :

$(a_n)_{n\geq 1}$ is convergent with limit a in (A,d_p)

$\Leftrightarrow \forall \varepsilon > 0 \, \exists n \in \mathbb{N}_1 \, \forall k \in \mathbb{N}_1 \; d_p(a, a_{n+k}) < \varepsilon$

$\Leftrightarrow \forall m \in \mathbb{N}_1 \, \exists n \in \mathbb{N}_1 \, \forall k \in \mathbb{N}_1 \; i(a, a_{n+k}) > m$

$\Leftrightarrow \forall m \in \mathbb{N}_1 \, \exists n \in \mathbb{N}_1 \, \forall k \in \mathbb{N}_1 \; p(m, a) = p(m, a_{n+k})$.

(3) follows from (1) . $\qquad\qquad\qquad\qquad\qquad\qquad\qquad\qquad\qquad\qquad\square$

1.5 <u>Remark</u> : To each Cauchy sequence in a projection space (A,p) there exists exactly one equivalent projective sequence; where $(a_n)_{n\geq 1}$ and $(b_n)_{n\geq 1}$ are equivalent Cauchy sequences if $\forall m \in \mathbb{N}_1 \, \exists n \in \mathbb{N}_1 \, \forall k \in \mathbb{N}_1 \; p(m, a_{n+k}) = p(m, b_{n+k})$, i.e. equivalent sequences have the same limits. Therefore the projective sequences form a complete system of representatives of equivalence classes of Cauchy sequences.

<u>Proof</u> : If $(a_n)_{n\geq 1}$ is Cauchy, then there is to each $n \in \mathbb{N}_1$ a minimal $m(n) \in \mathbb{N}_1$ so that for all $k \in \mathbb{N}_1 \; p(n, a_{m(n)+k}) = p(n, a_{m(n)})$. Furthermore $m(n+1) \geq m(n)$, since $p(n+1, a_{m(n+1)+k}) = p(n+1, a_{m(n+1)})$ implies $p(n, a_{m(n+1)+k}) = p(n, a_{m(n+1)})$.
Define $\underline{a}_n := p(n, a_{m(n)})$, then $p(n, \underline{a}_{n+1}) = p(n, p(n+1, a_{m(n+1)})) = p(n, a_{m(n+1)}) =$
$= p(n, a_{m(n)}) = \underline{a}_n$, i.e. $(\underline{a}_n)_{n\geq 1}$ is projective ; let $j := \max(n, m(n))$ for given $n \in \mathbb{N}_1$,
then $p(n, \underline{a}_{j+k}) = p(n, \underline{a}_j) = p(n, a_j) = p(n, a_{j+k})$ for all $n, k \in \mathbb{N}_1$, i.e. $(\underline{a}_n)_{n\geq 1}$ is equivalent to $(a_n)_{n\geq 1}$; now let $(b_n)_{n\geq 1}$ be any projective sequence equivalent to $(\underline{a}_n)_{n\geq 1}$, then for all n and $k \geq k_0 \; b_n = p(n, b_{n+1}) = p(n, b_{n+k}) = p(n, \underline{a}_{n+k}) = \underline{a}_n$;i.e.
$(\underline{a}_n)_{n\geq 1}$ is unique. $\qquad\qquad\qquad\qquad\qquad\qquad\qquad\qquad\qquad\qquad\square$

1.6 <u>Examples</u> (for Projection Spaces) :
(1) $\mathbb{N}$, the natural numbers with $p(n,k) = \min(n,k)$ for all $n,k \in \mathbb{N}$.
(2) The free monoid A^+ over a set A together with $p(n,w) = w[1,n]$, the first n letters of the word w.
(3) The set T of all finite or infinite trees with nodes from a given set A, where the n´th projection of a tree t is t, cut to level n.

We give an algebraic specification of this example (the **pnat1** specification is given at the beginning of chapter two) :

tree = **pnat1** +

    <u>formal sorts</u>   nodes

    <u>sorts</u>   tree

    <u>opns</u>   _:nodes $\to$ tree

                  +,·:tree tree $\to$ tree

                  p:nat1 tree $\to$ tree

    <u>eqns</u>   {+,· associative,+ commutative}

                  (t1+t2)·t3=t1·t3+t2·t3

                  p(1,a)=p(1,a·t)=a

                  p(n+1,a·t)=a·p(n,t)

                  p(n,t+t´)=p(n,t)+p(n,t´)

(Deletion of + yields a specification of strings.)

(4) $C^\infty(\mathbb{R})$, the set of infinitely often differentiable functions on $\mathbb{R}$, with $p(n,f)=n´$th Taylor polynomial $T_{n,x_0,f}$, for fixed $x_0\in\mathbb{R}$. The projection space axiom holds, since the $n´$th Taylor polynomial of a polynomial $p=$

$=\Sigma^\infty_{k=0} a_k\cdot(x-x_0)^k$ is given by $T_{n,x_0,p}(x)=\Sigma^n_{k=0} a_k\cdot(x-x_0)^k$.

(5) The set of real numbers $\mathbb{R}$, with $p(n,x)=\Sigma^n_{k=-m}a_k\cdot 10^{-k}$, if $x=\Sigma^\infty_{k=-m}a_k\cdot 10^{-k}$ $a_k\in\{0,1,..,9\}$ is the unique representation of x, excluding sums of the form $\Sigma^\infty_{k=-m} 9\cdot 10^{-k}$.

(6) Let M be any set, then

    (i) $p(n,x)=x$ for all $x\in M$ is called <u>discrete</u> projection;

    (ii) $p(n,x)=$const. if $M\neq\varnothing$, $p=\varnothing$ if $M=\varnothing$ is called <u>indiscrete</u> projection.

The set of morphisms of two projection spaces with a projection induced by the projection of the codomain forms a projection space.

<u>1.7 Fact</u> : Given projection spaces (A,p) and (B,q) , then

(1) $(B,q)^{(A,p)}:=\{f:(A,p)\to(B,q) :$ f is projection compatible$\}$ together with

$\pi:\mathbb{N}_1\times(B,q)^{(A,p)}\to(B,q)^{(A,p)}$, $\pi(n,f)=q(n,\_)\circ f$, is a projection space;

(2) $((B,q)^{(A,p)},\pi)$ is separated (complete) if and only if (B,q) is separated (complete).

<u>Proof</u> :

(1) $q(n,\_)\circ(q(n,\_)\circ f)=q(n,\_)\circ(q(n,\_)\circ f)\circ p(n,\_)$, thus $q(n,\_)\circ f$ is projection compatible and π is well defined;

furthermore $\pi(n,\pi(m,f))=q(n,\_)\circ q(m,\_)\circ f=q(\min(n,m),\_)\circ f=\pi(\min(n,m),f)$.

(2) (i) Let (B,q) be separated, then $\forall n\in\mathbb{N}_1$ $\pi(n,f)=\pi(n,g)$ is equivalent to $\forall a\in A\forall n\in\mathbb{N}_1$ $q(n,f(a))=q(n,g(a))$, i.e. $f=g$. If (B,q) is not separated, then there

are $b, b' \in B$, $b \neq b'$ with $q(n,b) = q(n,b')$ for all $n \in \mathbb{N}_1$. Then $f, g : (A,p) \to (B,q)$ with $f \equiv b$, $g \equiv b'$ are projection compatible and $q(n, f(a)) = q(n, g(a))$ for all $a \in A, n \in \mathbb{N}_1$, thus $\pi(n,f) = \pi(n,g)$ for all $n \in \mathbb{N}_1$ and neither $((B,q)^{(A,p)}, \pi)$ nor $(\{f : A \to B\}, \pi)$ is separated.

(ii) "$\Rightarrow$" Let $(f_k)_{k \geq 1}$ be a Cauchy sequence in $((B,q)^{(A,p)}, \pi)$, i.e.

$\quad \forall m \exists k \forall j \ \ \pi(m, f_{k+j}) = \pi(m, f_k)$, then

$\quad \forall m \exists k \forall j \forall a \ \ \pi(m, f_{k+j})(a) = \pi(m, f_k)(a)$, and

$\quad \forall m \exists k \forall j \forall a \ \ q(m, f_{k+j}(a)) = q(m, f_k(a))$ $\qquad (*)$.

Thus $(f_k(a))_{k \geq 1}$ is a Cauchy sequence in (B,q) for each $a \in A$. Since (B,q) is complete each $(f_k(a))_{k \geq 1}$ converges to an element $b := f(a) \in B$ with

$\forall m \exists k \forall j \forall a \ \ q(m, f_{k+j}(a)) = q(m, f(a))$ by $(*)$

Thus $\forall m \exists k \forall j \ \pi(m, f_{k+j}) = \pi(m, f)$ and $\lim_k f_k = f$.

It remains to be shown that f is projection compatible :

$\quad q(k, f(a)) = q(k, (\lim_m f_m)(a)) = q(k, (\lim_m (f_m(a)))) = \lim_m q(k, f_m(a)) =$

$\quad = \lim_m q(k, f_m(p(k,a))) = q(k, (\lim_m f_m)(p(k,a))) = q(k, f(p(k,a)))$.

"$\Leftarrow$" : If (B,q) is not complete, then there is a nonconvergent Cauchy sequence $(b_n)_{n \geq 1}$ in (B,q). The sequence $(f_n)_{n \geq 1}$ with $f_n(a) = b_n$ for all $a \in A$, $n \in \mathbb{N}_1$ then is a nonconvergent Cauchy sequence in $((B,q)^{(A,p)}, \pi)$. $\qquad \square$

Since projection compatible functions are closed under composition and the identities are projection compatible projection spaces together with projection compatible functions form a category.

<u>1.8 Definition</u> (the Category PRO_c) :

The category PRO_c is defined by

$\quad Ob(\mathrm{PRO}_c)$ = class of all projection spaces

$\quad Mor_{\mathrm{PRO}_c}((A,p),(B,q)) = \{f : (A,p) \to (B,q) : f \text{ is projection compatible}\}$.

In the last chapter of this paper recursive equation systems are solved using a fixed-point theorem corresponding to Banach's fixed-point theorem for complete metric spaces. As one of the presuppositions contractions in projection spaces have to be defined, compatible with the definition of a contraction in the corresponding pseudo-ultrametric space.

Banach's fixed-point theorem states that if X is a complete metric space and $f : X \to X$ is a contraction, then the sequence $(f^n(x))_{n \geq 1}$ for any $x \in X$ converges to a fixed point x^* of f (i.e. $f(x^*) = x^*$) and this fixed point is unique. A translated version of this theorem is given below.

1.9.Definition (Contraction) :
Given a projection spaces (A,p) and (B,q) a function $T{:}(A,p){\to}(B,q)$ is a *contraction*, if $\forall a,b \in A \ \forall n \in \mathbb{N}_1$
(1) $q(1,T(a))=q(1,T(b))$
(2) $p(n,a)=p(n,b) \to q(n+1,T(a))=q(n+1,T(b))$
(3) $A \neq \varnothing$.

1.10 Remark :
(1) $T{:}(A,p){\to}(B,q)$ is a contraction if and only if $d_q(T(a),T(b)) \leq 1/2 \cdot d_p(a,b)$ for all $a,b \in A$, where d_p and d_q are the standard pseudo-ultrametrics of (A,p) and (B,q) respectively.
(2) If T is a contraction then T is projection compatible.
(3) $A \neq \varnothing$ in the definition ensures that every contraction in a complete projection space has a unique fixed point, whereas $\varnothing{:}\varnothing{\to}\varnothing$ has none.

Proof :
(1) is obvious ;   (2) $q(1,T(a))=q(1,T(p(1,a)))$ and $p(n,a)=p(n,b)$ implies
$q(n,T(a))=q(n,p(n+1,T(a)))=q(n,q(n+1,T(b)))=q(n,T(b))$.   □

1.11 Theorem (Fixed-Point Theorem for Projection Spaces) :
Given a complete projection space (A,p) and a contraction $T{:}(A,p){\to}(A,p)$,
(1) T has a fixed point a^*, i.e. $T(a^*)=a^*$, and all fixed points of T are equivalent; i.e. if a and b are fixed points of T, then $p(n,a)=p(n,b)$ for all $n \in \mathbb{N}_1$. If a^* is a fixed-point of T, then $a^*=\lim T^n(a)$ ($a \in A$ arbitrary) and $p(n,a^*)=p(n,T^n(a))$ for all $n \in \mathbb{N}_1$.
(2) If (A,p) is separated, then T has a unique fixed point.

The category PRO_c , defined in 1.8, provides an appropriate framework for the constructions needed for the applications of the theory to algebraic specifications. The subcategory of complete separated projection spaces is a reflexive subcategory, that means that the separation and completion of a projection space (by factorization and adjunction of limits) yields a left adjoint to the inclusion functor, and is therefore universal. Combination of these two functors yields the separated completion of a projection space, which in the sequel will be used to define the semantics of a projection specification. A nice representation of the separated completion of a projection space, called standard construction, is given in terms of projective sequences.

1.13 Remark : The category PRO_c is complete, cocomplete, cartesian closed and a topos. The underlying sets of the corresponding limits, colimits, exponential objects and classifiers are as in SET (the category of sets and functions). The explicit construction of the projections concerned is in most cases straightforward, such that we do not present them here. The obvious forgetful functor has a right adjoint (indiscrete projection on the given set) and a left adjoint (discrete projection).

The standard construction of a projection space defined below will be used to define the semantics of a projection specification.

<u>1.14 Definition</u> (Standard Construction) :

Given a projection space (A,p) the standard construction (A^∞,p^∞) of (A,p) is defined by

$A^\infty=\{(a_n)_{n\geq 1}|(a_n)_{n\geq 1}$ is a projective sequence in $(A,p)\}$

$p^\infty(k,(a_n)_{n\geq 1})=(p(k,a_n))_{n\geq 1}$

and for a projection compatible morphism $f:(A,p)\to(B,q)$, f^∞ is defined by

$f^\infty((a_n)_{n\geq 1})=(q(n,f(a_n)))_{n\geq 1}$.

<u>1.15 Remark</u> : The definition above yields a functor $SC:PRO_c\to PRO_{Compl}{}^{Sep}$, defined by $SC(A,p)=(A^\infty,p^\infty)$ and $SC(f)=f^\infty$. The definition of the standard construction of morphisms includes $SC(p(n,\_))=p^\infty(n,\_):(A^\infty,p^\infty)\to(A^\infty,p^\infty)$, $p^\infty(k,(a_n)_{n\geq 1})=(p(k,a_n))_{n\geq 1}$.

<u>Proof</u> : $p^\infty(n,p^\infty(m,(a_k)_{k\geq 1}))=(p(n,p(m,a_k)))_{k\geq 1}=p^\infty(\min(n,m),(a)_{k\geq 1})$,

thus (A^∞,p^∞) is a projection space.

$\quad q(n,q(n+1,f(a_{n+1})))=q(n,f(a_{n+1}))=q(n,f(p(n,a_{n+1})))=q(n,f(a_n))$,

thus $(q(n,f(a_n)))_{n\geq 1}$ is projective and $f^\infty:(A^\infty,p^\infty)\to(B^\infty,q^\infty)$ is well defined.

$\quad q^\infty(n,f^\infty((a_k)_{k\geq 1}))=(q(n,q(k,f(a_k))))_{k\geq 1}=(q(n,q(k,f(p(n,a_k)))))_{k\geq 1}=$

$\quad =q^\infty(n,f^\infty((p(n,a_k))_{k\geq 1}))=q^\infty(n,f^\infty(p^\infty(n,(a_k)_{k\geq 1})))$,

i.e. f is projection compatible.

At last $SC(f\circ g)=SC(f)\circ SC(g)$. $\hfill\square$

<u>1.16   Theorem</u> : SC is a left adjoint to the inclusion functor $I:PRO_{Compl}{}^{sep}\to PRO_c$, i.e. and $SC\circ I\cong ID_{PRO_{Compl}{}^{Sep}}$.

The proof for the general case of projection algebras will be given in 2.12.

<u>1.17 Corollary</u> :
(1) The universal mapping $u_{(A,p)}:(A,p)\to SC(A,p)$, $u_{(A,p)}(a)=(p(n,a))_{n\geq 1}$ is isometric w.r.t. the standard pseudo-ultrametrics. If (A,p) is separated, then $u_{(A,p)}$ is injective.

(2) (i) (A^∞, p^∞) is a universal separated completion of (A,p) ;

   (ii) (A^∞, p^∞) is a universal completion of $(A_{fin}, p|A_{fin})$ and $Sep(A,p)$;

(3) (A^∞, p^∞) together with $\underline{p}(n,\_):(A^\infty, p^\infty) \to (A_n, p_n)$, $\underline{p}(n,(a_k)_{k\geq 1}) = a_n$ $(n \in \mathbb{N}_1)$ is projective limit of the diagram

$$(A_1, p_1) \xleftarrow{\quad p_1 \quad} (A_2, p_2) \leftarrow \ldots \leftarrow (A_n, p_n) \xleftarrow{\quad p_n \quad} (A_{n+1}, p_{n+1}) \leftarrow \ldots$$

where $A_i = p(i,A)$ and $p_i = p|A_i$.

<u>Proof</u> :

(1) $p(k,a) = p(k,a') \Leftrightarrow (p(n,p(k,a)))_{n\geq 1} = (p(n,p(k,a')))_{n\geq 1} \Leftrightarrow$

   $\Leftrightarrow p^\infty(k, u_{(A,p)}(a)) = p^\infty(k, u_{(A,p)}(a'))$;

but $u_{(A,p)}(a) = u_{(A,p)}(a') \Leftrightarrow \forall n\ p(n,a) = p(n,a')$;

(2) (i) follows from 1.16 and (ii) by 1.2(3) $(A_{fin}, p|A_{fin})$ is separated; since $SC(A,p)$ by defintion depends only on the finite elements $p(n,a)$ $SC(A,p) = SC(A_{fin}, p|A_{fin})$.

(3) is a special case of the corresponding theorem for projection algebras; see 2.21. $\qquad\qquad\qquad\square$

<u>1.18 Examples</u> (for standard constructions) :

(Compare the examples 1.6)

(1) $SC(\mathbb{N}, \min) \cong (\mathbb{N} \cup \{\infty\}, \min)$;

(2) $SC(A^+, p)$ is the set of all finite and infinite strings over A, with $p(n,w) = w[1,..,n]$;

(3) $SC(T,p)$ is the set of all finite and infinite trees with the same projection p ;

(4) $SC(C^\infty(\mathbb{R}), p)$ is the set of all formal power series $\Sigma^\infty_{k=0} a_k (x-x_0)^k$, with

   $p(n, \Sigma^\infty_{k=0} a_k (x-x_0)^k) = \Sigma^n_{k=0} a_k (x-x_0)^k$.

Note that $SC(C^\infty(\mathbb{R}), p)$ is no real function space and neither $(C^\infty(\mathbb{R}), p) \subseteq SC(C^\infty(\mathbb{R}), p)$ nor $SC(C^\infty(\mathbb{R}), p) \subseteq (C^\infty(\mathbb{R}), p)$.

(5) $SC(\mathbb{R}, p) \cong (\mathbb{R}, p) + \{\Sigma^\infty_{k=-m} a_k \cdot 10^{-k} : a_k = 9\ \forall k \geq k_0\}$. This unwanted side effect rests upon the fact that the addition $+:(\mathbb{R}, p) \times (\mathbb{R}, p) \to (\mathbb{R}, p)$ is not projection compatible (and also not uniformly continuous).

(6) (i) the discrete projection space : $SC(M,p) \cong (M,p)$

(ii) the indiscrete projection space : $SC(M,p) \cong (\{const.\}, p)$ if $M \neq \emptyset$ and
   $SC(\emptyset, \emptyset) = (\emptyset, \emptyset)$.

2. PROJECTION SPECIFICATIONS AND PROJECTION ALGEBRAS

The results from the first chapter will now be applied to algebraic specifications of combined data type and process algebras.

The idea of projection specifications is that the process sorts in the specification are enriched by non-discrete projection operations, while for the data sorts discrete projections are added. Thus a projection specification is a specification including a selected projection operation for each sort and therefore must also contain a specification of $\mathbb{N}_1$. Completion of the initial algebra then leaves the data sorts invariant, while the process sorts are enlarged by infinite processes, which are limits of the given finite processes.

Here (in-)finiteness is defined relative to the projections. The completion is done by standard construction leading to an initial complete separated projection algebra as semantics of the projection specification.

2.1. Syntax

Syntactically projection specifications differ from usual algebraic specifications only by the requirement, that there has to be a selected projection operation symbol for each sort. It is equivalent to define projections as a family of unary operations $(p_n)_{n\geq 1}$, $p_n:A\to A$, or as one operation $p:\mathbb{N}_1 xA\to A$. The latter version, which we have chosen yields finite signatures and makes calculations in the index explicit, i.e. dependencies between $p(n,\_)$ and $p(n+1,\_)$, which are likely to occure, are expressed in the specification. The disadvantage is, that each projection specification must contain a nat1-part which must be invariant under all constructions of specifications (extension, parameterization, renaming etc.) and must always be interpreted by $\mathbb{N}_1$, i.e. an (algebraic) initiality constraint is needed.

To make this nat1-part a projection specification **pnat1** the discrete projection is added:

pnat1 =
    <u>sorts</u> nat1
    <u>opns</u> 1: $\to$ nat1
           succ: nat1 $\to$ nat1
           min, p-nat1:nat1 nat1 $\to$ nat1
    <u>eqns</u> <u>for all</u> m,n <u>in</u> nat1 :
           min(n,1)=1
           min(1,n)=1
           min(succ(n),succ(m))=succ(min(n,m))
           p-nat1(m,n)=n

In the sequel $\mathbb{N}_1$ denotes the algebra $(\{1,2,...\},1,+1,\min,id)$, which is isomorphic to $T_{\mathbf{pnat1}}$. The following syntactic definition serves to ensure that each projection specification is a conservative extension of **pnat1**.

2.1 Definition : Given specifications SPEC=(S,OP,E) and SPEC1=(S1,OP1,E1) , SPEC *preserves* SPEC, if
(i) SPEC$\subseteq$SPEC1
(ii) if N:s1...sn$\rightarrow$s $\in$ OP1-OP then s$\notin$S ;
(iii)if $(t_1,t_2,X)\in$ E1-E , then $t_1\notin T_{(S,OP)}(X)$ ((t_1,t_2,X) means the equation $t_1=t_2$ where t_1, t_2 are terms with variables from X) .

2.2 Lemma : If SPEC1 preserves SPEC, then SPEC$\subseteq$SPEC1 is a conservative extension; i.e. $(T_{SPEC1})_{SPEC}\cong T_{SPEC}$.

Proof : Part (ii) in the definition ensures that SPEC$\subseteq$SPEC1 is complete, i.e. there are no new terms in $(T_{SPEC1})_{SPEC}$; part (iii) ensures that it is consistent, i.e. no SPEC-terms that are not equivalent w.r.t. E are identified in T_{SPEC1}. $\square$

Next projection specifications and constrained projection specifications are defined. The constraints correspond to the requirements of a projection algebra: Each base set together with its projection is a projection space and the operations are projection compatible.

2.3 Definition (Projection Specification) :
(1) A *projection specification* PS=(S,OP,E) is an algebraic specification with :
   (i) PS preserves **pnat1** ,
   (ii) for each sort s$\in$ S there is a selected operation symbol p-s:nat1 s$\rightarrow$s $\in$OP ,
(2) the *projection   constraints* C_{PS} for a given projection specification PS=(S,OP,E) are defined by C_{PS}=C1$\cup$C2$\cup$C3 , where

   $C1_s$: $\forall$n,m$\in$ nat1 $\forall$x$\in$ s   p-s(n,p-s(m,x))=p-s(min(n,m),x) $C1=\cup_{s\in S}\{C1_s\}$

   $C2_N$: $\forall$k$\in$ nat1 $\forall$x1$\in$ s1 ... $\forall$xn$\in$ sn

         p-s(k,N(x1,...,xn))=p-s(k,N(p-s1(k,x1),...,p-sn(k,xn)))

   $C2=\cup_{N\in OP}\{C2_N\}$

   C3={initial specification **pnat1**} (algebraic constraint) ;
(3) a *constrained projection specification* CPS is a projection specification PS=(S,OP,E) together with the corresponding projection constraints C_{PS} , i.e. CPS=(S,OP,E,C_{PS}) .

2.4 Definition (Category of Projection-PS-Algebras Cat(CPS)) :
Given a projection specification PS=(S,OP,E)
(1) a *projection-PS-algebra* is an algebra A=$((A_s)_{s\in S},(N_A)_{N\in OP})$ of the specification PS with
   (i) $(A_s,\text{p-s}_A)$ is a projection space for all s$\in$ S,

(ii) the operations N_A are projection compatible, i.e.

$\forall N: s1...sn \to s \ \forall k \geq 1 \ \forall a1 \in A_{s1}...\forall an \in A_{sn}$

$p\text{-}s_A(k,N_A(a1,...,an)) = p\text{-}s_A(k,N_A(p\text{-}s1_A(k,a1),...,p\text{-}sn_A(k,an)))$;

(iii) $A_{\textbf{pnat1}} \cong \mathbb{N}_1$;

(2) a *projection-PS-homomorphism* is a homomorphism $f: A \to B$ of projection-PS-algebras such that f_{nat1} is an isomorphism;

(3) Cat(CPS) is the category of projection-PS-algebras with projection-PS-homomorphisms.

Obviously Cat(CPS) is well defined.

<u>2.5 Fact</u> : A is a projection-PS-algebra if and only if A is a CPS-algebra.

<u>Proof</u> : follows from the definition of the projection constraints. $\square$

<u>2.6 Remark</u> :
The projections $p_A(k,\_): A \to A$ need not be projection-PS-homomorphisms, since in general $p\text{-}s_A(k,N_A(a1,...,an)) \neq N_A(p\text{-}s1_A(k,a1),...p\text{-}sn_A(k,an))$.

The following fact states that Cat(CPS) has free and initial algebras, using the Birkhoff construction of free algebras. The existence of free and initial algebras can also be shown using free functors (see 2.15,3.3).

<u>2.7 Fact</u> (Products and Subalgebras) :
Cat(CPS) is closed w.r.t. products and subalgebras; thus Cat(CPS) has free and initial algebras.

<u>Proof</u> : (1) Given $A_i \in$ Cat(CPS) $(i \in I)$ define $P = ((P_s)_{s \in S}, (N_P)_{N \in OP})$ by

$$P_s = \{(a_i)_{i \in I} \mid \forall i \in I \ a_i \in A_i \} \text{ if } s \neq nat1$$

$$P_{nat1} = \{(n)_{i \in I} \mid n \in \mathbb{N}_1 \}$$

and $N_P((a1_i)_{i \in I},...,(an_i)_{i \in I}) = (N_{A_i}(a1_i,...,an_i))_{i \in I}$.

Furthermore define the projections $\pi_j: P \to A$ by $\pi_j((a_i)_{i \in I}) = a_j \ (j \in I)$.

Then $\pi_j \circ N_P((a1_i)_{i \in I},...,(an_i)_{i \in I}) = N_{A_j}(a1_j,...,an_j) = N_{A_j} \circ \pi_j((a1_i,...,an_i))_{i \in I})$, $\pi_j((n)) = n$, i.e. the $\pi_j \ (j \in I)$ are projection-PS-homomorphisms. Given $f_i: B \to A_i \in$ Cat(SIG$_{init}$), $i \in I$, SIG=SIG(PS), where Cat(SIG$_{init}$) is the category of SIG-algebras with initial **pnat1**-part, then $f: B \to P$, $f_s(b) = (f_{si}(b))_{i \in I}$ is the unique factorization through P. f_{nat1} is an isomorphism, since each $f_{i,nat1}$ is an isomorphism. Thus P is a product of the $A_i \ (i \in I)$ in Cat(SIG$_{init}$). The product property shows that P is also a PS-algebra :

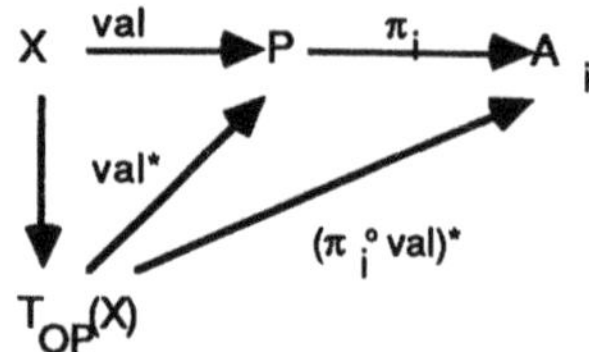

If (L,R,X) is an equation from PS, val:$X \to P$ an arbitrary valuation, then $\pi_i \circ val^*(L) = (\pi_i \circ val_i)^*(L) = (\pi_i \circ val_i)^*(R) = \pi_i \circ val^*(R)$ for all $i \in I$, thus val*(L)=val*(R). At last products of projection spaces and products of projection compatible functions are product spaces and projection compatible function respectively, thus $P \in Cat(CPS)$.

(2) (i) $\mathbb{N}_1$ has no proper subalgebras ;

(ii) each subalgebra A_0 of an algebra $A \in Cat(CPS)$ is closed w.r.t. the (restricted) projections $(p\text{-}s_A)_{s \in S}$ thus $(A_{0,s}, p\text{-}s_{A_0})$ is a projection space (for all $s \in S$);

(iii) the restriction of a projection compatible operation to a projection subspace is projection compatible. (i), (ii) and (iii) show that each SIG-subalgebra A_0 of a Cat(CPS)-algebra A is a Cat(CPS)-algebra.   □

2.2 Semantics

The semantics of a projection specification PS (which will be defined in 2.16) shall be a complete separated projection-PS-algebra which is initial in the category concerned. The separated completion (resp. standard construction) functor SC (see 1.14) is a left adjoint and therefore proper, but it cannot be defined on the whole category of PS-algebras Cat(PS), since in general it contains algebras which are not projection algebras. Thus a first (universal) step from PS-algebras to CPS-algebras is defined, which is described by a free functor. Since all projection specifications are conservative extensions of **pnat1** the considerations can be restricted to algebras with initial **pnat1**-part.

The following lemma shows that free functors induced by parameterized projection specifications preserve initial **pnat1**-parts.

<u>2.8 Lemma</u> : Given projection specifications PS, PS1 with PS⊆PS1 and a corresponding free functor F:Cat(PS)→Cat(PS1), then F preserves initial **pnat1**-parts, i.e. if $A \in Cat(PS_{init})$ then $F(A) \in Cat(PS1_{init})$, where $Cat(PS_{init})$ is the subcategory of Cat(PS) consisting of all PS-algebras with initial **pnat1**-part and all projection-PS-homomorphisms, ditto for $Cat(PS1_{init})$.

<u>Proof</u> : F(A) is isomorphic to the A-quotient term algebra $T_{PS}(A)$ (see [13], 7.14). Since $A_{pnat1}\cong\mathbb{N}_1$ and PS1 preserves **pnat1**, there are neither additional terms of sort nat1 in $T_{PS}(A)$ nor additional identifications between nat1-terms. Thus $(F(A))_{pnat1} \cong (T_{PS}(A))_{pnat1} \cong \mathbb{N}_1$. $\square$

Let PS=(S,OP,E) be a projection specification, CPS´=(S,OP,E$\cup$C1$\cup$C2), then (PS,CPS´) is a parameterized specification with corresponding free functor F:Cat(PS)$\rightarrow$Cat(CPS´). According to the lemma above F can be restricted to F*:Cat(PS$_{init}$)$\rightarrow$Cat(CPS) yielding the projection constraints functor Con$_{PS}$:

<u>2.9 Definition</u> (Projection Constraints Functor Con$_{PS}$) :
Given a projection specification PS the *projection constraints functor* Con$_{PS}$:Cat(PS$_{init}$)$\rightarrow$Cat(CPS) is the free functor (left adjoint to the inclusion functor I:Cat(CPS)$\rightarrow$Cat(PS$_{init}$)) .

As result we can state, that Cat(CPS) always has an initial algebra.

<u>2.10 Theorem</u> : (Initial CPS-algebra) : For each projection specification PS Cat(CPS) has an initial algebra T_{CPS} .

<u>Proof</u> : Cat(PS) has an initial algebra T_{PS}; by lemma 2.2 $(T_{PS})_{pnat1}\cong\mathbb{N}_1$, i.e. $T_{PS}\in$ Cat(PS$_{init}$). Thus T_{CPS}:=Con$_{PS}(T_{PS})$ is well defined and initial in Cat(CPS), since Con$_{PS}$ is free. $\square$

Separated completion by standard construction of the initial CPS-algebra yields the semantics of the projection specification PS. Since it is needed in the sequel the standard construction is defined for all projection algebras. It is a generalization of the standard construction for projection spaces and can also be extended to a functor SC on the category of projection algebras. Although there is nothing new in the definition, since operations and projection homomorphisms are (families of) projection compatible functions, a complete definition is given.

<u>2.11 Definition</u> (Standard Construction for Projection Algebras) :
Given a projection specification PS=(S,OP,E) and a projection-PS-algebra $A=((A_s)_{s\in S},(N_A)_{N\in OP})$ the *standard construction* $A^\infty =((A_s^\infty)_{s\in S},(N_A^\infty)_{N\in OP})$ is defined by

$$A_s^\infty=\{(a_n)_{n\geq 1} \mid \forall n\in \mathbb{N}_1 \ \ p\text{-}s_A(n,a_{n+1})=a_n\}$$
$$N_A^\infty((a1_n)_{n\geq 1},...,(ak_n)_{n\geq 1})=(p\text{-}s_A(n,N(a1_n,...,ak_n)))_{n\geq 1} \ \ .$$

For the following definition note that a projection-PS-algebra $A=((A_s)_{s\in S},(N_A)_{n\in OP})$ is called complete resp. separated if each base projection space $(A_s,\text{p-s}_A)$ is complete resp. separated; and $\text{Cat}_{Compl}\text{Sep(CPS)}$ is the category of complete separated projection-PS-algebras.

<u>2.12 Definition and Fact</u> : (Standard Construction Functor) :
The *standard construction functor* $SC:\text{Cat(CPS)}\to\text{Cat}_{Compl}\text{Sep(CPS)}$ is defined on algebras A by $SC(A)=A^\infty$ and on projection-PS-homomorphisms $h:A\to B$ by $SC(h)=h^\infty$, $h^\infty{}_s((a_n)_{n\geq1})=(h_s(a_n))_{n\geq1}$. SC is a left adjoint to the inclusion functor $I:\text{Cat}_{Compl}\text{Sep}\to\text{Cat(CPS)}$ and $SC\circ I$ is naturally isomorphic to the identity $ID_{\text{Cat}_{Compl}\text{Sep(CPS)}}$.

<u>Proof</u> : In 1.15 it has been shown that $SC(A_s,\text{p-s}_A)$ is a projection space; simple calculation shows that SC preserves products, i.e. $SC(\Pi(A_i,p_i))=\Pi SC(A_i,p_i)$, thus $SC(N_A)$ is a projection compatible function on the product concerned. Furthermore completeness and separation follow from 1.15. Since projection homomorphisms h are families of projection morphisms h_s ($s\in S$) and $(h_s(a_n))_{n\geq1}=(q\text{-s}_A(n,h_s(a_n)))_{n\geq1}$ for projective sequences $(a_n)_{n\geq1}$, also $SC(h)$ is well defined.
To show the universal property of SC consider the following diagram :

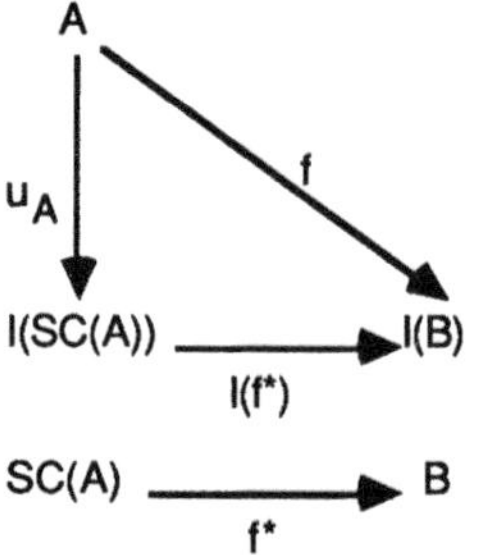

(i) First we show that the mapping $u_A:A\to A^\infty$ defined by $u_{A,s}(a)=(\text{p-s}_A(k,a))_{k\geq1}$ is a projection-PS-homomorphism :
$$u_{A,s}(N_A(a1,...an))=(\text{p-s}_A(k,N_A(a1,...,an)))_{k\geq1}=$$
$$=(\text{p-s}_A(k,N_A(\text{p-s}1_A(k,a1),...,\text{p-sn}_A(k,an))))_{k\geq1}=$$
$$=N_A^\infty((\text{p-s}1(k,a1))_{k\geq1},...,(\text{p-sn}(k,an))_{k\geq1})=N_A^\infty(u_{A,s1}(a1),...,u_{A,sn}(an))\ .$$

(ii) Now let B be any complete separated projection-PS-algebra and $f:A\to B$ a projection-PS-homomorphism. Define $f^*:A^\infty\to B$ by $f^*{}_s((a_k)_{k\geq1})=\lim_k f_s(a_k)$. Since $(a_k)_{k\geq1}$ is Cauchy in $(A_s,\text{p-s}_A)$, f is continuous and B_s is complete and separated, f^* is well defined and for all $a\in A_s$ we have $f^*{}_s(u_{A,s}(a))=$
$=\lim_k f_s(\text{p-s}_A(k,a))=f_s(\lim_k \text{p-s}_A(k,a))=f_s(a)$, hence $f^*\circ i_A=f$.
It remains to be shown that f^* is a projection-PS-homomorphism :

$$f^*{}_s(N_A{}^\infty((a1_j)_{j\geq 1},...,(an_j)_{j\geq 1}))=f^*{}_s((p\text{-}s_A(j,N_A(a1_j,...,an_j)))_{j\geq 1})=$$
$$=\lim_j f_s(p\text{-}s_A(j,N_A(a1_j,...,an_j)))=$$
$$=\lim_j p\text{-}s_B(j,N_B(f_{s1}(a1_j),...,f_{sn}(an_j)))=$$
$$=\lim_j p\text{-}s_B(j,N_B(f^*{}_{s1}\circ u_{s1}(a1_j),...,f^*{}_{sn}\circ u_{sn}(an_j)))=$$
$$=\lim_j p\text{-}s_B(j,N_B(f_{s1}((p\text{-}s1_A(k,a1_j)_{k\geq 1}),...,f_{sn}((p\text{-}sn_A(k,an_j)_{k\geq 1})))=$$
$$=N_B(f_{s1}((a1_j)_{j\geq 1}),...,f_{sn}((an_j)_{j\geq 1}))\ .$$

(iii) Next let $g:A^\infty\rightarrow B$ be any homomorphism of projection-SPEC-algebras with $g\circ u_A=f=f^*\circ u_A$. Since g_s and $f^*{}_s$ are continuous and $(a_n)_{n\geq 1}=\lim u_{A,s}(a_n)$ in $A_s{}^\infty$ for each $(a_n)_{n\geq 1}$ in $A_s{}^\infty$ and all $s\in S$ we have $g=f^*$, hence f^* is unique.

(iv) For $B\in Cat_{Compl}Sep(CPS)$ define $i:SC\circ I(B)\rightarrow B, i((b_n)_{n\geq 1})=\lim b_n$.

If $\lim b_n=\lim c_n$ and $(b_n)_{n\geq 1}$ and $(c_n)_{n\geq 1}$ are projective, then $(b_n)_{n\geq 1}=(c_n)_{n\geq 1}$ (see 1.5(3)), thus i is injective; surjectivity is obvious. Furthermore i_s is as an isomrophism for all $s\in S$, since

$$p\text{-}s_{B}\circ(n,(b_k)_{k\geq 1})=(p\text{-}s_B(n,b_k))_{k\geq 1}\ ,\ p\text{-}s_B(n,i_s(b_k)_{k\geq 1}))=p\text{-}s_B(n,\lim b_k)=b_n\ ,\ \text{whence}$$

$$p\text{-}s_{B}\circ(n,(b_k)_{k\geq 1})=p\text{-}s_{B}\circ(n,(b'_k)_{k\geq 1})\ \Leftrightarrow\ p\text{-}s_B(n,i_s(b_k)_{k\geq 1}))=p\text{-}s_B(n,i_s(b'_k)_{k\geq 1}))\qquad\square$$

<u>2.13 Corollary</u> : Given a projection specification PS and a projection-PS-algebra A, then
(1) SC(A) is a universal separated completion of A;
(2) SC(A) is a universal completion of Sep(A).

<u>2.14 Remark</u> : Since $a\equiv b\ :\Leftrightarrow\ \forall n\in \mathbb{N}_1\ p\text{-}s_A(n,a)=p\text{-}s_A(n,b)$ is a congruence relation, $Sep(A)=((A_s/\equiv)_{s\in S},((N_{A,\equiv})_{N\in OP})$ is well defined. On the other hand an algebra of finite elements cannot be defined, since the image of a finite element under a projection compatible function need not be finite.

The following theorem, stating that $Cat_{Compl}Sep(CPS)$ always has an initial algebra, is a corollary of 2.12 and leads to the definition of the semantics of a projection specification.

<u>2.15 Theorem</u> (Initial Complete Separated Projection Algebra) :
Given a projection specification PS, $SC\circ Con_{PS}(T_{PS})$ is initial in $Cat_{Compl}Sep(CPS)$, the category of complete separated projection-PS-algebras.

<u>2.16 Definition</u> (Semantics of a Projection Specification) :
Given a projection specification PS the (*initial complete separated algebra*) *semantics* CT_{PS} of PS is $SC\circ Con_{PS}(T_{PS})$; i.e.$CT_{PS}=SC\circ Con_{PS}(T_{PS})$. Furthermore the functor $PS:Cat(PS_{init})\rightarrow Cat_{Compl}Sep(CPS)$ defined by $PS=SC\circ Con_{PS}$ is called *projection semantics functor* .

2.17 Remark : If T_{CPS} is separated, then the universal mapping $u_{T_{CPS}}:T_{CPS}\rightarrow CT_{PS}=SC(T_{CPS})$ defined in the proof of 2.12 is injective, i.e. an embedding. It is in general impossible, however, to reconstruct T_{CPS} or even T_{PS}, given only CT_{PS}.

2.18 Remark : If all projections of a projection specification are discrete, then the projection constraints are always satisfied; i.e. $Con_{PS}=ID$. Furthermore the standard construction is naturally isomorphic to the identity (see 2.12). Thus $CT_{PS}=SC\circ Con_{PS}(T_{PS})\cong T_{PS}$, i.e. projection specifications contain usual algebraic specifications as special case. Therefore it is possible to specify combined data type and process algebras in one projection specification, taking discrete projections for the sorts incorporating data and non-discrete projections for the sorts incorporating processes.

2.19 Example : This is the standard example for processes : Elementary actions a1,...,an are given and can be executed sequentially ($\cdot$) or in indeterministic choice (+), c:action$\rightarrow$proc is a coercion. **probpa0** stands for **pro**jection specification of a **b**asic **p**rocess **a**lgebra, the **0** indicates that it is a preliminary version. In the sequel it will be extended to a projection specification of a process algebra, then an algebra of communicating processes and the action part will be replaced by a formal parameter projection specification in the third section of this paper.

probpa0 = pnat1 +

    <u>sorts</u> action, proc

    <u>opns</u> a1,...,an: $\rightarrow$action   {elementary actions}

           c:action$\rightarrow$proc   {coercion}

           +,$\cdot$:proc proc$\rightarrow$proc   {choice,sequence}

           p-action:nat1 action$\rightarrow$action

           p-proc:nat1 proc$\rightarrow$proc

    <u>eqns</u> <u>for all</u> x,y ,z <u>in</u> proc :

           x+x=x   (idempotent)

           x+y=y+x   (commutative)

           (x+y)+z=x+(y+z)   (associative)

           (x$\cdot$y)$\cdot$z=x$\cdot$(y$\cdot$z)   (associative)

           (x+y)$\cdot$z=x$\cdot$z+y$\cdot$z   (left distributive)

           <u>for all</u> n <u>in</u> nat1; <u>for all</u> a <u>in</u> action :

           p-action(n,a)=a

           <u>for all</u> n <u>in</u> nat1; <u>for all</u> a <u>in</u> action; <u>for all</u> x, y <u>in</u> proc :

           p-proc(n,c(a))=c(a)

           p-proc(1,c(a)$\cdot$x)=c(a)

           p-proc(succ(n),c(a)$\cdot$x)=c(a)$\cdot$p-proc(n,x)

           p-proc(n,x+y)=p-proc(n,x)+p-proc(n,y)

Deletion of the projection operation symbols and the corresponding equations yields a specification **bpa** of the basic process algebra BPA by Bergstra&Klop (see [6]). Note that **bpa** $\subseteq$ **probpa0** is a conservative extension.

Another helpful definition for the discussion of semantics and correctness of a projection specification is the projection functor, which generalizes the projections to the whole algebra. Fact 2.21 below shows that the semantics CT_{PS} of a projection specification PS is completely determined by the finite parts of the initial algebra T_{CPS} and generalizes the representation of the standard construction for projection spaces given in 1.17(3) to projection algebras.

<u>2.20 Definition and Fact</u> (Projection Functor) :
Given a projection specification PS the *projection functor*
P_n:Cat(CPS)$\to$Cat(CPS) ($n\in \mathbb{N}_1$) is defined on algebras
$A=((A_s)_{s\in S},(N_A)_{N\in OP})$ by
$P_n(A)=A^n=((A_s{}^n)_{s\in S},(N_A{}^n)_{N\in OP})$,
where $A_s{}^n=$p-$s_A(n,A_s)$ and $N_A{}^n=$p-$s_A(n,\_)\circ N_A$ if N:s1...sk$\to$s ;
and on projection-PS-homomorphisms h:A$\to$B by $P_n(h)=h|A^n$.

<u>Proof</u> : For each N:s1...sk$\to$s$\in$ OP we have : p-$s_A(n,\_)\circ N_A=$
    =p-$s_A(n,_)\circ N_A\circ($p-$s1_A(n,_)$x...xp-$sk_A(n,\_))=$
    =$N_A{}^n\circ($p-$s1_A(n,_)$x...xp-$sk_A(n,\_))$,
thus $p_A(n,\_)$:A$\to A^n$, $p_A(n,_)=($p-$s_A(n,\_))_{s\in S}$, is a homomorphism;
since p-nat$1_A=$id$_{nat1_A}$, $p_A(n,\_)$ is also a projection-PS-homomorphism, thus
$A^n=p_A(n,A)$ is a projection-PS-algebra. Since $h_s\circ N_A{}^n=h_s\circ$p-$s_A(n,\_)\circ N_A=$
=p-$s_B(n,_)\circ h_s\circ N_A=$p-$s_B(n,_)\circ N_B\circ(h_{s1}$x...x$h_{sk})=N_B{}^n\circ(h_{s1}$x...x$h_{sk})$
$P_n(h)$ is well defined. At last P_n is a functor, since $P_n(h)=h$. $\qquad\qquad$ $\square$

<u>2.21 Fact</u> : Given a projection-PS-algebra A, then SC(A) together with
$\underline{p}(n,\_)$:SC(A)$\to A^n$, $\underline{p}(n,(a_k)_{k\geq 1})=a_n$ ($n\in \mathbb{N}_1$) is a projective limit of the diagram

$$P_1(A)\longleftarrow\!P_2(A)\longleftarrow\!...\!\longleftarrow P_n(A)\longleftarrow\!P_{n+1}(A)\,...$$
$$\qquad p_A(1,\_)\qquad\qquad\qquad\qquad\qquad\qquad p_A(n,\_)$$

<u>Proof</u> :
(i) $\underline{p}(n,\_)$ is a projection-PS-homomorphism :
$\underline{p}(n,N_A{}^\infty((a1_k)_{k\geq 1},...,(am_k)_{k\geq 1}))=\underline{p}(n,($p-$s_A(k,N_A(a1_k,...,am_k)))_{k\geq 1})=$
=p-$s_A(n,N_A(a1_n,...,am_n))=N_A{}^n(\underline{p}(n,(a1_k)_{k\geq 1}),...,\underline{p}(n,(am_k)_{k\geq 1}))$
(ii) Since $a_n=p(n,a_{n+1})$ for all $(a_k)_{k\geq 1}\in A^\infty$ we have $\underline{p}(n,\_)=p_A(n,\_)\circ\underline{p}(n+1,\_)$

(iii) Let $(B,q)\in Cat(CPS)$, $f_n:B\to A^n \in Cat(CPS)$ with $p\text{-}s_A(n,\_)\circ f_{n+1,s}=f_{n,s}$ $(n\geq 1, s\in S)$. Define $f:B\to A^\infty$ by $f_s(b)=(f_{n,s}(b))_{n\geq 1}$ $(b\in B_s)$.

We have: $p\text{-}s_A(n,f_{n+1,s}(b))=f_{n,s}(b)$, hence f is well defined ;

$\underline{p}(n,f_s(b))=f_{n,s}(b)$, hence $\underline{p}(n,\_)\circ f=f_n$;

$f_s(N_B(b1,...,bm))=(f_{n,s}(N_B(b1,...,bm)))_{n\geq 1}=(N_A{}^n(f_{n,s1}(b1),...,f_{n,sm}(bm)))_{n\geq 1}=$

$=(p\text{-}s_A(n,N_A(f_{n,s1}(b1),...,f_{n,sm}(bm))))_{n\geq 1}=N_A{}^\infty((f_{n,s1}(b1))_{n\geq 1},...,(f_{n,sm}(bm))_{n\geq 1})=$

$=N_A{}^\infty(f_{s1}(b1),...,f_{sm}(bm))$ and $f_{nat1}=id_{nat1}$, hence f is a projection-PS-homomorphism.

(iv) If $g:B\to A^\infty\in Cat(PS)$ with $\underline{p}(n,\_)\circ g=f_n$ $\forall n\geq 1$ and $g(b)=(a_k)_{k\geq 1}$, then $a_n=\underline{p}(n,g(b))=f_n(b)$, hence g=f and f is unique.

To complete the proof also for corollary 1.17(3) first note that for one sorted projection-PS-algebras each projection-PS-homomorphism is a projection morphism and projection compatible.

It remains to be shown that if the given functions $f_n:(B,q)\to(A_n,p_n)$ for $(B,q)\in PRO_c$ are projection compatible, then also $f:(B,q)\to(A^\infty,p^\infty)$, $f(b)=(f(b_n))_{n\geq 1}$ is projection compatible : $p^\infty(k,f(b))=p^\infty(k,(f_n(b))_{n\geq 1})=(p(k,f_n(b)))_{n\geq 1}=$

$=(p(k,f_n(q(k,b))))_{n\geq 1}=p^\infty(k,f(q(k,b)))$,hence f is projection compatible.   □

2.3. Correctness

There are "two directions" of correctness for a projection specification PS. First the internal correctness, demanding that the initial algebra T_{PS} satisfies the projection constraints C_{PS} and is separated. This guarantees that the initial PS-algebra T_{PS} is a dense subalgebra of the semantics CT_{PS}. Secondly there is initial extension correctness w.r.t. a given model projection algebra. The latter is the usual notion of correctness of an algebraic specification.

<u>2.22 Definition</u> : Given a projection specification PS=(S,OP,E),
(1) PS is *correct*, if $T_{PS}=T_{CPS}$, (i.e. the initial algebra T_{PS} satisfies the projection constraints C_{PS}) and T_{PS} is separated.
(2) Given a SIG_0-algebra A, where SIG_0 is a subsignature of $SIG=(S,OP)$ containing all projection operation symbols, PS is *initial extension correct w.r.t.* A if $(CT_{PS})_{SIG_0}\cong A$; i.e. the SIG_0-reduct of the semantics of PS is isomorphic to A.

<u>2.23 Remark</u> : (1) In 1.17 it has been shown that the universal mapping $u_A:A\to SC(A)$ is an (injective) embedding if A is separated. Thus if PS is correct, then the initial algebra T_{PS} is isomorphic to a dense subalgebra of the

semantics CT_{PS}.

(2) A condition stronger than separation of T_{PS}, but often satisfied in applications, is the following: If each base set in T_{PS} together with its projection is a finite projection space, then T_{PS} is separated (see 1.2(3)).

2.21 showed that the semantics of a projection specification is completely determined by the finite parts $P_n(T_{CPS})$ of the inital projection algebra T_{CPS}. Also the correctness w.r.t. a model algebra A depends only on the finite parts of the semantics algebra CT_{PS} and A.

<u>2.24 Fact</u> : Given a projection specification PS,

(1) $\forall n \in \mathbb{N}_1 \ P_n(CT_{PS}) \cong P_n(T_{CPS})$;

(2) PS is inital extension correct w.r.t. a complete separated model projection algebra $A \in \text{Cat}(SIG_0)$ if and only if $\forall n \in \mathbb{N}_1 \ P_n((T_{CPS})_{SIG_0}) \cong P_n(A)$.

<u>Proof</u> : (1) Simple calculations show that $h:P_n(CT_{PS}) \to P_n(T_{CPS})$,

$h_s([p\text{-}s]^\infty(n,(a_k)_{k\geq 1})) = [p\text{-}s(n,a_n)] = a_n \ ((a_k)_{k\geq 1} \in CT_{PS})$ is an isomorphism (in $\text{Cat}(CPS)$) .

(2) Let $V:\text{Cat}(PS) \to \text{Cat}(SIG_0)$ be the forgetful functor.

" $\Rightarrow$ ": Since SIG_0 contains all projection operation symbols $P_n \circ V = V \circ P_n$, thus by

(1) $V(CT_{PS}) \cong A$ implies $P_n \circ V(T_{CPS}) = V \circ P_n(T_{CPS}) \cong V \circ P_n(CT_{PS}) = P_n \circ V(CT_{PS}) \cong$

$\cong P_n(A) \ (\forall n \in \mathbb{N}_1)$.

" $\Leftarrow$ ": Since V is a right adjoint and CT_{PS} is a limit of the diagram $(P_n(T_{CPS}) \leftarrow\!\!\!- P_{n+1}(T_{CPS}))_{n \geq 1}$, $V(CT_{PS})$ is a limit of the diagram $(V \circ P_n(T_{CPS}) \leftarrow\!\!\!- V \circ P_{n+1}(T_{CPS}))_{n \geq 1}$. On the other hand A is complete and separated, thus $A \cong SC(A)$ is a limit of the diagram $(P_n(A) \leftarrow\!\!\!- P_{n+1}(A))_{n \geq 1}$ (see 2.21). Since isomorphisms in a category of projection algebras commute with projections and isomorphic diagrams have isomorphic limits the assertion follows. $\square$

2.4. Extensions

Usually algebraic specifications are developed stepwise, i.e. in a basic specification generating operations are specified and in a second step further operations are defined.

Concerning projection specifications each extension step has to respect the projection constraints. The projection space axiom holds whenever the extension is complete, i.e. no new data is added in the extended initial (quotient term) algebra. If there occurs new data in the extended algebra the axiom has to be shown for the additional part. Furthermore projection compatibility of the old operations on the new parts of the base spaces has to be shown. Projection compatibility of the extension operations do not hold in general, even if the extension is an enrichment. I.e. there are more operations which can be defined by equations on a given data type than nonexpansive operations. But since projection compatibility is respected by composition at least derived operations are projection compatible.

Extensions of projection specifications by recursive equation systems are considered in the fourth section of this paper, where also sufficient criteria for projection compatibility of enrichment operations will be given.

2.25 Fact : Given projection specifications PS=(S,OP,E), PS1=(S1,OP1,E1) with PS⊆PS1 and PS is correct, then
(1) if PS⊆PS1 is a complete extension, then the projection space axiom C1 holds in $(T_{PS1})_{PS}$ i.e. $(T_{PS1,s},[p\text{-}s])$ is a projection space for all s∈S;
(2) if N*∈OP1 is derived from operations in OP, then N* is projection compatible

2.26 Remark : (1) N*:s1...sk→s ∈OP1 is derived from N,N1,...,Nk∈OP if for all xi∈ si (i=1,...,k) N*(x1,...,xk)=N(N1(x11,...,x1m1),...,Nr(xr1,...,xrmr)) , xij∈ {x1,...,xk} .
(2) Since PS1 is a projection specification by presupposition the initiality constraint C3 always holds in T_{PS1}.

Proof of 2.25 : (1) Since the unique homomorphism h:T_{PS}→$(T_{PS1})_{PS}$∈ Cat(PS) is surjective and T_{PS} satisfies [p-s](n,[p-s](m,_))=[p-s](min(n,m),_) for all s∈S the equation also holds in $(T_{PS1})_{PS}$.
(2) Composition of projection compatible funtions yields a projection compatible function.

2.27 Examples :
(i) (**propa0** extending **probpa0** , **propa0** = **pro**jection specification of a **p**rocess **algebra**) :
This is a specification of the arbitrary interleaving concept for parallel processes used by Bergstra&Klop and others.

propa0 = probpa0 +
    <u>opns</u> ‖,‖L:proc proc→proc   {merge,left merge}
    <u>eqns</u> <u>for all</u> a <u>in</u> action; <u>for all</u> x,y,z <u>in</u> proc:
        x‖y = x‖Ly+y‖Lx
        c(a)‖Lx = c(a)x
        c(a)x‖Ly = c(a)(x‖y)
        (x+y)‖Lz = (x‖Lz)+(y‖Lz)

(ii) (example for an enrichment operation which is not projection compatible) :

natmin = pnat1 +
   <u>sorts</u> nat
   <u>opns</u> 0:→nat
          succ:nat→nat
          p-nat:nat1 nat→nat
   <u>eqns</u> <u>for all</u> n1 <u>in</u> nat1; <u>for all</u> n <u>in</u> nat :
          p-nat(1,0)=0
          p-nat(1,succ(n))=succ(0)
          p-nat(succ(n1),0)=0
          p-nat(succ(n1),succ(n))=succ(p-nat(n1,n))

natmin is a specification of the natural numbers M with the minimum as projection (see 1.6). Now extend **natmin** by the subtraction sub(n,m)=n-m if n≥m and sub(n,m)=0 if n<m.

natsub = natmin +
   <u>opns</u> sub:nat nat→nat
   <u>eqns</u> <u>for all</u> n,m <u>in</u> nat :
          sub(n,0)=n
          sub(0,n)=0
          sub(succ(n),succ(m))=sub(n,m)

Let n1∈ nat1, n∈ nat with 1≤n1≤n , then p-nat(n1,sub(succ(n),n))=
=p-nat(n1,succ(0))=succ(0)   and p-nat(n1,sub(p-nat(n1,succ(n)),p-nat(n1,n)))=
=p-nat(n1,sub(n1,n))=0 ,
thus p-nat(n1,_)∘sub≠p-nat(n1,_)∘sub∘(p-nat(n1,_) x p-nat(n1,_))
and sub is not projection compatible.

(iii) The operation τ_I in ACP$_\tau$ (Algebra of Communicating Processes with internal action τ, see [6]), that hides all actions a∈ I, i.e. turns every a∈ I into τ , is not projection compatible, although it is an enrichment operation. Thus τ_I cannot be defined consistently for infinite processes. In the papers of the group in Amsterdam (see [6], [34]) *fair abstraction* is introduced at this point to avoid this inconsistency.

<u>2.28 Remark</u> : Trying to extend the **natmin** specification given in example 2.27(iii) by the equality between natural numbers leeds to difficulties : First note, that a projection p$_\mathbb{B}$ on the boolean algebra $\mathbb{B}$ =({T,F},T,F,¬,∧,∨) must be eventually discrete, since the semantics of a projection specification is always a separated projection algebra. Now let k∈ $\mathbb{N}_1$; n,m∈ $\mathbb{N}$ with m>n>k, then
   F = p$_\mathbb{B}$ (k,F) = p$_\mathbb{B}$ (k,n=m) = p$_\mathbb{B}$ (k,min{k,n}=min{k,m}) = p$_\mathbb{B}$ (k,k=k) =
   =p$_\mathbb{B}$(k,T)=T .

Thus each specification of the equality in the specification **natmin** leeds to an inconsistent projection semantics. In general it will be difficult to find non-constant projection compatible operations from a process sort (= sort with non-discrete projection) into a data type sort (= sort with a discrete projection).

3. PARAMETERIZED PROJECTION SPECIFICATIONS

Parameterization is one of the most expressive tools for algebraic specifications, making possible reusability and compositionality of specifications. So one of the main concepts of this paper is to introduce parameterized projection specifications.

Parameterized projection specifications are, syntactically, defined as pairs of projection specifications - a formal and a target specification - like usual algebraic specifications. Also the semantics of a parameterized projection specification is defined as a class of free functors, but source and target of these free functors must be categories of complete separated projection algebras. It is shown that this free functor always exists and can be defined with the help of the projection semantics functor defined in the previous chapter (see 2.16) , and the free functor of the parameterized projection specification considered as usual parameterized algebraic specification. To distinguish the latter and the free functors of complete separated projection algebras, we say free functor and projection free functor respectively, since completeness and separation are defined relative to the projections given in the specification.

A special accent is laid on the relation between free functors and projection free functors, i.e. the relation between usual (free functor) semantics and projection (free functor) semantics. A parameterized projection specification will be called correct, if these semantics are compatible, i.e. the free construction of a complete separated projection algebra A is isomorphic to a subalgebra of the projection free construction of A.

The remaining parts of this chapter introduce projection parameter passing, which is also defined analogously to usual algebraic specifications. Also the semantical constructions of parameter passing and the correctness results can be carried over to projection parameter passing.

For sake of readability first simple projection parameter passing is discussed; the results then can easily be stated for the general case of parameterized projection parameter passing. The whole chapter is developed along chapter 7 and 8 of the book [13] and uses its results.

3.1. Parameterized Projection Specifications

Since all notions which are needed to define syntax, semantics and correctness of parameterized projection specifications are already introduced, we can at once give the corresponding definitions.

<u>**3.1 Definition**</u> (Parameterized Projection Specification) :
(1) A *parameterized projection specification* (PS,PS1) is a pair of projection specifications PS,PS1 with PS$\subseteq$PS1; i.e. a parameterized specification where formal and target specification are projection specifications.
(2) The *semantics* of a parameterized projection specification (PS,PS1) is the class of *projection free functors* $PF:Cat_{Compl}^{Sep}(CPS)\to Cat_{Compl}^{Sep}(CPS1)$ given by PF=PS1∘F∘I, where

$I:Cat_{Compl}^{Sep}(CPS)\to Cat(PS_{init})$	is the inclusion functor
$F:Cat(PS_{init})\to Cat(PS1_{init})$	is the restriction of a free functor
	$F:Cat(PS)\to Cat(PS1)$
$PS1=SC1\circ Con_{PS1}$	is the projection semantics functor (see 2.16)
$Con_{PS1}:Cat(PS1_{init})\to Cat(CPS1)$	is the projection constraints functor (see 2.9)
$SC1:Cat(CPS)\to Cat_{Compl}^{Sep}(CPS1)$	is the standard construction functor (see 2.12) .

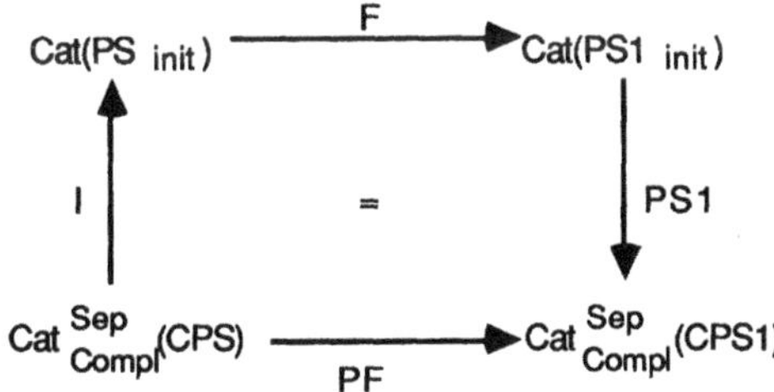

(In 3.3 it will be shown that projection free functors PF are actually free.)

(3) A parameterized projection specification (PS,PS1) is
(i) *correct*, if the free functor F preserves separated projection algebras, i.e. if
$\forall A\in Cat^{Sep}(CPS)$ $F(A)\in Cat^{Sep}(CPS1)$;
(ii) *correct w.r.t. a model functor*
$PM:Cat_{Compl}^{Sep}(CMPS)\to Cat_{Compl}^{Sep}(CMPS1),$
where   MPS$\subseteq$PS, MPS1$\subseteq$PS1 are model projection specifications
and $U:Cat_{Compl}^{Sep}(CPS)\to Cat_{Compl}^{Sep}(CMPS)$
and $U1:Cat_{Compl}^{Sep}(CPS1)\to Cat_{Compl}^{Sep}(CMPS1)$ are the corresponding forgetful functors,
if PM∘U$\cong$U1∘PF (natural).

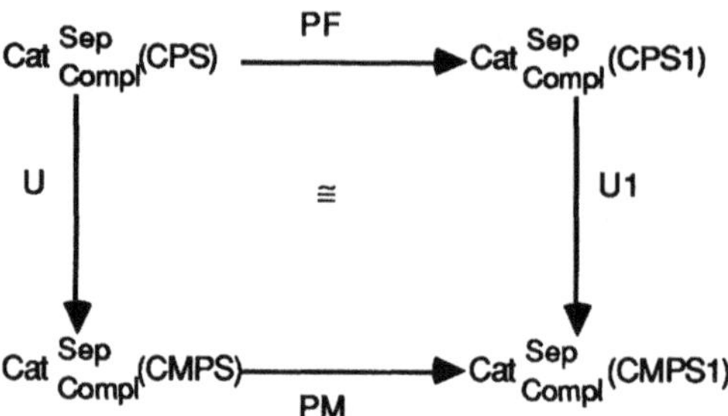

3.2 Remarks :

Syntax :

(1) The formal specification PS of a parameterized projection specification (PS,PS1) is a projection specification and thus **pnat1**⊆PS is always formal. It is of course intended to actualize **pnat1** always by **pnat1** .

Semantics :

(1) Since free functors are unique up to natural isomorphisms, also the projection free functors PF are naturally isomorphic. In slight abuse of language we will say "the projection free functor" instead of class of projection free functors.

(2) If A∈Cat(PS) has an initial **pnat1**-part then also F(A) has an initial **pnat1**-part (see 2.8); i.e. the restriction $F:Cat(PS_{init})\rightarrow Cat(PS1_{init})$ of the free functor is well defined .

(3) If PS=**pnat1**, then $Cat_{Compl}Sep(CPS)$ consists (up to isomorphic copies) of the algebra $\mathbb{N}_1$ and the identity $id_{\mathbb{N}_1}$ alone. Then $PF(\mathbb{N}_1)$ is initial in $Cat_{Compl}Sep(CPS1)$, since PF is free. Thus projection specifications may be considered as special case of parameterized projection specifications.

Correctness :

(1) If (PS,PS1) is correct, then F(A) is isomorphic to a subalgebra of PF(A) for all A∈$Cat_{Compl}Sep(CPS)$ (compare 2.23(1)).

(2) Even if the (restricted) free functor F is strongly persistent it need neither preserve projection algebras nor separation or completeness. Consider e.g. the following parameterized projection specifications (PS,PSi), i=1,2,3, with corresponding free functors Fi:Cat(PS)→Cat(PSi) :

PS = **pnat1**

PS1 = **pnat1** +

    sorts s

    opns c:→s

          f:s→s

          p-s:nat1 s→s

PS2 = PS1 +
    eqns <u>for all</u> n <u>in</u> nat1; <u>for all</u> x <u>in</u> s:
        p-s(n,c)=c
        p-s(1,x)=c
        p-s(succ(n),f(x))=f(p-s(n,x))
PS3 = PS2 +
    eqns <u>for all</u> n <u>in</u> nat1; <u>for all</u> x <u>in</u> s:
        p-s(n,x)=c

Then $F1(\mathbb{N}_1)$ is no projection algebra, since $[\text{p-s}(n,\text{p-s}(m,x))] \neq [\text{p-s}(\min(n,m),x)]$,

$F2(\mathbb{N}_1)$ is not complete, since the Cauchy sequence $([f^{(n)}(c)])_{n \geq 1}$ does not converge, and

$F3(\mathbb{N}_1)$ is not separated, since $[\text{p-s}(n,c)]=[c]=[\text{p-s}(n,f(c))]$ for all $n \in \mathbb{N}_1$, but $[c] \neq [f(c)]$;

although F1, F2 and F3 are strongly persistent.

(3) In 3.7 it will be shown that general forgetful functors w.r.t. "projection specification morphisms" preserve projection algebras and complete separated projection algebras; i.e. $U:\text{Cat}_{\text{Compl}}{}^{\text{Sep}}(\text{CPS}) \to \text{Cat}_{\text{Compl}}{}^{\text{Sep}}(\text{CMPS})$ and $U1:\text{Cat}_{\text{Compl}}{}^{\text{Sep}}(\text{CPS1}) \to \text{Cat}_{\text{Compl}}{}^{\text{Sep}}(\text{CMPS1})$ are the restrictions of $U:\text{Cat}(\text{PS}) \to \text{Cat}(\text{MPS})$ and $U1:\text{Cat}(\text{PS1}) \to \text{Cat}(\text{CMPS1})$ respectively and $V:\text{Cat}_{\text{Compl}}{}^{\text{Sep}}(\text{CPS1}) \to \text{Cat}_{\text{Compl}}{}^{\text{Sep}}(\text{CPS})$ is the restriction of the forgetful functor $V:\text{Cat}(\text{PS1}) \to \text{Cat}(\text{PS})$.

To complete the definition of the projection free functors it is now shown that projection free functors are actually free and that the projection semantics of a free construction of an arbitrary algebra A is isomorphic to the projection semantics of the free construction of the projection semantics of A.

<u>3.3 Fact</u> (Semantics): Given a parameterized projection specification (PS,PS1),
(1) the projection free functor $PF:\text{Cat}_{\text{Compl}}{}^{\text{Sep}}(\text{CPS}) \to \text{Cat}_{\text{Compl}}{}^{\text{Sep}}(\text{CPS1})$ is a left adjoint to the forgetful functor $V:\text{Cat}_{\text{Compl}}{}^{\text{Sep}}(\text{CPS1}) \to \text{Cat}_{\text{Compl}}{}^{\text{Sep}}(\text{CPS})$, i.e. PF is free;
(2) the projection semantics is compatible with the free construction, i.e. $PF \circ PS \cong PS1 \circ F$, especially $PS1 \circ F \cong PS1 \circ F \circ I \circ PS$.

<u>Proof</u> : (1) Since $PF(A)=PS1 \circ F(A)$ $(A \in \text{Cat}_{\text{Compl}}{}^{\text{Sep}}(\text{CPS}))$, $PS1 \circ F$ is a left adjoint to the forgetful functor $V:\text{Cat}_{\text{Compl}}{}^{\text{Sep}}(\text{CPS1}) \to \text{Cat}(\text{PS}_{\text{init}})$ and $V(A1) \in \text{Cat}_{\text{Compl}}{}^{\text{Sep}}(\text{CPS})$, $V(f1) \in \text{Cat}_{\text{Compl}}{}^{\text{Sep}}(\text{CPS})$ for all $A1 \in \text{Cat}_{\text{Compl}}{}^{\text{Sep}}(\text{CPS1})$, $f1 \in \text{Cat}_{\text{Compl}}{}^{\text{Sep}}(\text{CPS1})$, PF is a left adjoint to the forgetful functor $V:\text{Cat}_{\text{Compl}}{}^{\text{Sep}}(\text{CPS1}) \to \text{Cat}_{\text{Compl}}{}^{\text{Sep}}(\text{CPS})$.
(2) $PF \circ PS$ is a left adjoint to $I \circ V$, $PS1 \circ F$ is a left adjoint to $V \circ I1$ and $I \circ V = V \circ I1$.

Thus $PF \circ PS \cong PS1 \circ F$.

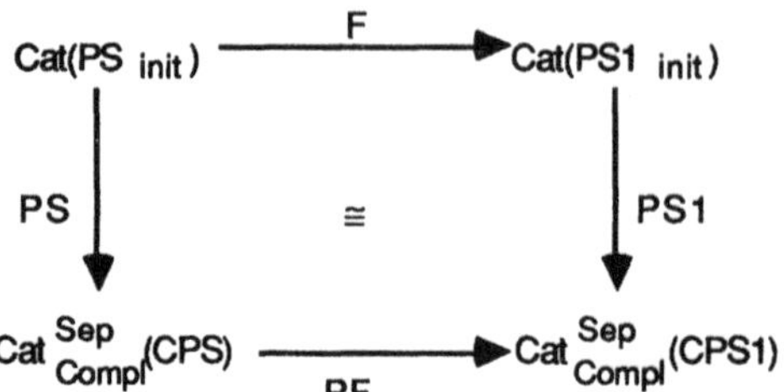

□

The following fact states how correctness w.r.t. a model functor relates to correctness w.r.t. a model functor of the underlying algebraic specification.

<u>3.4 Fact</u> (Correctness):
Given a parameterized projection specification (PS,PS1) and projection specifications MPS$\subseteq$PS, MPS1$\subseteq$PS1, a functor M:Cat(CMPS)$\rightarrow$Cat(CMPS1) induces a model functor

PM:Cat$_{Compl}$^{Sep}(CMPS)$\rightarrow$Cat$_{Compl}$^{Sep}(CMPS1) by PM=I$\circ$M$\circ$SCM1, where

    SCM1:Cat(CMPS1)$\rightarrow$Cat$_{Compl}$^{Sep}(CMPS1)   is the standard construction
                                             functor ,

    I:Cat$_{Compl}$^{Sep}(CMPS)$\rightarrow$Cat(CMPS)   is the inclusion functor.

Then (PS,PS1) is correct w.r.t PM, if Con$_{PS1}$$\circ$F is correct w.r.t. M, i.e.

M$\circ$U$\cong$U1$\circ$Con$_{PS1}$$\circ$F.

<u>Proof</u> :
The back rectangle in the diagram on the following page commutes by presupposition, commutativity of the left diagram is obvious, the right diagram commutes by 3.7, bottom and top diagram commute by definition.

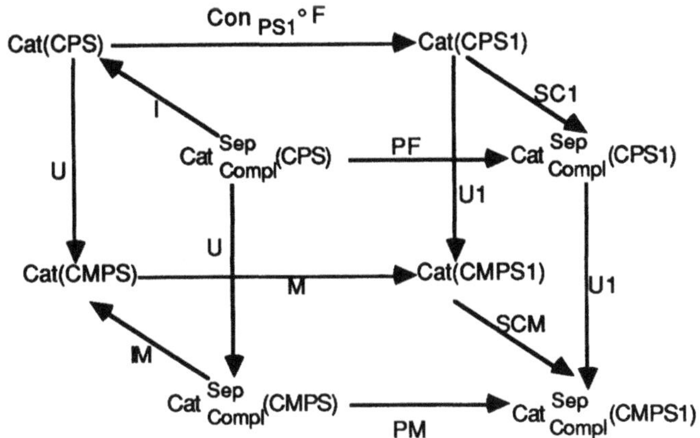

□

The following example takes the "action part" of the projection specification
propa0 given in 2.27(i) as formal part, whence the processes become
parameterized processes. A problem arises here, if the projection p-action of
actions is not discrete, i.e. also processes or other infinite objects are allowed as
actions. The coercion c:action→proc then is not projection compatible, unless
the projection p-proc of processes is adopted to this more general case: The n´th
projection of a process x now cuts x and all actions in x to length n.

<u>3.5 Example</u> : **probpa(action) = (action,probpa)**

action = pnat1 +
    <u>sorts</u> action
    <u>opns</u> p-action:nat1 action→action

probpa = action +
    <u>sorts</u> proc
    <u>opns</u> c:action→proc
           +,·:proc proc→proc
           p-proc:nat1 proc→proc
           cut:nat1 nat1 proc→proc
           { cut(n,m,x) cuts the process x to length m and each action in x to
                                              length n }
    <u>eqns</u> <u>for all</u> x,y ,z <u>in</u> proc :
           x+x=x
           x+y=y+x
           (x+y)+z=x+(y+z)
           (x·y)·z=x·(y·z)
           (x+y)·z=x·z+y·z

<u>for all</u> n,m <u>in</u> nat1; <u>for all</u> a <u>in</u> action; <u>for all</u> x, y <u>in</u> proc :
p-proc(n,x)=cut(n,n,x)
cut(n,m,c(a))=c(p-action(n,a))
cut(n,1,c(a)·x)=c(p-action(n,a))
cut(n,succ(m),c(a)·x)=c(p-action(n,a))·cut(n,m,x)
cut(n,m,x+y)=cut(n,m,x)+cut(n,m,y)

The basic parameterized process specification **probpa(action)** is now extended to a parameterized projection specification of processes with communication. It is an adapted version of a specification of Bergstra&Klop´s process algebra PA, where, however, the action part and the communication function are also specified. The deadlock δ is needed to express unsuccessful communication.

propa(action) = (action-with-comm,propa)

action-with-comm = action +

    <u>opns</u> δ:→action   {deadlock}
          | :action action→action   {communication function}
    <u>eqns</u> <u>for all</u> a,b,c <u>in</u> action :
          δ|a = δ
          a|b = b|a
          (a|b)|c = a|(b|c)

propa = action-with-comm + probpa +

    <u>opns</u> |,||,|L:proc proc→proc   {communication, merge, left merge}
    <u>eqns</u> <u>for all</u> a,b <u>in</u> action; <u>for all</u> x,y,z <u>in</u> proc :
          c(δ)x = c(δ)
          c(δ)+x = x
          c(a)|c(b) = c(a|b)
          (c(a)x)|c(b) = c(a|b)x
          (c(a)x)|(c(b)y) = c(a|b)(x||y)
          (x+y)|z = (x|z)+(y|z)
          x|y = y|x
          (x|y)|z = x|(y|z)
          x||y = x|Ly + y|Lx +x|y
          c(a)|Lx = c(a)x
          c(a)x|Ly = c(a)(x||y)
          (x+y)|Lz = (x|Lz)+(y|Lz)

3.2. Projection Parameter Passing

Projection parameter passing is defined analogously to usual parameter passing. To make the concept clear first simple projection parameter passing is introduced; then in the fourth part of this section the results are carried over to parameterized projection parameter passing.

The actualization of the formal part of a parameterized projection specification is expressed by a projection specification morphism, that ensures that the **pnat1**-part is actualized by **pnat1**.

<u>3.6 Definition</u> (Projection Specification Morphism) :
(1) Given projection specifications $PS=(S,OP,E)$, $PS'=(S',OP',E')$ a *projection specification morphism* $h:PS \rightarrow PS'$ is a specification morphism with

(i)　　$h|\textbf{pnat1}=id_{\textbf{pnat1}}$

(ii)　　$h(s)=nat1 \Rightarrow s=nat1$, for all $s \in S$

(iii)　for each projection operation symbol $p\text{-}s:nat1\ s \rightarrow s \in OP$　$h(p\text{-}s):nat1$　$h(s) \rightarrow h(s)$ is the projection operation symbol of sort $h(s)$ in OP'.

(2) Given a projection specification morphism $h:PS \rightarrow PS'$, PS' is called *h-secure* if the projection constraint C2 (projection compatibility) holds for all operations from $OP'\text{-}h(OP)$.

A basic fact for the following results is, that forgetful functors preserve projection algebras, separation and completeness. We call a parameterized projection specification (PS,PS') (or more generally a projection specification morphism $h:PS \rightarrow PS'$) secure (h-secure), if all operations $N \in OP\text{-}OP'$ (resp. $N \in OP\text{-}h(OP')$) are projection compatible. Security ensures that the corresponding forgetful functor is compatible with the projection semantics functor.

<u>3.7 Lemma</u> (Forgetful Functors) : Given projection specifications $PS=(S,OP,E),PS'=(S',OP',E')$ and a projection specification morphism $h:PS \rightarrow PS'$ the corresponding forgetful functor $V_h:Cat(PS') \rightarrow Cat(PS)$ preserves projection algebras, separation and completeness:
(1) Given $A' \in Cat(CPS')$, then
(i)　$V_h(A') \in Cat(CPS)$;
(ii) if A' is separated, then $V_h(A')$ is separated;
(iii)　if A' is complete, then $V_h(A')$ is complete.
(2) $V_h:Cat(CPS') \rightarrow Cat(CPS)$ is compatible with the standard construction functors $SC:Cat(CPS) \rightarrow Cat_{Compl}Sep(CPS)$ and
$SC':Cat(CPS') \rightarrow Cat_{Compl}Sep(CPS')$, i.e. $V_h \circ SC'=SC \circ V_h$.

(3) If PS' is h-secure, then $V_h:Cat(PS'_{init})\to Cat(PS_{init})$ is compatible with the projection constraints functors Con_{PS} and $Con_{PS'}$, i.e. $V_h \circ Con_{PS'} = Con_{PS} \circ V_h$.

(4) If (PS,PS') is a parameterized projection specification, h the inclusion, and the free functor $F:Cat(PS_{init})\to Cat(PS'_{init})$ preserves projection algebras, i.e. $F(A)\in Cat(CPS1)$ for all $A\in Cat(CPS)$, then $V_h:Cat(PS'_{init})\to Cat(PS_{init})$ is compatible with the projection constraints functors Con_{PS} and $Con_{PS'}$, i.e. $V_h \circ Con_{PS'} = Con_{PS} \circ V_h$.

<u>Proof</u> : (1) (i) Since h is a projection specification morphism $V_h(A')_{\textbf{pnat1}}=A'_{\textbf{pnat1}}\cong I\!N_1$. The projection constraints C1 and C2 (projection space axiom and projection compatibility of the operations) for PS' are the translated equations of the projection constraints C1 and C2 for PS. Since $V_h(A')$ satisfies an equation e if and only if A' satisfies the translated equation $h^{\#}(e)$ (see [13], 8.3) the assertion follows.

(ii) and (iii) $(V_h(A')_s, p\text{-}s_{V_h(A')})=(A'_{h(s)}, p\text{-}h(s)_{A'})$ $(s\in S)$, whence the assertions follow.

(2) Let U be the underlying set functor, then
$SC \circ V_h(A')=SC((A'_{h(s)})_{s\in S},(h(N)_{A'})_{N\in OP})=$

$\quad =((U \circ SC(A'_{h(s)}, p\text{-}h(s)_{A'}))_{s\in S},(SC(h(N)_{A'}))_{N\in OP})$

$V_h \circ SC'(A')=V_h((U \circ SC'(A'_s, p\text{-}s_{A'}))_{s\in S'},(SC'(N_{A'}))_{N\in OP'})=$

$\quad =((U \circ SC'(A'_{h(s)}, p\text{-}h(s)_{A'}))_{s\in S},(SC'(h(N)_{A'}))_{N\in OP})$

Since the standard construction is identically defined on single sorts and operations $SC \circ V_h = V_h \circ SC'$.

(3) First V preserves C3 (initial **pnat1**-part). By definition of the projection constraints functors (= free functors w.r.t. the equations C1 and C2) it is to be shown that $V_h \circ Con_{PS'}$ identifies the same elements as $Con_{PS} \circ V_h$ and vice versa. Thus let $A'\in Cat(PS'_{init})$ and $a,b\in A'_{s'}$ two elements which are identified in $Con_{PS'}(A')$. Then

(i) $a=p\text{-}s'_{A'}(n,p\text{-}s'_{A'}(m,c))$ and

$\quad b=p\text{-}s'_{A'}(m,p\text{-}s'_{A'}(n,c))$ or $b=p\text{-}s'_{A'}(\min(n,m),c)$ for a $c\in A'_{s'}$;

or

(ii) $a=p\text{-}s'_{A'}(n,N'_{A'}(c1,...,ck))$ and

$\quad b=p\text{-}s'_{A'}(n,N'_{A'}(p\text{-}s1'_{A'}(n,c1),...,p\text{-}sk'_{A'}(n,ck)))$

$\quad$ for an $N':s1'...sk'\to s'\in OP'$ and $ci\in A'_{si'}$, $i=1,...,k$.

Ad (i) If $s'=h(s)$ for an $s\in S$, then

$\quad\quad a=p\text{-}s_{V_h(A')}(n,p\text{-}s_{V_h(A')}(m,c))\in V_h(A')_s$ and

$\quad\quad b=p\text{-}s_{V_h(A')}(m,p\text{-}s_{V_h(A')}(n,c))\in V_h(A')_s$ or $b=p\text{-}s_{V_h(A')}(\min(n,m),c))\in V_h(A')_s$

thus a and b are also identified in $Con_{PS} \circ V_h(A')$.

If $s\notin h(S)$, then $a,b\notin Con_{PS} \circ V_h(A')$ and $a,b\notin V_h \circ Con_{PS'}(A')$.

Ad (ii) If $N':s1'...sk' \to s' = h(N):h(s1)...h(sk) \to h(s)$, then

$\quad a = p\text{-}s_{V_h(A')}(n, N_{V_h(A')}(c1,...,ck)) \in V_h(A')_s$ and

$\quad b = p\text{-}s_{V_h(A')}(n, N_{V_h(A')}(p\text{-}s1_{V_h(A')}(n,c1),...,p\text{-}sk_{V_h(A')}(n,ck))) \in V_h(A')_s$

thus a and b are also identified in $Con_{PS} \circ V_h(A')$.

If $N' \notin h(OP)$, then a=b in $A'_{s'}$, since $N'_{A'}$ is projection compatible by presupposition.

The reverse direction is obvious: If $a,b \in A'_{s'}$ are identified in $Con_{PS} \circ V_h(A')$, then they are also identified in $V_h \circ Con_{PS'}(A')$.

(4) It is to be shown that all operations in OP'-h(OP) are projection compatible, i.e. the second projection constraint C2 holds.

Let $X = U_{s \in S} X_s$ be a family of countably many variables. Since F is correct $F(T_{CPS}(X)) \in Cat(CPS')$. Thus for all $N:s1...sn \to s \in OP'\text{-}h(OP)$ the equation

$\quad p\text{-}s(k, N(x1,...,xn)) = p\text{-}s(k, N(p\text{-}s1(k,x1),...,p\text{-}sn(k,xn)))$

holds, i.e. N is projection compatible. $\qquad\qquad\qquad\qquad\qquad\qquad\square$

Projection parameter passing is defined analogously to usual parameter passing, where each part is replaced by the corresponding projection construction.

<u>3.8 Definition</u> (Projection Parameter Passing) :
Given a parameterized projection specification (PS,PS1), a projection specification PS' and a projection specification morphism $h:PS \to PS'$ the *projection parameter passing* is defined as follows:
(1) <u>Syntax</u> : the syntax of the projection parameter passing is given by the following pushout diagram, where PS2 is the (projection) *value specification* :

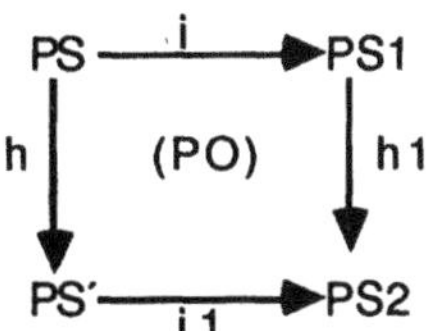

(2) <u>Semantics</u> : the semantics of the projection parameter passing is the triple of the semantics of the three components :
$PF:Cat_{Compl}{}^{Sep}(CPS) \to Cat_{Compl}{}^{Sep}(CPS1)$, $CT_{PS'}$ and CT_{PS2} .
(3) <u>Correctness</u> : projection parameter passing is correct, if
(i) $V_{i1}(CT_{PS2}) \cong CT_{PS'}$ (natural)   (actual parameter protection)
(ii) $V_{h1}(CT_{PS2}) \cong PF \circ V_h(CT_{PS'})$ (natural)   (passing compatibility) .

<u>3.9 Remark</u> : The pushout in 3.8(1) is always defined and PS2 is a projection specification.

<u>Proof</u> : First the pushout is always defined (see [13], 8.9). Secondly PS2 preserves **pnat1** , since neither PS′ nor PS1 have additional operation symbols with target nat1 or equations between nat1-terms. At last there is a selected projection operation symbol for each sort in S2=S′+h1(S1-S) :

i1(p-s:nat1 s→s)=p-s:nat1 s→s for all s∈S′ and

h1(p-s:nat1 s→s)=p-h(s):nat1 h(s)→h(s) for all s∈S1-S.   □

3.3 Amalgamation, Persistency and Extension of Persistent Functors

For the discussion of the semantics of projection parameter passing amalgamation, persistency and extension of persistent functors are the central notions.

The *amalgamation lemma* states how the category of the value projection specification can be represented in terms of the given parameterized projection specification and the actual projection specification. A *persistent projection specification* always preserves the actual parameter and the corresponding projection free functor can be *extended* to the *projection free functor* of the value projection specification.

The amalgamation lemma for algebraic specifications ([13], 8.11) states that the category of a value specification SPEC2 given by a parameter passing diagram

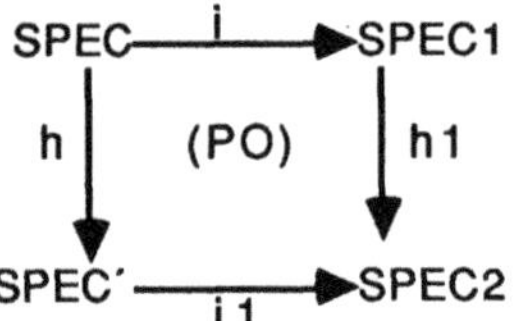

has the representation $Cat(SPEC2)=Cat(SPEC′)+_{Cat(SPEC)}Cat(SPEC1)$, i.e.

(1) each Cat(SPEC2)-algebra A2 has a unique representation $A2=A′+_A A1$, where

$A′=V_{i1}(A2)\in Cat(SPEC′)$, $A1=V_{h1}(A2)\in Cat(SPEC1)$ and

$A=V_h(A′)=V_i(A1)\in Cat(SPEC)$.

Vice versa given $A′\in Cat(SPEC′)$, $A1\in Cat(SPEC1)$ and

$A=V_h(A′)=V_i(A1)\in Cat(SPEC)$, then $A2=A′+_A A1$ is the unique Cat(SPEC2)-algebra satisfying $V_{i1}(A2)=A′$ and $V_{h1}(A2)=A1$.

(2) the same representation holds for Cat(SPEC2)-morphisms.

(3) Given $A2\in Cat(SPEC2)$ $A2\cong A′+_A A1$ if and only if $V_{i1}(A2)\cong A′$ and $V_{h1}(A2)\cong A1$.

Since forgetful functors preserve projection algebras, completeness and separation, a similar representation of the corresponding subcategories holds.

If the parameterized projection specification (PS,PS1) is correct, then the projection semantics of a $PS2_{init}$-algebra is the amalgamated sum of the projection semantics of its components.

<u>3.10 Fact</u> (Amalgamation):
Given a projection parameter passing diagram 3.8(1), then
(1) $Cat_{Compl}Sep(CPS2)=Cat_{Compl}Sep(CPS')+_{Cat_{Compl}Sep(CPS)}Cat_{Compl}Sep(CPS1)$,
   $Cat(CPS2)=Cat(CPS')+_{Cat(CPS)}Cat(CPS1)$.
(2) If (PS,PS1) is correct and $A2 \in Cat(PS2_{init})$ has the representation
   $A2=A'+_A A1$
   with $A' \in Cat(PS'_{init})$, $A1 \in Cat(PS1_{init})$ and $A=V_h(A')=V_i(A1) \in Cat(PS_{init})$,
then
   $PS2(A2)=PS'(A')+_{PS(A)}PS1(A1)$;

<u>Proof</u> : (1) By definition of amalgamated sums it is clear, that the amalgamated sum of (complete, separated) projection algebras is a (complete,separated) projection algebra. Thus the assertion follows from the amalgamation lemma ([13], 8.11) stated above and lemma 3.7 (1) and (2) (forgetful functors);

(2) follows from the amalgamation lemma and 3.7 (3) and (4). $\square$

Persistency of a parameterized (projection) specification ensures that the actual parameter of the (projection) parameter passing is respected by the semantics of the value (projection) specification. Persistency of a projection free functor is defined like persistency of other free functors. Note, however, that persistency of the projection free functor and persistency of the free functor of the underlying algebraic specification are in general independent of each other.

<u>3.11 Definition</u> (Persistency) :
A parameterized projection specification (PS,PS1) is called *persistent*, if the projection free functor $PF:Cat_{Compl}Sep(CPS) \to Cat_{Compl}Sep(CPS1)$ is persistent w.r.t. the forgetful functor $V:Cat_{Compl}Sep(CPS1) \to Cat_{Compl}Sep(CPS)$;
i.e. $V \circ PF \cong ID_{Cat_{Compl}Sep(CPS)}$ (natural).

<u>3.12 Fact</u> (Persistency) : Given a parameterized projection specification (PS,PS1) with corresponding free functor $F:Cat(PS) \to Cat(PS1)$, projection free functor $PF:Cat_{Compl}Sep(CPS) \to Cat_{Compl}Sep(CPS1)$ and forgetful functors $V:Cat(PS1) \to Cat(PS)$ and $V:Cat_{Compl}Sep(CPS1) \to Cat_{Compl}Sep(CPS)$.
If (PS,PS1) is correct, then :
(1) If F is persistent and (PS,PS1) is secure, then PF is persistent.
(2) F also preserves the projection semantics, i.e. $I \circ PS \cong V \circ I1 \circ PS1 \circ F$.

__Proof__ : (1) $V \circ PF = V \circ PS1 \circ F \circ I = PS \circ V \circ F \circ I = PS \circ I \cong ID_{Cat_{Compl}Sep(CPS)}$, where the last isomorphism is natural (see 3.7 for the second equation).
(2) $I \circ PS \cong I \circ PS \circ V \circ F = I \circ V \circ PS1 \circ F = V \circ I1 \circ PS1 \circ F$, the isomorphism is natural by definition. $\square$

__3.13 Remark__ : (1) Strong persistency of F alone is not sufficient for persistency of PF. Consider e.g. the parameterized projection specification (__natmin,natsub__), where __natmin__ is a projection specification of the natural numbers with minimum as projection and __natsub__ is __natmin__ extended by the subtraction (see 2.27(ii)). F is strongly persistent, but PF is not persistent, since sub is not projection compatible, thus $I1' \circ Con_{PS1} \neq ID$.
(2) Persistency of PF does not imply persistency of $F \circ I$ (or even F). Consider e.g. the following parameterized projection specification (PS,PS1) with corresponding free functor $F : Cat(PS) \rightarrow Cat(PS1)$ and projection free functor $PF : Cat_{Compl}Sep(CPS) \rightarrow Cat_{Compl}Sep(CPS1)$.

    PS = __pnat1__ +
       <u>sorts</u>   s
       <u>opns</u> $c : \rightarrow s$
                 p-$s : nat1 \; s \rightarrow s$
       <u>eqns</u>   <u>for all</u> x <u>in</u> s:
                 p-$s(n,x) = c$
    PS1 = PS +
       <u>opns</u> $d : \rightarrow s$

Let $A \in Cat_{Compl}Sep(CPS1)$, then $A \cong (\{a\},a,\pi)$, $\pi(n,a) = a$, since A is separated. Now $F(A) \cong (\{a,b\},a,b,\pi)$, $\pi(n,a) = \pi(n,b) = a$, is a complete non-separated projection algebra, but F is not persistent. Since PF(A) is the separation of F(A) $PF(A) \cong (\{a\},a,\pi)$, i.e. PF is persistent.
The following lemma shows that, similar to the case of algebraic specifications, each persistent projection free functor can be replaced by a (naturallly isomorphic) strongly persistent projection free functor.

__3.14 Lemma__ (Strong Persistency) : Given a persistent parameterized projection specification (PS,PS1) there is a strongly persistent projection free functor $PF : Cat_{Compl}Sep(CPS) \rightarrow Cat_{Compl}Sep(CPS1)$.

__Proof__ : By fact 3.12 (persistency) $Sep \circ Con_{PS1} \circ F \circ I =: G \circ I$ is persistent. Since G is a free functor corresponding to a parameterized algebraic specification (with one conditional equation with inifinite premise, respectively a congruence relation) it can be replaced by a (naturally isomorphic) free functor G´ ([13], 8.14). Now $PF \cong Compl* \circ G \circ I$ (natural) and $Compl*(A) \cong A$ for all $A \in Cat_{Compl}Sep(CPS1)$, whence the assertion follows. $\square$

A projection parameter passing diagram 3.8(1) induces a free functor $F':\mathrm{Cat}(PS')\to\mathrm{Cat}(PS2)$. If the free functor $F:\mathrm{Cat}(PS)\to\mathrm{Cat}(PS1)$ is strongly persistent, then F' is the extension of F via h ($F'=\mathrm{Ext}(F,h)$), i.e. $F'(A')=A'+_A F(A)$, whith $A=V_h(A')$, for all $A'\in\mathrm{Cat}(PS')$. A similar representation of the induced projection free functor $PF':\mathrm{Cat}_{\mathrm{Compl}}{}^{\mathrm{Sep}}(PS')\to\mathrm{Cat}_{\mathrm{Compl}}{}^{\mathrm{Sep}}(PS2)$ can be given :

<u>3.15 Fact</u> (Extension):
(1) Given a projection parameter passing diagram 3.8(1) and a strongly persistent functor $PF:\mathrm{Cat}_{\mathrm{Compl}}{}^{\mathrm{Sep}}(CPS)\to\mathrm{Cat}_{\mathrm{Compl}}{}^{\mathrm{Sep}}(CPS1)$ there is a strongly persistent functor $\mathrm{Ext}(PF,h):\mathrm{Cat}_{\mathrm{Compl}}{}^{\mathrm{Sep}}(CPS')\to\mathrm{Cat}_{\mathrm{Compl}}{}^{\mathrm{Sep}}(CPS2)$ given by the amalgamated sums

$\mathrm{Ext}(PF,h)(A')=A'+_A PF(A)$, with $A=V_h(A')$ $\qquad(A'\in\mathrm{Cat}_{\mathrm{Compl}}{}^{\mathrm{Sep}}(CPS'))$,

$\mathrm{Ext}(PF,h)(f')=f'+_f PF(f)$, with $f=V_h(f')$ $\qquad(f'\in\mathrm{Cat}_{\mathrm{Compl}}{}^{\mathrm{Sep}}(CPS'))$.

$\mathrm{Ext}(PF,h)$ is uniquely determined by the equations

$\qquad V'\circ\mathrm{Ext}(PF,h)=\mathrm{ID}_{\mathrm{Cat}_{\mathrm{Compl}}{}^{\mathrm{Sep}}(CPS')}$ and $V_{h1}\circ\mathrm{Ext}(PF,h)=PF\circ V_h$.

If in addition PF is a left adjoint to the forgetful functor $V:\mathrm{Cat}_{\mathrm{Compl}}{}^{\mathrm{Sep}}(CPS1)\to\mathrm{Cat}_{\mathrm{Compl}}{}^{\mathrm{Sep}}(CPS)$, then $\mathrm{Ext}(PF,h)$ is a left adjoint to the forgetful functor $V':\mathrm{Cat}_{\mathrm{Compl}}{}^{\mathrm{Sep}}(CPS2)\to\mathrm{Cat}_{\mathrm{Compl}}{}^{\mathrm{Sep}}(CPS')$; especially $CT_{PS2}\cong CT_{PS'}+_C PF(C)$, where $C=V_h(CT_{PS'})$.

(2) If the free functor $F:\mathrm{Cat}(PS)\to\mathrm{Cat}(PS1)$ is strongly pesistent, $(PS,PS1)$ is secure and PS' is h-secure, then the projection free functor $PF':\mathrm{Cat}_{\mathrm{Compl}}{}^{\mathrm{Sep}}(CPS')\to\mathrm{Cat}_{\mathrm{Compl}}{}^{\mathrm{Sep}}(CPS2)$ is persistent and has the representation $PF'=PS2\circ\mathrm{Ext}(F,h)\circ I'$ (see the diagram below).

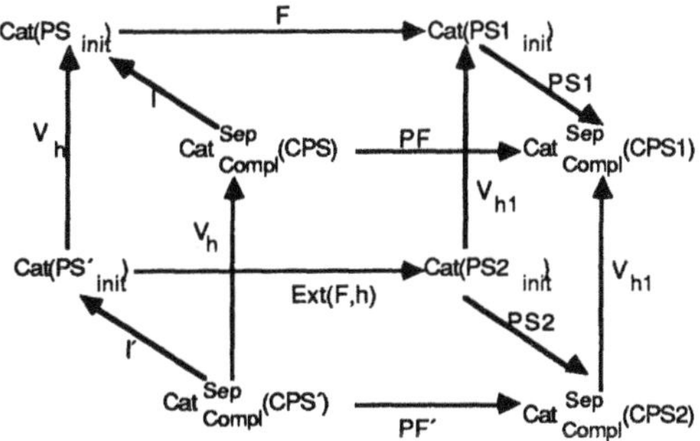

<u>Proof</u> : (1) is a corollary of the extension lemma [13], 8.15 and fact 3.10 (amalgamation).
(2) Since $\mathrm{Ext}(F,h):\mathrm{Cat}(PS'_{init})\to\mathrm{Cat}(PS2_{init})$ is a free functor, $PS2\circ\mathrm{Ext}(F,h)\circ I':\mathrm{Cat}_{\mathrm{Compl}}{}^{\mathrm{Sep}}(CPS')\to\mathrm{Cat}_{\mathrm{Compl}}{}^{\mathrm{Sep}}(CPS2)$ is projection free (see 3.3).
By 3.12-(1) PF is (strongly) persistent, whence (1) yields that PF' is persistent. $\square$

As a result of the previous facts we obtain : if a parameterized projection specification is persistent, then projection parameter passing is always correct and the semantics of the value projection specification is the amalgamated sum of the semantics of the actual parameter and the projection free construction of its h-reduct, where h is the projection specification morphism of the projection parameter passing.

3.16 Theorem (Correct Projection Parameter Passing): Given a parameterized projection specification (PS,PS1) the following statements are equivalent :
(1) Projection parameter passing is correct for all projection specification morphisms $h:PS \to PS'$.
(2) (PS,PS1) is persistent.
(3) The amalgamated sum $CT_{PS'} +_A PF(A)$ with $A = V_h(CT_{PS'})$ is well defined for all projection specification morphisms $h:PS \to PS'$ and the projection semantics CT_{PS2} of the value specification is given by $CT_{PS2} \cong CT_{PS'} +_A PF(A)$ (compositionality) where $CT_{PS'}$ and PF are the semantics of PS' and (PS,PS1) respectively.

Proof : $(2) \Rightarrow (3)$ by fact 3.15 (extension) and the fact that to each persistent projection free functor $PF:Cat_{Compl}{}^{Sep}(CPS) \to Cat_{Compl}{}^{Sep}(CPS1)$ there is a naturally isomorphic projection free functor that is strongly persistent (see 3.14).
$(3) \Rightarrow (1)$ by fact 3.10 (amalgamation).
$(1) \Rightarrow (2)$ Following the proof of the corresponding theorem for algebraic specifications ([13], 8.16) let $PS' = CPS(A) = CPS + (\varnothing, Const(A), Eqns(A))$ for given $A \in Cat_{Compl}{}^{Sep}(CPS)$. Then $A = V_h(T_{PS'}) \cong V_h(CT_{PS'})$, since $PS' = CPS'$ and $A = V_h(T_{PS'})$ is complete and separated. Using correctness of the projection parameter passing we have

$$V_i \circ PF(A) \cong V_i \circ PF \circ V_h(CT_{PS'}) \cong V_i \circ V_{h1}(CT_{PS2}) = V_h \circ V_{i1}(CT_{PS2}) \cong V_h(CT_{PS'}) \cong A . \quad \square$$

Also correctness of a persistent parameterized projection specification w.r.t. a strongly persistent model functor induces correctness of the value projection specification w.r.t. the induced model :

3.17 Theorem (Induced Correctness of Value Specifications) :
Given
(1) a persistent parameterized projection specification (PS,PS1) which is correct w.r.t. a strongly persistent model functor $PM:Cat_{Compl}{}^{Sep}(CMPS) \to Cat_{Compl}{}^{Sep}(CMPS1)$;
(2) an actual projection specification PS' which is (initial extension) correct w.r.t. a model algebra $A' \in Cat_{Compl}{}^{Sep}(CPS')$;

(3) a projection parameter passing morphism $h:PS \to PS'$ which is compatible with the model projection specifications MPS and MPS', i.e. $h(MPS) \subseteq MPS'$;
then we have
(4) the value projection specification PS2 is (initial extension) correct w.r.t. the PM-extension $PM'(A') \in Cat_{Compl}^{Sep}(CPS2)$ of A' given by the amalgamated sum
(5) $PM'(A') = A' +_A PM(A)$, with $A = V_k(A')$,

where $k:MPS \to MPS'$ is the restriction of h and MPS2 is the value specification in the projection parameter passing diagram

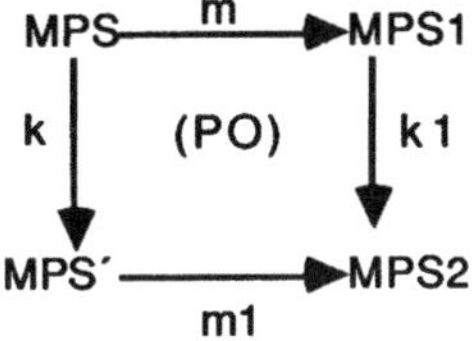

<u>Proof</u> : This fact is a special case of fact 3.21, where a proof will be given. □

3.4. Parameterized Projection Parameter Passing

In the last part of this section the results about simple projection parameter passing are stated for the more general case of parameterized projection parameter passing : the definition of parameterized projection parameter passing and the correctness results, which are generalizations of 3.16 and 3.17 respectively.

<u>3.18 Definition</u> (Parameterized Projection Parameter Passing) :
Given parameterized projection specifications (PS,PS1), (PS',PS1') and a projection specification morphism $h:PS \to PS1'$ the parameterized projection parameter passing is defined as follows:
(1) <u>Syntax</u> : the syntax of the parameterized projection parameter passing is given by the following pushout diagram, where (PS',PS2) is the (*parameterized projection*) *value specification* :

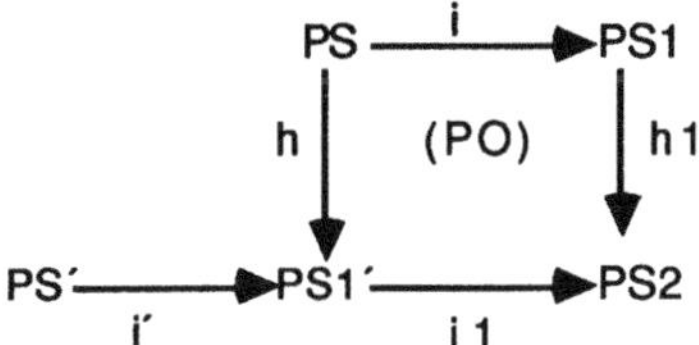

(2) <u>Semantics</u> : the semantics of the parameterized projection parameter passing is the triple of the semantics of the three components :

$$PF:Cat_{Compl}^{Sep}(CPS) \to Cat_{Compl}^{Sep}(CPS1),$$

$$PF':Cat_{Compl}^{Sep}(CPS') \to Cat_{Compl}^{Sep}(CPS1')$$

and

$$PF*_h PF':Cat_{Compl}^{Sep}(CPS') \to Cat_{Compl}^{Sep}(CPS2)$$, where $PF*_h PF'$ is the projection free functor induced by the free functor $F*_h F':Cat(PS') \to Cat(PS2)$.

(3) <u>Correctness</u> : parameterized projection parameter passing is correct, if

(i) $V_{i1} \circ (PF*_h PF') \cong PF'$ (natural)   (actual parameter protection)

(ii) $V_{h1} \circ (PF*_h PF') \cong PF \circ V_h \circ PF'$ (natural)   (passing compatibility) .

<u>3. 19 Remark</u> : Let $PS'=$**pnat1** , then parameterized projection parameter passing is equivalent to standard projection parameter passing (compare 3.2(3)).

<u>3.20 Theorem</u> (Correct Parameterized Projection Parameter Passing) :
Given persistent parameterized projection specifications (PS,PS1), (PS',PS1') and a projection parameter passing morphism $h:PS \to PS1'$ we have :
(1) The parameterized projection value specification (PS',PS2) is persistent.
(2) The projection free functor $PF*_h PF':Cat_{Compl}^{Sep}(CPS') \to Cat_{Compl}^{Sep}(CPS2)$
of (PS',PS2) is given by $PF*_h PF' \cong Ext(PF,h) \circ PF'$ (compositionality)

i.e. $(PF*_h PF')(A') \cong PF'(A')+_A PF(A)$ with $A=V_h \circ PF'(A')$,

where PF, PF' are the projection free functors of (PS,PS1) and (PS',PS1') respectively.
(3) Parameterized projection parameter passing is correct.

<u>Proof</u> : By lemma 3.14 we may assume that the projection free functors PF and PF' are strongly persistent. By fact 3.15 (extension) $Ext(PF,h):Cat_{Compl}^{Sep}(CPS') \to Cat_{Compl}^{Sep}(CPS2)$ is a persistent (projection) free functor, thus $Ext(PF,h) \circ PF' \cong PF*_h PF'$ (natural) and (PS',PS2) is persistent.

At last $V_{h1} \circ (PF*_h PF') \cong V_{i1} \circ Ext(PF,h) \circ PF' \cong PF'$, since $Ext(PF,h)$ is persistent,

and $V_{h1} \circ (PF*_h PF') \cong V_{h1} \circ Ext(PF,h) \circ PF' = PF \circ V_i \circ PF'$, by fact 3.15 (extension),

showing correctness of the parameterized projection parameter passing   □

<u>3.21 Theorem</u> (Induced Correctness of Value Specifications) :
Given
(1) a persistent parameterized projection specification (PS,PS1) which is correct w.r.t. a strongly persistent model functor $PM:Cat_{Compl}^{Sep}(CMPS) \to Cat_{Compl}^{Sep}(CMPS1)$;
(2) a correct and persistent parameterized projection specification (PS',PS1') which is correct w.r.t. a strongly persistent model functor $PM':Cat_{Compl}^{Sep}(CMPS') \to Cat_{Compl}^{Sep}(CMPS1')$;
(3) a projection parameter passing morphism $h:PS \to PS1'$ which is compatible

with the model projection specifications MPS and MPS1´, i.e. h(MPS)⊆MPS1´ ;
we have
(4) the value projection specification (PS´,PS2) is correct w.r.t. the model
functor $PM*_hPM´:Cat_{Compl}^{Sep}(CMPS´) \to Cat_{Compl}^{Sep}(CMPS2)$ given by the
amamlgamated sums
(5) $PM*_hPM´(A´)=PM´(A´)+_APM(A)$, with $A=V_k\circ PM´(A´)$,

where $k:MPS \to MPS1´$ is the restriction of h and MPS2 is the value specification
in the projection parameter passing diagram

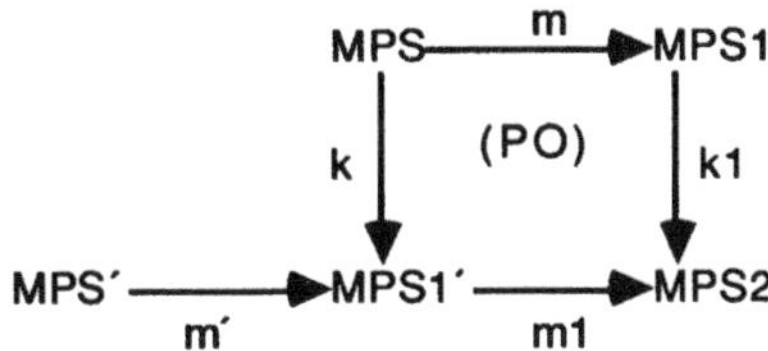

<u>Proof</u> : Let j,j1,j´,j1´ be inclusions and j2:MPS2→PS2 be the unique projection
specification morphism making the following diagram commute.

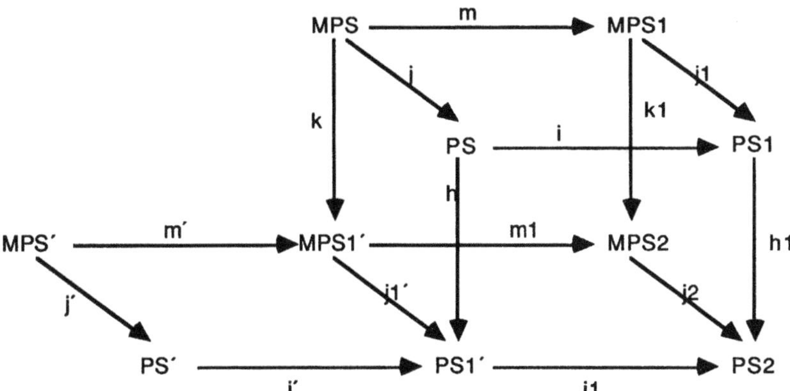

j1´ and j2 are welldefined by (3) and (5) respectively. $PM*_hPM´$ is well defined
and strongly persistent because PM and PM´are strongly persistent.
To show correctness of (PS´,PS2) w.r.t. $PM*_hPM´$ means to show
$V_{j2}\circ(PF*_hPF´) \cong (PM*_hPM´)\circ V_{j´}$; using 3.14 (amalgamation) it is therfore
enough to show the following equations :
$V_{m1}\circ V_{j2}\circ(PF*_hPF´) \cong PM´\circ V_{j´}$ and $V_{k1}\circ V_{j2}\circ(PF*_hPF´) \cong PM\circ V_k\circ PM´\circ V_{j´}$.
Since (PS,PS1) is persistent, parameterized projection parameter passing is
correct, i.e.
(a) $V_{i1}\circ(PF*_hPF´) \cong PF´$

(b) $V_{h1}\circ(PF*_hPF')\cong PF\circ V_h\circ PF'$;

correctness of (PS,PS1) w.r.t. PM and correctness of (PS',PS2) w.r.t. PM' yields

(c) $V_{j1}\circ PF\cong PM\circ V_j$

(d) $V_{j1'}\circ PF'\cong PM'\circ V_{j'}$.

Thus

$$V_{m1}\circ V_{j2}\circ(PF*_hPF')=V_{j1'}\circ V_{i1}\circ(PF*_hPF')\cong V_{j1'}\circ PF'\cong PM'\circ V_{j'},$$

$$V_{k1}\circ V_{j2}\circ(PF*_hPF')=V_{j1}\circ V_{h1}\circ(PF*_hPF')\cong V_{j1}\circ PF\circ V_h\circ PF'\cong PM\circ V_j\circ V_h\circ PF'=$$

$$=PM\circ V_k\circ V_{j1'}\circ PF'\cong PM\circ V_k\circ PM'\circ V_{j'}$$; where all isomorphisms are natural. $\square$

4. RECURSIVE PROJECTION SPECIFICATIONS

4.1. Recursive Projection Specifications

At the end of the second section extensions of projection specifications have been introduced and it has been shown that even enrichment operations need not be projection compatible. This leads to the definition of continuous enrichments, which are enrichments relative to the projection semantics (see definition 4.1 below). Extensions of projection specifications by recursive equation systems are of special interest for process specifications, since a large class of processes occuring in praxis can be (or are naturally) defined recursively. In this chapter recursive projection specifications are introduced, where the solution of the recursive equation system defining the recursive processes is computed using the fixed point theorem for projection spaces (1.11). The semantics of a recursive projection specification defined this way coincides with the projection semantics defined in the previous chapters, i.e. we have initial algebra resp. projection free functor semantics and compositionality, provided the specifications are correct. A recursive projection specification is correct, if the operator defined by the equation system is a contraction. Furthermore it will be shown that correct recursive projection specifications are continuous enrichments.

<u>4.1 Definition</u> (Continuous Enrichment) :
Given parameterized projection specifications (PS,PS1), (PS',PS1') with
$PS\subseteq PS'$, $PS1\subseteq PS1'$, $S(PS)=S(PS')$, $S(PS1)=S(PS1')$,
forgetful functors $V:Cat_{Compl}Sep(CPS')\to Cat_{Compl}Sep(CPS)$,
$V1:Cat_{Compl}Sep(CPS1')\to Cat_{Compl}Sep(CPS1)$ and projection free functors
$PF:Cat_{Compl}Sep(CPS)\to Cat_{Compl}Sep(CPS1)$,
$PF':Cat_{Compl}Sep(CPS')\to Cat_{Compl}Sep(CPS1')$,

then $(PS,PS1)\subseteq(PS',PS1')$ is a *continuous enrichment*, if the projection free functors are compatible with the forgetful functors, i.e. $PF \circ V \cong V1 \circ PF'$.

<u>4.2 Remark</u> : Let $PS=PS'=$**pnat1** , then the definition above includes : $PS1 \subseteq PS1'$ is a continuous enrichment, if $CT_{PS1} \cong V1(CT_{PS1'})$.

Next recursive projection specifications are defined. In the case of parameterized recursive projection specifications there shall only be a recursive equation system for the target specification, since there is no need to solve formal recursive equation systems. This does not exclude recursive equations for formal operations, which are included in the general theory given in chapter three.

<u>4.3 Definition</u> (Recursive Projection Specification) :
(1) Given a projection specification PS, a projection specification RECPS of the shape
RECPS = PS +
    <u>opns</u> proci:si1...simi$\rightarrow$si
    <u>eqns</u> <u>for all</u> xi1 <u>in</u> si1;...;<u>for all</u> ximi <u>in</u> simi :
        proci(xi1,...,ximi)=Ti $(i \in I)$
where $Ti \in T_{SIG(RECPS)}(Xi)$, $Xi=\{xi1,...,ximi\}$,
is a *recursive projection specification* based on PS with equation system
proci(xi1,...,ximi)=Ti $(i \in I)$.
(2) A parameterized projection specification (PS,RECPS1) is a *parameterized recursive projection specification* based on a parameterized projection specification (PS,PS1) (with equation system proci(xi1,...,ximi)=Ti $(i \in I)$), if RECPS1 is a recursive projection specification based on PS1 (with equation system proci(xi1,...,ximi)=Ti $(i \in I)$).

Obviously (parameterized) recursive projection specifications can be considered as parameterized projection specifications with projection free functor semantics defined in the previous chapters. In the sequel a condition will be defined, how to decide whether a recursive projection specification is a continuous enrichment and how the semantics can be computed from the semantics of the (parameterized) projection specification the recursive projection specification is based on, and the recursive equation system. I.e. projection compatibility of the processes proci need not be prooved like in general extensions defined in the second section.
Before we proceed a lemma about product projection spaces and contractions in product projection spaces is needed.

<u>4.4 Lemma</u> : Given projection spaces (A,p), (A_i,p_i) $(i \in I)$,
(1) a function $T:(A,p) \rightarrow \prod(A_i,p_i)$, $T(a)=(T_i(a))_{i \in I}$ is a contraction if and only if

each $T_j:(A,p)\to(A_j,p_j)$ $(j\in I)$ is a contraction ;

(2)   (i) if each (A_j,p_j) $(j\in I)$ is complete, then $\Pi(A_i,p_i)$ is complete;

(ii)   if $\Pi(A_i,p_i)$ is complete and none of the (A_j,p_j) $(j\in I)$ is empty, then each (A_j,p_j) $(j\in I)$ is complete;

(iii) if each (A_j,p_j) $(j\in I)$ is separated, then $\Pi(A_i,p_i)$ is separated;

(iv) if $\Pi(A_i,p_i)$ is separated and none of the (A_j,p_j) $(j\in I)$ is empty, then each (A_j,p_j) $(j\in I)$ is separated.

<u>Proof</u> : (1) First note that by definition of a contraction $A\neq\varnothing$ and $A_j\neq\varnothing$ $(\forall j\in I)$. Since $\Pi p_i(n,T(a))=(p_i(n,T_i(a)))_{i\in I}$ $(\forall n\in \mathbb{N}_1, \forall a\in A)$ the assertion follows.

(2) (i) Let $((a_i{}^n)_{i\in I})_{n\geq 1}$ be a Cauchy sequence in $\Pi(A_i,p_i)$, then we have the following equivalencies :

$\forall n\exists m\forall k\quad \Pi p_i(n,(a_i{}^m)_{i\in I})=\Pi p_i(n,(a_i{}^{m+k})_{i\in I})$

$\forall n\exists m\forall k\forall i\quad p_i(n,a_i{}^m)=p_i(n,a_i{}^{m+k})$

$\forall n\exists m\forall k\forall i\quad p_i(n,\lim_r a_i{}^r)=p_i(n,a_i{}^{m+k})$ (the limits exist since each (A_i,p_i) is complete)

$\forall n\exists m\forall k\quad \Pi p_i(n,(\lim_r a_i{}^r)_{i\in I})=\Pi p_i(n,(a_i{}^{m+k})_{i\in I})$

$\lim_n (a_i{}^n)_{i\in I} = (\lim_n a_i{}^n)_{i\in I}$

(ii) Let $(a_j{}^n)_{n\geq 1}$ be a Cauchy sequence in (A_j,p_j), then $((a_i{}^n)_{i\in I})_{n\geq 1}$ with $a_i{}^n=p(n,a_i{}')$ for an $a_i{}'\in A_i$ $(i\neq j)$ is a Cauchy sequence in $\Pi(A_i,p_i)$ and the j'th component of its limit is $\lim_n a_j{}^n\in (A_j,p_j)$.

(iii) and (iv) $\Pi p_i(n,(a_i)_{i\in I})=\Pi p_i(n,(b_i)_{i\in I}) \Leftrightarrow \forall i\ p_i(n,a_i)=p_i(n,b_i)$. $\square$

To solve the equation system of a recursive projection specification for a given complete separated projection-PS-algebra A we first take the space of all projection compatible operations on A defined by the signature of the processes proci $(i\in I)$ alone. If the equation system has a solution, then it must be contained in this process space. Next the operator $T(A)$ on the process space is defined, which replaces the evaluated left hand sides of the equations by their evaluated right hand sides.

<u>4.5 Definition</u> (Process Space) :

Given a parameterized recursive projection specification (PS,RECPS1) based on a parameterized projection specification (PS,PS1) with equation system proci$(xi1,...,ximi)=Ti$ $(i\in I)$ and a complete separated projection algebra $A\in \mathrm{Cat}_{\mathrm{Compl}}{}^{\mathrm{Sep}}(CPS1)$, the *process space* $(\mathrm{Hom}(A),pH)$ is the product projection space of the morphism spaces $(\{fi:A_{si1}x...xA_{simi}\to A_{si} \mid fi$ is projection com -patible$\},pHi)$; i.e.

$(\mathrm{Hom}(A),\mathrm{pH})=\prod_{i\in I}((A_{si},\text{p-si}_A)^{\prod_{j=1,..,mi}}(A_{sij},\text{p-sij}_A),\mathrm{pHi})$, with

$\mathrm{pH}:\mathbb{N}_1\times\mathrm{Hom}(A)\rightarrow\mathrm{Hom}(A)$,

$\mathrm{pH}(n,(fi)_{i\in I})(ai1,...,aimi)_{i\in I}=(\text{p-si}_A(n,fi(ai1,...,aimi)))_{i\in I}$.

<u>4.6 Remark</u> : Since $(A_{si},\text{p-si}_A)$ is a complete separated projection space (for each $i\in I$), $\mathrm{Hom}(A)$ is a complete separated projection space for all $A\in\mathrm{Cat}_{\mathrm{Compl}}{}^{\mathrm{Sep}}(\mathrm{CPS1})$ (see 1.7;4.4).

<u>4.7 Definition</u> (Operator T):
Given a parameterized recursive projection specification (PS,RECPS1) based on a parameterized projection specification (PS,PS1) with equation system $\mathrm{proci}(xi1,...,ximi)=Ti$ $(i\in I)$ and a complete separated projection algebra $A\in\mathrm{Cat}_{\mathrm{Compl}}{}^{\mathrm{Sep}}(\mathrm{CPS1})$, define the *operator* $T(A):(\mathrm{Hom}(A),\mathrm{pH})\rightarrow(\mathrm{Hom}(A),\mathrm{pH})$ by

$T(A)((fi)_{i\in I})(ai1,...,aimi)_{i\in I}=(\mathrm{vali}_A{}^*(Ti))_{i\in I}$

where $\mathrm{vali}_A{}^*:T_{\mathrm{SIG(RECPS1)}}(Xi)\rightarrow(A,(fi)_{i\in I})$ is the unique extension of the valuation $\mathrm{vali}_A:Xi\rightarrow(A,(fi)_{i\in I})$, $\mathrm{vali}_A(xij)=aij$, $j=1,...,mi$.

The operator $T(A)$ depends on the given $\mathrm{Cat}_{\mathrm{Compl}}{}^{\mathrm{Sep}}(\mathrm{CPS})$-algebra A and operates on its process space. If $T(A)$ has a fixed point, then it is a solution of the equation system and vice versa.

<u>4.8 Fact</u> : The operator $T(A)$ is well defined for all $A\in\mathrm{Cat}_{\mathrm{Compl}}{}^{\mathrm{Sep}}(\mathrm{CPS1})$.

<u>Proof</u> : It is to be shown, that

$\mathrm{pH}(n,T(A)(f))((ai1,...,aimi)_{i\in I}) =$

$= \mathrm{pH}(n,T(A)(f))((\text{p-si1}_A(n,ai1),...,\text{p-simi}_A(n,aimi)_{i\in I})$.

Let vali_A be given like in definition 4.7 and let $\mathrm{valin}_A:Xi\rightarrow A$, $\mathrm{valin}_A=(\text{p-s}_A(n,\_)\circ\mathrm{vali}_A)_{s\in S}$ be the valuation corresponding to the right hand side of the equation above. By uniqueness of the extension of valuations to the term algebra $T_{\mathrm{SIG(RECPS)}}(Xi)$ we have $\text{p-s}_A(n,\_)\circ\mathrm{vali}_A{}^*=\mathrm{valin}_A{}^*$, i.e.

$(\text{p-si}_A(n,\mathrm{vali}_A{}^*(Ti)))_{i\in I}=(\text{p-si}_A(n,\mathrm{valin}_A{}^*(Ti)))_{i\in I}$,q.e.d.

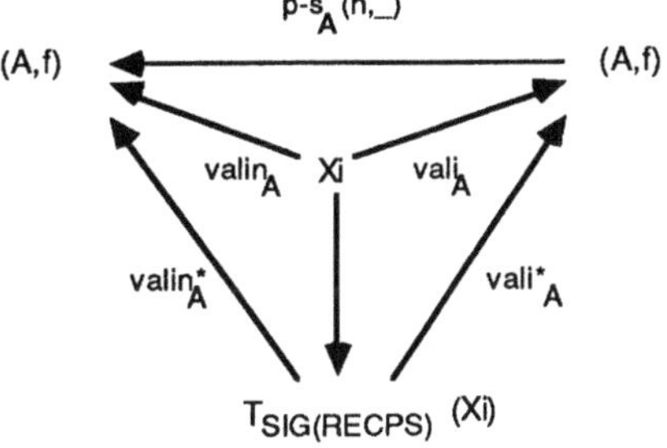

□

Next the correctness condition for recursive projection specifications is defined :

4.9 Definition : A parameterized recursive projection specification (PS,RECPS1) based on a parameterized projection specification (PS,PS1) is *contracting* , if the operator $T(A)$ is a contraction for each projection free algebra $A \in Cat_{Compl}{}^{Sep}(CPS1)$, i.e. each algebra $A=PF(A')$ with $A' \in Cat_{Compl}{}^{Sep}(CPS)$, PF the projection free functor $Cat_{Compl}{}^{Sep}(CPS) \to Cat_{Compl}{}^{Sep}(CPS1)$.

The fixed point in Hom(A) of the operator $T(A)$ is denoted by $(PROCi(A))_{i \in I}$.

4.10 Remark : If $Hom(A)=\emptyset$ for an algebra $A=PF(A') \in Cat_{Compl}{}^{Sep}(CPS1)$, $A' \in Cat_{Compl}{}^{Sep}(CPS)$, then $T(A)$ is no contraction. But also $(PS,PS1) \subseteq (PS,RECPS1)$ cannot be a continuous enrichment if $Hom(A)=\emptyset$:
If $Hom(A)=\emptyset$, then there is an $i \in I$ with $A_{si}=\emptyset$ but $A_{sij} \neq \emptyset$ for all $j \in \{1,...,mi\}$, i.e. one of the factors $((A_{si},p\text{-}si_A)^{\Pi_{j=1,..,mi}}(A_{sij},p\text{-}sij_A),pHi)$ is empty.
Let $RECPF:Cat_{Compl}{}^{Sep}(CPS) \to Cat_{Compl}{}^{Sep}(CRECPS1)$,
$PF:Cat_{Compl}{}^{Sep}(CPS) \to Cat_{Compl}{}^{Sep}(CPS1)$ be the projection free functors, then $RECPF(A')$ contains an element $[proci(ai1,...,aimi)]$, $aij \in A_{sij}$, $j=1,...,mi$, since $RECPS1 \supseteq PS1$ is an extension. Thus $RECPF(A')_{si} \neq \emptyset$, $PF(A')_{si}=A_{si}=\emptyset$ and $(PS,PS1) \subseteq (PS,RECPS1)$ is no continuous enrichment.

The following fact gives a syntactic criterion how to prove that $T(A)$ is a contraction for all $A \in Cat_{Compl}{}^{Sep}(CPS1)$.

4.11 Fact :Given a parameterized recursive projection specification (PS,RECPS1) based on a parameterized projection specification (PS,PS1) with equation system $proci(xi1,...,ximi)=Ti$ $(i \in I)$, the operator $T(A)$ is a contraction for all $A \in Cat_{Compl}{}^{Sep}(CPS1)$, if
(1) for each $i \in I$ there is an operation symbol $Ni:s1'...sm' \to si \in OP(PS1)$ with
  $\{s1',...sm'\} \subseteq \{si1,...,simi\}$;
(2) for each $i \in I$ there is a term $t \in T_{SIG(PS1)}(Xi)$ with $p\text{-}si(1,Ti) \equiv_{E(RECPS1)} t$;
(3) $p\text{-}si(succ(n),Ti) \equiv_{E(RECPS1)} p\text{-}si(succ(n),Ti^n)$ (for all $i \in I$),
where the terms $Ti^n \in T_{SIG(RECPS1)}(Xi)$ are recursively defined for arbitrary terms $Ti \in T_{SIG(RECPS1)}(Xi)$ by
(i) if $Ti=x$ (variable $x \in Xi$) or $Ti=c$ (constant symbol $c \in T_{OP(RECPS1)}$),
  then $Ti^n=Ti$;
(ii) if $Ti=procj(tj1,...,tjmj)$ with terms $tjk \in T_{SIG(RECPS1)}(Xi)$,

then $Ti^n = p\text{-}sj(n,Ti)$;

(iii) if $Ti = N(t1,...,tk)$ with $N \neq procj$ $(j \in I)$ and terms $tm \in T_{SIG(RECPS1)}(Xi)$,
   then $Ti^n = N(t1^n,...,tk^n)$.

(2) and (3) state that $p\text{-}si(1,Ti)$ is congruent to a term without process operation
symbol $procj$ by the equations of RECPS1 and that $[p\text{-}si(succ(n),Ti)] =$
$= [p\text{-}si(succ(n),Ti^n)]$ in $T_{RECPS1}(X)$.

<u>Proof</u> : Let $A \in Cat_{Compl}{}^{Sep}(CPS1)$. Condition (1) implies, that if
$\Pi_{j=1,...,mi}A_{sij} \neq \varnothing$ then $\Pi_{j=1,...,m}A_{sj} \neq \varnothing$, and $A_{si} \neq \varnothing$, i.e. $Hom(A) \neq \varnothing$.
Let $T(A)_i$ denote the i'th component of the operator $T(A)$. If (2) holds, then for all
$f \in Hom(A)$ $pHi(1,T(A)(f)_i)$ is independent of f, i.e. $pHi(1,T(A)(f)_i)=pHi(1,T(A)(g)_i)$
for all $f,g \in Hom(A)$. To prove this let $vali:Xi \to (A,f)$ and
$vali^*:T_{SIG(RECPS1)} \to (A,f)$ be given like in definition 4.7 and let
$vali^{**}:T_{SIG(PS1)} \to A$ be the unique extension of vali to $T_{SIG(PS1)}$.

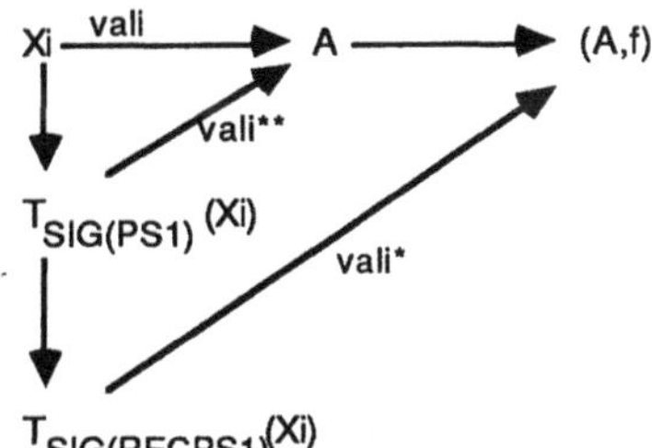

Then $pHi(1,T(A)(f)_i)(ai1,...,aimi)=p\text{-}si_A(1,vali^*(Ti))=vali^*(p\text{-}si(1,Ti))=vali^{**}(t)$,
which is independent of f.
To prove the last condition first the following equivalence is shown :

$pHi(n,T(A)(f)_i)=pHi(n,T(A)(g)_i) \to pHi(n+1,T(A)(f)_i)=pHi(n+1,T(A)(g)_i)$

for all $f,g \in Hom(A)$
is equivalent to

$pHi(n+1,T(A)(f)_i)=pHi(n+1,T(A)(pH(n,f))_i)$ for all $f \in Hom(A)$.

Proof : "$\Rightarrow$" : $pH(n,f)=pH(n,pH(n,f))$ implies
   $pHi(n+1,T(A)(f)_i)=pHi(n+1,T(A)(pH(n,f))_i)$;
"$\Leftarrow$" : $pHi(n,T(A)(f)_i)=pHi(n,T(A)(g)_i)$ implies
   $pHi(n+1,T(A)(f)_i)=pHi(n+1,T(A)(pH(n,f))_i)=pHi(n+1,T(A)(pH(n,g))_i)=$
   $=pHi(n+1,T(A)(g)_i)$

Next let $valin^*:T_{SIG(RECPS1)} \to (A,(p\text{-}si_A(n,\_)\circ fi)_{i \in I})$ be the unique extension of
$vali:Xi \to (A,(p\text{-}si_A(n,\_)\circ fi)_{i \in I})$ (Actually vali is a family of set functions $Xi \to A$;
$vali^*$ interprets the operation symbol proci by fi, $valin^*$ interprets it by
$p\text{-}si_A(n,\_)\circ fi$.)

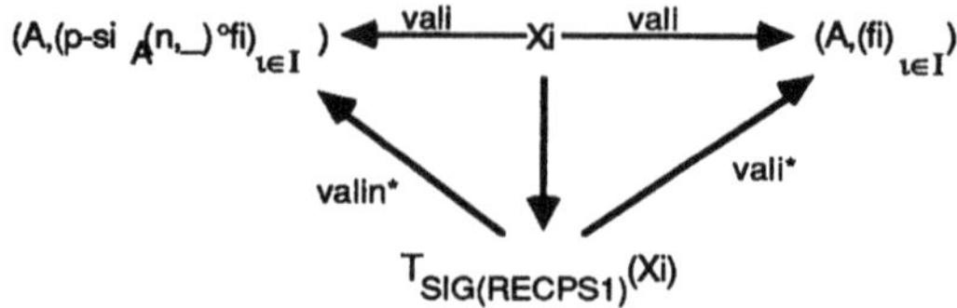

By definition of the terms Ti^n it is clear that vali*(p-si(succ(n),Ti))=
=vali*(p-si(succ(n),Ti^n))=valin*(p-si(succ(n),Ti)) ,i.e. $pHi(n+1,T(A)(f)_i)$=
=pHi(n+1,T(A)(pH(n,f))$_i$) , whence it is shown that T(A) is a contraction   □

The following theorem summarizes the results of this chapter. If the
recursive projection specification is contracting, then, by the fixed point
theorem 1.11, the operator T(A1) has a unique fixed point for each projection
free A1∈ Cat$_{Compl}$Sep(CPS1). This fixed point serves as semantics for the
processes proci and the extended algebra coincides with the projection
semantics of the recursive projection specification.

<u>4.12 Theorem</u> :
Given a contracting parameterized recursive projection specification
(PS,RECPS1) based on a parameterized projection specification (PS,PS1) with
equation system proci(xi1,...,ximi)=Ti (i∈ I) , a projection free algebra
A1∈ Cat$_{Compl}$Sep(CPS1) and the fixed point (PROCi(A1))$_{i∈ I}$ of the operator T(A1),
then
(1) (A1,(PROCi(A1))$_{i∈ I}$) is a complete separated projection RECPS1 algebra, i.e.
(A1,(PROCi(A1))$_{i∈ I}$)∈ Cat$_{Compl}$Sep(CRECPS1).
(2) Let PF:Cat$_{Compl}$Sep(CPS)→Cat$_{Compl}$Sep(CPS1),
RECPF:Cat$_{Compl}$Sep(CPS)→Cat$_{Compl}$Sep(CRECPS1) be the corresponding
projection free functors and A∈ Cat$_{Compl}$Sep(CPS) ,
then RECPF(A)≅(PF(A),(PROCi(PF(A)))$_{i∈ I}$) .

<u>Proof</u> : (1) is clear by the definition of the fixed point (PROCi(A1))$_{i∈ I}$∈ Hom(A1).
(2) Let V1:Cat$_{Compl}$Sep(CRECPS1)→Cat$_{Compl}$Sep(CPS1),
V:Cat$_{Compl}$Sep(CPS1)→ Cat$_{Compl}$Sep(CPS) be the forgetful functors,
B∈ Cat$_{Compl}$Sep(CRECPS1) and f:A→V∘V1(B)∈ Cat$_{Compl}$Sep(CPS) arbitrary.Then
it is to be shown, that the unique projection-PS1-homomorphism
f*:PF(A)→V1(B) with V(f*)∘u$_A$=f is compatible with (PROCi(PF(A)))$_{i∈ I}$ and
(proci$_B$)$_{i∈ I}$ respectively ; i.e. the diagram (*)

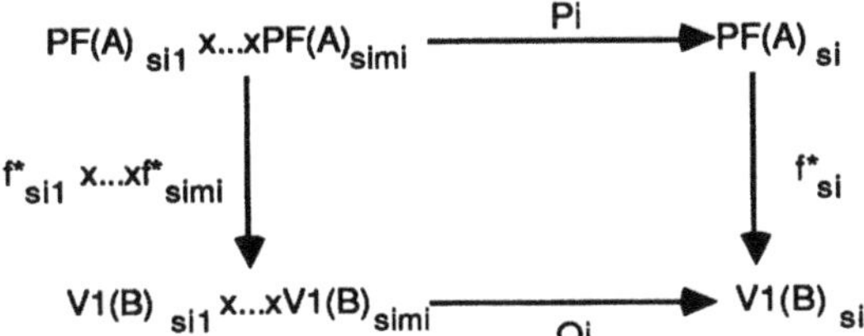

with $Pi := PROCi(PF(A))$, $Qi := proci_B$ commutes for all $i \in I$.

Let $P0, Q0$ be constant functions with values $p0, q0$ and $f^*_{si}(p0) = q0$. If Pi is replaced by $P0$ and Qi by $Q0$ in $(*)$, then $(*)$ commutes. Furthermore we have :

Lemma: If $(*)$ commutes with $P \in Hom(PF(A))$, $Q \in Hom(V1(B))$, then $(*)$ commutes with $T(PF(A))(P), T(V1(B))(Q)$.

Proof of the lemma:

In the diagram

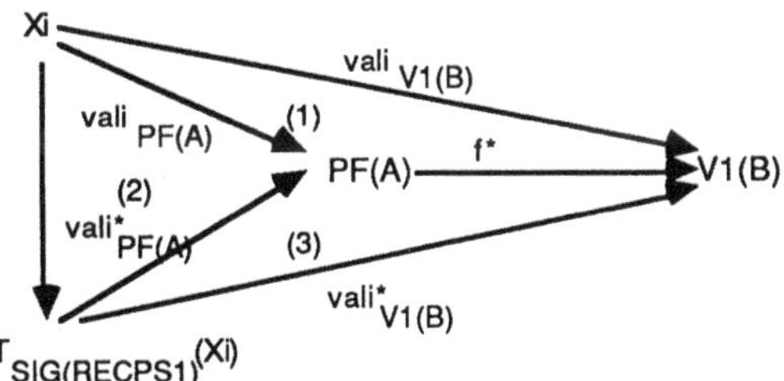

f^* is a homomorphism by the assumption of the lemma,

(1) commutes by definition of $T(PF(A))$ and $T(V1(B))$: $vali_{PF(A)}(xij) = aij$, $vali_{V1(B)}(xij) = f^*(aij)$

(2)&(2)$\cup$(3) commute by definition of $vali_{PF(A)}^*$ and $vali_{V1(B)}^*$

hence

$$T(PF(A))(P)(ai1,...,aimi) = vali_{PF(A)}^* (Ti)$$
$$T(V1(B))(Q)(f^*_{si1}(ai1),...,f^*_{simi}(aimi)) = vali_{V1(B)}^*(Ti) = f^*(vali_{PF(A)}^*(Ti))$$

i.e. $(*)$ commutes with $T(PF(A))(P), T(V1(B))(Q)$ q.e.d.

Hence $(*)$ commutes with $T(PF(A))^k(P0), T(V1(B))^k(Q0)$ for all $k \in \mathbb{N}$ and

$$f^*_{si} \circ Pi = f^*_{si} \circ (\lim_k T(PF(A))^k(P0)) =$$
$$\lim_k (f^*_{si} \circ T(PF(A))^k(P0)) =$$
$$\lim_k (T(V1(B))^k(Q0) \circ (f^*_{si1} x...xf^*_{simi})) =$$
$$\lim_k (T(V1(B))^k(Q0)) \circ (f^*_{si1} x...xf^*_{simi}) =$$
$$Qi \circ (f^*_{si1} x...xf^*_{simi})$$

$\square$

Reducing the statement of the theorem to the case of non-parameterized recursive projection specifications yields:

<u>4.13 Corollary</u> : Given a contracting recursive projection specification RECPS based on a projection specification PS, $(CT_{PS},(PROCi(CT_{PS}))_{i \in I})$ is initial in $Cat_{Compl}Sep(CRECPS)$, i.e. $(CT_{PS},(PROCi(CT_{PS}))_{i \in I}) \cong CT_{RECPS}$.

Another corollary or theorem 4.12 states, that a contracting recursive projection specification is a continuous enrichment, as indicated in the introduction.

<u>4.14 Corollary</u> :
A contracting parameterized recursive projection specification (PS,RECPS1) based on a parameterized projection specification (PS,PS1) is a continuous enrichment (PS,PS1)$\subseteq$(PS,RECPS1).

<u>Proof</u> : Let $A \in Cat_{Compl}Sep(CPS)$, then

$V1 \circ RECPF(A) \cong V1(PF(A),(PROCi(PF(A))_{i \in I}) = PF(A)$.   □

5. CONCLUSION

Let´s first give a summary of the results concerning projection specifications:

1. A projection specification PS is an algebraic specification preserving the projection specification **pnat1** (see the introduction to chapter two) with a selected projection operation symbol for each sort (see definition 2.3). The projections induce pseudo-ultrametrics on the base sets of a PS-algebra. Roughly speaking they are discrete for the data types and non discrete for the process types. This distinction is only preliminary, since there may be data types containing process subtypes or elements, such as (finite) sets of (infinite) processes.

A parameterized projection specification (PS,PS1) is a pair of projection specifications with PS$\subseteq$PS1, PS is the formal- or requirement specification, PS1 the target specification (see definition 3.1(1)).

A recursive projection specification RECPS based on a projection specification PS is an extension of PS by operation symbols proci and equations proci(xi1,...,ximi)=Ti (i$\in$ I); a parameterized projection specification (PS,RECPS1) is a parameterized recursive projection specification based on (PS,PS1), if RECPS is a recursive projection specification based on PS1 (see definition 4.3).

2. The semantics of a projection specification PS is the class of initial complete separated projection PS-algebras. It can be obtained from the initial quotient term algebra T_{PS} (which has an initial **pnat1** part, see theorem 2.10 and lemma 2.2) as follows:

First apply the projection constraints functor Con_{PS} (see definition 2.9); $T_{CPS}:=Con(T_{PS})$ is an initial projection PS-algebra. Then take the standard construction SC (see definition 2.11); $CT_{PS}\cong SC\circ Con_{PS}(T_{PS})$, the semantics of PS is initial in the category of complete separated projection PS-algebras $Cat_{Compl}{}^{Sep}(CPS)$. Con_{PS} and SC are left adjoints to the corresponding inclusion and forgetful functors, thus also the projection semantics functor PS is a left adjoint (see definition 2.16).

The semantics of a parameterized projection specification (PS,PS1) is the class of free functors $PF:Cat_{Compl}{}^{Sep}(CPS)\rightarrow Cat_{Compl}{}^{Sep}(CPS1)$, called projection free functors, and is given by $PF\cong PS1\circ F\circ I$, the projection semantics functor PS1 of the target specification, the free functor F induced by the underlying algebraic specification and the inclusion functor I of complete separated projection algebras (see definition 3.1(2)).

The semantics of a parameterized recursive projection specification (PS,RECPS1) based on (PS,PS1) can be constructed as follows:

On the process (projection) space (Hom(A),pH) (where the algebra $A\in Cat_{Compl}{}^{Sep}(CPS)$ is given), that consists of all projection compatible functions fi (i$\in$ I) that are given by the signature of the processes proci alone (see definition 4.5), the operator T(A) is defined, that maps fi on the evaluated term Ti, the right hand side of the equation for proci (see definition 4.7). If T(A) is a contraction, then it has a unique fixed point, which serves as semantics for the processes proci. This construction leads to the same result as the projection free functor $RECPF:Cat_{Compl}{}^{Sep}(CPS)\rightarrow Cat_{Compl}{}^{Sep}(CRECPS1)$, but saves its explicit construction (see theorem 4.12). Furthermore (PS,PS1)$\subseteq$(PS,RECPS1) is a continuous enrichment if T is a contraction.

Concerning further research there are two main directions at the moment. One is the application of the theory to existing specification languages, such as eg LOTOS (see [24], [25]). Here the assumption of projection compatible operations should be checked, and methods how to prove correctness of a given specification (internal and external) should be developed. Concerning recursive processes/process specifications some results have been worked out (see [10]).

The other direction is the investigation of recursively definable processes. It is possible to redefine the semantics of a simple projection specification to consist only of such processes, instead of all (Cauchy-) finitely approximable processes. This also constitutes a reflexive subcategory. However, this semantics fails to be compositional; the actualization of a parameterized projection specification may generate new syntax, such that in the target specification some processes are recursively definable, which are neither recursively definable in the parameterized projection specification nor in the actual parameter. (More technically : the amalgamation lemma does not hold.) The solution of this problem is still open.

6. REFERENCES

[1] J.A.Goguen, J.W.Thatcher, E.G.Wagner, J.B.Wright : Initial Algebra Semantics as Continuous Algebras, Journal ACM 24, 68-95, 1977

[2] J.B.Wright, E.G.Wagner, J.W.Thatcher : A Uniform Approach to Inductive Posets and Inductive Closure, Theoretical Computer Science 7, 57-77, 1978

[3] A.Arnold, M.Nivat : The metric space of infinite trees. Algebraic and topological properties , Societatis Mathematicae Polonae , Series IV:Fundamenta Informatica III,4 p. 445-476 , 1980

[4] E.K.Blum, H.Ehrig,F.Parisi-Presicce : Algebraic Specifications of Modules and Their Basic Interconnections, Journal of Computer and System Sciences Vol 34, April/June 1987

[5] J.A. Bergstra, J.W. Klop, 1983 : The Algebra of recursively defined processes and the algebra of regular processes, Report IW 235/83, Math. Centrum, Amsterdam 1983

[6] J.A. Bergstra, J.W. Klop: Algebra of Communicating Processes, in: CWI Monographs I Series, Proceedings of the CWI Symposium Mathematics and Computer Science, North-Holland, Amsterdam 1986, p. 89-138

[7] J.W. De Bakker, J.I. Zucker : Denotational semantics of concurrency Proc. 14th. ACM Symp. on Theory of Computing, p.153-158, 1982

[8] J.W. De Bakker, J.I. Zucker : Processes and the denotational semantics of concurrency, Information and Control, Vol.54, No.1/2, p.70-120, 1982

[9] C.Dimitrovici, H.Ehrig, M.Große-Rhode, C.Rieckhoff : Projektionsräume und Projektionsalgebren: Eine Algebraisierung von ultrametrischen Räumen , Technical Report No. 87-7, TU Berlin, 1987

[10] C.Dimitrovici : Methoden zur Lösung rekursiver Gleichungssysteme in Projektionsräumen und -algebren, Diplomarbeit ,TU Berlin 1988

[11] H.Ehrig,W.Fey,H.Hansen : ACT ONE: An Algebraic Specification Language with Two Levels of Semantics, TUB Bericht Nr.83-01

[12] H.Ehrig, W.Fey, F.Parisi-Presicce, E.K.Blum : Algebraic Theory of Module Specifications with Constraints, invited paper for MFCS´86, LNCS 233 (1986), 59-77

[13] H.Ehrig, B.Mahr : Fundamentals of Algebraic Specifications 1 : Equations and Initial Semantics , Springer Verlag , Berlin-Heidelberg-New York-Tokyo 1985

[14] H.Ehrig, B.Mahr : Fundamentals of Algebraic Specifications 2 : Modules and Constraints, Springer Verlag , Berlin-Heidelberg-NewYork-Tokyo 1990

[15] H.Ehrig, F.Parisi-Presicce, P.Boehm, C.Rieckhoff, C.Dimitrovici, M.Große- Rhode : Combining Data Type and Recursive Process Specifications using Projection Algebras, Theoretical Computer Science 71 (1990), 347-380

[16] M.Große-Rhode, H.Ehrig : Transformation of Combined Data Type and Process Specifications Using Projection Algebras , in: Stepwise Refinement of Distributed Systems, Springer LNCS 430, pp.301-339, 1990

[17] M.Große-Rhode : Specification of Projection Algebras , Diploma Thesis, TU Berlin, 1988

[18] M.Große-Rhode : Parameterized Data Type and Process Specifications Using Projection Algebras, in: Categorical Methods in Computer Science with Aspects from Topology, H.Ehrig, M.Herrlich, H.J.Kreowski G.Preuß (eds.), Springer LNCS 393, 1989

[19] H.Herrlich, H.Ehrig : The Construct PRO of Projection Spaces: Its Internal Structure, in: Categorical Methods in Computer Science with Aspects from Topology, H.Ehrig, M.Herrlich, H.J.Kreowski G.Preuß (eds.), Springer LNCS 393, 1989

[20] M.Hennessy, R.Milner : Algebraic Laws for Nondeterminism and Concurrency, University of Edingburgh, Department of Computer Science, Internal Report CSR-133-83, June 1983

[21] M.Hennessy, G.Plotkin : A term model for CCS, Proc. 9th MFCS, LNCS 88 (1980), 261-274

[22] E.Kranakis : Approximating the Projective Model , Report CS-R8607, Centre for Mathematics and Computer Science, Amsterdam 1980, 122-133,179-191

[23] E.Kranakis : Fixed point equations with parameters in the projective model, CWI Report CS-R8606, Amsterdam 1986, to be published in Information and Computation

[24] ISO-documents and draft proposals on the Specification Language LOTOS since 1983

[25] H.Ehrig, J.Buntrock, P.Boehm, K.P.Hasler, F.Nürnberg, C.Rieckhoff, J.deMeer : Towards an Algebraic Semantics of the ISO-Specification Language LOTOS, draft version, Technische Universität Berlin, May 1986

[26] Information processing systems-Open systems interconnection-LOTOS-A Formal Description Technique Based on the Temporal Ordering of Observational Behaviour,ISO DIS 8807 (ISO/TC97/SC21N), July 20, 1987

[27] R.Milner : A Calculus of Communicating Systems, Springer LNCS 92, 1980

[28] R.Milner : Communication and Concurrency, Prentice Hall 1989.

[29] B.Möller, W.Dosch : On the Algebraic Specification of Domains, in Recent Trends in Data Type Specification (e.d. H.J.Kreowski), Informatik Fachberichte 116, Springer Verlag 1986, 178-195

[30] B.Möller : Unendliche Objekte und Geflechte, Fakultät für Mathematik und Informatik der TU München, Dissertation, TUM-18213, 1982

[31] B.Möller : On the Algebraic Specification of Objects - Ordered and Continuous Models of Algebraic Types, Acta Informatica 22, 537-578, 1985

[32] M.Nivat : On the Interpretation of Recursive Polyadic Program Schemes, Istituto Nazionale di Alta Mathematica XV, 255-281, Academic Press, London 1975

[33] A.Tarlecki, M.Wirsing : Continuous abstract data types, Fundamenta Informaticae IX (1986) 95-126, North-Holland

[34] F.W.Vaandrager : Verification of two Communication Protocols by means of Process Algebra, Report CS-R8608, Center for Mathematics and Computer Sciences, Amsterdam 1986

Tree Automata and Languages
M. Nivat and A. Podelski (editors)
1992 Elsevier Science Publishers B.V.

TREES AND ALGEBRAIC SEMANTICS

Irène GUESSARIAN *
C.N.R.S. - L.I.T.P. Université Paris 6
4, Place Jussieu, 75252 Paris Cedex 05, France
e-mail: ig@litp.ibp.fr

I INTRODUCTION

Trees are used in most areas of computer science, both practical and theoretical, such as for instance sorting, parsing, compiling, complexity theory, semantics. In the present paper, we will focus on the use of trees in algebraic semantics.

The trees which are used in algebraic semantics represent unfoldings of recursive program schemes, or unfoldings of control structures of iterative program schemes. The algebraic semantics of a program scheme S is defined by first describing the meaning $T(S)_I$ of the program scheme S in a free initial algebra I, then deducing from $T(S)_I$ the meaning $T(S)_A$ of the program scheme S in any other algebra A, via a canonical morphism. If Σ is the set of operation symbols which are used in S, then, the free initial algebra I will consist of the algebra CT_Σ of all finite and infinite terms, well-formed on the signature Σ : such terms are represented by ranked labeled trees. Once S is solved in the initial algebra CT_Σ, then, the universal algebra tools enable us to derive for free the solution of S in any other algebra A, together with its properties. Let D be a domain in which S can be interpreted : then in D all the symbols of Σ must be interpreted, hence D can be endowed with a Σ-algebra structure, and there exists a unique canonical morphism μ from the free algebra CT_Σ into D, $\mu :\ CT_\Sigma \to D$. If S_D denotes the program corresponding to S on the domain D, we will show (Theorem III.1) that its meaning $m(S_D)$ is the image of the meaning $T(S)$ of S in CT_Σ under the morphism μ, i.e. $m(S_D) = \mu(T(S))$. This enables us to prove properties of $m(S_D)$ by proving them abstractly, i.e. independently of any particular model, at the level of $T(S)$. In fact, we can obtain a simple proof system which is complete for the logics of inequalities, i.e. for proving all valid inequalities $T \leq T'$ between infinite trees. This proof system contains the standard rules for inequalities to which it suffices to add a very simple infinitary axiom for inductive proofs about infinite trees. (Note that at least one infinitary axiom is unavoidable). This proof system can also be generalized to complete proof systems for proving inequalities valid only in restricted classes of models, see [Guessarian 85, Guessarian 90].

---

* Support from the PRC Mathématiques-Informatique and Esprit BRA3230 is gratefully acknowledged.

The present paper is organized as follows : we first define the trees that we will use, we then show how they are used in computer science to represent the semantics of program schemes. We study rational (resp. algebraic) forests which represent iterative (resp. recursive) program schemes, and show their various characterizations, via grammars, automata or algebra. We finally conclude briefly with some problems.

II TREES

The trees which are of use in semantics can be considered:
- either from an operational viewpoint, as labeled graphs
- or from an algebraic viewpoint, as terms of a free algebra. We will briefly present both viewpoints, as both will be useful in the sequel.

II.1 Operational viewpoint

Let Σ (resp. Φ) be a finite ranked alphabet of base function symbols (resp. of variable function symbols). The rank of a symbol s in $\Sigma \cap \Phi$ is denoted by $r(s)$. Symbols in Σ are denoted $f, g, h, \ldots$ if they have rank ≥ 1, and $a, b, \ldots$ if they have rank 0; Σ_i denotes the symbols of rank i in Σ. Symbols in Φ are denoted $\phi, \psi, \ldots$. Let X be a set of variables: the variables have rank 0 and are denoted by $u, v, w, \ldots$ possibly with indices.

The notions of tree, node in a tree, occurrence of a variable or a subtree in a tree, substitution, etc... are supposed to be known (see [Guessarian]). We recall however those notions we shall use in the paper in order to fix the notations.

We use the Dewey notation for trees (see also [Gorn]): nodes in a tree are denoted by finite words over the alphabet $\mathbb{N}$, i.e. elements of the free monoid $\mathbb{N}^*$.

Formally, a *tree* on Σ is a pair (D_t, t) consisting of:

(i) a tree domain D_t which is a finite subset of $(\mathbb{N} - \{0\})^*$ such that if $o = n_1 \ldots n_p$ is in D_t then (a) every left factor $o' = n_1 \ldots n_q, q \leq p$, also is in D_t, (b) for all $i \leq n_p$, $o'' = n_1 \ldots n_{p-1} i$ also is in D_t.

(ii) a total mapping t from D_t into Σ such that, for any o in D_t, if $t(o) = f \in \Sigma$ and f is of rank p then o has exactly p "sons" (i.e. nodes o' of the form $o' = on_i$) in D_t.

Note that the restriction that a node labeled by a symbol of rank p have p sons exactly is needed for the purpose of using trees for algebraic semantics. General trees can have nodes labeled by symbols of arbitrary ranks, or even unlabeled nodes. A slightly different viewpoint, the partial algebra viewpoint, would require that nodes labeled with symbols of rank p have $q \leq p$ sons.

In the sequel, D_t will be omitted and a tree will be denoted by t. A *node* in a tree is an element of D_t; $t(o)$ is called the *label* of node o; o is also called an *occurrence* of $t(o)$ in t; nodes or occurrences will be denoted by o. Note that D_t contains the empty word ε: the root of t. A node without sons is called a *leaf*.

Let t, t' be trees on Σ and o be a node in t; $t(t'/o)$ denotes the tree obtained by substituting t' for the node o in t and is defined by:

(a) $t(t'/o)(oo') = t'(o')$ for any o' in $D_{t'}$, and

(b) $t(t'/o)(o'') = t(o'')$ if o is not a left factor of o'' (recall that o is a left factor of o'' iff there exists an o' with $o'' = oo'$).

The subtree $t_{|o}$ of t (at occurrence o) is the tree t'' defined by: $t''(o') = t(oo')$ for any oo' in D_t.

The depth $d(t)$ of a tree t is defined by: $d(t) = sup\{|o| + 1 \ / \ o \in D_t\}$, where $|o|$ is the length of o considered as an element of $\mathbb{N}^*$.

II.2 Algebraic viewpoint

Let T_Σ denote the set of trees over Σ. A tree with variables in X is a tree over $\Sigma \cup X$, i.e. an element of $T_{\Sigma \cup X}$: intuitively, the variables are intended to range over a set of trees, e.g. T_Σ or $T_{\Sigma \cup X}$. T_Σ (resp. $T_{\Sigma \cup X}$) is endowed with a Σ-algebra structure: for f in Σ, f of rank p, and $t_1, \ldots, t_p$ in T_Σ (resp. $T_{\Sigma \cup X}$), the operation f_{T_Σ} (resp. $f_{T_{\Sigma \cup X}}$) is defined by: $(t_1, \ldots, t_p) \mapsto f(t_1, \ldots, t_p)$. T_Σ (resp. $T_{\Sigma \cup X}$) is the *domain*, or *carrier set*, of the free initial Σ-algebra, or Σ-magma in the terminology of [Nivat], (resp. the free Σ-algebra over generators X, $T_{\Sigma \cup X}$, which is also denoted by $T_\Sigma(X)$). This will be justified by the fact that T_Σ and $T_{\Sigma \cup X}$ have the free property, which we will recall below in full generality. We will in the sequel identify T_Σ (resp. $T_{\Sigma \cup X}$) with the corresponding free algebra.

Let us first recall the notions of ordered and complete Σ-algebras, which we will use later. A Σ-*algebra* A, or algebraic structure of similarity type Σ, consists of a carrier set A and for each function symbol σ in Σ, a function $\sigma_A : A^w \to A$ where $A^w = A \times \cdots \times A$ with $w = rank(\sigma)$, and σ_A is a constant in A if $w = 0$.

An ordered Σ-algebra is a Σ-algebra such that the carrier set A is endowed with an ordering $\leq_A$ and a least element $\perp_A$, and the operations σ_A are order preserving. Ordered Σ-algebras are actually a special case of algebraic structures of similarity type $\Sigma' = \Sigma_\perp \cup \{\leq\}$, where $\Sigma_\perp = \Sigma \cup \{\perp\}$. The class of all ordered Σ-algebras forms a quasivariety, i.e. is axiomatisable by quasi-atomic formulae, in the sense of [Mal'cev]. See also [Grätzer, Guessarian 90] for quasi-equational logic.

A Σ-algebra A is said to be *complete* iff, all directed subsets of A have a lub (least upper bound) in A, and the σ_A's are continuous, i.e. preserve lub's of directed sets. The category of Σ-algebras (resp. ordered Σ-algebras, or complete Σ-algebras) is defined as follows: objects are Σ-algebras (resp. ordered Σ-algebras, or complete Σ-algebras); morphisms are Σ-homomorphisms (resp. order-preserving Σ-homomorphisms, i.e. Σ'-homomorphisms, or continuous Σ-homomorphisms that is: homomorphisms which, in addition to preserving the Σ-structure, also preserve lub's of directed sets, which implies that they preserve $\perp$ and the order also). Recall that a Σ-homomorphism from a Σ-algebra A to a Σ-algebra B is a function $f : A \to B$ that satisfies: for σ in Σ with $rank(\sigma) = n$, and $a_i \in A$ for $i = 1, \ldots, n : f(\sigma_A(a_1, \ldots, a_n)) = \sigma_B(f(a_1), \ldots, f(a_n))$; and for σ in $\Sigma_0: f(\sigma_A) = \sigma_B$.

It was proved in [Lehmann-Pasztor] that the category of complete Σ-algebras is reducible to a variety of partial algebras in the sense of [Andréka-Németi].

A (complete, ordered) $\Sigma \cup X$-algebra F is said to be *free over generators* X iff for any (complete, ordered) $\Sigma \cup X$-algebra A there exists a unique Σ-homomorphism (in the category of complete or ordered Σ-algebras) $\bar{v} : F \to A$ making the following diagram commute:

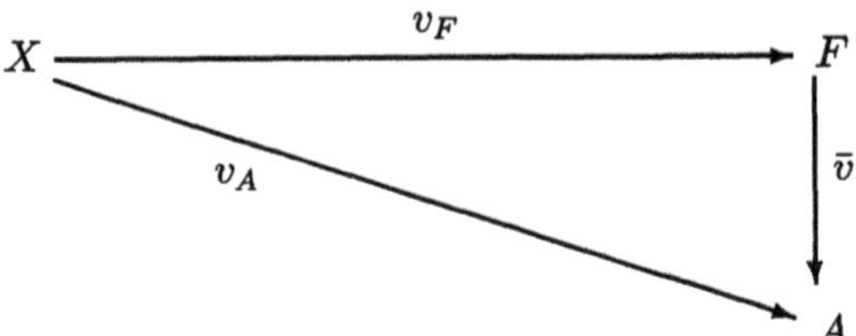

where, for any $\Sigma \cup X$-algebra J, $v_J(x) = x_J$, for all x in X. F is also called the *free (complete, ordered)* Σ-algebra generated by X.

The free Σ-algebra generated by X, also called the algebra of Σ-terms with variables in X, and denoted by $T_\Sigma(X)$, exists and can be constructed as follows: its carrier set is the set of well-formed terms (or trees) on the alphabet $\Sigma \cup X$, and is defined inductively by:

(i) $\Sigma_0 \cup X \subseteq T_\Sigma(X)$,

(ii) $\sigma(t_1, \ldots, t_n) \in T_\Sigma(X)$ for each σ in Σ and t_i in $T_\Sigma(X)$ for $i = 1, \ldots, n$, and $rank(\sigma) = n$.

The Σ-algebra structure of $F = T_\Sigma(X)$ is defined by: $\sigma_F = \sigma$ if $\sigma \in \Sigma_0$, and otherwise: $\sigma_F(t_1, \ldots, t_n) = \sigma(t_1, \ldots, t_n)$, for t_i in $T_\Sigma(X)$ for $i = 1, \ldots, n$.

The free ordered Σ-algebra $F' = T_{\Sigma_\perp}(X)$ generated by X has carrier sets the sets of well-formed terms on the alphabet $\Sigma \cup X \cup \{\perp\}$; its Σ-algebra structure is defined as the one of F; its carrier set F' is endowed with the least ordering such that: (i) $\perp$ is the least element, and (ii) the Σ-algebra operations are order-preserving.

Finally, the free complete Σ-algebra $H = CT_{\Sigma_\perp}(X)$ generated by X is the ideal completion of F' (see [Birkhoff]). $H = CT_{\Sigma_\perp}(X)$ is the set of ideals of F', ordered by inclusion, and the Σ-algebra operations are extended by continuity; namely for σ in Σ and I_i ideal of F' for $i = 1, \ldots, n$, $\sigma_H(I_1, \ldots, I_n)$ is the ideal generated by $\{\sigma_{F'}(i_1, \ldots, i_n)/i_j \in I_j$ for $j = 1, \ldots, n\}$. H can be viewed as the set of well-formed finite and infinite trees generated by X. Its ordering $\leq$ extends the ordering on F' and can be intuitively described by: $T \leq T'$ iff T' can be deduced from T by substituting some occurrences of $\perp$ by terms different from $\perp$.

By the freeness of $H = CT_{\Sigma_\perp}(X)$, we know that for each valuation $a : X \to A$ in an arbitrary complete Σ-algebra A, there exists a unique morphism $\bar{a}$ making the following diagram commute:

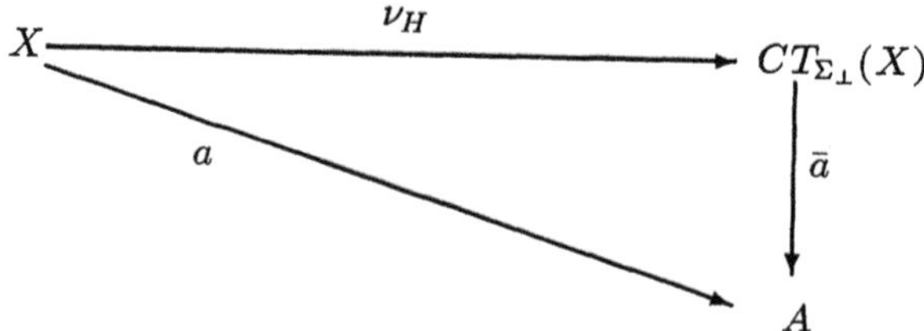

For T in $CT_{\Sigma_\perp}(X)$, having variables $\{x_1,\ldots,x_n\}$ we will denote by T_A the application $A^n \to A$ defined by $a = (a_1,\ldots,a_n) \mapsto \overline{a}(T)$, i.e. every n-tuple $a \in A^n$ determines a valuation $X \to A$, and we define $T_A(a) = \overline{a}(T)$.

Let CT_Σ (resp. $CT_\Sigma(X)$) be the subset of $CT_{\Sigma_\perp}(X)$) consisting of terms without occurrences of $\perp$. T_Σ (resp. CT_Σ) is the set of finite (resp. finite and infinite) trees on the alphabet Σ; similarly, $CT_\Sigma(X)$ is the set of finite and infinite trees on $F \cup X$. Among these trees, some are of particular interest for semantics. We will study them in the next section. We first introduce some notations.

II.3 Some notations

$T_{\Sigma,i}(X)$ and $CT_{\Sigma,i}(X)$ denotes the trees having at most i variables; $T_\Sigma^i(X)$ denotes the trees of depth at most i.

Throughout this paper we will use the word notation for trees. A tree is represented by a word over the alphabet $\Sigma \cup \{(,),c\}$, where c is the comma, as follows:

a stands for $a \in \Sigma_o$

$f(t_1,\ldots,t_r)$ stands for the tree having root $f \in \Sigma_r$ and direct subtrees $t_1,\ldots,t_r$ and would be pictured as:

$$
\begin{array}{c}
f \\
\diagup\ \diagdown \\
t_1 \quad \cdots \quad t_r
\end{array}
$$

A tree t in $T_{\Sigma,p}(X)$ will be denoted by $t(x_1,\ldots,x_p)$ in order to point out the variables; the shorthand vector notation $t(\vec{x})$ will be used: $t(t_1,\ldots,t_p)$ denotes the tree obtained by simultaneously substituting t_i for each occurrence of x_i in $t(x_1,\ldots,x_p)$, for $i = 1,\ldots,p$; formally, $t(t_1,\ldots,t_p)$ is the image of $t(x_1,\ldots,x_p)$ by the $\Sigma \cup X$-algebra morphism h defined by: $h(x_i) = t_i$ for $i = 1,\ldots,p$, and $h(f) = f$ for f in Σ. We will also use a vector shorthand notation $t(\vec{t})$ for $t(t_1,\ldots,t_p)$. Alternatively, when wanting to point out which trees are being substituted for which variables, we will note: $t(t_1/x_1,\ldots,t_p/x_p)$ (or $t(\vec{t}/\vec{x})$).

A *prefix* of $t \in T_\Sigma(X)$ is a tree $\underline{t} \in T_\Sigma(\{w_1,\ldots,w_s\})$, with $\{w_1,\ldots,w_s\}$ disjoint from X, $D_{\underline{t}} \subset D_t$, and such that there exist subtrees $t_1,\ldots,t_s$ of t with $\underline{t} = t(t_1/w_1,\ldots,t_s/w_s)$.

Finally, $t(g/f)$ denotes the tree where all occurrences of g are substituted for by f, and, if $rank(g) = p < n = rank(f)$, the last $n - p$ arguments of f are dropped. We will most often use $t(\perp/f)$, where all subtrees with root f are substituted for by $\perp$. In fact, $\perp$ being of rank 0, it suffices to substitute only outermost subtrees whose root is f by $\perp$. These notations are illustrated by the example below.

EXAMPLE II.1 Let t be defined by:

$$t = t(v_1, v_2) =$$

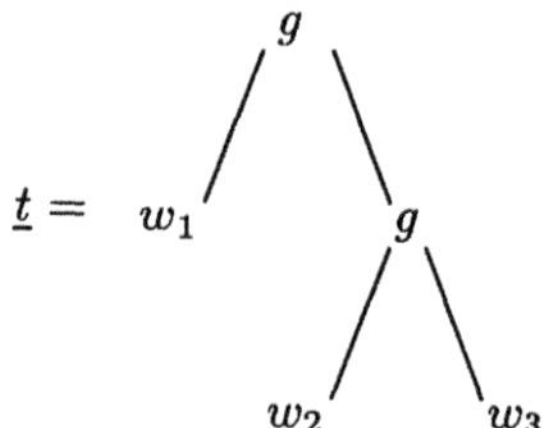

$t \in T_\Sigma(\{v_1, v_2\})$ with $\Sigma = \{a, h, g\}$, $r(a) = 0$, $r(h) = 1$, $r(g) = 2$

$D_t = \{\varepsilon, 1, 2, 21, 22, 211\}$
$d(t) = 4$, $t(\varepsilon) = t(2) = g$, $t(1) = v_1$, $t(21) = h$, $t(211) = v_2$, $t(22) = a$.

$t' = t(t/1) = t(t, v_2)$ can be drawn as:

$$t' =$$

Let:

$$\underline{t} =$$

$\underline{t}$ is a prefix of t and t'.

Figure 1.

III RATIONAL AND ALGEBRAIC TREES AND SETS OF TREES.

We study in the present section those trees which are most often used in semantics.

III.1 Rational and algebraic trees

Rational (resp. algebraic) trees are trees which are solutions of iterative (resp. recursive) schemes, and correspond intuitively to unfoldings of iterative, i.e. WHILE-loop programs (resp. LISP-like recursive programs). Formally, we have:

A *recursive scheme* on Σ is a pair (S, t), where S is a system of n equations:

$$S: \quad \phi_i(x_1^i, \ldots, x_{n_i}^i) = t_i \quad , \quad i = 1, \ldots, n \tag{1}$$

where, for $i = 1, \ldots, n$, $\phi_i \in \Phi$, $x_j^i \in X$ for each j, $t_i \in T_{\Sigma \cup \Phi}(X)$, and $t \in T_{\Sigma \cup \Phi}(X)$.

Each scheme is associated with a term rewriting system (or context free tree grammar) which is denoted by $S_\perp$, and defined by : $\phi_i(x_1^i, \ldots, x_{n_i}^i) \to t_i + \perp$. Let $t' \Longrightarrow_S t''$ denote the rewriting relation associated with $S_\perp$, i.e. t'' is deduced from t' by substituting t_i or $\perp$ for one occurrence of some ϕ_i in t'. Let $\overset{*}{\Longrightarrow}_S$ denote the reflexive and transitive closure of $\Longrightarrow_S$, and $L(S, t) = \{t'/t' \in T_{\Sigma_\perp}(X) \quad t \overset{*}{\Longrightarrow}_S t'\}$ be the tree-language generated by S with axiom t (cf. [Guessarian]). It is well-known that $L(S, t)$ is a directed subset of $CT_{\Sigma_\perp}(X)$, let $T(S, t) = lubL(S, t)$ in $CT_\Sigma(X)$. Note that there are no longer occurrences of $\perp$ in $T(S, t)$. Formally, immediate rewritings $\Longrightarrow_S$ and derivations $\overset{*}{\Longrightarrow}_S$ according to S are defined as usual. Let us recall the definitions here for the reader's convenience. The immediate rewriting according to S is the relation $\Longrightarrow_S$ defined on $T_{\Sigma \cup \Phi} \times T_{\Sigma \cup \Phi}$ by: $t \Longrightarrow_S t'$ iff there exists some prefix $\underline{t}$ of t, an occurrence o in t labeled by the function variable ϕ_i of rank n (i.e. $t(o) = \phi_i$), and subtrees $t_1, \ldots, t_n$ of t such that: $t = \underline{t}(\phi_i(t_1, \ldots, t_n)/o)$ and $t' = \underline{t}(t_i(t_1, \ldots, t_n)/o)$, or $t' = \underline{t}(\perp/o)$.

Intuitively, at occurrence o, ϕ_i is macroexpanded, i.e. replaced by the right hand-side t_i of production $\phi_i(x_1, \ldots, x_n) \to t_i$; simultaneously, $t_1, \ldots, t_n$ are substituted for $x_1, \ldots, x_n$. Whenever no ambiguity can occur, the subscript S is omitted and $\Longrightarrow_S$ (resp. $\overset{*}{\Longrightarrow}_S$) is denoted by $\Longrightarrow$ (resp. $\overset{*}{\Longrightarrow}$), as will be done in the sequel.

A recursive scheme is said to be *iterative* iff all function variables in Φ are of rank 0, or, from a programming viewpoint, iff all the t_i's are left-linear: intuitively, an iterative scheme corresponds to left-linear or terminal recursion, which is well-known to be equivalent to iteration. Left linear terms of $T_\Sigma(X)$ are defined inductively by:

− terms in $T_\Sigma \cup \Phi_0$ are left-linear
− for f in Σ_n, and $t_1, \ldots, t_n$ left linear terms, $f(t_1, \ldots, t_n)$ is left-linear.

In a left-linear term t, function variables from Φ can label only the leaves of t, since $\Phi = \Phi_0$, the set of function variables of rank 0.

The solution of a recursive (resp. iterative) scheme is called an *algebraic* (resp. rational, or regular) tree. The terminology comes from language theory, because algebraic trees are obtained as languages generated by context-free tree grammars, and rational trees are generated by regular tree grammars. (see section IV.2 below).

EXAMPLE III.1 1) A recursive scheme and its algebraic tree

Having defined a binary operation *mult* on a data type, we want to extend that operation to an operation *lmult* on list of elements of that data type by means of the recursive LISP like program P:

$$mult(L, L') = \text{if } L = NIL \text{ then } NIL \text{ else}$$
$$\text{if } L' = NIL \text{ then } NIL \text{ else}$$
$$cons(mult(car L, car L'), lmult(cdr L, cdr L'))$$

where $car L$ (resp. $cdr L$) is the first element (resp. the rest) of the list L.

To this recursive program, corresponds a program scheme, naturally obtained by abstracting the meanings of the functions if-then-else, car, etc..., and replacing as much as possible of the right hand side of the above equation by a single function name. We obtain here the scheme S, where $\Phi = \{lmult\}, X = \{L, L'\}, \Sigma = \{\bot, cdr, g\}$.

$$S: \quad lmult(L, L') = g(L, L', lmult(cdr L, cdr L'))$$

The solution of scheme S (and the corresponding program) is then computed by successive approximations. The nth approximation σ_n to that solution is defined by:

1) unwinding first the scheme S n times, i.e. replacing n times all occurrences of $lmult$ by the righthand side of S, starting with the term $lmult(L, L')$. This gives here, after n unwindings: $s_n(L, L') = g(L, L', g(cdr L, cdr L', g(\ldots, g(cdr^{n-1} L, cdr^{n-1} L', lmult(cdr^n L, cdr^n L'))\ldots)))$.

2) replacing then all occurrences of $lmult$ in s_n by $\bot$ (the totally undefined function) yielding here: $\sigma_n(L, L') = g(L, L', g(cdr L, cdr L', g(\ldots, g(cdr^{n-1} L, cdr^{n-1} L', \bot)\ldots)))$.

σ_n belongs to $L(S, lmult(L, L'))$, but not s_n. In the present case, evaluating σ_n yields:

$$\sigma_n(a_1 a_2 \ldots a_p, a_1' a_2' \ldots a_{p'}') = mult(a_1, a_1') mult(a_2, a_2') \ldots mult(a_q, a_q')$$
$$\text{if } q = inf(p, p') < n$$
$$= \bot \quad \text{(the undefined element) otherwise.}$$

We choose here the simpler but ambiguous notation of christening of the same name $\bot$ the undefined elements, or equivalently least elements, of the various domains, syntactic or semantic (data, lists of data), that we consider.

Thus the σ_n make their domain of definition grow larger and larger. That results in a chain $\sigma_0 < \sigma_1 < \ldots < \sigma_n < \sigma_{n+1} < \ldots$ where the ordering $<$ represents the relation: "to be less defined than".

A tree like representation of S, the s_n's and σ_n's which might help the intuition is given in Figure 2.

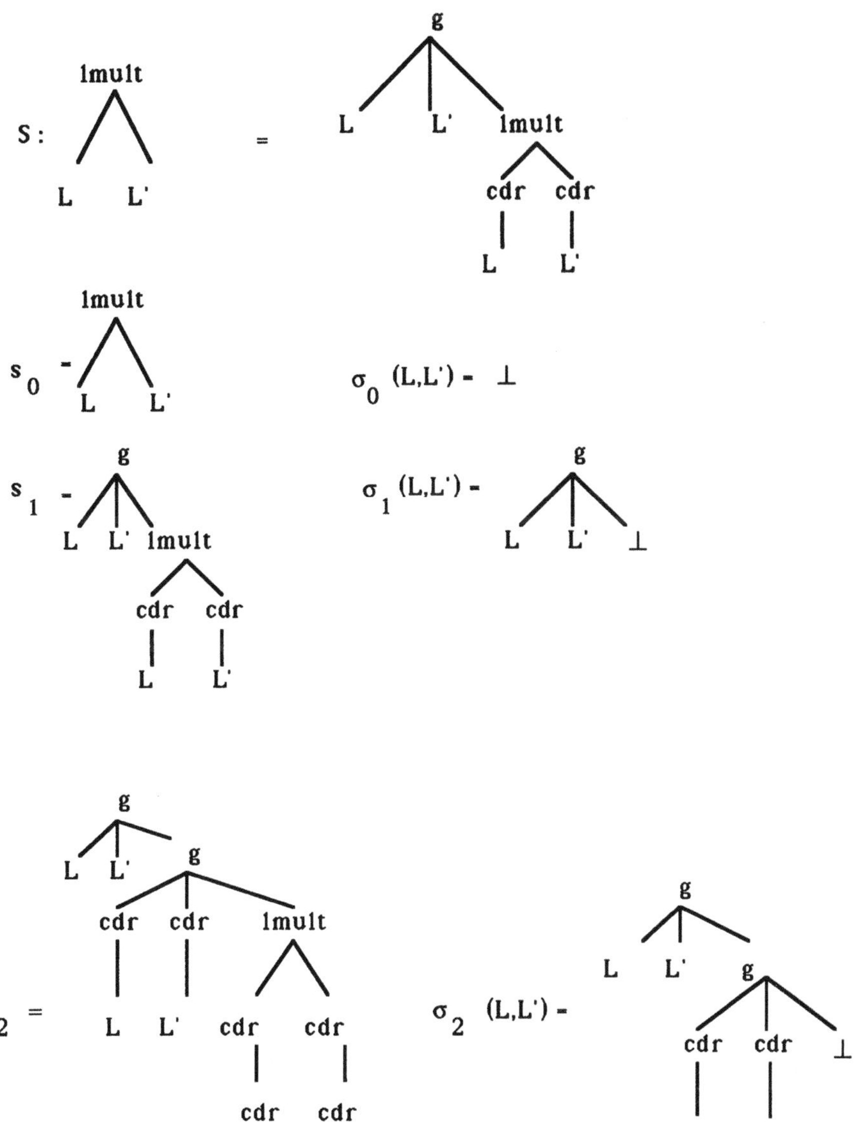

Figure 2.

Finally, the algebraic tree $T(S, lmult(L, L')) = lub\{\sigma_n / n \in \mathbb{N}\}$ can be depicted as:

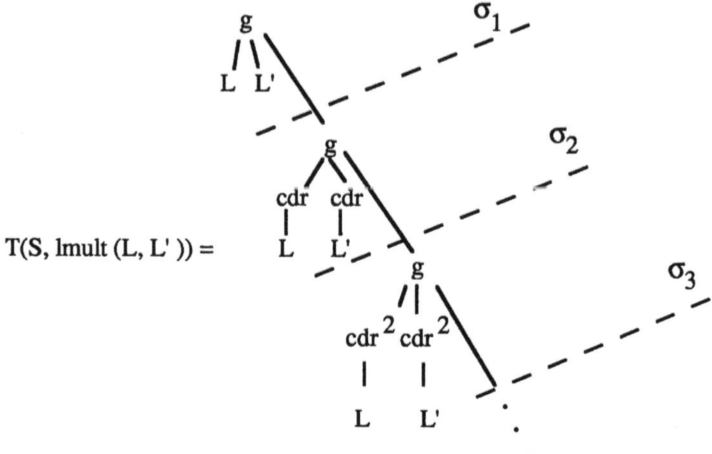

Figure 3.

2) An iterative scheme and its rational tree

Define a simple semaphore for the mutual exclusion of two processes A and B by ([Milner]):

$$P\begin{cases} sem = \overline{get};\overline{release};sem \\ X = get;A;release;X \\ Y = get;B;release;Y \\ Z = sem\|X\|Y \end{cases}$$

where $\Phi = \{sem, X, Y, Z\}$, $\|$ is the parallel composition where $\overline{get}$ and get, $\overline{release}$ and $release$ are synchronized, and all non-variable function symbols are considered to be of rank 1. If we denote by a the composition of $(\overline{get}\|get)$; A; $(\overline{release}\|release)$ and similarly for b, then, the above program P can be represented, after some computations, by the iterative scheme:

$$S\begin{cases} Z = or(X,Y) \\ X = a(Z) \\ Y = b(Z) \end{cases}$$

and its solution is represented by the rational tree:

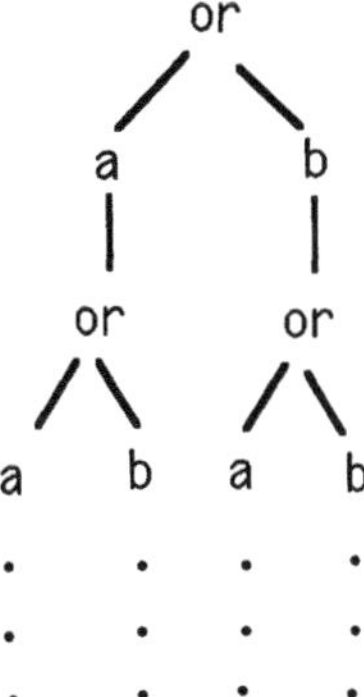

This example introduces the notion of non deterministic choice, which we represented here by an explicit operation "or", with the advantage of determinizing a nondeterministic program; the nondeterministic choice can also be represented somehow more naturally by a set theoretic union, or a "+" operation. In that case we will obtain sets of rational or algebraic trees, also called forests. Before studying forests, we will briefly sketch how trees are used in semantics.

III.2 Semantics

The basic idea of algebraic semantics is to characterize the meaning of a scheme in the free algebra $CT_\Sigma(X)$ via a tree, and to deduce from that tree the meaning of the scheme, and the corresponding programs, in all other possible models, or interpretations.

An *interpretation* A of Σ is a complete Σ-algebra; a valuated interpretation is an interpretation together with a valuation $a : X \to A$. An n-tuple $(\gamma_1, \ldots, \gamma_n)$ of operations $\gamma_i : A^{n_i} \to A$ is said to be a solution of S iff the equations of S are satisfied in A endowed with the $\Sigma \cup \{\phi_1, \ldots, \phi_n\}$-algebra structure defined by $\phi_{iA} = \gamma_i$ for $i = 1, \ldots, n$; equivalently, $(\gamma_1, \ldots, \gamma_n)$ is a fixpoint of the system of equations S in A. Note that $H = CT_{\Sigma_\perp}(X)$ is a particular interpretation of Σ, called the Herbrand model. We then have (see section IV.1 for the proof):

THEOREM III.2 *The n-tuple* $(T_1, \ldots, T_n) = (T(S, t_1), \ldots, T(S, t_n))$ *is the least solution of* S *in* H.

Note that: (i) we identify $T \in H = CT_{\Sigma_\perp}(X)$ with T_H, and (ii) for any t, $L(S, t)$ can be obtained by substituting $L(S, t_i)$ to all occurrences of ϕ_i in t, for $i = 1, \ldots, n$. By definition, $T(S, t)$ is the function computed by the scheme (S, t) in the free model, or interpretation, H. If A is now an arbitrary interpretation, the function defined by scheme S in A will be defined as $T(S, t)_A$. The adequacy of this definition follows from the:

THEOREM III.3 *Let* A *be a complete* Σ-*algebra, and* (S, t) *and* (S', t') *be two schemes:*
(i) $T(S, t)_A \leq T(S', t')_A$ *for all* A *iff* $T(S, t) \leq T(S', t')$
(ii) *for all* A, $T(S, t)_A = lub\{t_{kA} \ / \ t_k \leq T(S, t) \text{ and } t_k \in T_{\Sigma_\perp}(X)\}$

(iii) $(T_{1A}, \ldots, T_{nA})$ is the least solution of S in A.

Theorem III.3 is an immediate consequence of theorem III.2, together with the freeness of H.

(i) shows that $T(S, t)$ characterizes the behavior of (S, t) with respect to all models, namely that what happens in the model H suffices to describe what will happen in all other models; the name of Herbrand model comes from that property, together with the fact that H is a term model.

(ii) says that the function computed by (S, t) in A can be defined as a lub of finite computations, by successive approximations.

(iii) finally expresses the link between the algebraic semantics above described and the denotational semantics of [Scott].

Theorem III.3 (i) is fundamental in the study of abstract data types [Goguen-Meseguer]: it expresses the fact, that, due to its initiality, the free complete algebra provides an abstract (in the sense that it is representation independent and unique up to isomorphism) way of studying data types. The initial model $CT_{\Sigma_\perp}(X)$ is the abstract data type, wherefrom all other data types can be deduced by homomorphism.

III.3 Forests

A *non deterministic* (as opposed to the deterministic ones considered up to now) recursive scheme on Σ is a pair (S, t), where S is a system of n equations:

$$S: \quad \phi_i(x_1^i, \ldots, x_{n_i}^i) = T_i \quad , \quad i = 1, \ldots, n \tag{2}$$

where, for $i = 1, \ldots, n$, $\phi_i \in \Phi$, $x_j^i \in X$ for each j, $T_i \subseteq T_{\Sigma \cup \Phi}(X)$, and $t \in T_{\Sigma \cup \Phi}(X)$.

(S, t) is said to be iterative if all the trees in the T_i's and t are left-linear.

Similarly, each scheme S (recursive or iterative) is associated with a tree grammar defined by:

$$S_\perp: \quad \phi_i(x_1^i, \ldots, x_{n_i}^i) \to T_i + \perp \quad , \quad i = 1, \ldots, n$$

As previously, $\Longrightarrow_S$ and $\overset{*}{\Longrightarrow}_S$ denote the immediate rewriting according to $S_\perp$ and its reflexive and transitive closure.

The solution of the scheme (S, t) in $CT_\Sigma(X)$, where X is the set of variables occurring in t is the forest:

$$F(S, t) = Fin(S, t) \cup Inf(S, t)$$

where:

$$Fin(S, t) = \{t'/t' \in T_\Sigma(X) \text{ and } t \overset{*}{\Longrightarrow}_S t'\} = L(S_\perp, t)$$

and

$$Inf(S, t) = \{T'/T' \in CT_\Sigma(X) \text{ and } T' = lub\{t_n/n \in \mathbb{N}\} \qquad \text{for some sequence}$$

$$t_n = t'_n(\vec{\perp}/\vec{\phi}) \text{ with } t'_0 = t \text{ and } \forall n \ t'_n \overset{*}{\Longrightarrow}_S t'_{n+1}\}$$

$Fin(S, t)$ (resp. $Inf(S, t)$) correspond to the finite (resp. infinite) trees over Σ generated by the scheme (S, t), i.e. the finite (resp. infinite) unfoldings of t according to S. Note that no occurrences of $\perp$ appear in the trees of $F(S, t)$.

Note that:

$$Inf(S,t) \neq \{T'\,/\,T' \in CT_\Sigma(X) \text{ and}$$
$$T' = lub\{t_n/n \in \mathbb{N}\} \text{ for some sequence } t_n \in L(S_\perp,t)\}$$
$$= adh(L(S_\perp,t))$$

where $adh(L)$ is the set of limits of sequences of trees in L. For word grammars we would have $Inf(S,w) = adh(L(S,w))$; this shows one respect in which tree languages can be more complex than word languages.

For instance, let S be the iterative scheme:

$$S: \quad \begin{cases} X = g(Y,g(Z,X)) + g(Y,g(Y,g(Z,X))) \\ Y = b(Y) \\ Z = c(Z) \end{cases}$$

Then, letting:

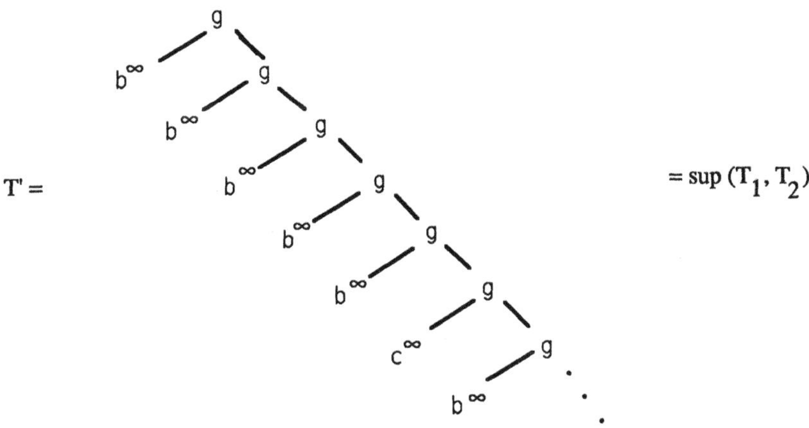

with

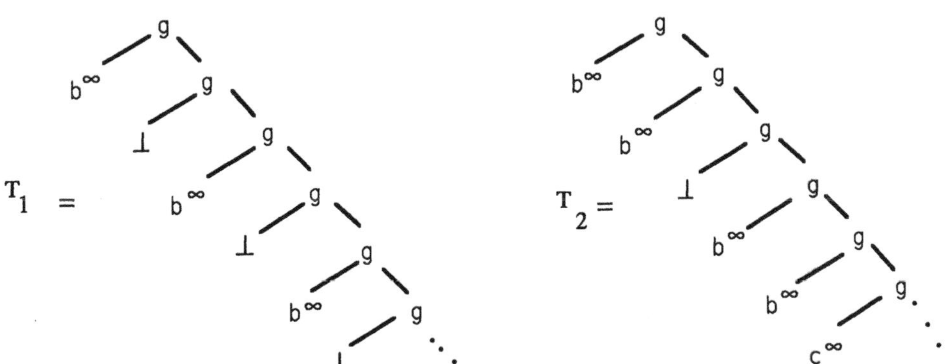

$T' \in adh(L(S_\perp, X))$, but $T' \notin Inf(S, X)$ since T' is not a limit of a computation sequence of X according to scheme S, even though $T' = sup(T_1, T_2)$ with T_1 and T_2 partial limits of (non terminated) computation sequences of X according to S, i.e. $\exists T_i' \in Inf(S, X), T_i \leq T_i', i = 1, 2$.

EXAMPLE III.4

1) Consider the recursive scheme:

$$S: \phi(x) = x + a(\phi(b(x)))$$

$(S, \phi(x))$ generates the algebraic forest, which happens here to be identifiable with a word language: $F(S, \phi(x)) = \{a^n b^n(x)/n \geq 0\} \cup \{a^\omega\}$, where a^ω denotes the infinite tree $a^\omega = lub\{a^n(\perp)/n \geq 0\}$. Here $Inf(S, \phi(x)) = a^\omega$.

2) The example of the semaphore can be written more naturally as the iterative scheme: $S: Z = aZ + bZ$, which generates the rational forest, reduced to its infinitary part: $F(S, Z) = \{w/w \in \{a, b\}^\omega\} = Inf(S, Z)$.

3) Consider finally:

$$S: \phi(x) = x + \phi(b(x))$$

then:

$$F(S, \phi(x)) = Fin(S, \phi(x)) = \{b^n(x) \; / \; n \in \mathbb{N}\}$$

Here the forest generated by $(S, \phi(x))$ is reduced to its finitary part.

We can thus see that the case of the deterministic program schemes generating a single infinite tree corresponds to schemes whose finitary part is empty and whose infinitary part is reduced to a single element.

We will now briefly survey the various characterizations of algebraic and iterative forests, and this will constitute the goal of the next section.

IV CHARACTERIZATIONS OF ALGEBRAIC AND ITERATIVE FORESTS.

As in the case of word languages, tree languages have equational, grammatical and automaton-theoretical characterizations. In the case of arbitrary tree languages, this study is quite complex because of the richness and flexibility of the tree structure. Fortunately, in the case of the tree languages we are considering, namely trees labeled by ranked symbols, the problems are more tractable, even whilst there still remains a richer structure than in the case of words, source to quite a few unexpected problems.

IV.1 Equational characterization

In the case of a deterministic (recursive or iterative) scheme (S, t), we have a nice characterization of the tree $T(S, t)$ which follows from the theorem III.2 of section III.2, which we recall here:

Let S be a deterministic recursive scheme, then $(T(S, \phi_1(\vec{x})), \ldots, T(S, \phi_n(\vec{x})))$ is the least solution, i.e. the least fixpoint of the system of equations S in the free algebra $CT_{\Sigma_\perp}(X)$.

Before proving this theorem, note that $T(S, t)$ then is the first component of the least solution of the system $S' = S \cup \{\phi'(\vec{x}') = t\}$ in $CT_{\Sigma_\perp}(\vec{x}')$, where $\vec{x}'$ is the vector of the variables occurring in t.

Sketch of proof of theorem III.2 (see [Guessarian,Nivat] for a detailed proof): we begin by a few remarks which we will use later.

Let A be a complete algebra, and S a system of recursive equations:

$$S : \quad \phi_i(x_1^i, \ldots, x_{n_i}^i) = t_i \quad , \quad i = 1, \ldots, n \tag{1}$$

Let $A_i = [A^{n_i} \to A]$ be the set of continuous applications from A^{n_i} into A, for $i = 1, \ldots, n$, and $D = A_1 \times \cdots \times A_n$.

Fact 1: $A_i, i = 1, \ldots, n$, and D are complete Σ-algebras.

Each $\vec{\phi} = (\gamma_1, \ldots, \gamma_n) \in D$ determines a $\Sigma \cup \Phi$-algebra structure on A, where $\Phi = \{\phi_1, \ldots, \phi_n\}$ and the γ_i's are intended to represent the interpretation of the ϕ_i's in the $\Sigma \cup \Phi$-algebra, whence a mapping: $t_{i\,A(\vec{\phi})} : A^{n_i} \to A$ associating to each valuation $v = (a_1, \ldots, a_{n_i}) \in A^{n_i}$ the unique value of $\overline{v}(t_i)$ given by the commutative diagram:

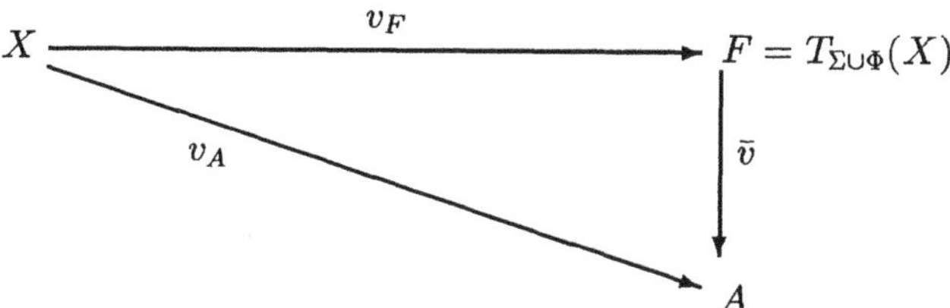

$t_{i\,A(\vec{\phi})}$ is continuous, hence belongs to A_i. One can thus define a mapping $S_A : D \to D$ by $S_A(\vec{\phi}) = (t_{1\,A(\vec{\phi})}, \ldots, t_{n\,A(\vec{\phi})})$.

Fact 2: S_A is continuous.

Note that Fact 2 is the basic step in the definition of denotational semantics.

Hence S_A has a least fixpoint μS_A on D, and moreover, μS_A can be computed by $\mu S_A = lub\{S_A^n(\vec{\perp})/n \in \mathbb{N}\}$, where $\perp \in A_i$ is the function everywhere equal to $\perp$. (Recall that every continuous function f on a complete lattice has a least fixpoint $\mu f = lub\{f^n(\perp)/n \in \mathbb{N}\}$).

When $A = CT_{\Sigma_\perp}(X)$, one can check that $lub\{S_A^n(\vec{\perp})/n \in \mathbb{N}\} = (T(S, \phi_1(\vec{x})), \ldots, T(S, \phi_n(\vec{x})))$, which proves theorem III.2.

This theorem is no longer true in the case of non deterministic schemes. For instance, in example III.4.1, $\{a^n b^n(x)/n \geq 0\} \cup \{a^\omega\}$ is the greatest fixpoint of the system of equations S in $CT_\Sigma(X)$; note that $CT_\Sigma(X) = \{a, b\}^\infty(x)$. In example III.4.2 also,

$F(S, Z)$ is the greatest fixpoint of S in CT_Σ. Examples can be found where $F(S, \phi(\vec{x}))$ is neither the greatest nor the least fixpoint of S in the corresponding free algebras [Guessarian 89].

This point shows the advantage of the approach of section III.1, where all schemes were "determinized" by considering "+" and "or" as just formal uninterpreted symbols of rank 2 in Σ, and where we solved the schemes in the free algebra $CT_{\Sigma_\perp}(X)$, by comparison to the "more natural" approach in the non-deterministic case, where the non-deterministic "or" is explicitly interpreted as set union. In the latter case, the natural free domains are $\mathcal{P}(CT_\Sigma(X))$, equipped with set inclusions as ordering, or the closed subsets thereof [Arnold-Nivat], and then, $F(S, \phi(\vec{x}))$ is no longer necessarily a component of the least fixpoint of S in the free Σ-algebra, as shown by the above examples.

IV.2 Grammatical characterization

This characterization coincides with the definition which has already been given to describe our trees. Namely, we have the

DEFINITION IV.1   *A context-free tree grammar $\mathcal{G}$ is a 4-tuple (Σ, Φ, P, G_0), where:*
(a) *Σ is a finite ranked alphabet (of terminal symbols)*
(b) *$\Phi = \{G_0, G_1, \ldots, G_n\}$ is a finite ranked alphabet of function variables (or nonterminal symbols)*
(c) *P is a finite set of pairs $< G_i(x_1, \ldots, x_{r(G_i)}), t_i^j >$ where, for $i = 0, \ldots, n$, $j = 1, \ldots, n_i$, $t_i^j \in T_{\Sigma \cup \Phi}(x_1, \ldots, x_{r(G_i)})$*
(d) *G_0, the axiom or initial nonterminal, is a distinguished symbol of rank 0 in Φ.*

We will use a vector shorthand notation and abbreviate $G(x_1, \ldots, x_{r(G)})$ in $G(\vec{x})$. Each pair $< G_i(\vec{x}), t_i^j >$ in P is called a *production* of $\mathcal{G}$, and is denoted by $G_i(\vec{x}) \to t_i^j$. Productions of $\mathcal{G}$ turn into rewriting rules: consider the $x_k's$ as dummy variables (or parameters) standing for trees in T_Σ. The triple (Σ, Φ, P) can thus be viewed as a nondeterministic recursive program scheme, or a tree rewriting system. Grouping together all right hand sides of productions having the same nonterminal as left hand side, we get the following synthetic notation for a tree rewriting system:

$$G_i(\vec{x}) \to T_i = t_i^1 + \ldots + t_i^{n_i} \subset T_{\Sigma \cup \Phi}(x_1, \ldots, x_{r(G_i)})$$

where + denotes the set theoretical union.

A context-free tree grammar is said to be *deterministic* iff the root of the right hand side of each production is a terminal symbol in Σ, i.e. the grammar is in Greibach form, and moreover, for each left hand side G there exists at most one production whose right hand side has a given root $f \in \Sigma$. A non deterministic recursive scheme (S, t) on Σ is thus a context-free tree grammar $\mathcal{G} = (\Sigma', \Phi', P, G_o)$, where $\Sigma' = \Sigma \cup X$, X being the set of variables occurring in t, $\Phi' = \Phi \cup \{G_o\}$ and $P = S \cup \{G_o \to t\}$, modulo a possible renaming of the variables of S. Conversely, a context-free tree grammar $\mathcal{G}$ is a nondeterministic recursive scheme $(\mathcal{G}, G_o)$. Finally, a context-free tree grammar is said to be *regular* iff all the righthand sides of its productions are left linear. Deterministic recursive schemes can be considered as deterministic context free tree grammars, but the converse is false.

The finitary context free tree language generated by $\mathcal{G}$ is $L(\mathcal{G}) = \{t \in T_\Sigma / G_0 \overset{*}{\Longrightarrow}_\mathcal{G} t\} = Fin(\mathcal{G}, G_0)$, where $(\mathcal{G}, G_0)$ is the nondeterministic scheme associated with $\mathcal{G}$. Similarly, the infinitary tree language generated by $\mathcal{G}$ is $F(\mathcal{G}, G_0)$.

We thus obtain the following characterizations:
- a rational tree is a $T(S,t) = F(S,t)$ for a deterministic iterative scheme, and is thus generated by a deterministic regular tree grammar of a special type
- an algebraic tree is a $T(S,t) = L(S,t)$ for a deterministic recursive scheme, i.e. a deterministic context-free tree grammar of a special type
- a rational forest is an $F(S,t)$ for a non deterministic iterative scheme (S,t), i.e. a regular tree grammar
- an algebraic forest is an $F(S,t)$ for a non deterministic recursive scheme (S,t), i.e. a context free tree grammar.

IV.3 Automaton theory characterization

The finitary forest generated by a non-deterministic recursive (resp. iterative) scheme can be recognized by a pushdown tree automaton, in short a PDTA, (resp. a finite tree automaton). The forest $F(S,t)$ generated by a (non)-deterministic iterative scheme can be recognized by a finite top-down tree automaton with a Büchi (or Muller) acceptance condition for infinite branches and an acceptance by final state for finite branches. Finally, no good characterization is known in full generality for the algebraic infinitary forests generated by non deterministic recursive schemes. We will not detail these results, but survey them briefly and refer the reader to the literature.

Since iterative schemes coincide with regular tree grammars it is well-known that the finitary languages, i.e. forests, that they generate can be recognized by top-down, or bottom-up, finite tree automata (see e.g. [Thatcher, Engelfriet]).

Finitary forests generated by non deterministic recursive schemes can be recognized by PDTA's which extend the usual string pushdown automata by allowing trees instead of strings in both the input and the pushdown [Guessarian 83].

The PDTA reads its input like a finite top down tree automaton [Thatcher] while scanning the top (root) of the store (the root is the only stack symbol accessible at a given moment). PDTA's recognize context-free tree languages. A different model, called TPDA is investigated in [Gallier-Schimpf]: it stresses the parsing standpoint. Henceforth, TPDA's process trees in a bottom-up manner. The advantage of TPDA's is that a wide subclass of context-free tree languages can be recognized by deterministic TPDA's. The trade-off is that TPDA's are of course much more complicated than PDTA's. The latter stick more closely to the behaviour of grammars and their pushdown consists of a single tree; whereas the former's pushdown consists of arbitrary sets of trees whose roots must be simultaneously accessible and they consult appropriate shift-reduce tables. PDTA's thus provide a parallel LL-parsing method for context free tree languages, as opposed to the parallel LR-parsing of TPDA's.

That parsing can be improved and made really operational by restricting the pushdown and allowing only words instead of trees on it: i.e. we can linearize the pushdown;

this would be difficult to realize with TPDA's and is also one of the advantages of PDTA's. Last but not least, the advantage of PDTA's is that they can be generalized to accept some infinite algebraic trees, which is impossible with a bottom-up model. See [Guessarian 83] (resp. [Gallier-Schimpf]) for more details on the PDTA (resp. TPDA) construction.

Infinitary forests generated by iterative program schemes can be recognized by generalized finite automata, which behave as finite automata on finite branches, and have a Büchi (or Muller) type acceptance condition for infinite branches [Saoudi]. Similarly, the PDTA construction can be extended to generalized PDTA's accepting infinite trees, which behave like normal PDTA's (with a linear stack) on finite branches, i.e. accept by final state or empty stack, and which have a Büchi or Muller type acceptance condition for infinite branches. It has been shown [Saoudi] that, for some program schemes S, $F(S,t)$ can indeed be recognized by such a generalized PDTA.

V COMPLEMENTS AND CONCLUSIONS

V.1 Classes of interpretations

Usually, we do not consider all models (or interpretations) but only subclasses of models subject to some constraints, e.g. that some operation σ is a test, that some other operations are associative, or commutative, etc... Given a class $\mathcal{C}$ of models, we would thus like to find an "abstract" or Herbrand model which characterizes the behavior of the whole class $\mathcal{C}$.

This leads to considering factor algebras of the algebras $T_{\Sigma_\perp}, CT_{\Sigma_\perp}$, and usually is the source of many problems, because the factor algebra is no longer complete, or, even if it is complete, continuity is not preserved, etc... See [Courcelle, Guessarian 85, Courcelle-Guessarian] for some of the problems, and references to related literature.

However, there is one case when we obtain a nice characterization of the free and Herbrand algebra relative to the class $\mathcal{C}$, generalizing the construction of CT_Σ, this is the case of relational classes.

DEFINITION V.1   *A class $\mathcal{C}$ of models (i.e. interpretations or Σ-complete algebras) is said to be relational iff it is of the form:*

$$\mathcal{C}_R = \{A/A \models t \leq t', \text{ for all } (t,t') \text{ in } R\},$$

where $R \subseteq T_{\Sigma_\perp}(X)^2$. $\mathcal{C}_R$ is said to be equational if R is an equivalence relation. In other words, an equational class is of the form $\mathcal{C}_R^e = \{A/A \models t = t', \text{ for all } t,t' \text{ in } R\}$, where R is as above.

Relational classes form semi-varieties and have been considered by different authors under various names (see [Guessarian 85] for references concerning the subject). Let $\prec_R$ be the least substitution-closed (or fully invariant) preordering containing R on $T_{\Sigma_\perp}(X)$ (i.e. $t \prec_R t'$ implies that for any homomorphism $h : T_{\Sigma_\perp}(X) \to T_{\Sigma_\perp}(X)$, we also have $h(t) \prec_R h(t')$). Let $\sim_R$ be the associated equivalence relation. Define $H_R = T_{\Sigma_\perp,R}(X)^\infty$, where $T_{\Sigma_\perp,R}(X)$ is the ordered Σ-algebra obtained by factoring $T_{\Sigma_\perp}(X)$ through $\sim_R$ and ordering it with $\prec_R / \sim_R$, and $T_{\Sigma_\perp,R}(X)^\infty$ is the ideal completion of $T_{\Sigma_\perp,R}(X)$.

THEOREM V.2 *Let $R \subseteq T_{\Sigma_\perp}(X)^2$, and $H_R = T_{\Sigma_\perp,R}(X)^\infty$. Then H_R is an initial and Herbrand model for the class C_R.*

For the proof, see [Guessarian].

V.2 Conclusion

We have in the present survey studied the trees which are most often used in algebraic semantics, namely the algebraic and rational ranked trees. We showed how to characterize them (algebraically, grammatically, or via automata), and how they are used to describe the semantics of a program scheme through its semantics in an initial algebra, the algebra of trees.

There remain of course quite a few interesting problems to be studied in connection with such trees. We will cite here three directions for research:
- the automaton theory characterization of forests of algebraic infinitary trees
- the study of factor algebras of algebras of trees, in particular those corresponding to factors of CT_Σ, which characterize $\equiv_C$ for classes of interpretations C which are not semi-varieties ($\equiv_C$ is defined by $T \equiv_C T' \iff T_A = T'_A \quad \forall A \in C$).
- the transformation of solutions to program schemes, or more generally sets of equations, under factoring morphisms.

VI REFERENCES

[**Andréka-Németi**] , H. Andréka, I. Németi, Generalization of variety and quasi variety concepts to partial algebras through category theory, *Dissertationes Mathematicae (Rosprawy Math.)*, 204 (1982).

[**Arnold-Nivat**] A. Arnold, M. Nivat, The metric space of infinite trees: Algebraic and Topological properties, *Fund. Inform.* 3 (1980), 445-476.

[**Birkhoff**] G. Birkhoff, *Lattice Theory*, 3rd edition, *AMS Coll.*, New York (1979).

[**Courcelle**] B. Courcelle, Arbres infinis et systèmes d'équations, *RAIRO Info. Théor.* 13 (1979), 31-48.

[**Courcelle-Guessarian**] B. Courcelle, I. Guessarian, On some classes of interpretations, *J. Comput. and Syst. Sci.* 17 (1978), 388-413.

[**Engelfriet**] J. Engelfriet, Some open questions an recent results on tree transducers and tree languages, in *Formal Languages: Perspectives and open problems*, R. Book ed. Academic Press, London (1980), 241-286.

[**Gallier-Schimpf**] J. Gallier, K. Schimpf, Tree pushdown automata, *J. of Comput. and Syst. Sci.* 30 (1986).

[**Goguen-Meseguer**] J. Goguen, J. Meseguer, Initiality, Induction and Computability, in *Algebraic Methods in Semantics*, M. Nivat and J. Reynolds eds., Cambridge Univ. Press (1985), 459-540.

[**Gorn**] S. Gorn, Explicit definitions and linguistic dominoes, in *Systems and Computer Science*, J. Hart, S. Takasu eds., (1965).

[**Grätzer**] G. Grätzer, *Universal Algebra*, 2nd edition, Springer-Verlag, Berlin (1979).

[**Guessarian**] I. Guessarian, *Algebraic Semantics*, Springer-Verlag, *LNCS* 99, Berlin (1981).

[**Guessarian 83**] I. Guessarian, Pushdown tree automata, *Math. Systems Theory* 16 (1983), 237-263.

[**Guessarian 85**] I. Guessarian, Classes of interpretations and some of their applications, in *Algebraic Methods in Semantics*, M. Nivat and J. Reynolds eds., Cambridge Univ. Press (1985), 383-409.

[**Guessarian 89**] I. Guessarian, Improving fixpoint tools for computer science, *IFIP'89* Proc., G. Ritter ed., North-Holland, Amsterdam (1989), 1109-1114.

[**Guessarian 90**] I. Guessarian, Many sorted logics and algebraic semantics, to appear in Proc. *Many-sorted Logics and its Applications*, Leeds (1989).

[**Lehmann-Pasztor**] , D. Lehmann, A. Pasztor, Epis need not be dense, *TCS* 17 (1982), 151-162.

[**Mal'cev**] A.I. Mal'cev, *Algebraic systems*, North-Holland, Amsterdam (1973).

[**Milner**] R. Milner, *A calculus of communication systems*, LNCS 92, Springer-Verlag, Berlin (1980).

[**Nivat**] M. Nivat, On the interpretation of recursive polyadic program schemes, *Symp. Mathematica* 15 (1975) 255-281.

[**Saoudi**] A. Saoudi, Contribution la théorie des automates d'arbres infinis et applications la logique et la programmation, thèse de 3ème cycle, Univ. Paris 7 (1987).

[**Scott**] D. Scott, Data types as lattices, *SIAM Jour. Comput.* 5 (1976), 522-587.

[**Thatcher**] J.W. Thatcher, Tree automata: An informal survey, in *Currents in Theory of Computing*, A. Aho ed., Prentice-Hall, London (1973).

Tree Automata and Languages
M. Nivat and A. Podelski (editors)
© 1992 Elsevier Science Publishers B.V. All rights reserved.

311

A SURVEY OF

TREE TRANSDUCTIONS

Jean-Claude Raoult

IRISA, Campus de Beaulieu,
F-35042 RENNES CEDEX
email: raoult@irisa.fr

Abstract: We describe the various approaches to tree-transformations, and their behaviour with respect to a few criteria: stability under composition, inverse, iteration, rationality of their domains and images, finite or infinite branching, restriction to the case of words.

I. INTRODUCTION AND NOTATIONS. RATIONAL FORESTS

Since the beginning (around 1965, cf. Thatcher & Wright [1965]) the study of sets of trees, or tree-languages, has followed roughly the same route as the theory of word-languages. The same classification has been established, and the following families of trees have been defined: regular or rational (see Gecseg & Steinby [1984]), algebraic (program schemes, cf. Nivat [1973]) recursive, and recursively enumerable (standard definitions). And the following strict hierarchy has been established:

$$\text{regular} \subset \text{algebraic} \subset \text{recursive} \subset \text{recursively enumerable}$$

Many results carry over from sets of words to sets of trees. In particular, the richest class, the one having the most characterizations: the regular or rational languages, can also be defined for trees. To fix ideas, trees will be identified with terms built over an alphabet F of 'function symbols', considered as a terminal alphabet. The languages $T(F) = T$ of trees or terms over F and T^* of sequences of trees over F are defined (mutually) recursively by the following equations:

$$T = FT^* \qquad\qquad t = av \text{ or } a(v)$$
$$T^* = \{\varepsilon\} + TT^* \qquad\qquad v = t_1 \ldots t_n \text{ or } (t_1, \ldots, t_n)$$

where the cartesian product of sets is denoted by concatenation, as is customary in language theory. Note that sequences of trees may be empty (ε), but trees may not. The alphabet F is often graded by an arity $\alpha : F \to \mathbf{N}$, and the trees are subject to the restriction that for all subtree $v = av_1 \ldots v_n$ the arity of a equals n: this is a purely syntactic restriction which can be expressed equationally if needed. In this case, one defines a F-algebra as a set D together with a function $f_D : D^n \to D$ where n is the arity of f for all f in F. A F-morphism $\mu : D \to D'$ between two F-algebras is a mapping satisfying $\mu(f_D(d_1, \ldots, d_n)) = f_{D'}(\mu(d_1), \ldots, \mu(d_n))$, and F-morphisms make a category in which T(F) itself with functions $f(t_1, \ldots, t_n) = ft_1 \ldots t_n$, is an initial object.

Frequently, one wants to single out some occurrence of a subtree s into a tree t. For this purpose, we write $t = c(s)$ where $c(x)$ is a term containing x (of arity o) only once, called a context: the context of s in t. And t is got from $c(x)$ by replacing x by s. We shall also assume that the reader is aware of the representation of terms by trees considered as oriented graphs with nodes labelled by F, and use freely the graph-theoretic terminology, like root, node, leaf, depth, etc.

The equivalence, true with words, between right-linear grammars, finite automata, finite index congruences and rational expressions still holds, with the following definitions:

Definition. 1°) *A tree-grammar is a finite set of productions of the form* $A \to t$ *where A belongs to a finite set* N *of non-terminals of arity* o, *and* $t \in T(F \cup N)$.

2°) *A descending automaton is a finite set of transitions of the form* $q \xrightarrow{f} q_1 \ldots q_n$ *where q belongs to a finite set* Q *of states.*

3°) *An ascending automaton is a finite set of transitions of the form* $q_1 \ldots q_n \xrightarrow{f} q$ *where q belongs to a finite set* Q *of states.*

4°) *A congruence is an equivalence relation over* T(F) *such that for all* $f \in F$
$$s_i \equiv t_i \text{ for } i = 1, \ldots, n \Rightarrow f(s_1, \ldots, s_n) \equiv f(t_1, \ldots, t_n) \text{ with } n = \alpha(f)$$

5°) *The rational operations over the subsets of* T(F) *are the following:*
$$L + M = L \cup M$$
$$L \cdot_a M = \{s(t_1, \ldots, t_p) ; s(x_1, \ldots, x_p) \in T(F - \{a\}) \wedge s(a, \ldots, a) \in L \wedge t_1, \ldots, t_n \in M\}$$
$$L^{*a} = \bigcup_{n \geqslant o} L \cdot_a \cdots \cdot_a L \ (n \text{ times})$$

1°) One step of derivation according to a tree-grammar G is the relation $c(A) \to_G c(t)$ where $A \to t$ is a production, and G is omitted if no confusion results. A derivation is a sequence of such steps. The language L(G,A) generated by the grammar G starting from the axiom A is defined as for words: $L(G,A) = \{t \in T(F) ; A \to^* t\}$.

2°) A run of a descending automaton on a tree t is a labelling of the nodes of t by states, in a way compatible with the transitions: if a node labelled by q has function symbol f, and if its successors are labelled by $q_1, \ldots, q_n$ then $q \xrightarrow{f} q_1 \ldots q_n$ must be a transition. The tree t is accepted by the automaton if it admits a run starting at some given initial state.

3°) A run of an ascending automaton on t is defined similarly, except that $q_1 \ldots q_n \xrightarrow{f} q$ must be a transition at each node. The tree t is accepted if the state at the root belongs to some given set of states: the final states. Note that there is no difference between an ascending run and a descending run, nor really between the ascending and the descending transitions. At each node with function symbol f of arity n, with state label q and state labels $q_1 \ldots q_n$ for the successors, the tuple $q f q_1 \ldots q_n \in QFQ^*$ must belong to the transition relation. The difference is in the acceptance condition. The words ascending and descending refer to the drawings of the graphs representing the trees: The tradition has evolved to draw the root at the top and the leaves at the bottom.

4°) If $s \equiv t$ is a congruence, the factor set $T(F)/\equiv$ is 'naturally' a F-algebra for general algebraic reasons. The index of the congruence is the cardinality of the factor set.

These few definitions (!) provided, the next result is welcome (see Gecseg & Steinby [1984]).

Theorem. *For a set* L *of trees, the following conditions are equivalent:*
1°) L = L(G,A) *for some tree-grammar* G *and axiom* A.
2°) L *is the set of trees accepted by some descending automaton.*
3°) L *is the set of trees accepted by some ascending automaton.*
4°) L = $\mu^{-1}(\mu(L))$ *for some F-morphism* $\mu:T(F) \to D$ *where* D *is finite.*
5°) L *is got by applying a finite number of rational operations to a finite number*
of finite subsets of T(F).

The sets of trees satisfying the equivalent conditions of the proposition are called rational forests. Therefore, everything goes as smoothly as with words, with a small exception: Ascending tree automata have equivalent deterministic counterparts, but deterministic descending automata are strictly weaker than general non deterministic descending automata. More precisely $\{f(a,a), f(b,b)\}$ is a rational forest, but no deterministic descending automaton can accept it, since the state labelling the left leaf must recognize a and b, and likewise for the state labelling the right leaf. Therefore, the automaton should also accept $f(a,b)$ and $f(b,a)$, which are not in the set.

One remark is in order here: the counterpart of concatenation has been chosen. Its role is played by tree substitution (see the definition above, point 5°). Therefore, the iteration of concatenation, the 'star' operation is played by iterated substitution. Another point of view has been adopted by a few authors (for instance Steinby [1983]). For instance, following Steinby, we may consider the concatenation as a binary operator; therefore its generalization is the prefixing of two − or more − trees by a function symbol. Likewise, the generalization of L*, which is also the sub-monoid generated by L, is in this setting the sub-algebra ⟨L⟩ generated by L. Rational subsets in this sense have some good properties, for instance they are preserved by boolean operations, but are much more restricted than in the definition above.

II. TREE TRANSDUCTIONS

Regarding tree transformations, results do not flow so easily. Several definitions are candidate for the label 'tree transductions', with irrelated properties. People will keep in mind how gracefully behave rational (word) transductions. These admit several equivalent definitions: by grammars of relations with right linear productions, by finite automata with output, by pairs of morphisms of monoids, and others.

Example 1: The relation $R = \{(a^p b^n, b^n a^q)\,;\ n,p,q \geqslant 0\}$ is rational: $R = (a,\varepsilon)^* (b,b)^* (\varepsilon,a)^*$. It can be defined by this grammar:

$$A \to (a,\varepsilon)A \qquad B \to (b,b)B \qquad C \to (\varepsilon,a)C$$
$$A \to B \qquad\qquad B \to C \qquad\qquad C \to (\varepsilon,\varepsilon)$$

Or it can be generated by this automaton, where state C is final:

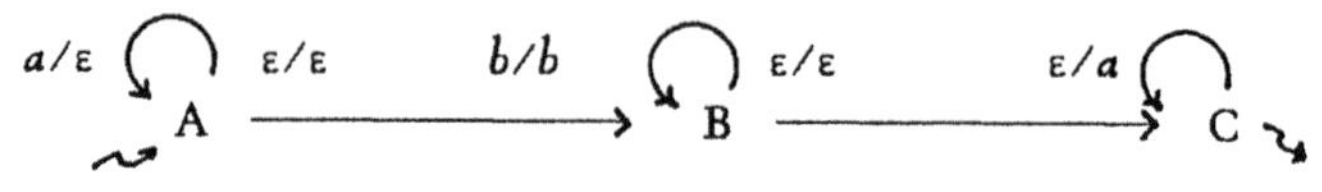

$$a/\varepsilon \ \circlearrowright A \xrightarrow{\ b/b\ } B \xrightarrow{\ \varepsilon/a\ } C$$

with loops ε/ε on A, ε/ε on B, and ε/a on C.

Or it can be defined by this pair of morphisms (the bimorphism), applied to the rational language $x^* y^* z^*$, which represents the relation actually, while the morphisms represent the first and second projection:

$$\alpha(x) = a \qquad\qquad \beta(x) = \varepsilon$$
$$\alpha(y) = b \qquad\qquad \beta(y) = b$$
$$\alpha(z) = \varepsilon \qquad\qquad \beta(z) = a$$

All these means have been tried in the case of trees, and tested against several touchstones:

- Do these transformations contain the identity? This is a minimal requirement.
- Are they preserved by composition? To make a monoid.
- Are they preserved by inverse? To make a group.
- Are they locally finite (given s only a finite number of $s \to t$)?
- What are the images of rational forests? of algebraic forests?
- Can they be decomposed into a succession of 'simple' transformations, like relabelling, morphisms, inverse morphisms, etc.?
- When restricted to words, do they coincide with rational transductions?
- Is the accessibility relation $s \to^* t$ decidable? It is decidable in the case of words.

Preserved by composition, and including the identity, the tree homomorphisms have been known since the beginning. They are not only locally finite, but deterministic: one and only one image for a tree; and the set of images need not be rational:

Example 2: Define $\mu(ax) = b(\mu(x),\mu(x))$, $\mu(c) = c$. The image of $T(\{a,c\}) = a^* c$ is the set of perfect binary trees, which is not even algebraic.

The point of view of automata has been investigated first (cf. J. Engelfriet [1975]) and been developped enough to be included in a book (cf. Gecseg & Steinby [1984]). It is developed in the following section. Section IV is devoted to the approach using two morphisms (cf. Dauchet [1977], Arnold & Dauchet [1982]). Section V describes the transformations generated by tree rewriting systems which do not contain variables: ground tree transducers. Section VI looks at the relations generated by grammars.

III. TRANSDUCTIONS USING AUTOMATA

The descending transducer looks like a descending automaton with output:

Definition. A *descending transducer is a finite set Q of states of arity one together with a finite set of transitions of the form: $qf(x_1,\ldots,x_n) \to t(q_1 x_{\sigma(1)},\ldots,q_p x_{\sigma(p)})$, where t contains no variable other than explicitly written and σ is a mapping $[1,p] \to [1,n]$.*

To be fair, let us define immediately the ascending transducers, which are exactly finite automata plus an output:

Definition. An *ascending transducer is a finite set Q of states of arity one together with a finite set of transitions of the form: $f(q_1 x_1,\ldots,q_n x_n) \to qt(x_{\sigma(1)},\ldots,x_{\sigma(p)})$ where t contains no variable other than explicitly written, and σ is a mapping $[1,p] \to [1,n]$.*

If states are considered as relations, the transitions can be seen as a definition of these relations in terms of each other. The variables x_i represent the subtrees hanging under the node f. If σ is injective in all transitions, no subtree is duplicated, and the transducer is called *linear*. If σ is surjective in all transitions, no subtree is erased, and the transducer is called *non erasing*.

Here, the distinction between ascending and descending is more relevant. More precisely, following J. Engelfriet [1975], ascending automatic transducers can perform actions AC and AE below:

(AC) Copy, duplicate, an output tree.

(AE) Erase, delete, an input tree (after processing it).

Descending automata can instead perform actions DC and DE:

(DC) Copy an input tree, then translate copies independently.

(DE) Erase, delete, an input tree (even before looking at it).

Operations (AC) and (DC) prevent the image to be rational: tree-morphisms are particular cases of transducers. Indeed they are deterministic, total transducers with a single state.

Proposition. *A tree-morphism can be realized by a descending transducer, or an ascending transducer. Both have only one state and are deterministic: there is at most one right-hand side corresponding to a given left-hand side and a given function symbol.*

Proof: The definition of a morphism has the format of a descending transducer: If μ is a morphism, it satisfies for all function symbol f:

$$\mu(f(x_1,\ldots,x_n)) = t[\mu(x_1),\ldots,\mu(x_n)]$$

where the square brackets indicate that t may contain several occurrences, or none, of each variable x_i, but no other variable.

Corresponding descending transition: $\mu(f(x_1,\ldots,x_n)) \to t[\mu(x_1),\ldots,\mu(x_n)]$.

Corresponding ascending transition: $f(\mu(x_1,\ldots,x_n) \to \mu(t[x_1,\ldots,x_n])$.

See the proof in Gecseg & Steinby [1984].

Now, we showed in example 2 a morphism having for image the set of all perfect binary trees, which is not rational, and not even algebraic.

The main results concerning these transducers are gathered in the following proposition, using the following notations: A for ascending, D for descending, L for linear, N for non-deleting, H for homomorphic (see the proposition above), and F^n indicates the class of n-fold composition of transducers of class F.

Theorem. *The following inclusion relations (represented from left to right) are strict, and circled families are closed by composition*

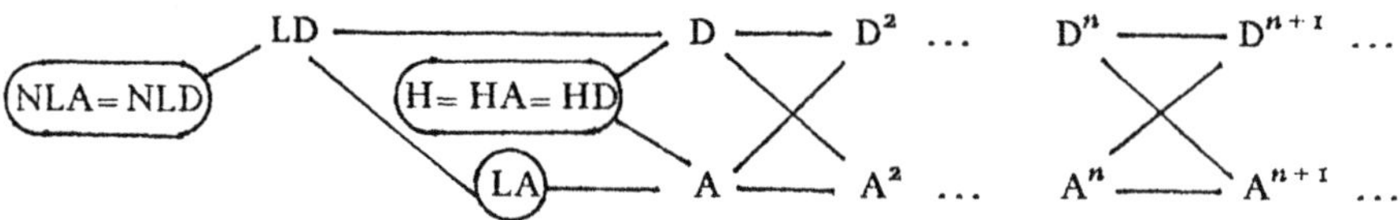

The strict inclusions $D^n \subset A^{n+1}$ and $A^n \subset D^{n+1}$ are deduced from the equalities

$$A = LA \circ H = LD \circ H$$
$$D = H \circ LD \subset H \circ LA$$

Example 3: Consider the following transducer with $Q = \{q,p,s,i\}$

$$
\begin{array}{ll}
qb(x,u) \to b(b(ix,pu),su) \qquad & ib(x,y) \to b(ix,iy) \\
pb(y,z) \to iy & ic \to c \qquad\qquad \text{(ND–A)} \\
sb(y,z) \to iz &
\end{array}
$$

The state i produces the identity, states p and s yield respectively the first and second argument of b and state q is chosen as the initial state. The reader can check that the tree transformation produced by this transducer is the left rotation of balanced trees. It cannot be realized by an ascending automaton. It is not linear, but it is non-deleting: it belongs to ND–A.

Example 4: Consider the following transducer with $Q = \{q\}$ and $F = \{b,a,c\}$ of arities respectively $2,1,0$:

$$
\begin{array}{l}
b(qx,qy) \to qa(x) \\
c \to qc \qquad\qquad\qquad \text{(LA–D)}
\end{array}
$$

This transducer is almost homomorphic: it has a unique state, is linear, deterministic, but not total. It copies the left branch of a binary tree containing only b's, and replaces everywhere b by a. This cannot be accomplished by a descending transducer, because deleting the right branch in a descending transducer allows the right subtree to be any tree − including trees containing a's. This cannot be checked before inspection. It is an example of property (AE) compared with property (DE). This transducer is in LA−D.

All these transduction have locally finite images. On the other hand, if erasing is allowed, the converse image of a tree may be an infinite tree language. Therefore, it is not surprising that none of the classes considered so far are preserved by taking the inverse. Restricted to words, all yield ordinary rational transductions, but not all the transductions. No class is preserved by iteration. Their behaviour with respect to rational forests is described in the following proposition where we say that a class preserves rational forests when all transducers in the class give of a rational forest a rational image.

Proposition. 1°) *The domain of any transducer is a rational forest.*
2°) *Neither* H *nor* D *nor* A *preserve rational forests.*
3°) LA *preserves rational forests, hence* LD *also preserves rational forests.*

Proof: 1°) We noticed that an ascending transducer is really a finite automaton with output. Suppressing the output yields the fact that its domain is a rational forest.
2°) Example 2 shows a homomorphism, and the image of $T(\{a,c\})$ is the set of perfect binary trees, which is not rational, nor even algebraic.
3°) The range of a linear ascending transducer is rational: build the following grammar

$$q \to t(q_{\sigma(1)},\ldots,q_{\sigma(p)}) \in G \quad \Leftrightarrow \quad f(q_1 x_1,\ldots,q_n x_n) \to qt(x_{\sigma(1)},\ldots,x_{\sigma(p)})$$

It can be checked with no pain that a tree is in the range of the transducer if and only if it is generated by the associated grammar. The result now comes from the fact that linear ascending transducer are preserved by composition, QED.

IV. BIMORPHISMS

Originally, the two morphisms used for representing a relation R are nothing else than the two projections of this relation (see section I):

$$x \to y \quad \Leftrightarrow \quad x = \pi(r) \wedge y = \rho(r) \text{ for some element } r = (x,y) \in R.$$

In the case of monoids, morphisms, which preserve the rational languages, should certainly be particular cases of rational transductions. The intersection with a rational set is also rational. Finally, word transductions are preserved by taking the inverse. Since the composition of two word transductions is again a transduction, an inverse morphism, followed by an intersection with a rational set, followed by a direct morphism still is a transduction. Nivat [1968] proved that every transduction can be decomposed in this way: The class of rational transductions is the least class containing the morphisms and the identity id_R with rational domain (this accounts for the intersection with a rational set R), and closed by composition.

In the case of trees, this point of view has been investigated by Dauchet [1977]. Here, arbitrary morphisms do not preserve the rationality, due to possible duplication of branches (cf. example 1). The least condition to be satisfied is the linearity:

$$\mu(f(x_1,\ldots,x_n)) = t[\mu x_1,\ldots,\mu x_n] \;\Rightarrow\; (\forall i)\; |t[\mu x_1,\ldots,\mu x_n]|_{x_i} \leqslant 1 \quad (L)$$

That is, right-hand sides never contain two occurrences of the same variable. It turns out that this condition is not enough to ensure stability under inversion and composition.

Example 5: consider the following morphism.

$$\begin{aligned}
\varphi f(x,y,z)) &= b(b(\varphi x,\varphi y),\varphi z)\\
\varphi b(x,y)) &= b(\varphi x,\varphi y)\\
\varphi a(x)) &= a(\varphi x)\\
\varphi c &= c
\end{aligned}$$

And consider the identity restricted to the following rational forests.

$$K = f(x,y,y)^{*x}\cdot_x b(y,y)\cdot_y a(z)^{*z}\cdot_z c \text{ and } K' = b(x,y)\cdot_x f(x,y,y)^{*x}\cdot_y a(z)^{*z}\cdot_z c$$

Then the composition of the two bimorphisms $(\mathrm{id}_K \circ \varphi) \circ (\varphi^{-1} \circ \mathrm{id}_{K'})$ cannot be expressed by any bimorphism, because of the constant delay between the depth of the arguments of f and the depth of the arguments of their images (in the figure, corresponding arguments have the same numbers $1,2,\ldots$).

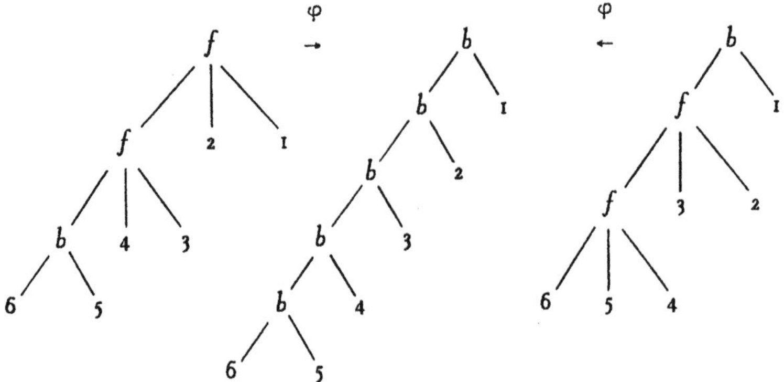

In the framework of trees, some other properties need be added: morphisms will be non erasing and strict. Recall that a morphism is non erasing when $\mu(f(x_1,\ldots,x_n))$ contains all the variables $x_1,\ldots,x_n$: no subtree can be deleted.

Definition. A *morphism* μ *is strict when* $\mu(f(x_1,\ldots,x_n))$ *is not reduced to a variable, for all* $f \in F$.

Denoting by LNSB the class of all linear, non erasing, strict bimorphisms, a rather unsatisfactory result can be obtained (Arnold & Dauchet [1982]).

Proposition. *The class LNSB is not preserved by composition. The class $LNSB^2$ is preserved by composition. More precisely, we have the following inclusions:*
$$LNSB \subset LNSB^2 = LNSB^n \text{ for } n > 2.$$

Proof: It can be found in Arnold & Dauchet [1982] and is neither short nor simple.

Notice however that the product of two monoids still is a monoid, while the direct product of two term algebras is not a term algebra. This might account for the awkward results using trees. Arnold & Dauchet investigated the case of direct products of term algebras, following in fact a trend initiated by Pair & Quéré [1968] who considered words on trees, with an extra operation: prefixing a root to a word of trees. Arnold & Dauchet also consider p-tuples of trees containing variables only in $\{x_1, \ldots, x_q\}$: Define the set $T_q^p = T(F \cup \{x_1, \ldots, x_q\})^p$. Considered as words on $T(F)$, they inherit a version of the product of concatenation uv denoted here $u \times v$ with an explicit variable renaming for v: if $u \in T_q^p$, then all variables x_i in v become variables x_{i+q} in $u \times v$. But instead of prefixing by a root, they are considered as morphisms $T(F)^q \to T(F)^p$ and inherit the composition of functions. Endowed with these two operations, the set T_q^p is an algebraic structure called a 'magmoid' and now, relations between two magmoids T_q^p and $T_{q'}^{p'}$ are included in the magmoid $T_{q+q'}^{p+p'}$. The morphisms of magmoids are mappings that respect the algebraic structure.

Definition. *A morphism of magmoids is a mapping compatible with the product and the composition, and satisfying $\mu(f(x_1, \ldots, x_n)) = (t_1, \ldots, t_k) \in T_{kn}^k$ for some fixed $k > 0$.*

Ordinary tree morphisms correspond to total deterministic descending transducers having only one state. Allowing several states bring in the morphisms of magmoids.

Proposition. *Associate with any k-morphism the following total deterministic descending transducers :* $q_i f(x_1, \ldots, x_n) \to t[q_{\tau(1)} x_{\sigma(1)}, \ldots, q_{\tau(p)} x_{\sigma(p)}]$ *if and only if*
$$\mathrm{pr}_i \varphi(f(x_1, \ldots, x_n)) = t[x_{(\sigma(1)-1)k+\tau(1)}, \ldots, x_{(\sigma(p)-1)k+\tau(p)}].$$
Then this correspondence is bijective and satisfies $q_i t \to^* u_i \Leftrightarrow \varphi t = (u_1, \ldots, u_k)$.

Proof: by induction on t.

In this case again, we shall restrict to morphisms linear non erasing, 'regulated' (or 'strict with delay') and 'separated with delay'. These last two properties are extensions respectively of strict and separated morphisms. Strictness for magmoids means strictness at all components. Separation is related to magmoids only: restricted to linear morphisms, it means that the morphism would remain linear even if the variables occurring in different components had not been renamed.

Definition. A *morphism* μ *is strict when it satisfies for all function symbol* f:
$$\mu(f(x_1,\ldots,x_n))=(t_1,\ldots,t_k) \;\Rightarrow\; (\forall i)\, t_i \notin X \;\; (\text{no } t_i \text{ is a variable}).$$
A morphism is separated when it satisfies for all function symbol f:
$$\mu(f(x_1,\ldots,x_n))=(t_1,\ldots,t_k) \;\Rightarrow\; \text{no variable } x_i \text{ has been renamed twice, in } t_j$$
and t_k *for* $j \neq k$.

Morphisms satisfy the same property with delay p if the same relations hold not for elementary tree of the form $f(x_1,\ldots,x_n)$ but for all trees of depth p in which all variables are at depth exactly p. And they hold 'with delay' if they hold with some delay p.

Theorem. *The class of bimorphisms linear non erasing, strict with delay and separated with delay contains the identity and is preserved by composition and inverse.*

These morphisms are more general than the linear ascending transducer. For instance, the transformation of example 3, which cannot be generated by an ascending transducer is defined by the following bimorphism.

Example 6: Consider the morphisms (linear non erasing, strict):

$$\varphi f(x_1,x_2,x_3) = b(b(\varphi x_1,\varphi x_2),\varphi x_3) \qquad \psi f(x_1,x_2,x_3) = b(\psi x_1,b(\psi x_2,\varphi x_3))$$
$$\varphi b(x_1,x_2) = b(\varphi x_1,\varphi x_2) \qquad\qquad\quad \psi b(x_1,x_2) = b(\psi x_1,\psi x_2)$$
$$\varphi c = c \qquad\qquad\qquad\qquad\qquad\qquad \psi c = c$$

Then the left rotation realized in example 3 is given by (ψ, K, φ) where K is the forest $\{f(s_1,s_2,s_3)\,;\; s_1,s_2,s_3 \in T(\{b,c\})\}$. This transformation is not in A, *a fortiori* not in LA.

Somewhat curiously, more general results hold for binary trees.

Proposition. *If no function symbol in* F *has an arity more than* 2, *then the class of linear bimorphisms contains the identity and is preserved by composition and inverse.*

Proof: It can be found in Arnold & Dauchet [1982] but their proof is older. This result is quoted in Takahashi [1977] who proved it independently, using a trick which has been generalized since (see Dauchet & Tison [1990]): When two trees have the same shape, *viz.* when both are binary, they can be superposed to give a unique binary tree labelled with the product of the alphabets augmented by a symbol ω meaning 'not defined' (Takahashi uses λ). For an instance of this trick, see below where the two tree on the left are projections of the unique tree on the right.

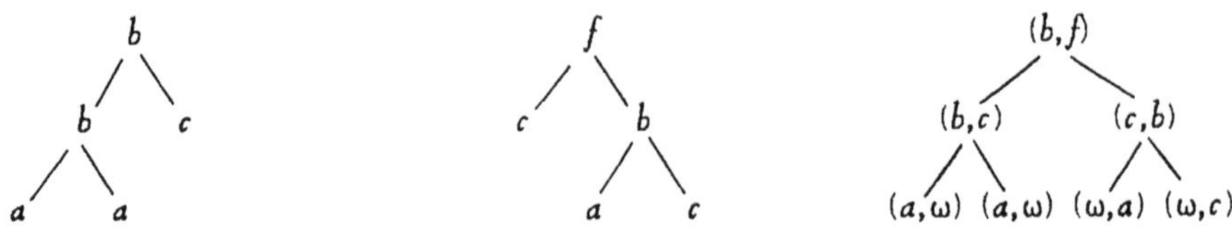

V. GROUND TRANSDUCERS

There is, after all, a simple way to define tree transformations: term rewriting systems. Given a term rewriting system, *i.e.* a finite set R of pairs $g \to d$ with $g, d \in T(F \cup X)$, the relation generated by this system using contexts and substitutions is defined in the usual way:

$$s \to t \quad \text{if} \quad s = c(g\sigma) \;\wedge\; t = c(d\sigma) \;\wedge\; (g \to d) \in R$$

This is an extension of semi-Thue systems in the case of words: The semi-Thue systems are finite sets of couples of words $g \to d$ and the relation generated is the set of all couples of the form $ugv \to udv$, for arbitrary words u and v and couple $g \to d$ in the system. Unfortunately, it is well-known that about everything concerning the transitive closure of this relation is undecidable, since semi-Thue systems can represent the computations of a Turing machine. The same undecidability results clearly hold also for term rewriting systems.

Let us modify the generation of the relation. Suppose, for instance, that we forbid the contexts and allow only substitutions σ:

$$s \to t \quad \text{if} \quad s = g\sigma \;\wedge\; t = d\sigma \;\wedge\; (g \to d) \in R$$

In the case of words, this may reduces more or less to the definition of Post rewriting systems: A Post rewriting system is a finite set of pairs $g \to d$ of words $g, d \in (A \cup X)^*$ containing letters and variables. The generated relation is the set of all pairs $g\sigma \to d\sigma$ for all substitutions σ of variables by words. Unfortunately, it is well-known that these systems can simulate a Turing machine, and again almost everything concerning the transitive closure still is undecidable.

Actually the analogy here points to something that we discussed in section II: what is the analogue of word concatenation in the case of trees? Using Steinby's definition again, we define the family Rat of rational sets in $T(F) \times T(F)$ as follows:

(1) Any finite subset of $T(F) \times T(F)$ is rational.
(2) $R, R' \in \text{Rat} \Rightarrow R \cup R' \in \text{Rat}$.
(3) $R_1, \ldots, R_n \in \text{Rat} \;\wedge\; f \in F \Rightarrow \{f(s_1, \ldots, s_n), f(t_1, \ldots, t_n)\;;\; \forall i \;(s_i, t_i) \in R_i\} \in \text{Rat}$.
(4) $R \in \text{Rat} \;\wedge\; c(x_1, \ldots, x_n) \in T(F \cup \{x_1, \ldots, x_n\}) \Rightarrow \{c(s_1, \ldots, s_n), c(t_1, \ldots, t_n)\;;\; \forall i \;(s_i, t_i) \in R\} \in \text{Rat}$.

Then Steinby proves the following proposition.

Proposition. *The family* Rat *in* $T(F \cup X) \times T(F \cup X)$ *is closed by composition and inverse and contains the graph of any morphism* $T(F \cup X) \to T(F \cup X)$.

Proof: in Steinby [1983].

The family Rat is also preserved by intersection. This property is interesting in itself, but is not satisfied by ordinary transductions on words. Also the transductions thus defined have locally finite image: This notion of rationality does not extend really the corresponding notion on words.

In fact, in section II, we assumed that concatenation was to be replaced by substitution. In this setting, another modification of the way a system generates a relation is to forbid substitutions:

$$s \to t \quad \text{if} \quad s = c(g) \;\wedge\; t = c(d) \;\wedge\; (g \to d) \in R$$

This amounts to forbidding variables in the system: all terms g and d are 'ground' terms, and we have a ground term rewriting system. Supposing that letters are unary symbols, words are represented by filiform trees. What we get in this case is the notion of suffix rewriting: the relation $\{(ug, ud); u \in A^* \wedge (g \to d) \in R\}$. This notion has been studied first by Büchi [1964] who proved the rationality of the language of all words accessible from a given axiom, and by Caucal [1988] who constructs a rational transduction R such that $u R v \Leftrightarrow u \to^* v$. The transductions thus constructed are not general, since they are preserved by inverse, composition, but also iteration (star composition): They are the concatenations ΔR, where Δ is the identity of A^* and R is a recognizable subset of $A^* \times A^*$. In the case of trees, Dauchet & Tison [1985] prove a similar result. They define ground tree transducers by considering trees in which leaves may be labelled also by states, considered as constant function symbols.

Definition. *Given an ascending tree automaton with state set Q, consider the transition $q_1 \ldots q_n \xrightarrow{f} q$ as a ground rewriting rule $f(q_1, \ldots, q_n) \to q$ and let $s \to t$ be the generated relation over $T(F \cup Q)$. A ground tree transducer is the relation $\{(s,t); (\exists u)\, s \xrightarrow[Q]{*} u \;\wedge\; t \xrightarrow[Q']{*} u\}$ for two given ascending tree automata Q and Q'.*

The procedure described in this definition is reminiscent of the ΔR decomposition of the transduction representing suffix rewritings: the ascending automaton Q recognizes some subtree of s, stopping short before completing the recognition, hence leaving some prefix u untouched. Then Q' considered as a descending automaton recognizes the remaining suffixes of t that are not in u. Actually Q and Q' may be united into $Q \cup Q'$ modulo state renaming and a few ε-transitions, so that a unique automaton is good enough. These ground tree transducers have several qualities:

Theorem. *1°) The ground tree transducers contain the identity and are closed by composition, inverse and iteration. They give of a rational forest a rational image. 2°) A ground tree transducer can be associated with any ground term rewriting system R in such a way that $s \to t$ for R if and only if $s \to t$ for the transducer.*

Proof: There are several proofs of these results. For the most up to date, cf. Dauchet & Tison [1990], where they use the tree-superposition technique described above.

These ground tree transducers share with ordinary word transductions another pleasant property: accessibility is decidable (see Deruyver & Gilleron [1989]). Since ground tree transducers are preserved by iteration, this result comes from the fact that the ground tree transducers generate decidable relations. Only drawback: reduced to filiform trees, *i.e.* to words, they correspond to the recognizable transductions, of the form ΔR, and not to all rational transductions.

VI. GRAMMARS OF RELATIONS

There is still another way to define tree transductions: by a grammar generating the relation, in the same vein as for words (see example 1). This has been tried with concatenation replaced by prefixing a function symbol to two or more trees. It has not been tried using tree substitution instead. Let us define recursively relations as follows.

Example 7: Let $I(x,y)$ be defined over $T(\{b,a,c\})$, with $\alpha(b)=2$, $\alpha(a)=1$ and $\alpha(c)=0$, by

$$I = (b(x,y),b(u,v)) \text{ where } I(x,u) \wedge I(y,v) \quad \text{or} \quad (ax,ay) \text{ where } I(x,y) \quad \text{or} \quad (c,c)$$

Then I clearly is the identity on $T(\{b,a,c\})$. The definition above may be written slightly more concisely:

$$I = (b(x,y),b(u,v)), Ixu,Iyv + (ax,ay), Ixy + (c,c)$$

The binary relations have been written as binary hyperarcs: Ixy instead of $I(x,y)$, to tell them at first sight from function symbols.

This example suggests that this sort of recursive definition should characterize 'rational' relations. Unfortunately, these binary relations contain the identity and are preserved by inverse, but not by composition.

Example 8: Example 5 can be redefined in this way:

$$R = (f(x,y,z),b(b(u,v),w)), Rxu,Iyv,Izw + (b(x,y),b(u,v)), Ixu,Iyv$$
$$S = (b(x,y),b(u,v)), R'ux,Iyv$$
$$R' = (f(x,y,z),b(b(u,v),w)), R'xu,Iyv,Izw + I$$

It can be drawn as below (see next page).

It can be proved that the composition $R \circ S$ cannot be defined recursively using only binary relations. The idea of the proof is to notice that the second projection of R contains an even number of b's until the end, when a $2n+1^{st}$ symbol b is added; similarly, the first projection of R' contains an even number of b's, but the first projection of S starts with a b. Therefore, S is always one b late, or early. The composition can be defined using a ternary relation:

$$R \circ S = (f(x,y,z),b(u,v)), Txyu,Izv$$
$$T = (f(x,y,z),x',f(u,v,w)), Txyu,Izv,Ix'w + (b(x,y),z,f(u,v,w)), Ixu,Iyv,Izw$$

As soon as n-ary relations are allowed $(n>2)$, interesting problems arise. The composition of two such relations need no longer have a rational range, nor even an algebraic one.

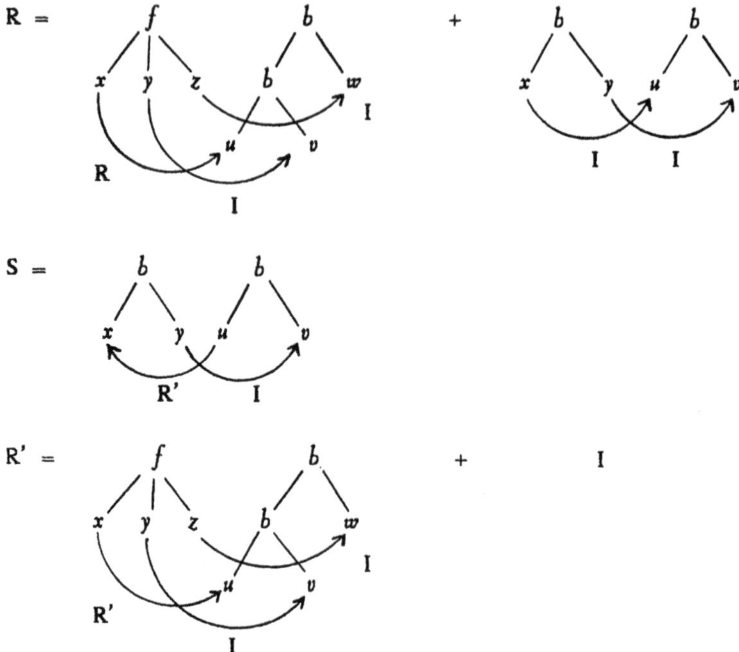

Example 9: Consider the alphabet $F = \{a,b,c,\varepsilon\}$ with arities 1, 1, 1 and 0, and relations $R \subset T(F) \times T(F)^3$ and $A \subset T(F)^3 \times T(F)$ defined below:

$$
\begin{aligned}
R &= (au, a^4 x, a^4 y, a^4 z),\ Ruxyz &+&\quad (\varepsilon, \varepsilon, \varepsilon, \varepsilon) \\
A &= (a^2 x, ay, az, au),\ Axyzu &+&\quad B \\
B &= (ax, a^2 y, az, bu),\ Bxyzu &+&\quad C \\
C &= (ax, ay, a^2 z, cu),\ Cxyzu &+&\quad (\varepsilon, \varepsilon, \varepsilon, \varepsilon)
\end{aligned}
$$

Then it can be checked that the composition $R \circ A$ is the relation $\{(a^n, a^n b^n c^n);\ n \geqslant 0\}$ where ε has not been written, to emphazise the analogy with words. Actually the example could have been written entirely in the framework of words. The image $R \circ A(a^*)$ is the language $\{a^n b^n c^n;\ n \geqslant 0\}$ which is not even algebraic!

If the recursive definitions are viewed as grammars, then the rewriting of several occurrences is synchronized. This synchronization is described by the hyperarcs like Ixy or $Ruxyz$, the last one, for instance, meaning that the productions at occurrences u, x, y and z must be synchronized and chosen among the productions of R. Understandably, too much synchronization forbids the generated language to be rational. Grazon & Raoult [1990] have designed a condition of 'desynchronization with delay' similar to the 'separation with delay' of Arnold & Dauchet [1982], which ensures that the generated language is rational in the usual sense. In this case, all the stability results of rational tree languages carry over to relations. But a condition ensuring the stability under composition remains yet to be found.

CONCLUSION

The simplicity of rational word transductions seems lost among the trees, but the analogy between words and trees is probably misleading. Tree relations should rather be compared with relations between tuples of words: $R \subseteq (A^*)^* \times (A^*)^*$, a topic on which results are not plenty. The reader also is bound to get lost in the jungle of tree transformations. In almost all directions, non trivial results have been obtained. The most general definition so far seems to be through bimorphisms (cf. Arnold & Dauchet [1982]) and linear ascending transducers (cf. Engelfriet [1975]), which are incomparable in strength. Grammars of relations seem promising, but no condition has been found yet to ensure the stability under composition. Another recent tentative defines a transduction by a formula in second order monadic logic (cf. Courcelle [1990]). These transductions contain the identity and are preserved by composition and inverse. But they have a locally finite image. There may still be discussions as to what constitutes the exact analogue of word transductions in the case of trees.

Tree Automata and Languages
M. Nivat and A. Podelski (editors)
© 1992 Elsevier Science Publishers B.V. All rights reserved.

Structural Complexity of Classes of Tree Languages

M. Dauchet and S. Tison

LIFL (URA 369-CNRS), University of Lille-Flandres-Artois, UFR IEEA,
59655 VILLENEUVE D 'ASCQ Cedex FRANCE.   e-mail: {dauchet, tison}@lifl.lifl.fr

Abstract
In a comparition framework, we study three basic classes of tree languages :
recognizable, context-free (or, algebraic) and recursively enumerable. In particular,
we investigate closure (resp; non-closure) properties with respect to elementary
transformations of tree languages : morphisms, inverse morphisms and
bimorphisms.

1. INTRODUCTION

In the case of languages of words (or, strings), the algebraic properties of
various classes have been extensively studied. The most standard classes are the
recognizable, context-free (or, algebraic) and recursively enumerable languages.
The most important tranformations of languages are morphisms, inverse
morphisms and those defined by rational transductions. In the case of trees, those
classes of languages and transformations have been investigated, too. The
straightforward generalizations is, however, not alway possible, or leads to a theory
with complex results.

The Table below summarizes the situation and the content of this paper. Some
results are well known, other ones are new. Some exercises are devoted to
complementary results, those which are difficult and marked with a star *. Our
bibliography is short, we give only typical references in which the reader can find
larger bibliographies.

In the following, we refeer to the concepts (0), (1),... in the Table bellow.

(0) Formally, in the case of finite automata, the formal definitions extend from the
word to the tree case in a straightforward fashion, preserving elementary
properties and yelding a fruitful theory. Finite automata have been extensively
studied (for a survey, see [20]), in section 2, we recall only elementary definitions
and properties .

(1)   A bimorphism B is defined by a triple $B = (f,K,g)$, where f and g are morphisms
and K a recognizable (tree) language; B defines a transformation: $B(t) = g(K \cap f^{-1}(t))$.
In the word case, the Nivat theorem identifies a-transducers with bimorphisms. This
result is crucial in theory of languages and is the basis of the study of cones and
trios. It means that the algorithmic notion of finite state automaton with output is
equivalent to the structural notion of bimorphism (we call this result in Section 4.1).
Furthermore, bimorphisms are closed under composition and inversion;
recognizability and context-freeness of languages is preserved by these
transformations.

Table: words and trees: a comparison

The property	... holds in the word case	... in the tree case
Finite automata: Homomorphic characterisation	(0) YES	(0) YES
Homomorphic characterization of sequential transducers	(1) YES bimorphisms = rational transductions	(2) NO simple correspondence. Very intricate situation
Closure properties of bimorphisms	(1) YES bimorphisms are closed under composition and inverse	(3) $B^3(L,L) \neq B^4(L,L) = B^5(L,L)$. Infinite hierarchy of $B^n(L,M)$. $B(L,M) \circ B(M,L)$ simulates any computable transformation
REC closed under bimorphisms	(1) YES (**REC**)* is the smallest class closed under bimorphisms	(4) YES *in the linear case:* **REC** is the smallest class closed under linear bimorphisms (5) NO *in the general case* (see (3))
ALG closed under bimorphism	(1) YES	(5) NO, neither in the **IO** nor the **OI** case, even in the case of linear morphisms
Parsing **ALG** with pushdown automata	(6) YES	(6) Very intricate situation
For any computable L, there are algebraic sets A, A' such that $L = h(A \cap A')$	(7) YES	(7) YES in the **IO** and in the **OI** case (A and A' homomorphic images of recognizable forests in the IO case)
Existence of "rational bijections" on rational sets	(8) YES	(8) We conjecture that YES

(2)   All these foundations of classical language theory listed under (1) do not hold in the tree case. To design linguistic transformations or compilers, a lot of tree transformations have been introduced and studied: syntax directed translators, bottom up tree transducers, top down (with or without look ahead) tree transducers... [16]. The properties of these classes are very intricate: One obtains infinite hierarchies with respect of iteration of composition.

(3) **B(X,Y)** denotes the class of bimorphisms with f of class **X** and g of class **Y**. In the tree case, **M** denotes the class of any morphisms and **L** the class of linear

morphisms (morphisms which do not duplicate subtrees; i.e., every variable in the argument tree occurs only once). Engelfriet studied connections between bimorphisms and tree transducers [16]. There are no elegant results as in the word case; the non linearity of morphisms is the main difficulty. More precisely, it is possible to describe composition of transducers namely by composition of morphisms, for the inverse image, it suffices to take on kind of morphisms : letter to letter morphisms (also called delabelling). The reason is that the inverse morphisms are associated with the inputs, and a transducer reads only one letter at each step. If transducers do not duplicate, it is sufficient to use only linear morphisms; then the corresponding hierarchies collapse. But a more surprising phenomena arises [3,11] if we consider only linear morphisms: $B^3(L,L) \neq B^4(L,L) = B^5(L,L)$. It is not possible to sketch the difficult proof in this paper, but Figure 3 illustrates the fact that, even in the linear case, unexpected and complicated phenomena appear when we generalize from words to trees.

(4) We prove that the smallest (non trivial) class closed under morphism and inverse morphism is the class of recursively enumerable sets of trees (Section 6, Corollary 4), and that the smallest (non trivial) class closed under linear morphisms and inverse linear morphisms is the class of recognizable sets of trees (Section 3, Corollary 1). These two results illustrate the gap between the linear and the general case.

(5) The natural generalization of the notion of algebraic grammar leads to a class which is closed neither under linear morphisms, nor under inverse linear morphisms (Section 5). If we derive from Outside to Inside ("call by name"), we obtain the whole class of **Alg**; if we derive from Inside to Outside ("call by value"), we get **IO**-algebraic tree languages, constituting a proper subclass of **Alg**. A nice theory of **IO** algebraic languages has been developed [17, 18, 23]. The class **IO** is not yet closed under linear bimorphisms.

(6) In the same way, there is no correspondance between algebraic tree grammars and classes of tree pushdown automata considered in [9, 21, 27]. We don't approach this study here.

(7) This is one of the rare cases where we get the same kind of result as in the word case (Theorems 6.2, 6.3).

(8) Maurer and Nivat got a nice characterization of bijective correspondence of two recognizable languages (they defined bimorphisms called bitransductions [24]). Here, the non linearity make the problem very difficult; but we prove in a particular case that if we can do a bijection between recognizable tree languages by the mean of a bimorphism, we can get a linear bimorphism [14]. We conjecture that this result is always true (see Section 4.3).

Conclusion The set of algebraic properties is confused in the tree case and very different from the word case. It seems that the situation is more difficult than in the word case: this feeling is half true and half false. The comparison above illustrates that the situation is indeed intricate in trees where it is clear in words, but a lot of new phenomena are due to the non linearity and are very deep. Many of them are easy to understand and lead to simulate "anything" (i.e. any recursively enumerable set) with classical techniques. Let us illustrate our purpose with an example: we know that we can simulate any Turing machine with only a rewrite rule [13]; so termination and halting are undecidable for a (tree) rewrite rule. We conjecture that termination is decidable in the case of a linear rule as well as in the case of a word rule (semi-Thue rule). This decidability result seems difficult to obtain in the word case, and the linear tree case should not be difficult to deduce from the word case. The techniques developed for the decision result have to be radically new.

To conclude, let us "classify" the main results of the paper.

The smallest (non trivial) class closed under linear morphisms and inverse linear morphisms is the class of recognizable sets of trees (Corollary 1 - see a preliminary version in [12]): Here, we illustrate how classical word techniques can be adapted to the linear case.

The smallest (no trivial) class closed under morphisms and inverse morphisms is the class of recursively enumerable sets of trees (Corollary 4); any recursively enumerable tree language is the homomorphic image of the intersection of two algebraic tree languages (Theorem 6.2), or the homomorphic image of the intersection of two homomorphic images of recognizable tree languages (Theorem 6.3). In this way, we exhibit the power of non linearity. The proof technique is the same as in the word case, namely the simulation of any rewrite system.

We obtain a gap in the hierarchy of bimorphisms with respect to eterate compositions : $B^3(L,L) \neq B^4(L,L) = B^5(L,L)$ (Section 4.2 and [11]).

Any bijective bitransduction of recognizable domain and image is equivalent to a linear one (Section 4.3) (proved in a particular case, conjectured in the general case).

These phenomena do not find corresponding ones in the word case. In our opinion, a general framework which could...
- illuminate the property $B^3(L,L) \neq B^4(L,L) = B^5(L,L)$
- permit a (comprehensive) proof of the conjecture "any bijective bitransduction of recognizable domain and image is equivalent to a linear one"
- provide a decision algorithm for recognizability of irreducible forms associated to a rewrite system
... would be a valuable contribution to a work leading to a deep understanding of tree structures.

2. PRELIMINARIES

2.1. Generalities

We denote $\mathbb{N}$ the set of non negative integers, $\subset$ the *strict* inclusion ($\subseteq$ means $\subset$ or $=$).

For two binary relations R and R', the *composition* of R with R' is denoted RoR' or just R R'= $\{(t,v) \mid \exists\, u,\, (t,u) \in R$ and $(u,v) \in R'\}$. The *inverse* R^{-1} of R is defined by $R^{-1} = \{(u,t) \mid (t,u) \in R\}$. We consider functions as particular relations, thus $f \circ g(t) = g(f(t))$.

Conventions Usually, "grammar" abbreviates "tree grammar", automata abbreviates "tree automata", etc... We precise only if we consider simultaneously trees and words. **Alg** (respectively **IO-Alg, OI-Alg, Rec, Reg** etc...) denotes the class of algebraic tree languages (respectively, **OI**-algebraic, **IO**-algebraic,

Recognizable, Regular,etc...) For any class **C**, **M(C)** (respectively **M⁻¹(C)**) denotes the class of homomorphic images of languages of the class **C** under morphisms of the class **M** (respectively, under inverse morphisms of the class **M**). We only consider *non trivial* classes of languages, i.e. classes which contain at least a non empty language.

2.2. Trees
Trees (or terms) over a ranked alphabet Σ

$T_\Sigma(X)$ denotes the set of *terms (i.e.trees) over some finite ranked alphabet Σ*. For simplicity, we suppose that the rank of a letter is unique, but the results of this paper are true even without this restriction. Σ_n denotes the set or letters of rank n. X denotes an infinite set of substitution variables, these variables are usually denoted x_1, x_2,..., x_n, y_1,...., y_p, x, x', x", y, z, ... A letter of rank k and of label a is denoted (a,k) or $a(x_1, x_2,..., x_k)$. A letter a of rank 2 is often denoted a(x,y), etc...

An alphabet is *monadic* if any letter is of rank 0 or 1. We can obviously identify words with monadic trees. For example, if Σ = {a,b} is a word alphabet, we define the tree alphabet $\Sigma = \Sigma_0 \cup \Sigma_1$ with Σ_0 is the special symbol # and Σ_1 = {a(x), b(x)} and identify the word abbab with the tree a(b(b(a(b(#))))), also denoted abbab. A word rule (*semi-Thue rule*) l $\rightarrow$ r is then identified with the monadic rule l(x)$\rightarrow$ r(x).

A term is *ground* if it does not contain any variable. T_Σ denotes the set of ground terms over Σ. It is the *free algebra* generated by Σ.

A term is *linear* (or *no copying*) if it does not contain more than one occurrence of each variable. To specify all the occurrences of variables of a term t, we sometimes denote $t(x_{i1}, x_{i2},..., x_{ip})$ instead of t.

Example Σ = {c, c', a(x), a'(x), b(x,y), d(x_1,x_2,x_3)} means that Σ_0 = {c,c'} (it is the set of *constants* of the alphabet); Σ_1 = {a,a'}; Σ_2 = {b}; Σ_3 = {d}.

t_1 = b(d(aa'aa(x_3), x_1, aac),a(x_3)) contains two occurrences of the variable x_3 and one occurrence of x_1 ; we denote aa'aa(x_3) instead of a(a'(a(a(x_3)))) and aac instead of a(a(c)). To specify all the occurrences of variables, we note t_1 = $t_1(x_3,x_1, x_3)$; if t_2 = b(d(aa'aaa(y),b(c',c),aac),aa(y)), we write t_2 = t_2(y,y)
The term t_3 = b(d(aa'aaac,b(c',c),aac),aac) is ground.
Remark We call variables *substitution variables* whenever we want to distinguish them from syntactical variables, which are associated with grammars in the other parts of the paper.

Context and substitution A *substitution* σ is an application from a (finite) set of syntactical variables to terms.When we apply a substitution σ to a term t, we substitute σ(x) to every occurrence of the variable x in t. We denote t.σ or σ(t), or t σ the result. A *context* is a term which contains exactly one occurrence of a variable. A tree t' is a *subtree* of a ground tree t iff there is a context u and a substitution σ such that t = u.t'.σ (it is easy to extend this notion to a non ground term t, but it is not useful here).
Example (continuation) Let σ : $\sigma(x_1)$ = b(c',c) ; $\sigma(x_3)$ = a(y) and σ'(y) = c

then $\sigma(t_1)$ = $t_1.\sigma$ = t_2 ; $\sigma'(t_2)$ = $t_2.\sigma$ = t_3 ; $\sigma'(\sigma(t_1))$ = $t_1.\sigma.\sigma'$ = t_3 . There are 4 occurrences of the tree aa(x) in t_3 ; they correspond to the three decompositions t_1 =

$u_1 . aa(x) . \sigma_1 = u_2 . aa(x) . \sigma_2 = u_3 . aa(x) . \sigma_3 = u_4 . aa(x) . \sigma_4$ with $u_1 =$ b(d(aa'a(x),b(c',c),aac),aac) ; $\sigma_1 = cu_2 = $ b(d(aa'(x),b(c',c),aac),aac); $\sigma_2 = ac$; $u_3 = $ b(d(aa'aaac,b(c',c),x),aac); $\sigma_3 = c$ $u_4 = $ b(d(aa'aaac,b(c',c),aac),x); $\sigma_4 = c$

Canonical decomposition of a term To explain the occurrences of variables in a term t, we denote $t(x_{i1}, x_{i2},..., x_{ip})$ instead of t. The *torsion* Θ_t of t is the substitution from $\{x_1, x_2,..., x_p\}$ to $\mathbb{N}$ defined by $\Theta_t(j) = ij$ $(1{\leq}j{\leq}p)$; we also denote Θ_t by $(x_{i1}, x_{i2},..., x_{ip})$. A term is *torsion-free* if Θ_t is identified, for some p, with the identity over $\{1,2,...,p\}$. If t is ground, we consider that Θ_t is the application defined on the empty set. So, the following fact obviously holds.

Canonical decomposition: For any term t, *there is one and only one torsion free term, denoted* [t], *and one and only one torsion, denoted* Θ_t, *such that* $t = [t]\Theta_t$

Example (continuation) $[t_1] = $ b(d(aa'aa(x_1), x_2, aac),a(x_3)); $\Theta_{t1} = (x_3, x_1, x_3)$

Classification of morphisms Let f be a morphism from $\mathbf{T}_\Sigma$ to $\mathbf{T}_\Delta$ and $a(x_1,...,x_n)$ a letter of Σ. Then, the following definitions hold:

$f(a(x_1,...,x_n))$ is *complete* (or no *deleting*) if $x_1,..., x_n$ occur in $f(a)$.

$f(a(x_1,...,x_n))$ is *no-collapsing* if it contains at least a letter of Δ

A morphism f is *linear* (respectively *complete, torsion-free, no-collapsing*) if, for every letter a of Σ, $f(a)$ is linear (resp. complete, torsion-free, no-collapsing).

Example Let f be a morphism from $\mathbf{T}_\Sigma$ to $\mathbf{T}_\Delta$. $f(a(x)) = $ b(c,x) is torsion-free; $f(b(x,y)) = $ c(y,b(x,c),y) is complete but not linear; $f(d(x,y)) = $ y is linear but not complete, it is collapsing.

2.3. Finite tree automata

Finite tree automata are a natural generalization of finite word automata. We only sketch definitions and summarize the main properties; the proofs of these properties are easy extensions of the word case. A tree automaton can be seen as a signature and a recognizable tree language as a sort.

Definition A *bottom-up tree automaton* M is
A finite input ranked alphabet Σ
A finite set Q of states and a subset F of final states

A finite set of rules of the form $a(q_1,...,q_n) \rightarrow q$, where a is a letter of rank n and $q_1,...,q_n$, q are states. If we identify states with constants, we get a ground rewrite system, and a tree t is *recognized* (or *accepted*) by M iff it can be reduced to a final state q. Formally, we denote t =*M=> q. A tree language L is *recognizable* if it is the set of terms accepted by some tree automaton M, that we denote L = **L**(M).

Examples

1/ The states are q, q' and qf, qf is final; the rules are a $\rightarrow$ q ; a' $\rightarrow$ q' ; b(q,q) $\rightarrow$ qf; b(q',q') $\rightarrow$ qf . The automaton recognizes the language {b(a,a), b(a',a')}

2/ The states are c, m, p, π, p is final. The rules are A $\rightarrow$ c ; B $\rightarrow$ c ; C $\rightarrow$ c ; c $\rightarrow$ m ; X $\rightarrow$ π ; *(π,π) $\rightarrow$ π ; *(c,π) $\rightarrow$ m ; m $\rightarrow$ p ; + (m,p) $\rightarrow$ p

t = +(*(A,X),*(B,*(*(X,X),*(X,X)))) is accepted by the move t =*=> +(*(c,π),*(c,*(*(π,π),*(π,π)))) =*=> +(m,*(c,*(π,π))) ==> +(m,*(c,π)) ==> +(m,m) ==> +(m,p) ==> p
*(+(A,B),X)) is not accepted, its irreducible forms are *(p,π) and *(+(p,p),π))

Basic properties [20]
> *i/We can reduce non determinism.*
> *ii/ The class **REC** of recognizable tree languages is effectively closed under boolean operations.*
> *iii/ Emptiness, finiteness are decidable for tree automata.*

3. MORPHISMS AND RECOGNIZABILITY

3.1. Proposition
Any morphism f can be decomposed in $f_1 \circ f_2$ where f_1 is letter-to-letter and f_2 torsion-   free.

Proof Let f be a morphism from T_Σ to T_Δ and $\Sigma' = \{a' \mid a \in \Sigma$, with rank(a') = number of occurrences of variables in $f(a)\}$. For any a, we define $f_1(a) = a'(\Theta_{f(a)})$, where $\Theta_{f(a)}$ denotes the torsion of $f(a)$, and $f_2(a') = [f(a)]$. Then $f_2(f_1(a)) = f_2(a'(\Theta_{f(a)})) = f_2(a')(\Theta_{f(a)}) = [f(a)](\Theta_{f(a)}) = f(a)$.
Example $f(c) = b(c,a(c)); f_1(c) = c'; f_2(c') = b(c,a(c)); f(a(x))=b(c,x); f_1(a(x))=a'(x); f_2(a'(x))=b(c,x); f(b(x,y))=c(y,b(x,c),y); f_1(b(x,y))=b'(y,x,y); f_2(b'(x,y,z))=c(x,b(y,c),z); f(d(x,y))=y; f_1(d(x,y))=d'(y); f_2(d'(x))=x$

3.2. Proposition
The class of recognizable tree languages is (effectively) closed under linear morphisms and inverse morphisms.

Proof Let R be a tree language recognized by a deterministic automaton M and let f a morphism from T_Σ to T_Δ. It suffices (proposition 3.1) to prove that recognizability is preserved by linear morphisms, inverse letter-to-letter morphism and inverse torsion-free morphism. In the three cases, we build an automaton M'. Final states of M' are those of M. Q is the set of states of M.

i/ **REC** *is closed under linear morphism*   For any letter $a(x_1,..., x_d)$, let $f(a(x_1,..., x_d)) = [f(a)](x_{i1},..., x_{ip})$. For any rule $a(q_1,..., q_d) \to q$, we introduce rules using new intermediate states such that $f(a)(q_{i1},..., q_{ip}) =*M'=> q$. For example, if $f(a(x_1,x_2,x_3)) = b(b(c,x_3),e(x_1))$ and $a(q,q', q'') \to q$, we define the rules $c \to k; b(k,q'')\to k'; e(q)\to k''$ and $b(k',k'')\to q$, where k,k' and k" are new states. It is easy to check that, M' recognizes $f(R)$.

ii/ **REC** *is closed under inverse letter-to-letter morphism* Let q_Σ be the final state of some automaton M_Σ which recognizes T_Σ. Rules of M' are those of M_Σ and those defined as follows. For any letter a of Σ, there is some letter α of Δ such that $f(a(x_1,..., x_d)) = \alpha(x_{i1},..., x_{ip})$. For every rule $\alpha(q_1,..., q_p) \to q$ of M such that there exists a substitution σ from $\{x_1,..., x_d\}$ to Q which satisfies $\propto (\sigma(x_{i1}),..., \sigma(x_{ip})) \rightarrow q$, we build the rule $a(\sigma(x_1),..., \sigma(x_d)) \rightarrow q$ where $\sigma(x_i) = q_\Sigma$ if x_i does not appear in $\propto (x_{i1},..., x_{ip})$. It is easy to check that M' recognizes $f^{-1}(R)$. Remark that our construction uses determinism of M: if two occurrences of a variable occur in $f(a)$, the same term is substituted and so the same state is reach. For example, if $f(a(x,y)) = \propto (x,x)$, we associate to the rule $\propto (q',q') \rightarrow q$ of M the rule $a(q', q_\Sigma) \rightarrow q$ of M'; but to the rule $\propto (q',q'') \rightarrow q$ of M we don't associate any rule and indeed, there is no term t such that $t =^*M \Rightarrow q'$ and $t =^*M \Rightarrow q''$ because M is deterministic.

    iii/ **REC** *is closed under inverse torsion-free morphism* For any letter a of Σ, there is some term $\propto$ of $T_\Delta(X)$ such that $f(a(x_1,...,x_d)) = \propto (x_1,..., x_d)$. For any move let a rule $\propto (q_1,..., q_d) =^*M \Rightarrow q$ we built the M' rule $a(x_1,..., x_d) \rightarrow q$. It is easy to check that, M' recognizes $f^{-1}(R)$.

3.3. Theorem

For any recognizable tree language L, there exists a ranked alphabet Σ, a projection π and two linear no collapsing morphisms φ_1 and φ_2 such that

$$L = \pi(\varphi_2^{-1}(\varphi_1(T_\Sigma))).$$

Proof Let us assume that $L \subseteq T_\Delta$ is recognized by an automaton M; Q is the set of states of M, Q_F the subset of final states. #(x) is a new symbol.

$\Sigma = \{a_{q;q1,...,qd}(x_1,...,x_d) \mid a(x_1,...,x_d) \in \Delta \ \& \ q,q1,...,qd \in Q\} \cup \{\#_q(x_1) \mid q \in Q_F\}$

$\Sigma' = \Delta \cup \{q(x) \mid q \in Q\} \cup \{\#(x)\}$

φ_1 is the morphism from T_Σ to $T_{\Sigma'}$ defined, for any a,q,q1,...,qn, by $\varphi_1(\#_q(x_1)) = \#(q(x))$; $\varphi_1(a_{q;q1,...,qd}(x_1,...,x_d)) = q(a(q1(x_1),..., qd(x_d)))$. φ_2 is the morphism from $T_{\Sigma'}$ to $T_{\Sigma'}$ defined by $\varphi_2(q(x)) = q(q(x))$; φ_2 is the identity otherwise. π is the morphism from $T_{\Sigma'}$ to T_Δ defined, for any q, by $\pi(q(x)) = x$; $\pi(\#(x)) = x$; π is the identity otherwise. We leave the reader prove that $t =^*M \Rightarrow qf$ iff $t \in \varphi_1(T_\Sigma) \cap \varphi_2(T_{\Sigma'})$. Figure 1 draws an example.

Remark and definition If Σ does not contain any constant, T_Σ is empty. To avoid this trivial case, we consider implicitly only alphabets Σ containing at least a constant. In the same way, it will be convenient to say that *a class is empty* not only if it contains no language but too if it is reduced to the empty language.

Corollary 1 *The class of recognizable tree languages is the smallest (non empty) class closed under linear morphisms and inverse morphisms.*

Proof If a class **C** is not empty, it contains a language L0 which contains a term t (see 2.1). Let Σ be a ranked alphabet (containing at least a constant); considering the morphism f such that $f(c) = t$ for every constant c and f collapsing otherwise, we get $f^{-1}(L0) = \mathbf{T}_\Sigma$. Then, we deduce from 3.3 that if a class **C** is closed under linear morphisms and inverse morphisms, it contains **REC**. We conclude using 3.2.

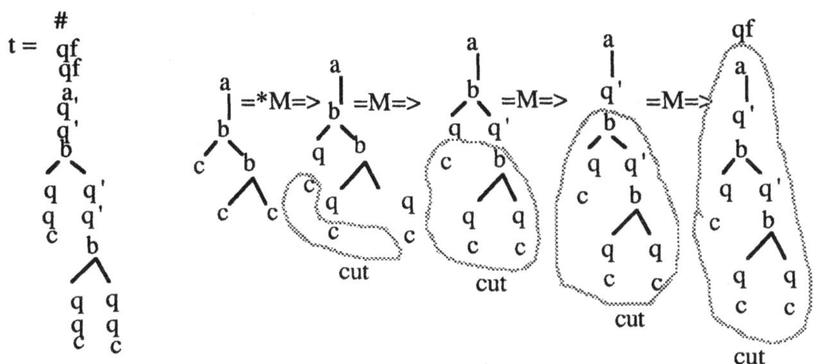

Figure 1. t= a(b(c,b(c,c))) and rules of M are c→q; b(q,q)→ q'; b(q,q')→ q'; a(q') → qf

4. BIMORPHISMS

4.1. Reminder: the word case.

Let Σ^* be the free monoid generated by some finite set Σ of letters (Σ^* is the set of words over Σ). A *transduction* τ from Σ^* to Δ^* is a part of $\Sigma^* \times \Delta^*$. Let us remind that the three following properties a equivalent

i/ *Algorithmic point of view:* τ is defined by some *a-transducer*, i.e. some finite automaton with ε-moves and output.

ii/ *Structural (or specification) point of view:* τ is defined by some *rational expression* over $(\Sigma \times \Delta)^*$. A rational expression is defined by + , . and * interpreted as alternative, sequence and iteration over some alphabet. Here letters are elements of $(\Sigma \cup \{\varepsilon\}) \times (\Delta \cup \{\varepsilon\})$.

iii/ *Algebraic (or analysis-synthesis) point of view:* [&] τ is defined by some *bimorphism*, i.e. there exists some triple (f,K,g) where f , g are morphisms and K is a recognizable language, such that $\tau = \{(f(m),g(m)) \mid m \in K\}$. We only sketch this result with an example (Figure 2): in this example, τ is computed by the a-transducer on the left-hand side of Figure 2; it is the interpretation of the rational relation $(a,\varepsilon).(ba,a)^* \cup (ab,bbb)^*$ and it is structured by the bimorphism on the right-hand side of Figure 2.

This class of transduction is called the class of *rational transductions*. We usually confuse the a-transducer or the bimorphism with the relation that it defines. The main result is

Theorem *The class of rational transduction is (effectively) closed under composition and inverse; the classes of recognizable and algebraic languages are (effectively) closed under rational transductions.*

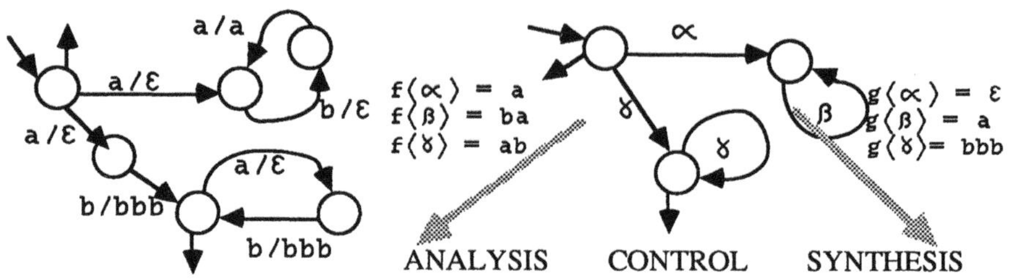

Figure 2. $\tau = \{(a(ba)^n, a^n) \mid n \in \mathbb{N}\} \cup \{((ab)^n, b^{3n}) \mid n \in \mathbb{N}\}$.

4.2. The tree case

The situation is much more intricate in the tree case. No one of the properties above holds. We focus on the homomorphic approach.

Notations A tree bimorphism is obviously defined by a triple $B = (f,K,g)$, where f and g are morphisms and K a recognizable tree language. B defines a transformation - also denoted B - as in the tree case; so, for any term t, $B(t) = g(K \cap f^{-1}(t))$; the composition is the composition of relations. $B(X,Y)$ denotes the class of bimorphisms, the first morphism of which is in class **X** and the second one is in class **Y**. **L** denotes linear, **C** complete, **S** non erasing, **M** general morphisms.

The main results are sketched in the following theorem
Theorem of bimorphisms
i/ Any computable transduction can be effectively simulated by a transformation of **B(L,M) B(M,L)**
ii/ The classes **B(L,M)**[n] *and* **B(M,L)**[n] *define an infinite hierarchies of transformations*
iii/ For any n, **B(M,L)**[n]**(REC) = REC**
iv/ For any n, **B(L,M)**[n]**(REC)** *is a class of recursive tree languages*
v/ **B(LCS,LCS)**[3] = **B(LCS,LCS)**[2] ≠ **B(LCS,LCS)**
vi/ **B(L,L)**[5] = **B(L,L)**[4] ≠ **B(L,L)**[3]
vii/ If there is no letter of rank >2, then **B(L,L)**[2] = **B(L,L)**
About the proof i/ Is an easy consequence of theorem 6.3 (see exercise 1). ii/ We omit the proof [20] . iii/, iV/Easy exercise. v/ and vi/ are very difficult to prove (see [5] for v/, [11] for vi/ and vii/).
Figure 3 illustrates why **B(L,L)**[4] ≠ **B(L,L)**[3]. We define **B1, B2, B3, B4** as following and leave the proof to the reader that **B1 B2 B3 B4** ∉ **B(L,L)**[3]. For i=1,2,3,4 **Bi** = $(\varphi i, Fi, \phi i)$ with F1 defined by the automaton $c \to q$; $a(q) \to q$; $b(q,q)$

→ q' ; a'(q') → q' ; b(q',q') → q" ; b(q",q') → q" ; q" final. φ1 is the identity; φ1 is the
identity except that φ1(a'(x)) = x. F2 is defined by the automaton c → q ; a(q) → q ;
b(q,q) → q' ; d(q',q,q) → q' ; q' final. φ2 is the identity except that φ2(d(x,y,z)) =
b(x,b(y,z)) ; φ2 is the identity except that φ2(d(x,y,z)) = b(b(x,y),z). F3 is defined by
the automaton c → q ; a(q) → q ; d(q,q,q) → q' ;   d(q',q,q) → q'; b(q',q) → q" ; q"
final. φ3 = φ2 ; φ3 = φ2. F4 is defined by the automaton c → q ; a(q) → q ; b(q,q) →
q' ; a'(q') → q' ; b(q,q') → q" ; b(q",q') → q" ; b(q",q) → q''' ; q''' final. φ4 = φ1 ; φ4 is
the identity.

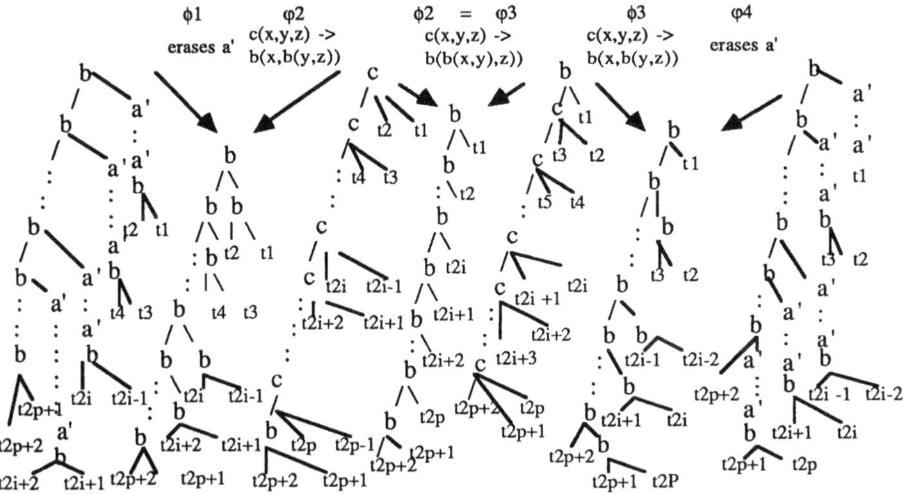

Figure 3. A transformation in **B(L,L)⁴ - B(L,L)³**.

Exercise 1: We say that a transduction T is computable if {\$(t,u) | (t,u) ∈ T} is recursively
enumerable. Use theorem 6.3 to prove i/.
Exercise 2: Prove iii/ and iv/.
Exercise 3: Prove vii/.

4.3. Recognizability and non linearity

Homomorphic image of a recognizable tree language is generally no
recognizable, except if the morphism is linear. The following lemma means that if
the homomorphic image of a recognizable tree language is recognizable, we can
in some sense suppose that the morphism is linear. Unfortunately, the lemma
actually holds only for a particular class of morphisms, the quasi-alphabetic
morphisms. In spit of its "academism", the generalisation of this lemma would be a
great improvement of knowledge about the structure of tree sets. We give as
exercises some consequences of this result.

Definition A morphism *f* is *quasi-alphabetic* if, for any letter a, *f*(a) is a
substitution variable or every occurrence of variable in *f*(a) is a son of the root.
Lemma of recognizability and duplication

For any quasi-alphabetic morphism f and any recognizable tree language K such that f(K) is recognizable, we can design a recognizable tree language R and a linear, complete, non erasing and quasi-alphabetic morphism π such that $\pi \circ f$ is linear, π(R) is included in K and $f(\pi(R)) = f(K)$.
We omit the proof [14]. It is very technical and tedious.

Example $K = K' \cup \{a'' \; a^n c \mid n \geq 0\}$ with $K' = \{b(a^{3n}a'^{2p}c, a^{3n'}a'^{2p'}c) \mid n,n',p,p' \geq 1\}$;
$f(b(x,y)) = b(x,y)$; $f(a''(x)) = b(x,x)$; $f(a'(x)) = f(a(x)) = a(x)$; $f(c) = c$; $R = K' \cup \{d,d'\}$;
$\pi = f$ over $\{b(x,y), a(x), a'(x), c\}$; $\pi(d) = b(c,c)$; $\pi(d') = b(ac,ac)$.

¿**Conjecture?** *For any morphism f and any recognizable tree language K such that f(K) is recognizable, we can design a recognizable tree language R and a linear, complete, non erasing morphism π such that $\pi \circ f$ is linear, π(R) is included in K and $f(\pi(R)) = f(K)$.*

¿**Conjecture?** *We can decide if the homomorphic image of a recognizable tree language is recognizable.*

Exercise 4*: *The cross-section theorem for trees.*
i/ Prove that for any linear morphism f and recognizable tree language K, there exists some recognizable tree language K' included in K such f is a bijection between K' and f(K).
ii/ Prove that i/ does not hold for $K = \{b(a^n c, a^{n'}c) \mid n,n' \geq 0\} \cup \{b'(a^n c, a^{n'}c) \mid n,n' \geq 0\}$ and
$f(b(x,y)) = d(x,x,y)$; $f(b'(x,y)) = d(x,y,y)$; $f(a(x)) = a(x)$; $f(c) = c$
Exercise 5: *Applications of the lemma of recognizability and duplication.*
i/ Prove the following generalization of the cross section theorem: " for any quasi-alphabetic morphism f and tree language K such that f(K) is recognizable, there exists some recognizable tree language K' included in K such f is a bijection between K' and f(K)".
ii/ Prove that if a quasi-alphabetic bimorphism is a bijection from T_Σ to T_Δ, then we can design an equivalent linear bimorphism.
iii/ Prove that if the domain and the image of a bijective quasi-alphabetic bimorphism is recognizable, then we can design an equivalent linear bimorphism.
Exercise 6: Prove that if we can decide if the homomorphic image of a recognizable tree language is recognizable, then we can decide if the set of terms which are irreducible for a rewrite system is recognizable.

5. TREE GRAMMARS

5.1. General definitions

Syntax *A tree grammar (or type 0-tree grammar)* G is a 4-uple (V, T, P, S) where V and T are finite sets of *grammatical variables* and *terminals,* respectively. We assume that V and T are disjoint. P is a finite set of *rules*; each rule is of the form $l \rightarrow r$, l and r are terms of $T_{V \cup T}(X)$, there is no substitution variable which occurs in r and not in l, at least one syntactical variable occurs in l.   l is called the *left hand-side* and r is the *right hand-side* of the rule. The *rank* of a rule is the number of variables which occur in this rule (*not* the number of occurrences). A is a special variable called the *axiom*, the rank of A is zero.

Semantic Let G be a tree grammar. We write t =G=> t' (or just t ==> t' if G is understood), and we say that " *t generates t' in one step*" or "*t is rewritten in t' in one step*", iff

t and t' are ground & there exists a rule $l \rightarrow r$ of G

& there exists a context u and a substitution σ such that $t = u.l.\sigma$ and $t'= u.r.\sigma$

u is called the *Out-context of the derivation (or rewriting) step;* σ is called the *In-context of the derivation (or rewriting) step* . $=^*=>$ denotes the reflexive and transitive closure of $==>$

$L(G, t)$ denotes $\{ u \mid t =^*G=> u$ and $u \in T_T \}$. $L(G) = L(G, A)$ is the language generated by the grammar G. If $L(G) = L$, we say that the grammar G generates the language L. G and G' are *equivalent* iff $L(G) = L(G')$.

Usual classification A rule *collapses* or is *an X-rule* iff its right hand-side is a substitution variable. A *rule is left-linear* (respectively *right linear, linear*) if its left hand-side (respectively right hand-side, both sides) is left-linear. The same definition hold for *complete* and *torsion-free*. A *grammar collapses* iff it contains <u>a</u> collapsing rule; a grammar is *left-linear*, etc... iff <u>every</u> rule is left-linear, etc... A rule or a grammar is *monadic* if it uses a monadic alphabet. A grammar is *algebraic* (or context-free) iff every left hand-side is a syntactical variable (a syntactical variable is implicitly torsion free). *An algebraic grammar is reduced* if, for every syntactical variable B, $L(G,B)$ is not empty and there is some t such that $A =^*=> t$, B occurs in t and the nodes of the branch leading to this B are terminals (A is the axiom). Note that *an algebraic grammar is always left-linear.* A grammar is *regular* iff every syntactical variable is of rank 0. In a regular grammar, if the right hand-side of a rule is reduced to a syntactical variable, the rule is called a *V-rule*; if the right hand-side of a rule is a terminal or of the form a(A1,...,An), where A1,...,An are syntactical variables, the rule is normal. A *regular grammar is in normal form* if it is reduced and all its rules are in normal form. A tree language L is algebraic (respectively regular), if there is some algebraic grammar G (respectively regular grammar) such that $L(G) = L$.

Usual notation The set of rules and the set of terminals are implicitly defined. Usually, A denotes the axiom. Capital letters denote syntactical variables. As usually, | denotes "or", i.e. $A(x) \rightarrow A(s(x)) \mid S(x)$ denotes the two rules $A(x) \rightarrow A(s(x))$ and $A(x) \rightarrow S(x)$.

In the algebraic and regular cases, syntactical variables are implicitly symbols which occur on the left hand-sides.

Exercise 7: Let the algebraic grammar $A \rightarrow B(C,a)$ $B(x,y) \rightarrow y$. Is it reduced? Criticize and justify the definition of reduce. Design an algorithm which reduce any algebraic grammar (see&).

Examples

(G1) $V = \{A, S(x), +(x,y) \}$; $A \rightarrow A(0)$; $A(x) \rightarrow A(s(x)) \mid S(x)$; $S(0) \rightarrow 0$

   $S(s(x)) \rightarrow s(+(S(x),+(x,x)))$; $+(s(x),y) \rightarrow s(+(x,y))$; $+(0,x) \rightarrow x$

(G2) $A \rightarrow B(S)$ $S \rightarrow s(S) \mid 0$ $B(x) \rightarrow b(x,x)$

(G3)   P is the axiom $P \rightarrow +(M,P) \mid M$ $M \rightarrow *(M, X) \mid I$ $;X \rightarrow x$; $I \rightarrow s(I) \mid 0$

(G4)   H is the axiom $H \rightarrow +(*(H, x),I) \mid I$; $I \rightarrow s(I) \mid 0$

G2 is algebraic, G3 and G4 are regular, G3 is normal. G1 collapses, G1 and G2 are X-complete but not linear. The rank of the rule $S(s(x)) \rightarrow s(+(S(x),+(x,x)))$ is 1. $\Diamond$

The following result is classical and very easy to check, it states that we can associate "in real time" a regular grammar which generates the tree language recognized by some finite automaton, and conversely. It is a trivial adaptation of the word case.

Theorem [20]*The notion of regular tree grammar and the notion of recognizable tree language are effectively equivalent. There is a linear time algorithm which associates with any regular tree grammar* G *a finite automaton* M *such that* $L(M) = L(G)$, *and conversely.*

Exercise 8: i/ Simplify the definition of "reduce" in the regular case.
ii/ Design an efficient normalization algorithm of regular grammars, apply it to G4.
Exercise 9: *Algebraic word grammars and regular tree grammars.*
Let A → aAbAc | d an algebraic word grammar. The regular (tree) language generated by the grammar A → A(a,A,b, A,c) | d is the set of derivation trees of the algebraic word grammar above.
i/ Generalize that construction and associate to any algebraic word grammar G without ε-rule, a normal regular tree grammar R such that L(R) is the set of derivation trees of G (ε denotes the empty word, an ε-rule is a rule such that A → ε).
ii/ The *yield* $\Phi(t)$ *(or frontier)* of a tree t is the word inductively defined as usually: $\Phi(b(t1,...,tn)) = \Phi(t1)... \Phi(tn)$ and, for every $a \in \Sigma_0 \cup X$, $\Phi(a) = a$. Prove that the frontier of a regular tree language is an algebraic word language.
iii/ Check that the language { c(ba,ba') } is not the set of derivation trees of any algebraic word grammar.
Exercise 10: *Algebraic word grammars and regular tree grammars* (continuation).
i/ Design a no regular tree language L such that $\Phi(L)$ is a regular word language.
ii/ Let Σ be a given ranked alphabet. Prove that, for any regular word language L of Σ_0^*, $\Phi^{-1}(L)$ is a regular tree language. Using the fact that the class of regular tree languages is closed under intersection, deduce the classical result "the intersection of an algebraic word language and a regular one is algebraic".
Exercise 11: *Branch languages* We can consider the sequence (from the root to the frontier) of the labels of nodes of any branch as a word. So, we associate a word language with any tree and to any tree language; we call this language the branch language B(F) of the tree language F. prove that B(F) is regular (respectively algebraic) if F is regular (respectively algebraic).

5.2. Grammars and rewrite systems

If we don't distinguish syntactical variables and terminals and if we start from any tree, the set of rules of any grammar defines a *rewrite system*. For example, the two rules $\cdot(\cdot(K, x), y) \to x$ and $\cdot(\cdot(\cdot(S, x), y), z) \to \cdot(\cdot(x, z), \cdot(y, z))$ of combinatory logic compute the result of any programme, the programme being the given tree. *A system is a finite set of rules.*

Termination and halting problems for one rule

Let us recall that the *halting problem* is decidable for some class R of systems if there is an algorithm which decides, for any term t and any system of R, if any derivation starting from t halts. The *termination problem* is decidable for some class R if there is an algorithm which decides if for any system of R, every derivation halts (so, termination means halting for any term).
Example A regular system S (i.e. the set of rules of some regular grammar) terminates iff, for every variable A, $L(S,A)$ is finite; S halts for some t iff no variable A such that $L(S,A)$ is infinite, appears in t. So halting and termination are decidable for the class of regular systems.

Termination and halting are generally undecidable because rewrite systems are a paradigm for general computation. Logical combinators prove that halting is undecidable for two rules. We sketch in the exercise 12 a construction which proves that termination and halting are undecidable only for one left linear rule. The following problems remain open. These problems are interesting because it seems that they are decidable, and that the proofs require new kind of pumping lemmas.

¿Open problems? *Is halting decidable for a monadic rule (i.e. in the word case)?*
Is termination decidable for a monadic rule ?
Is halting decidable for a (tree) linear rule ?
Is termination decidable for a (tree) linear rule ?

Exercise 12*: *The power of one rule.* We associate with every Turing machine M a rule R_M which "simulates" M. Intuitively, this result means that a single rewrite rule is as powerful as a Turing machine. Furthermore, this rule is left-linear, non-overlapping and complete. Particularly, the rule associated with an Universal Turing machine simulates any computations.

Let M be a *Turing machine* with a single tape. Let $Q = \{q_1, q_2, \dots q_n\}$ be the set of *states*. Letters of the *tape alphabet* are $\{\#L, a_1, a_2, \dots, a_p, \#R\}$. Without lack of generality, an *instantaneous description* ID of M is of the form $(c_m \dots c_1 \#L\$, q, a, d_1 \dots d_n \#R\$)$ where, $\#L$, $\#R$ and $\$$ are special tape symbols. This ID means that M is in state q, the head scans the symbol a, the non-blank left portion of tape is $\$\#Lc_1 \dots c_m$ (from the left end to the symbol preceding the scanhead) and right-portion is $d_1 \dots d_n \#R\$$ (from the symbol following the scanhead). An empty non-blank portion is denoted by NIL. (q, a, b, q', L) denotes a *left-moving M-instruction* ; it means "if M is in state q reading the symbol a, then replace on the tape a by the symbol b, move left and go into state q' ". Right-moving instructions are defined the same way. To increase the non-blank portion of a tape, there are *special-left moving instructions* of the form (q, #L, #Lb, q', L). They mean "if M is in state q reading the symbol #L, then insert the symbol b, move left to position the head on #L again, and go into state q' ". Special-right moving instructions are defined the same way.

The rule R_M is $\mathrm{Left}(x_h, x_t, z_s, z_c, y_h, y_t) \rightarrow \mathrm{Right}(x_h, x_t, z_s, z_c, y_h, y_t)$ defined as follows:

$\mathrm{Left}(_h, x_t, z_s, z_c, y_h, y_t) = A(Q(T(^*(x_h, x_t), z_s, z_c, ^*(y_h, y_t)), q_1, q_2, \dots q_n), a_1, a_2, \dots, a_p)$

$\mathrm{Right}(x_h, x_t, z_s, z_c, y_h, y_t) = V(I(t_I), \dots, J(t_J), \dots, K(t_K))$

where $A, Q, T, V, ^*, I, \dots, K$ are new letters; $I, \dots, J, \dots, K$ are monadic symbols associated with the names of the instructions of M; for any instruction J, t_J is defined as follows

If J is a right-moving instruction (q_i, a_j, b, q_k, R), then $t_J = A(Q(T(^*(b, ^*(x_h, x_t)), q_k, y_h, y_t), R_1, R_2, \dots R_n), B_1, B_2, \dots, B_p)$ with $B_j = z_c$ and $B_{j'} = a_{j'}$ if $j' \neq j$; $R_i = z_s$ and $R_{i'} = q_{i'}$ if $i' \neq i$. If J is a left-moving instruction (q_i, a_j, b, q_k, L), then $t_J = A(Q(T(x_t, q_k, x_h, ^*(b, ^*(y_h, y_t))), R_1, R_2, \dots R_n), B_1, B_2, \dots, B_p)$ with $B_j = z_c$ and $B_{j'} = a_{j'}$ if $j' \neq j$; $R_i = z_s$ and $R_{i'} = q_{i'}$ if $i' \neq i$. If J is a special right-moving instruction $(q_i, \#R, b\#R, q_k, R)$, then $t_J = A(Q(T(^*(b, ^*(x_h, x_t)), q_k, y_h, ^*(\$, NIL)), R_1, R_2, \dots R_n), B_1, B_2, \dots, B_p)$ with $B_p = z_c$ and $B_{j'} = a_{j'}$ if $j' \neq p$; $R_i = z_s$ and $R_{i'} = q_{i'}$ if $i' \neq i$. If J is a special left-moving instruction $(q_i, \#L, b\#L, q_k, L)$, then $t_J = A(Q(T(^*(\$, NIL), q_k, x_h, ^*(b, ^*(y_h, y_t))), R_1, R_2, \dots R_n), B_1, B_2, \dots, B_p)$ with $B_1 = z_c$ and $B_{j'} = a_{j'}$ if $j' \neq 1$; $R_i = z_s$ and $R_{i'} = q_{i'}$ if $i' \neq i$

Precise how R_M "simulates" M. Prove that halting and termination are undecidable for a rule.

5.3. The main classes of algebraic grammars

IO- and OI- grammars

Controls on the derivation are introduced to describe strategies of evaluation. An algebraic grammar is **IO-** (Innermost Outermost) if innermost variables are derived first; that control corresponds to a call by value evaluation. An algebraic grammar is **OI-** (Outermost Innermost) if outermost variables are derived first; that control corresponds to a call by name evaluation.

Example Let the grammar (G2) defined above. $L(G2) = \{b(s^n 0, s^{n'} 0) \mid n, n' \in \mathbb{N} \}$.

An **IO**-derivation is necessarily of the form $A ==> B(S) =n=> B(s^n(S)) ==> B(s^n 0) ==> b(s^n 0, s^n 0)$. An **OI**-derivation is necessarily of the form $A ==> B(S) ==> b(S, S)$

$=^* =>$ b(s^n0, $s^{n'}0$). So, we ge **L(IO-G2)** = {b(s^n0,s^n0) | $n \in \mathbb{N}$} and **L(OI-G2)** = {b(s^n0,$s^{n'}0$) | n, n' $\in \mathbb{N}$ }. For this example, **L(IO-G)** $\subset$ **L(OI-G)** = **L(G)**. The following theorem precises this observation.

Theorem i/ *For any algebraic grammar* G, **L(IO-G)** $\subseteq$ **L(OI-G)** = **L(G)**.

ii/ *If G is linear and reduced, then* **L(IO-G)** = **L(OI-G)** = **L(G)**.

Sketch of proof   i/ Let a current derivation and suppose that some variable B occurs on the path from the root to an other occurrence of some variable C. Let us consider the step ∂ deriving B and the step ∂' deriving C: we don't modify the term generated if we do ∂ before and ∂' after (or duplications of ∂' if ∂ applies a non linear rule).

ii/ G2 proves that the equalities are false if g is not linear. The grammar (A $\rightarrow$ B(C) ; B(x) $\rightarrow$ 0 ; C $\rightarrow$ C) proves that they are false too if G is not reduced. If G is linear and reduced, a dual reasoning of i/ is now valid. $\Diamond$

We know that the homomorphic image of a regular tree language (i.e. a recognizable tree language) is generally not regular if the morphism is not linear. We give in the exercise 15 an example which proves that this homomorphic image is usually nor algebraic. But the following property states that **IO**-languages are closed under any morphisms.

Proposition   *The class of* **IO**-*algebraic languages is effectively closed under morphisms.*

Construction   There is not any trap! Let G be an **IO**-grammar of terminal alphabet Σ and φ a morphism of source alphabet Σ and image alphabet Γ. We can suppose without loss of generality that $\Sigma \cap \Gamma$ = $\varnothing$. We define an **IO**-grammar G_φ such that **L(G_φ)** = φ(**L(G)**) as follows : Γ is the terminal alphabet of G_φ ;

{rules of G_φ } = {rules of G} $\cup$ {a(x_1,...,x_n) $\rightarrow$ φ (a(x_1,...,x_n)) | a $\in \Sigma$}

Corollary 2   *Homomorphic images of regular languages are* **IO**.

Exercise 13:   Prove the construction of the proposition . Precise the complexity of the algorithm.

Exercise 14:   Prove that **Alg** is closed under union, algebraic substitution, intersection with **Reg**.

Exercise 15: i/ Prove that **IO-Alg** $\cap$ **Reg** = **IO-Alg** (Hint: adapt the construction used in the word case). ii/ Prove that **IO-Alg** is not closed under inverse diadic alphabetic linear morphism. (Hint: consider the **IO-Alg** language {b(a^nca^n0, a^nca^n0) | $n \in \mathbb{N}$ }, the recogn. lang. {b($a^{'n}ca^{'n'}0$, $a^mca^{m'}0$) | m,m',n,n'$\in \mathbb{N}$ } and the morphism φ : φ (a(x)) = φ (a'(x)) = a(x) φ is the identity on b(x,y), c(x) and 0.)

Exercise 16:   i/ Prove that emptiness and finiteness of the lang. generated by an alg. tree grammar are decidable . ii/ Prove that any alg. tree language is recursive. Study the complexity of the analysis.

Exercise 17*: i/ Prove that it is possible to suppress collapsing rules for linear algebraic grammars (i.e. that it is possible to design an equivalent linear no collapsing algebraic grammar). ii/ Prove that it is possible to suppress collapsing rules for **IO-Alg**. iii/ Prove that it is not possible to get a non collapsing algebraic grammar equivalent to the following one:

$A \rightarrow X(0,\partial(0,0))$; $X(x,y) \rightarrow X(a(x),Y(a(x),y)) \mid Z(d(x,y))$; $Z(x) \rightarrow b(x,x) \mid Z(b(x,x))$; $Y(x,y) \rightarrow \partial(x,x) \mid y$

Exercise 18: *Semi-Greibach Form* The preceding exercise proves that we can't get in the tree case the analogous of the Greibach form of algebraic word languages. Nevertheless, prove that any algebraic tree grammar can be convert to a semi-Greibach grammar defined as following: A Greibach normal form contains only rules which ollaps or are of the form $X(x_1,...,x_n) \rightarrow a(x_{i_1},...,x_{i_p})$ for some terminal a, or $X(x_1,...,x_n) \rightarrow X'(x_1,...,x_n)$ for some variable X', or $X(x_1,...,x_n) \rightarrow X(Y1(x_1,...,x_n),..., Ym(x_1,...,x_n))$ for some variables $Y1(x_1,...,x_n),..., Ym(x_1,...,x_n)$

Exercise 19: *Algebraic grammars and morphisms* Let G be an algebraic tree grammar and φ a morphism. i/ Prove that if φ is linear then $\varphi(L(G))$ is algebraic. ii/* Prove that if φ is linear and Σ does not contain any letter of rank >2, then $\varphi^{-1}(L(G))$ is algebraic. iii/* Let φ and G defined as following

$\varphi(c(x,y,z))=b'(b''(x,y),z)$; $\varphi(b(x,y)) = b(x,y)$; $\varphi(a(x)) = a(x)$; $\varphi(\#) = \#$; $A \rightarrow B(b''(\#,\#)); B(x,y) \rightarrow B(S(x,y),a(y))$

$\mid D(b'(x,y))$; $D(x) \rightarrow D(b(x,x)) \mid b(x,x); S(x,y) \rightarrow b''(y,y) \mid x$. Prove that $\varphi^{-1}(L(G))$ is not alg.

iv/* Let Σ be some nonmonadic ranked alphabet. Prove that { $b(t,t) \mid t \in T_\Sigma$ } is not algebraic (see[2].

Exercise 20: *Coregular grammars* An alg. grammar is co-regular iff any right hand-side contains no syntactical variable or only one at the root. i/ Prove that T_Σ is not coregular. ii/ Prove that the class of coregular lang. is closed under morphisms. iii/ If you know EDTOL systems, compare with coregular.

5.4. Rational transformations

In the word case, the classes of recognizable and algebraic languages are obviously closed under the three rational operations . (concatenation), * (iteration) and + (union). In the tree case, it is necessary to introduce sorted concatenations to get a like Kleene theorem; the sorts are auxiliary constants of the ranked alphabet. Then, the fact that **REC** and **Alg** are closed under rational operations is obvious but less interesting than in the word case. In the **IO** case, it is necessary to define an **IO** concatenation (substitute the *same* term to each occurrence of a constant).

5.5. Deterministic grammars, programme schemes and infinite algebraic trees

These algebraic grammars are the main tool of algebraic semantic. We only sketch the notion. For the simplicity - but without essential restriction - we suppose that the root of any right hand-side is a terminal. An algebraic grammar is *deterministic* if, for any variable, there is only one right hand-side. In the classical sense, such a grammar generates only zero or one tree; let us see how we can always associate an infinite tree with this grammar.

The syntactical distance Let us denote $D(t,t')$ the greatest integer such that t and t' are the same up to the depth $D(t,t')$ ($D(t,t') = \infty$ if t = t'). More precisely,

$D(a(t_1,...,t_n), a'(t'_1,...,t'_{n'})) =$ if $a \neq a'$ then 0 else $1+\inf\{\infty, D(t_i,t'_i) \mid 1 \leq i \leq n\}$

$\Delta(t,t') = 1/(1+D(t,t'))$ defines a metric on T_Σ . Let us consider the set T^∞_Σ of finite and infinite trees on Σ. The above definition of Δ is available for T^∞_Σ and T^∞_Σ is complete. Then, it is easy to check that by fair infinite derivations we associate one and only one infinite tree $P(G)$ with any deterministic grammar G.

Example (S) $A \rightarrow$ "FAC"(F(X)); $F(x) \rightarrow$ if(x,*(x,F(a(x))))

(S') $A \rightarrow$ "FAC"(G(X)); $G(x) \rightarrow$ if(x, H(x)); $H(x) \rightarrow$ *(x,F(a(x)))

$A =S=>$ "FAC"(F(X)) $=S=>$ "FAC"(if(X,*(X,F(a(X))))) $=S=>$ "FAC"(if(X,*(X,F(a(X)))))
$=S=>$"FAC"(if(X,*(X, if(a(X),*(a(X),F(a(a(X))))))))
$=\infty S=>$ $P(S) =$ "FAC"(if(X,*(X, if(a(X),*(a(X),if(a(a(X)),*(a(a(X))),...)...)

Roughly speaking, such grammars are called programme schemes, two programme schemes are equivalent iff they compute the same function for any interpretation of the symbols. In other words, **P**(S) = **P**(S') iff the symbolic programmes S and S' are equivalent. An important problem arises: is the equivalence of programme schemes equivalent ? This problem is interreductible with the famous problem of equivalence of deterministic pushdown word automata. So, we can state

¿Open problem? *Design a decision algorithm for the equivalence of deterministic algebraic tree grammars.*

6. ALGEBRAIC CHARACTERIZATIONS OF RECURSIVELY ENUMERABLE TREE LANGUAGES

The following result states that any recursively enumerable tree language can be generated by some grammar. It is an easy extension of the word case, but the fact that a special kind of tree grammars suffices will simplify further results.

6.1 Proposition
Any recursively enumerable tree language can be generated by a grammar.
Furthermore, this grammar can be chosen torsion-free and such that only monadic rules collapse.
Sketch of proof
Let Σ be a ranked alphabet (let us recall that we agree that we can identify letters and their labels). Let n the greatest rank of letters of Σ. We associate to Σ the word alphabet $\bar\Sigma = \bar\Sigma_(\cup \bar\Sigma_)$ with $\bar\Sigma_(= \{a'(x) \mid a \in \Sigma\}$ and $\bar\Sigma_) = \{a''(x) \mid a \in \Sigma\}$. For any tree t, let us denote $\bar t$ the word which codes t using a special bracket notation that we define as follows: Let $t = a(t1,...,tn)$, with $a \in \Sigma_n$; then, $\bar t = a'\bar{t1}...\bar{tn}a''$. $\mathbb{F}$ denotes $\{\bar t \mid t \in F\}$. Obviously, if F is recursively enumerable, so is $\mathbb{F}$ too. The idea of the construction is the following

a/ First of all, we generate the word language $\mathbb{F}$. It is well known that it is possible, using a word grammar (which simulates a Turing machine). We can identify this word grammar to a monadic tree grammar and the word $\bar t$ to the monadic tree $\bar t\#$. Let a'' the last letter of $\bar t$, for technical reasons, we consider a# as a single constant $a_\#$.

b/ Secondly, we define special rules U which translate $\bar t\#$ into t "as we open an umbrella". The tree "grows from the leaves to the root" . We give the construction for $\Sigma = \{b(x,y), a(x), o\}$, it can be easily extended. We use new ranked letters #(x) and #(x,y). Rules U are $o'(o''(x)) \to \#(o,x)$; $o'(o''_\#) \to o$; $a'(\#(x,a''(y))) \to \#(a(x),y)$; $a'(\#(x,a''_\#)) \to a(x)$; $b'(\#(x,\#(y,b''(z)))) \to \#(b(x,y),z)$; $b'(\#(x,\#(y,b''_\#))) \to b(x,y)$. The theorem is a consequence of the following lemma (we omit the end of the proof).
Lemma of umbrella (see Figure 3) *For any* $\bar t\#$, $(\bar t\# =^*U\!\!=> u$ *and* $u \in T_\Sigma$) *iff* $t = u$

Exercise 21: Complete the proof of the th. for the same alphabet. Build U for any ranked alphabet.

In the word case, it is well known - by simulation of a Turing machine - that, for any recursively enumerable language L, we can design two (linear) algebraic languages A and A' and an alphabetic morphism π such that $L = \pi(A \cap A')$. We will get the same result in the **OI**-(i.e. standard-) case (theorem 6.2.) and the **IO**-case (corollary of theorem 6.3.). The fact that we can't generate any duplication in the **OI**-case [2] makes the proof more complicated in this case.

6.2. Theorem *For any recursively enumerable tree language* L, *there exists a projection* π *and two algebraic tree languages* A *and* A' *such that* $L = \pi(A \cap A')$

Proof Section 6.2. is devoted to the proof.

Let G such that $\mathbf{L}(G) = L$; we can suppose that G is torsion-free and such that only some monadic rules collapse (proposition 6.1). We will design two algebraic grammars H and H' and a projection π such that $L = \pi(\mathbf{L}(H) \cap \mathbf{L}(H'))$.

The *mirror* of a word $w = a_1...a_n$ is the word $w^r = a_n...a_1$

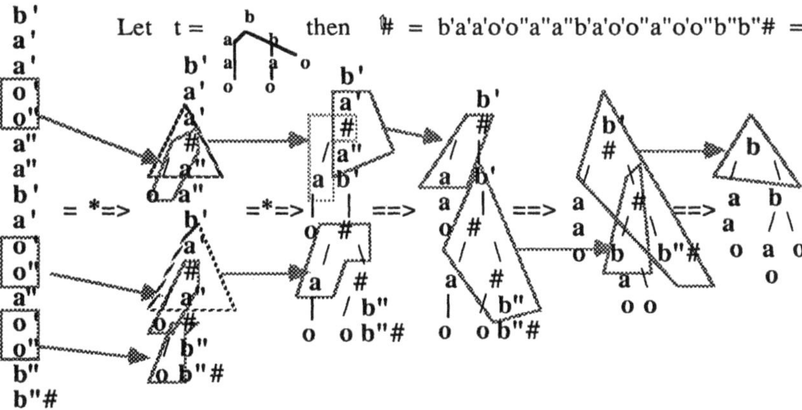

Figure 4. Illustration of the lemma of umbrella.

Notations and description of G. The axiom is σ, the variable alphabet is V, the terminal alphabet Σ. The rules are $\rho_1,..., \rho_g$ with $\rho_i: l_i \to r_i$ and $rank(\rho_i) = d_i$. Greek characters denote variables of G, capital letters denote variables of H and H'. As a projection acts on Σ, we can suppose that two different letters of Σ get different label, and so identify letters and labels. More precisely, we associate a monadic letter $\underline{a}(x)$ to any letter of $V \cup \Sigma$, and the word - i.e. monadic tree - $\underline{t}$ denotes the prefix coding of any tree t. For example, if $t = b(\beta(a,a), \alpha(a'))$ then $\underline{t}(x) = \underline{b}\underline{\beta}\underline{a}\underline{a}\underline{\alpha}\underline{a}'(x).\underline{V}$ and $\underline{\Sigma}$ denote the sets of underlined letters of V and Σ.

Definition and properties of H. (we can suppose that the variable alphabets are disjoint for G, H and H'). Axiom: A. Variables: those of the rules H_i defined below $\cup$ copy of(V) $\cup \underline{V} \cup \{A, B(x), C, C'(x),T, T_f, X(x)\}$. Terminals: $\Sigma \cup \underline{\Sigma} \cup \{\neq(x,y), \neq'(x), \$(x), \#(x)\}$.

First property   For every rule ρ_i of G, we can design a set of rules H_i, using special
  variables indexed with i, such that $R_i(x) =^* H_i => t(x)$ with $t \in \mathbf{T}^\circ_{(\underline{\Sigma} \cup \underline{V} \cup \{\neq(x,y),T\})}(x)$
  iff $\exists\ u = \rho_i => u'$ such that $t(x) = \underline{u}^r\ (\neq(\underline{u}(T),\underline{u}'(x))$.

  Proof of the first property (For the simplicity, we omit the indices: we consider
the rule $\rho: l \to r$ of rank d and define the set H of rules)

1/ There exists a finite family $\underline{l}_0(x),..., \underline{l}_d(x), \underline{r}_0(x),..., \underline{r}_d(x)$ associated with ρ such that
$u = \rho => u'$ iff $\exists\ t_0(x) \in \mathbf{T}^\circ_{(\underline{\Sigma} \cup \underline{V})}$, $t_1(x),..., t_d(x) \in \mathbf{T}_{(\underline{\Sigma} \cup \underline{V})}$, such that $\underline{u}(x) = \underline{t}_0\underline{l}_0\underline{t}_1...\underline{l}_{d-1}\underline{t}_d\underline{l}_d(x)$ and $\underline{u}'(x) = \underline{t}_0\underline{r}_0\underline{t}_1...\underline{r}_{d-1}\underline{t}_d\underline{r}_d(x)$. Instead of proving formally 1/, we give an
example: let $\rho: \propto (x,\beta(y,b,z),a) \to \beta(\delta(x,y,z))$,

then $u = \rho => u'$ iff $\underline{u}(x) = \underline{t}_0\underline{\propto}\underline{t}_1\underline{\beta}\underline{t}_2\underline{b}\underline{t}_3\underline{a}(x)$ and $\underline{u}'(x) = \underline{t}_0\underline{\beta}\underline{\delta}\underline{t}_1\underline{t}_2\underline{t}_3(x)$

i.e. $\underline{l}_0(x) = \underline{\propto}(x)$; $\underline{l}_1(x) = \underline{\beta}(x)$; $\underline{l}_2(x) = \underline{b}(x)$; $\underline{l}_3(x) = \underline{a}(x)$; $\underline{r}_0(x) = \underline{\beta}\underline{\delta}(x)$; $\underline{r}_1(x) = \underline{r}_2(x) = \underline{r}_3(x) = x$

2/   We introduce d+2 new variables $R(x)$, $R^0(x,y),..., R^d(x,y)$ and define the rules of
H, which use $\underline{l}_0(x),..., \underline{l}_d(x), \underline{r}_0(x),..., \underline{r}_d(x)$ defined in 1/: $R(x) \to \underline{l}_d{}^r(R^d(\underline{l}_d(T), \underline{r}_d(x))$;

  for every $1 \le i \le d$, for every $a(x_1,...,x_d) \in \Sigma \cup V$, $R^i(x,y) \to \underline{a}(R^i(\underline{a}(x),\underline{a}(y)))$;

  for every $1 \le i \le d$, $R^i(x,y) \to \underline{l}_{i-1}{}^r (R^{i-1}(\underline{l}_{i-1}(x),\underline{r}_{i-1}(x))$;

  for every $a(x_1,...,x_d) \in \Sigma \cup V$, $R^1(x,y) \to \underline{a}(R^1(\underline{a}(x),\underline{a}(y)))\ |\ \neq(x,y)$;

The first property is easy to prove by induction (left to the reader).

Rules of H (B\$C denotes B(\$(C)) : sets H_i defined in 1/; $A \to B\$C$ $C \to C'T_f$; for every
rule ρ_i of G , $B(x) \to A(x)\ |\ B\$R_i(x)$; $A(x) \to \neq (\sigma, \underline{l}_1(x))\ |...|\ \neq (\sigma, \underline{l}_g(x))$; for any $a \in$
$V \cup \Sigma$, $C'(x) \to \neq'(x)\ |\ \underline{a}C'\underline{a}(x)$; for any $a(x_1,...,x_d) \in \Sigma$, $T_f \to a(XT_f,..., XT_f)$; for any
$a(x_1,...,x_d) \in V \cup \Sigma$, $T \to a(XT,..., XT)$; for any $a \in V \cup \Sigma$, $X(x) \to \#(x)\ |\ \underline{a}X\underline{a}(x)$

The morphism φ : $\varphi(\#(x) = x$; for any $a \in V \cup \Sigma$, $\varphi(\underline{a}(x)) = x$; $\varphi(a(x)) = a(x)$

The language χ : χ is the smallest set of trees such that $V_0 \cup \Sigma_0 \subseteq \chi$ and

for any $a(x_1,...,x_d) \in V_d \cup \Sigma_d$, for any $t_1,...,t_d \in \chi$, for any $\underline{m}_1(x_1),..., \underline{m}_d(x_d) \in \mathbf{T}^\circ_{(\underline{\Sigma} \cup \underline{V})}$
  then $a(\underline{m}_1{}^r\#\underline{m}_1(t_1),..., \underline{m}_d{}^r\#\underline{m}_d(t_d)) \in \chi$

Then, we get the following property (proof left to the reader):
The χ-*property* $L(H, T) = \chi$.

The language L(H): It is the set of trees of the form described on Figure 5, for
any rule $\sigma \to u_0$ of G (σ is the axiom of G), for any $n \in \mathbb{N}$, $i \le n$ and $u_i = G => u'_i$, for
any $i \le n+1$ and $t_i \in \chi$, and with $t_{n+1} \in \mathbf{T}_{(\underline{\Sigma} \cup \Sigma \cup \{\#(x)\})}$. The proof is a straightforward
checking left to the reader.

Definition and properties of H': Axiom: A; variables: copy of$(V) \cup \underline{V} \cup \{A, B(x), Y\}$
$\cup \{V_a{}^i(x_1,...,x_d)\ |\ a \in V_d \cup \Sigma_d$ with d>0, $1 \le i \le d\}$; terminals: the same as for H, obviously.
The rules of H' : $A \to \neq(Y,B(A))\ |\ \neq(Y,B(\neq'(Y)))$; for any $a(x_1,...,x_d) \in V \cup \Sigma$ $B(x) \to \$(x)\ |$
$\underline{a}(B(\underline{a}(x)))$; for any $a \in V_0 \cup \Sigma_0$ $Y \to \underline{a}a$;   for any $a \in V_d \cup \Sigma_d$ with d>0

$Y \to \underline{a}V_a^1(\#(Y),...,\#(Y))$; for any $b \in V \cup \Sigma$, $a \in V_d \cup \Sigma_d$ with d>0 & $1 \leq i \leq d$, $V_a^i(x_1,...,x_d) \to$
$\underline{b}(V_a^i(x_1,...x_{i-1},\underline{b}(x_{i-1}),x_{i+1},...,x_d)$; for any $b \in V \cup \Sigma$, $a \in V_d \cup \Sigma_d$ with d>0 & $1 \leq i < d$,
$V_a^i(x_1,...,x_d) \to \underline{b}(V_a^{i+1}(x_1,...x_{i-1},\underline{b}(x_{i-1}),x_{i+1},...,x_d))$; for any $a(x_1,...,x_d) \in V \cup \Sigma$,
$V_a^d(x_1,...,x_d) \to a(x_1,...,x_d)$

The language ϑDefinition: ϑ is the smallest set of trees such that for any $a \in V_0 \cup \Sigma_0$
, $\underline{a}a \in \vartheta$; for any $v_1,...,v_d \in \vartheta$, any $a(x_1,...,x_d) \in V_d \cup \Sigma_d$ & any $\underline{t}_1(x),..., \underline{t}_d(x) \in T^\circ_{(\Sigma \cup V)}$
then $\underline{a}\underline{t}_1^r... \underline{t}_d^r a(\underline{t}_1(\#(v_1)),..., \underline{t}_d(\#(v_d))) \in \vartheta$

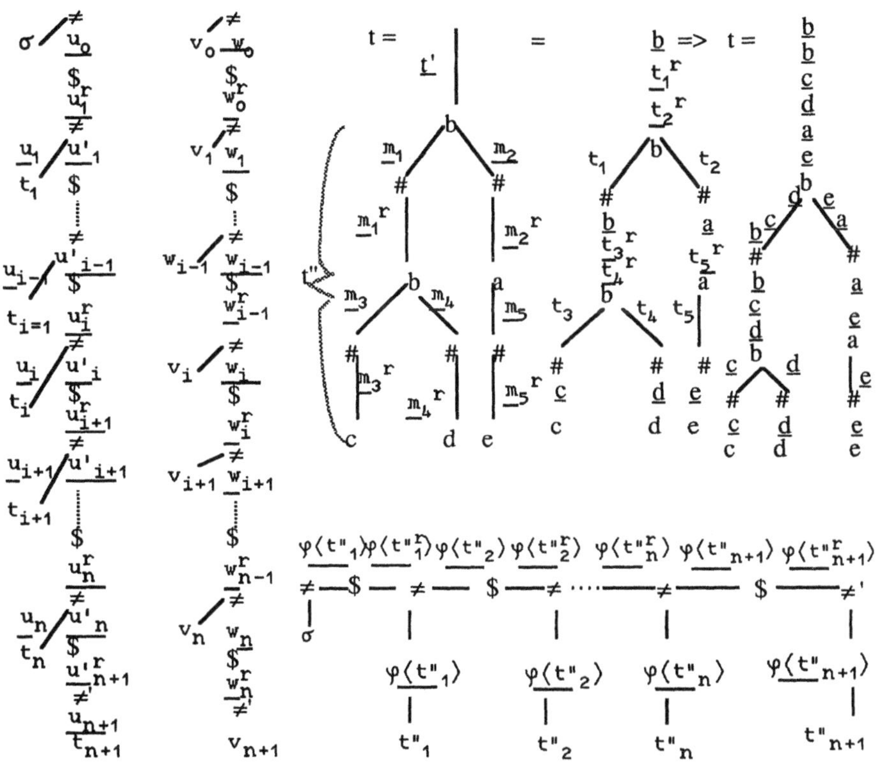

Figure 5. On the left side,   On the right side,

Description   Description   Above: the $\chi \cap \vartheta$-property with $\varphi(t) = b(b(c,d),a(e))$
of **L(H)**   of **L(H')**   Below: description of **L(H)** $\cap$ **L(H')**

Then, we get the following property (proof left to the reader):
The ϑ-*property* $L(H', Y) = \vartheta$.

The language **L(H')**. It is the set of trees of the form described on the left hand side of Figure 5 for any rule $\sigma \to u_0$ of G (σ is the axiom) and for any $n \in \mathbb{N}$, any $v_1,...,v_{n+1} \in \vartheta$. (Straightforward checking left to the reader).

Characterization of **L(H) $\cap$ L(H'):** **the** $\chi \cap \vartheta$-***property***

For any $t \in \vartheta \cap T^\circ_{(\Sigma \cup \underline{V})}(\chi)$, there is one and only one t" such that $t = \underline{\varphi\ (t")}(t")$.

Proof There are one and only one $\underline{t'} \in T^\circ_{(\Sigma \cup \underline{V})}$ and $t" \in \chi$ such that $t = \underline{t'}(t")$; then it suffice to prove $\underline{t'}(x) = \underline{\varphi\ (t")}$ by induction on the size of t". We leave it to the reader but we illustrate the structure of $\vartheta \cap T^\circ_{(\Sigma \cup \underline{V})}(\chi)$ with an example (Figure 5). Intuitively, $\varphi(t)$ keeps only the term t' of $T_{(\Sigma \cup V)}$ embedded in t, and the overlapping of the structure of ϑ and of the structure of χ are able to propagate this term t' of $T_{(\Sigma \cup V)}$ up to the root in monadic prefix form. In the example of Figure 5, we get $\varphi(t) =$ b(b(c,d),a(e)) and $\underline{t'}(x) = \underline{\varphi\ (t)} = \underline{bbcdae}(x)$. $\Diamond$

Using the form of **L(H)**, **L(H')** and the $\chi \cap \vartheta$-property, we get immediately the general form of terms of the intersection: for any $n \in \mathbb{N}$, any rule $\sigma \to \varphi(t"_1)$ of G (σ is the axiom of G) & any $1 \le i \le n$, Figure 5 describes terms of **L(H) $\cap$ L(H')**, these terms satisfy $t"_i \in \chi$; $\varphi(t"_i)$ =G=> $\varphi(t"_{i+1})$ and $t"_{n+1} \in T_{\{\Sigma \cup \Sigma \cup \{\#(x)\}\}}$

Definition of the projection π: $\pi(\neq(x,y))=y$; $\pi(\$(x))=x$; $\pi(\neq'(x))=x$;$\pi=\varphi$ otherwise

End of the proof of theorem 6.2.: With notations of Figure 5, it is easy to deduce of the preceding results that

$\pi(\mathbf{L(H)} \cap \mathbf{L(H')}) = \{\varphi(t"_{n+1})\ |\ \sigma$ =G=> $\varphi(t"_1)...=^*G=> \varphi(t"_{i+1}) \in T_\Sigma\}$.

6.3. Theorem *For any recursively enumerable tree language* L, *there exists a projection* π, *two morphisms* φ_1, φ_2 *and two recognizable tree languages* R_1, R_2 *such that* $L = \pi(\varphi_1(R_1) \cap \varphi_2(R_2))$

Proof We can suppose that **L(G)** = L for some grammar G which is torsion-free and such that only some monadic rules collapse (Proposition 6.1).

Notations and description of G: axiom is σ, the variable alphabet is V, the terminal alphabet Σ. The rules are $\rho_1,..., \rho_g$ with ρ_i: $l_i \to r_i$ and rank(ρ_i) = d_i. Greek characters denote variables of G. We add the following symbols: $\Sigma = \{a^p(x_1,...,x_{d+2})\ |\ 1 \le p \le d,\ a(x_1,...,x_d) \in \Sigma\}$;
$\mathbb{R} = \{R_i(x_1,...,x_{rank(\rho i)})\ |\ \rho_i$ rule of G}; $\mathbb{A} = \{A_i\ |\ \sigma \to \rho_i$ rule of G (σ axiom of G)};

$D_\Sigma = \{\neq_a(x_1,...,x_{d+1}) \mid a(x_1,...,x_d) \in \Sigma\};$
$\neq(x,y,z); \neq'(x,y); \partial(x_1,x_2,x_3,x_4); \partial'(x,y); \$(x,y,z); \$'(x,y); £(x,y,z); £'(x,y)$

Definition of the recognizable tree language R_1 : Using the tree language substitution notation, R_1 is described, for any p, etc..., by Figure 6. **T** denotes T_Σ , **M** denotes $T_{(\Sigma \cup V)}$ and **U1** is the recognizable tree language described by Figure 6, where **W** denotes $T_{(\Sigma \cup \Sigma \cup V)}$.

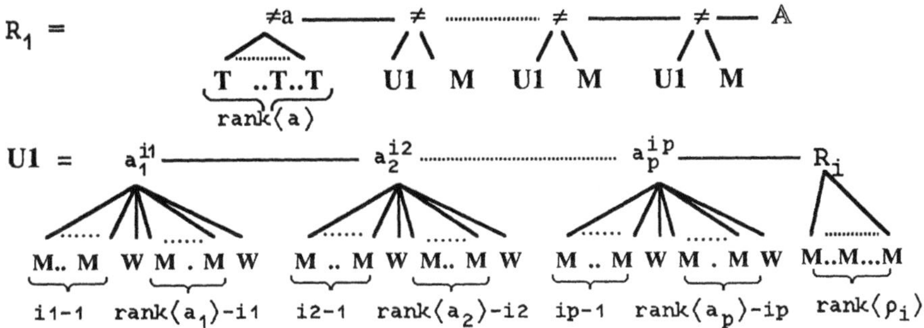

Figure 6. Description of R_1 and **U1**

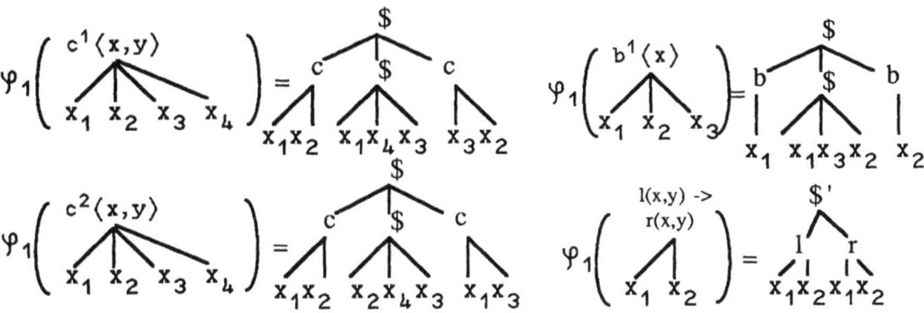

Figure 7. The meaning of φ_1 . (An example).

Definition of the morphism φ_1:

φ_1 is the identity on $\Sigma \cup V$; $\varphi_1 (\neq(x,y,z)) = \neq'(x,\neq'(y,\neq'(y,z)))$;

for any rule ρ_i of G, $\varphi_1 (R_i(x_1,...,x_{rank(\rho i)})) = \$'(r_i (x_1,...,x_{rank(\rho I)}), l_i (x_1,...,x_{rank(\rho i)}))$;

for any rule ρ_i such that the axiom σ is the left hand side, $\varphi_1 (A_i) = \$'(r_i ,\sigma)$;

for any $\neq_a \in D_\Sigma$, $\varphi_1(\neq_a(x_1,...,x_{d+1})) = \neq_a(x_1,...,x_d, \neq'(a(x_1,...,x_d),x_{d+1}))$; for any $a^p \in \Sigma$,

$\varphi_1(a^p(x_1,...,x_{d+2})) = \$(a(x_1,...,x_d),\$(x_p,x_{d+2},x_{d+1}),a(x_1,...,x_{p-1},x_{d+1},x_{p+1},...,x_d))$

Definition of the recognizable tree language R_2 : R_2 is described, for any p, etc..., by the Figure 8 (**T** denotes T_Σ , **M** denotes $T_{(\Sigma \cup V)}$ (as for R_1) and **U2** is the recognizable tree language described by figure 9, where **W** denotes $T_{(\Sigma \cup \overline{\Sigma} \cup V)}$).

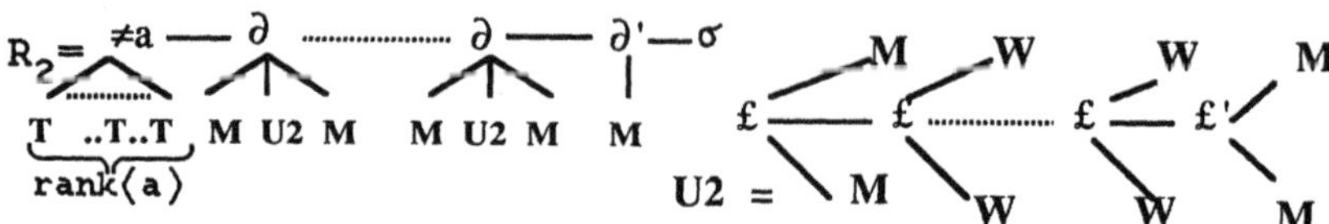

Figure 8. Description of R_2 and **U2**

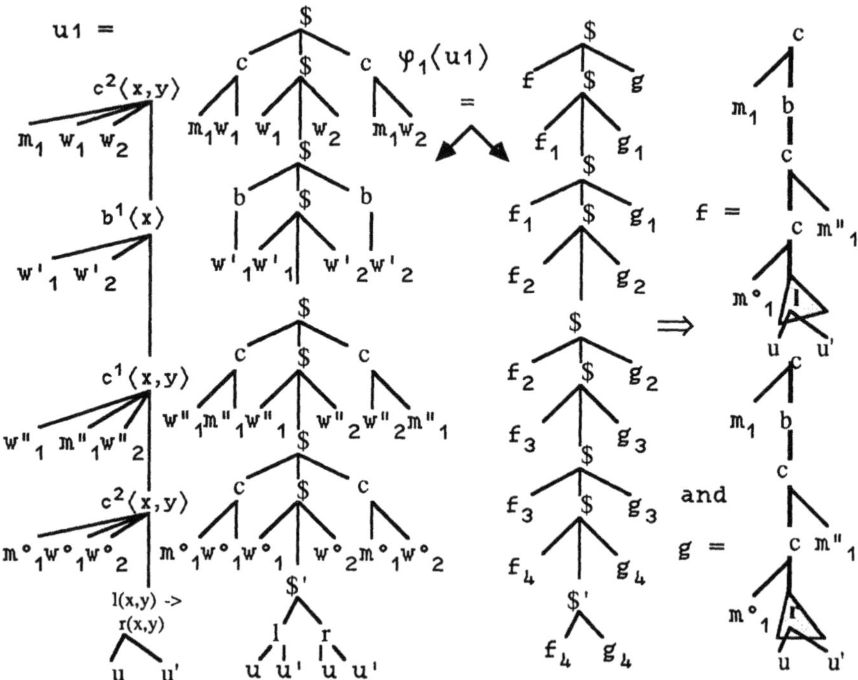

Figure 9. U-*property* : f =G=> g   iff $\exists$ h , $(f,h,g) \in \varphi_1(U1) \cap (M, \varphi_2(U2), M)$

Definition of the morphism φ_2 : φ_2 = identity on $\Sigma \cup V \cup \overline{\Sigma}$;

$\varphi_2(\partial(x_1, x_2, x_3, x_4)) = \neq'(x_1, \neq'((x_1, x_2, x_3), \neq'(x_3, x_4)))$; $\varphi_2(\partial'(x,y)) = \neq'(x, '(x,y))$;

$\varphi_2(\pounds(x,y,z)) = (x, (x,y,z),z)$; $\varphi_2(\pounds'(x,y)) = (x, '(x,y),y)$.

Description of $\varphi_1(U1)$ and $\varphi_2(U2)$: The U-*property* :

f =G=> g　iff there exists some h such that \$(f,h,g) $\in$ $\varphi_1(U1) \cap$ \$(M,$\varphi_2(U2)$,M).

Figure 9 sketches why the U-property holds. We leave the proof to the reader (straightforward induction).　　　　◊

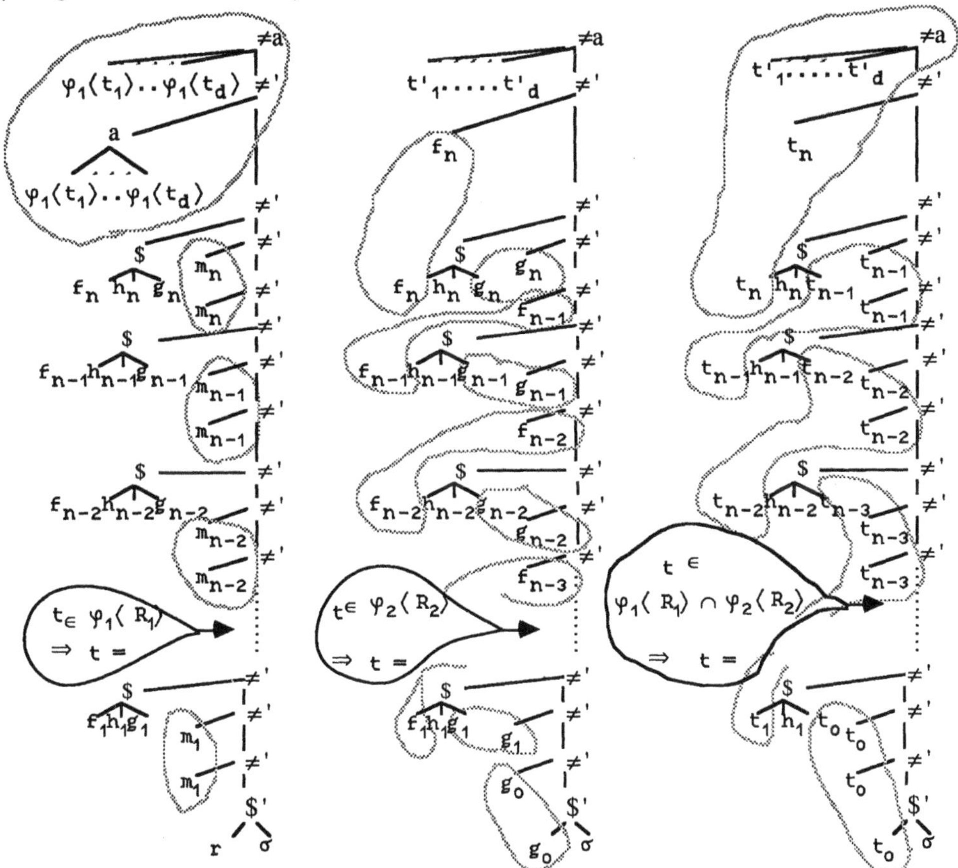

Figure 10.　　R-*property*

Description of $\varphi_1(R_1)) \cap \varphi_2(R_2)$:　The R-*property*

t $\in$ $\varphi_1(R_1) \cap \varphi_2(R_2)$ iff there exists some letter a and some term u

such that $t = \neq_a(t'_1,...,t'_{rank(a)},u)$ and $a(t'_1,...,t'_{rank(a)}) \in$ **L**(G)

Figure 10 sketches why this property holds. The crucial point is that, using the U-property, we get f_i =G=> g_i for any i in the term on the right hand side of figure 11. We leave the proof to the reader (straghtforward induction).　　　　◊

Definition of the projection π : Fo any $\neq_a(x_1,...,x_{d+1}) \in D_\Sigma$, $\pi(\neq_a(x_1,...,x_{d+1})) = a(x_1,...,x_d)$; π is the identity otherwise.

The property $\pi(\varphi_1(R_1) \cap \varphi_2(R_2)) = \mathbf{L}(G)$ is then an obvious consequence of the R-property.

Corollary 3 *For any recursively enumerable tree language* L, *there exists a projection* π *and two* IO-*algebraic tree languages* A *and* A' *such that*
$L=\pi(A \cap A')$
Proof Obvious consequence of Theorem 6.3 and of Proposition 5.3.

Corollary 4 *The class of recursively enumerable tree language is the smallest (non empty) class closed under morphisms and inverse morphisms.*

Proof Let $L = \pi(\varphi_1(R_1) \cap \varphi_2(R_2))$ be a recursively enumerable tree language. Corollary 1 proves that a class **C** closed under morphisms and inverse morphisms must contain any recognizable tree language, particularly $\$(R_1,R_2)$, where R_1 and R_2 are defined as in Theorem 6.3. We can suppose that R_1 and R_2 are defined on disjoint alphabets and consider a morphism φ which is the identity on the special symbol $\$(x,y)$, which is equal to φ_1 on the alphabet of R_1 is equal to φ_2 on the alphabet of R_2. So, **C** contains $\varphi_1(\$(R_1,R_2)) = \$(\varphi_1(R_1),\varphi_2(R_2))$. On the other hand, let ψ be the identity, except that $\psi(\$(x)) = \$(x,x)$, and let π of the theorem 6.3 extended with $\pi(\$(x)) = x$. Then $\pi(\psi^{-1}(\$(\varphi_1(R_1),\varphi_2(R_2)))) = \pi(\varphi_1(R_1) \cap \varphi_2(R_2)) = L$ belongs to **C**.

We thank Andréas Podelski for improving the introduction.

7. REFERENCES

1   A. Arnold, Le théorème de transversale rationnelle dans les arbres, *Math. Syst. Th.* **13** (1980) 275-282
2   A. Arnold, M. Dauchet, Un théorème de duplication pour les forêts algébriques, *J. Comput. System Sci.* **13** (1976) 223-244
3   A. Arnold, M. Dauchet, Morphismes et bimorphismes d'arbres, *Th. Comp. Sc.* **20** (1982) 33-93
4   A. Arnold, M. Dauchet,Théorie des magmoïdes I, *RAIRO Inform. Théorique* **12**-3 (1978) 235-257
5   A. Arnold, M. Dauchet,Théorie des magmoïdes II, *RAIRO Inform. Théorique* **13**-2 (1979) 135-154
6   A. Arnold, M. Dauchet, Caractérisation algébrique des ensembles d'arbres récursivement énumérables, colloque AFCET-SMF, Paris (1978), T2 1-14
7   B.S. Baker and R.V. Book, Reversal-Bounded multipushdown Machines, *J. Comput. System Sci.* **8** (1974) 315-332

8　W.S. Brainerd, Tree Generating Regular Systems, *Inform. Contr.* **14** (1969) 217-231

9　J.-L. Coquidé, M. Dauchet, R. Gilleron, S. Vagvölgyi, Tree pushdown automata and Rewrite Systems, CAAP'91 and Technical Report IT 190, LIFL, Lille

10　B. Courcelle, Fundamental properties of infinite trees, *Th. Comp. Sc.,* 25, (1983) 95-169.

11　M. Dauchet, Transduction de forêts. Bimorphismes de Magmoïdes, thèse d'Etat, Lille, 1977

12　Dauchet, M. , De Comité, F., A gap between linear and non-linear systems, 2nd Rewriting Techniques and Applications, Bordeaux, *LNCS* **256** (1987), 95-104.

13　Dauchet,M. Simulation of Turing Machines by a Left-Linear Rewrite Rule, Rewriting Techniques and Applications, 3rd International Conference, RTA-89, Chapel Hill, NC, USA, *LNCS* **355**, 109-120 (1989)

14　M. Dauchet and S. Tison, Réduction de la non-linéarité des morphismes d'arbres, Technical Report IT 196, LIFL, University of Lille I, (1990)

15　N. Dershowitz, Termination. *J. Symbolic computation,* **3** (1987) 69-116.

16　J. Engelfriet, Bottom-up and top-down tree transformations. A comparision, *Math. Syst. Th.* **9** (1975) 198-231

17　J. Engelfriet and E.M. Schmidt, IO and OI I, *J. of Comp. Syst. Sc.* **15** (1977) 328-353

18　J. Engelfriet and E.M. Schmidt, IO and OI II, *J. of Comp. Syst. Sc.* **16** (1978) 67-99

19　M.J. Fisher, Grammars with macro-like productions, *IEEE Symp.* **9** (1968), 131-142

20　F. Gécseg and M. Steinby, Tree Automata, *Akadémiai Kiado*, Budapest (1984)

21　Guessarian, I., Pushdowm Tree Automata, *Math. Systems Theory,* **16**, (1983), 237-264.

22　J.P. Jouannaud and N. Dershowitz, Rewrite systems, in: *Handbook of Theoretical Computer Sciences,* North Holland.

23　T.S.E. Maibaum, A generalized approach to formal languages, *J. of Comp. Syst. Sc.* **8** (1974) 409-439

24　H.A. Maurer and M. Nivat, Rational Bijections on Rational Sets, *Acta Informatica* **13** (1980) 365-378

25　M. Nivat, Transductions des langages de Chomsky, thèse d'Etat, Paris (1968)

26　M.Nivat, Langages algébriques sur le magma libre et sémantique des schémas de programmes, in: M. Nivat, ed., *Automata, Languages and Programming* (North-Holland, Amsterdam, 1973) 293-307

27　K. Salomaa, Deterministic Tree Pushdown Automata and Monadic Tree Rewriting Systems, *Journal of Comput. and Syst. Sci.,* **37**, (1988), 367-394

This work was supported in part by the ESPRIT Basic Research Working Group 3166 ASMICS

Tree Automata and Languages
M. Nivat and A. Podelski (editors)
1992 Elsevier Science Publishers B.V.

355

Ambiguity and Valuedness

Helmut Seidl

Fachbereich Informatik, Universität des Saarlandes, D-6600 Saarbrücken, Federal Republic of Germany

Abstract
This chapter surveys results on the ambiguity of finite tree automata and the valuedness of bottom-up finite state tree transducers.

0. Introduction

The ambiguity of an automaton is the maximal number of different accepting computations for a given input. The valuedness of a transducer gives the least upper bound on the number of outputs produced for any input. Both the interest in ambiguity of automata and valuedness of transducers is motivated by the equivalence problems of the corresponding devices. Automata are called equivalent iff they accept the same language; transducers are called equivalent iff they define the same translation. For finite word automata one finds that equivalence is PSPACE-complete in general [MeSto72, StoMe73], but decidable in polynomial time provided the ambiguities of the automata are bounded by some constant [SteHu85, Kui88]. Accordingly, equivalence of finite word transducers is in general undecidable [Gri68]; however if the transducers are finite-valued we obtain a decidable problem [CuKa86, We88b].
It turns out that similar results hold for tree automata and bottom-up tree transducers. However, due to the more complicated tree structure, nontrivial extensions of the proof techniques are necessary.
This paper surveys the results on ambiguity and valuedness of [Sei89, Sei90a, Sei90b, Sei90c]. It is organized in two sections each of which is subdivided into three subsections.
Section 1. is concerned with finite tree automata. Subsection 1.1. presents our notation concerning trees and automata. Moreover, it analyzes the complexities of fundamental decision problems such as the emptyness problem and the equivalence problem for

finite tree automata. Subsection 1.2. introduces the notion of ambiguity and argues that provided the ambiguities of finite tree automata are bounded by some constant equivalence among them can be decided in polynomial time. Subsection 1.3. investigates ambiguity of finite tree automata for its own sake. Especially, it gives a precise characterization of finitely ambiguous automata and derives an upper bound for the ambiguity provided it is finite. The first two subsections report results from [Sei90a] whereas the last one is based on [Sei89].

Section 2. is concerned with bottom-up finite state tree transducers (FST's for short). Subsection 2.1. gives methods to decide whether or not the valuedness of a given FST M is at least k. Subsection 2.2. gives a characterization of finite-valued FST's and derives an upper bound for the valuedness of an FST provided it is finite. The structural insights provided by this subsection allow to construct the announced decision test for finite-valued FST's (Subsection 2.3). Section 2. presents the results from [Sei90b]. Only the deterministic polynomial test of single-valuedness in Subsection 2.1. is taken from [Sei90c].

1. Finite Tree Automata

Finite tree automata were defined in the late sixties by Thatcher and Wright and Doner as generalizations of finite automata accepting word languages to finite state devices for tree languages [ThaWri68,Do70]. Their main interest in tree automata was a logical point of view. Finite tree automata can describe classes of (finite or infinite) models for formulas of monadic second order theories with multiple successors and therefore can be used to obtain effective decision procedures for these theories [ThaWri68,Do70,Ra69,Tho84]. For other possible applications see [GeStei84].

1.1. Basics

There are two dual views of trees, namely the algebraic one as a term over a signature, and the graph theoretical one as a special kind of node labeled ordered digraph. In fact, we use (and therefore introduce) both of them.

A *ranked alphabet* or *signature* is a pair (Σ,ρ) where Σ is a finite alphabet and $\rho : \Sigma \to \mathbb{N}_0$ is a function mapping symbols to their rank. The *rank* of (Σ,ρ) is the maximal value of ρ on Σ, i.e. $\max\{\rho(a) \mid a \in \Sigma\}$. Usually, if ρ is understood we write Σ for the signature (Σ,ρ) and define $\Sigma_j := \rho^{-1}(j)$. T_Σ denotes the free Σ-algebra of (finite ordered Σ-labeled) *trees*, i.e. T_Σ is the smallest set T satisfying (i) $\Sigma_0 \subseteq T$, and (ii) if $a \in \Sigma_m$ and $t_1,..,t_m \in T$, then $a(t_1,..,t_m) \in T$. Note: (i) can be viewed as the subcase of (ii) where $m = 0$.

The *depth* of a tree $t \in T_\Sigma$, depth(t), is defined by depth(t) := 0 if $t \in \Sigma_0$, and depth(t) := $1+\max\{$depth(t_1),...,depth(t_m)$\}$ if $t = a(t_1,..,t_m)$ for some $a \in \Sigma_m$, $m > 0$. Assume N

$\subseteq \Sigma$. the N-*size* of $t = a(t_1,..,t_m)$, $|t|_N$, is defined by $|t|_N := 1+\sum_{j=1}^{m}|t_j|_N$ if $a \in N$ and

$|t|_N := \sum_{j=1}^{m}|t_j|_N$ otherwise. If $N = \Sigma$, we skip the index N.

The set of *nodes* of t, $O(t)$ is the subset of $\mathbb{N}^*$ defined by $O(t) := \{\varepsilon\}\cup\bigcup_{j=1}^{m}j\cdot O(t_j)$ where

$t = a(t_1,..,t_m)$ for some $a \in \Sigma_m$, $m \geq 0$. Note that the cardinality of $O(t)$, $\#O(t)$, equals $|t|$.

t defines maps $\lambda_t(\_) : O(t)\rightarrow\Sigma$ and $\sigma_t(\_) : O(t)\rightarrow T_\Sigma$ mapping the nodes o of t to their labels or the subtree of t with root o, respectively. We have

$$\lambda_t(o) := \begin{cases} a & \text{if } o = \varepsilon \\ \lambda_{t_j}(o') & \text{if } o = j\cdot o' \end{cases} \quad \text{and} \quad \sigma_t(o) := \begin{cases} t & \text{if } o = \varepsilon \\ \sigma_{t_j}(o') & \text{if } o = j\cdot o' \end{cases}$$

Let X denote a set of variables of rank 0. Define $T_\Sigma(X) := T_{\Sigma\cup X}$. We use this different notation in order to indicate which variables are to be substituted. (Clearly, $T_\Sigma \subseteq T_\Sigma(X)$.) Assume $t \in T_\Sigma(X)$. For $x \in X$ the set $O_x(t)$ of *occurrences* of x is the set $\{o\in O(t) \mid \lambda_t(o) = x\}$. t is called X-*proper* iff every $x \in X$ occurs in t exactly once, i.e. $\#O_x(t) = 1$ for every x. If $X = \{x\}$ we write x-proper instead of $\{x\}$-proper, and if X is understood we skip the prefix X.

Every map $\theta : X\rightarrow T_\Sigma(X)$ can be uniquely extended to a map $\theta : T_\Sigma(X)\rightarrow T_\Sigma(X)$. θ is called X-*substitution* or simply substitution if X is understood. If $X = \{x_j,..,x_m\}$ (i.e. the variables are indexed by some interval of natural numbers) and $x_i\theta = t_i$, we denote $t\theta$ also by $t[t_j,..,t_m]$. Of special importance is the case where the set X of variables which are to be substituted consists of just one element x. Assume $x\theta = t_2$ and $t_1 \in T_\Sigma(x) := T_\Sigma(\{x\})$. Then we write $t_1\theta = t_1t_2$. The set $T_\Sigma(x)$ is a monoid w.r.t. x-substitution. (The neutral element is x). $T_\Sigma(x)$ is *not* a free monoid. Especially, $t_1t_2 = t_1$ if t_1 does not contain an occurrence of x.

Observe that in case a tree t may contain variables from X, its "size" is more adequately described by a polynomial over the variable set X of degree at most 1. Let us denote the set of all these polynomials with coefficients in the field of rational numbers $\mathbb{Q}$ by $\mathbb{Q}^{(1)}[X]$. Then, we define a map $\omega : T_\Sigma(X)\rightarrow\mathbb{Q}^{(1)}[X]$ by $\omega(t) := |t|_\Sigma + \sum_{x\in X}|t|_{\{x\}}\cdot x$, i.e. the constant is the Σ-size of t, whereas the coefficient of x is just the number of occurrences of x in t.

As for substitutions, a map $\theta : X\rightarrow\mathbb{Q}^{(1)}[X]$ can be extended to a map $\theta : \mathbb{Q}^{(1)}[X]\rightarrow\mathbb{Q}^{(1)}[X]$. For $f \in \mathbb{Q}^{(1)}[X]$, $f\theta$ is defined by function composition: $f\theta$ is the function obtained from f by first applying the functions $x\theta$ and then applying f to the results. As for trees, we call θ a X-*substitution* or substitution, if X is understood. If $X = \{x_1,..,x_m\}$ and $x_i\theta = f_i$ we denote $f\theta$ by $f[f_1,..,f_m]$ and write $f\ f_1$ for $f[f_1]$.

Clearly, the map ω commutes with substitution.

A *finite tree automaton* (FTA for short) is a 4-tuple $A = (Q,\Sigma,\delta,Q_F)$ where Q is a finite set of *states*, $Q_F \subseteq Q$ is the set of *final* states, Σ is the signature of *input trees*, $\delta \subseteq \bigcup_{m \geq 0} Q \times \Sigma_m \times Q^m$ is the set of *transitions* of A. The transitions in $\delta \cap \bigcup_{m \geq 0} \{q\} \times \Sigma_m \times Q^m$ are also called q-*transitions*.

In the sequel, we always consider automata with the same fixed input signature Σ.

We do not define a computation (which sometimes also is called "run") of an FTA by a reduction sequence but prefer a "static" view where the computation is given as a whole. This allows us to decompose, compare and classify computations more easily.

Let $t = a(t_1,..,t_m) \in T_\Sigma(X_k)$ and $q,q_1,..,q_k \in Q$. A $(q,q_1..q_k)$-*computation* of A for t starts at variables x_j in the states q_j and consists of $(p_j,q_1..q_k)$-computations of A for the subtrees t_j, $j \in \{1,..,m\}$, together with a transition $(q,a,p_1..p_m) \in \delta$ for the root. We write the state at the root to the left of the states at the variable leafs. This convention is chosen in accordance with our prefix notation of trees and the left-to-right order of substitutions. Formally, a $(q,q_1..q_k)$-computation of A for t is a map $\phi : O(t) \to \delta \cup Q \times X_k \times \{\varepsilon\}$ with

(i)   If $\lambda_t(\varepsilon) \in \Sigma$ then $\phi(\varepsilon)$ is a q-transition ; if $\lambda_t(\varepsilon) = x_j$ then $q = q_j$.

(ii)   If $\sigma_t(o) = x_j$, then $\phi(o) = (q_j,x_j,\varepsilon)$.

(iii)   If $\lambda_t(o) = a \in \Sigma_m$, and $\phi(oj)$ is a q_j-transition for $j = 1,..,m$, then $\phi(o) = (q',a,q_1...q_m)$ for some $q' \in Q$.

$\phi(o)$ is also called the transition *chosen* in ϕ at node o.

In abuse of the symbol δ we also write $(q,t,q_1...q_k) \in \delta$ iff there is a $(q,q_1...q_k)$-computation of A for t.

In [Sei89, Sei90a] a computation of a $(q,q_1..q_k)$-computation maps the nodes of an input tree to states. This is equivalent to our definition here. However, choosing transitions as images allows to view a $(q,q_1..q_k)$-computation as a term in a free Q-sorted algebra with signature δ. We do not introduce this notion formally. Nevertheless, we write $\phi = \tau (\phi_1,...,\phi_m)$ if τ is the transition chosen at the root and ϕ_j are the subcomputations of ϕ for the subtrees t_j. In case $t = x_j$ we simply write $\phi = x_j$. Also we say ϕ is X_k-proper iff every variable x_j of X_k occurs in t exactly once.

Assume $t \in T_\Sigma(X_k)$ and $t = t_0[t_1,..,t_k]$. Assume ϕ_0 is a $(q,p_1..p_k)$-computation for t_0, and ϕ_i are $(p_i,q_1..q_m)$-computations for t_i , $i = 1,..,k$. Then we can combine them to a $(q,q_1..q_m)$-computation ϕ of A for t. We denote ϕ by $\phi_0[\phi_1,...,\phi_k]$. If $k = 1$ we write $\phi_0\phi_1$ for short. Conversely, if t_0 contains exactly one occurrence of any x_j , $j = 1,...,k$ (i.e. is X_k-proper), then every $(q,q_1..q_m)$-computation ϕ for t uniquely can be decomposed into a $(q,p_1..p_k)$-computation ϕ_0 for t_0, and $(p_i,q_1..q_m)$-computations ϕ_i for t_i (for suitable states p_i) such that $\phi = \phi_0[\phi_1,...,\phi_k]$. ϕ_i is called *subcomputation* of ϕ on t_i.

A (q,ε)-computation is also called q-computation. A q-computation is called *accepting*, iff $q \in Q_F$. $L(A) := \{t \in T_\Sigma \mid$ there is an accepting computation of A for t$\}$ is the *language* accepted by A. The *size* of A, $|A|$, is defined by $|A| := \sum_{(q,a,q_1...q_m) \in \delta} (m+2)$.

Call a state $q \in Q$ *useless* if no accepting computation exists in which a q-transition occurs, and *useful* otherwise. An FTA A is *reduced* iff A has no useless states. Clearly, useless states can be removed without changing the "behavior" of A. We have:

Theorem 1

(1)   For every FTA $A = (Q,\Sigma,\delta,Q_F)$ there is an FTA $A_r = (Q_r,\Sigma,\delta_r,Q_{r,F})$ such that

- $Q_r \subseteq Q$, $\delta_r \subseteq \delta$ and $Q_{r,F} \subseteq Q_F$;

- A_r is reduced; and

- $L(A) = L(A_r)$.

   A_r can be constructed from A in polynomial time.

(2)   The emptyness problem for FTA's is complete in P (w.r.t. logspace reductions).

The method of reducing an FTA is analogous to the reduction of a contextfree grammar. Constructing the reduced FTA according to (1) clearly can be used to decide whether or not the language accepted by some FTA A is empty. This proves that the emptyness problem for FTA's is in P. The hardness part can be proven by reducing the Monotonic Circuit Value Problem to the emptyness problem for FTA's. $\square$

Observe in contrast that emptyness for finite word automata is in the complexity class LOGSPACE which is a subclass of P.

Two FTA's A_1, A_2 are called *equivalent*, iff $L(A_1) = L(A_2)$. We analyze the complexity of the inequivalence problems for unrestricted FTA's and FTA's accepting finite languages, respectively. For nondeterministic finite word automata the corresponding two problems are known to be PSPACE-complete and NP-complete (w.r.t. logspace reductions), respectively [MeySto72,StoMey73]. We find that for FTA's these problems are slightly more difficult.

Theorem 2

(1)   Equivalence for FTA's is complete in DEXPTIME w.r.t. logspace reductions.

(2)   Equivalence for FTA's accepting finite languages is complete in PSPACE w.r.t. logspace reductions. $\square$

1.2. Deciding Equivalence of m-Ambiguous FTA's

We saw in the last subsection that deciding equivalence for FTA's is provably difficult in general. Therefore, we are interested in (non-trivial) subclasses of FTA's which admit fast decision procedures. One structural parameter which attracts attention in this context is the ambiguity.

Assume $A = (Q,\Sigma,\delta,Q_F)$ is an FTA. For $t \in T_\Sigma$, define the *ambiguity vector* of t w.r.t. A as the Q-tuple $n_A(t) = (n_A(t)_q)_{q \in Q}$ where $n_A(t)_q$ denotes the number of different q-computations of A for t. The *ambiguity* of t w.r.t. A, $da_A(t)$, is the number of different accepting computations of A for t. The ambiguity of A, $da(A)$, is defined as $da(A) := \sup\{da_A(t) \mid t \in T_\Sigma\}$. Note that what we call ambiguity here, is sometimes called *degree of ambiguity*, e.g. in [Sei89, Sei90a]. For simplicity and in analogy to the "valuedness" of a transducer we decided to use the shorter form here. A is called

- *unambiguous*, if $da(A) \leq 1$;

- *ambiguous*, if $da(A) > 1$;

- m-*ambiguous*, if $da(A) \leq m$;

- *finitely ambiguous*, if $da(A) < \infty$; and

- *infinitely ambiguous*, if $da(A) = \infty$.

Two FTA's A_1, A_2 are called *ambiguity-equivalent* (written $A_1 \equiv A_2$), iff $da_{A_1}(t) = da_{A_2}(t)$ for all $t \in T_\Sigma$. Clearly, $A_1 \equiv A_2$ implies $L(A_1) = L(A_2)$. Moreover, if A_1 and A_2 are unambiguous, then $A_1 \equiv A_2$ iff $L(A_1) = L(A_2)$.

In [SteHu81,SteHu85] Stearns and Hunt III give a polynomial time algorithm which decides equivalence of m-ambiguous finite word automata for every fixed constant m. They employ difference equations for their decision procedure. Kuich in [Kui88] simplifies their constructions. Although not stated explicitly, Kuich's proof can be used to show that the equivalence problem for m-ambiguous finite word automata is even in NC. Kuich uses semiring automata and the formalism of formal power series. Semiring automata are obtained from ordinary automata by giving ambiguities in some semiring to transitions and initial states. In particular, Kuich shows how the equivalence problem for m-ambiguous word automata can be reduced to the ambiguity-equivalence problem for unambiguous automata with ambiguities in **Q**.

For tree languages the concept of formal power series seems to be much more involved than for word languages [BeReu82]. We avoid to use this concept, but show how some basic ideas of Kuich can be carried over to tree automata.

The key observation for this to work is that the ambiguity vector $n_A(t)$ of a tree $t = a(t_1,...,t_m)$ can be computed from the ambiguity vectors $n_A(t_j)$ of the subtrees by a *multilinear* mapping. Accordingly, the ambiguity $da_A(t)$ can be determined from $n_A(t)$ by means of a *linear* mapping. We have:

$$n_A(t)_q = \sum_{(q,a,q_1..q_m) \in \delta} n_A(t_1)_{q_1} \cdots n_A(t_m)_{q_m} \tag{1}$$

$$da_A(t) = \sum_{q \in Q_F} n_A(t)_q \tag{2}$$

To have more "freedom" for algebraic manipulations we would like to compute the ambiguities in some arbitrary (commutative) *semiring* (with 0 and 1). Such a semiring is a structure $(R,+,\cdot,0,1)$ where $(R,+)$ and $(R,\cdot)$ are commutative monoids with neutral elements 0 and 1 respectively such that $a\cdot(b+c) = (a\cdot b)+(a\cdot c)$ and $0\cdot a = 0$ hold for arbitrary semiring elements $a,b,c \in R$. If $+$, $\cdot$, 0 and 1 are understood, we write R instead of $(R,+,\cdot,0,1)$.

Assume R is a semiring. A finite tree automaton with ambiguities in R (short: R-FTA) is a quadruple $A = (Q,\Sigma,I,\delta)$ where Q is a finite set of states; Σ is a ranked alphabet; δ is a map $\delta : \bigcup_{m\geq 0} Q\times\Sigma_m\times Q^m \to R$ denoting the *transition ambiguity*; and $I = (I_q)_{q\in Q} \in R^Q$ is the Q-tuple of *final ambiguities*.

Taking equations (1) and (2) as definitions, we define, for every $t = a(t_1,...,t_m) \in T_\Sigma$, an R-*ambiguity vector* $n_A(t) = (n_A(t))_{q\in Q}$ where

$$n_A(t)_q := \sum_{q_1..q_m\in Q^m} \delta(q,a,q_1..,q_m)\cdot n_A(t_1)_{q_1}\cdot..\cdot n_A(t_m)_{q_m} .$$

The R-*ambiguity*, $da_A(t)$, of the tree t is given by $da_A(t) := \sum_{q\in Q} I_q\cdot n_A(t)_q$.

The language $L(A)$ accepted by A is the set $L(A) := \{t \in T_\Sigma \mid da_A(t) \neq 0\}$.

Similar to the case of FTA's, the size of the R-FTA A, $|A|$, is defined by

$$|A| := \sum_{\delta(q,a,q_1..q_m)\neq 0} (m+2) .$$

An R-FTA A is called *unambiguous*, iff $da_A(t) \in \{0,1\}$ for all $t \in T_\Sigma$.

Two R-FTA's A_1, A_2 are called equivalent, iff $L(A_1) = L(A_2)$. They are called *ambiguity-equivalent* (denoted: $A_1 \equiv A_2$), iff $da_{A_1}(t) = da_{A_2}(t)$ for every tree t.

Any FTA $A = (Q,\Sigma,\delta,Q_F)$ can be viewed as the definition of an R-FTA $A_R = (Q,\Sigma,\delta_R,1_{Q_F})$ where

$$1_{Q_F,q} := \begin{cases} 1 & \text{if } q \in Q_F \\ 0 & \text{otherwise} \end{cases} \quad \text{and} \quad \delta_R(q,a,q_1..q_m) := \begin{cases} 1 & \text{if } (q,a,q_1..q_m) \in \delta \\ 0 & \text{otherwise} \end{cases} .$$

The following holds for all $t \in T_\Sigma$:

(1) If $\mathbb{N}_0$ is a subsemiring of R, then

  (1.1) $(n_{A_R}(t)_q = n_A(t)_q)$ for all $q \in Q$;

  (1.2) $da_{A_R}(t) = da_A(t)$.

(2) If $p > 1$ is a natural number, then

  (2.1) $(n_{A_{\mathbb{Z}_p}}(t)_q = n_A(t)_q \bmod p)$ for all $q \in Q$;

  (2.2) $da_{A_{\mathbb{Z}_p}}(t) = da_A(t) \bmod p$.

(3) If $\mathbb{B}$ is the Boolean semiring defined by $1+1=1$, then

> (3.1)   ($n_{A_B}(t)_q = 1$ iff $n_A(t)_q > 0$) for all $q \in Q$;
>
> (3.2) $da_{A_B}(t) = 1$ iff $da_A(t) > 0$.

The properties Kuich needs to transform an m-ambiguous finite word automaton into an unambiguous **Q**-automaton A' are the presence of constructions for 1. the linear combination, 2. the product of **Q** automata and 3. of **Q**-automata accepting every word with a fixed ambiguity [Kui88]. These constructions generalize the usual constructions for union, intersection and a 1-state automaton accepting everything. It turns out that such constructions are available for R-FTA's as well.

Assume $A_i = (Q_i, \Sigma, \delta_i, I^{(i)})$, $i = 1, 2$, are two R-FTA's and $\mu_1, \mu_2 \in R$.

The *linear combination* $\mu_1 A_1 + \mu_2 A_2$ of A_1 and A_2 is the the R-FTA $(Q', \Sigma, \delta', I')$ where Q' is the disjoint union of Q_1 and Q_2;

$$I'_q := \begin{cases} \mu_1 I_q^{(1)} & \text{if } q \in Q_1 \\ \mu_2 I_q^{(2)} & \text{if } q \in Q_2 \end{cases} \quad ; \quad \delta'(q,a,q_1..q_m) := \begin{cases} \delta_1(q,a,q_1..q_m) & \text{if } q,q_1,...,q_m \in Q_1 \\ \delta_2(q,a,q_1..q_m) & \text{if } q,q_1,...,q_m \in Q_2 \\ 0 & \text{otherwise} \end{cases} .$$

The *product* $A_1 \times A_2$ of A_1 and A_2 is the R-FTA $(Q'', \Sigma, \delta'', I'')$ where $Q'' := Q_1 \times Q_2$;

$$I''_{<p,q>} := I_p^{(1)} \cdot I_q^{(2)} \text{ for } p \in Q_1, q \in Q_2 \text{ ; and}$$
$$\delta''(<p,q>,a,<p_1,q_1>..<p_m,q_m>) := \delta_1(p,a,p_1..p_m) \cdot \delta_2(q,a,q_1..q_m) .$$

For every $j \in R$ the *constant* R-FTA **j** is defined by $\mathbf{j} := (\{q\}, \Sigma, \delta^1, j)$ where

$$\delta^1(q,a,q^m) := 1 \text{ for all } a \in \Sigma_m , m \geq 0 .$$

The following proposition summarizes the algebraic properties of these constructions.

Proposition

For all $t \in T_\Sigma$:

(1) $n_{\mu_1 A_1 + \mu_2 A_2}(t)_q = \begin{cases} n_{A_1}(t)_q & \text{if } q \in Q_1 \\ n_{A_2}(t)_q & \text{if } q \in Q_2 \end{cases} ;$

$n_{A_1 \times A_2}(t)_{<p,q>} = n_{A_1}(t)_p \cdot n_{A_2}(t)_q$ for all $p \in Q_1$ and $q \in Q_2$;

$n_{\mathbf{j}}(t) = n_{\mathbf{j}}(t)_q = 1 .$

(2) $da_{\mu_1 A_1 + \mu_2 A_2}(t) = \mu_1 da_{A_1}(t) + \mu_2 da_{A_2}(t) ;$

$da_{A_1 \times A_2}(t) = da_{A_1}(t) \cdot da_{A_2}(t) ;$

$da_{\mathbf{j}}(t) = j . \quad \square$

Using Kuich's construction we define $U(A) := \sum_{j=1}^{m} \dfrac{(-1)^{j+1}}{j!} [A]_j$ where

$[A]_j := A \times (A-1) \times .. \times (A-(j-1)) .$

Moreover, to keep the numbers occurring in our algorithms small we relativize this construction modulo a suitable prime number.

Assume p is a prime number greater than m. By Bertrand's postulate (see [HaWri60, Theorem 418]) such a prime p exists in the range between m and 2m. Since $p > m$, the multiplicative inverse $(j!\ \mathrm{mod}\ p)^{-1}$ is defined in $\mathbb{Z}_p$ for all $j \in \{1,..,m\}$. Therefore, we can define $U_p(A) := \sum_{j=1}^{m}(-1)^{j+1}(j!\ \mathrm{mod}\ p)^{-1}[A_{\mathbb{Z}_p}]_j$. We find:

Theorem 1

Assume A is an m-ambiguous FTA. Then the following holds:

(1)   U(A) is an unambiguous **Q**-FTA with L(U(A)) = L(A).

(2)   For every prime number $p > m$, $U_p(A)$ is an unambiguous $\mathbb{Z}_p$-FTA with $L(A) = L(U_p(A))$.   □

Since for unambiguous R-FTA's equivalence coincides with ambiguity-equivalence, Theorem 1 can be used to reduce the equivalence problem for m-ambiguous FTA's to the ambiguity-equivalence problem for unambiguous **Q**- or $\mathbb{Z}_p$-FTA's. It remains to construct an algorithm deciding ambiguity-equivalence for R-FTA's, provided R is a *field*. This algorithm relies on a generalization of Eilenberg's equality theorem [Ei74, Theorem 8.1] to trees.

Theorem 2

If R is a field, and A_1 and A_2 are R-FTA's with n_1 and n_2 states respectively, then the following holds.

(1) $A_1 \equiv A_2$ iff $da_{A_1}(t) = da_{A_2}(t)$ for all $t \in T_\Sigma$ with $depth(t) < n_1+n_2$.

(2)   Whether or not $A_1 \equiv A_2$, can be decided in polynomial time on a RAM which can hold elements of R in its registers and performs the R-operations +, -, ·, : and R-tests for 0 in constant time.   □

An algorithm deciding ambiguity equivalence may work as follows. Assume $A_2-A_1 = (Q,\Sigma,\delta,I)$ and $n := \#Q = n_1+n_2$. The algorithm iteratively computes a basis B_k of sub-vectorspaces $V_k \subseteq R^n$, $k = 0,...,n-1$. The subvectorspace V_k is the linear hull of the R-ambiguity vectors of trees of depth at most k, i.e. $V_k := <\ n_{A_2-A_1}(t)\ |\ depth(t) \leq k\ >$. The algorithm outputs "$A_1 \equiv A_2$" if for every $v = (v_q)_{q \in Q} \in V_{n-1}$

$$I\cdot v = \sum_{q \in Q} I_q v_q = 0 \tag{3}$$

and outputs "$A_1 \not\equiv A_2$" otherwise.

Clearly, (3) only has to be verified for the basis B_{n-1}. Moreover, recall that every symbol $a \in \Sigma$ corresponds to a multilinear mapping of the ambiguity vectors. Let M_a

denote this mapping. Then, V_0 is generated from M_a, $a \in \Sigma_0$, whereas for $k > 0$, V_k is generated from the vectors $M_a(v_1,...,v_m)$, $m \geq 0$, $a \in \Sigma_m$, $v_1,...,v_m \in B_{k-1}$. From this generating system a basis B_k may be computed by using e.g. the Gaussian elimination method. Provided the rank of Σ is bounded by some constant (which w.l.o.g. always may be assumed), this algorithm runs in polynomial time. $\square$

Observe that the algorithm given in the proof of Theorem 2 (2) makes unrestricted use not only of R-addition and R-comparisons but also of R-multiplication. In fact, if we employ $R = \mathbf{Q}$, numbers may grow to double exponential size and hence exponential bit length. Therefore, this algorithm does not give a polynomial algorithm deciding ambiguity-equivalence of FTA's (viewed as $\mathbf{Q}$-FTA's). However, if $R = \mathbb{Z}_p$ for some fixed prime number p then addition, multiplication and comparison become polynomial time operations. Since $U_p(A)$ can be constructed from A in polynomial time, Theorems 1 and 2 together give a deterministic polynomial equivalence test for m-ambiguous FTA's:

Theorem 3
For every fixed constant $m > 0$, equivalence of m-ambiguous FTA's can be decided in polynomial time. $\square$

Recall that the emptyness problem even of unambiguous FTA's is P-complete. Since the emptyness problem can be reduced to the equivalence problem it follows from Theorem 3 that in fact, equivalence of m-ambiguous FTA's is complete in P (w.r.t. logspace reductions). For testing ambiguity-inequivalence of FTA's Theorem 2 at least allows to construct a randomized polynomial algorithm:

Theorem 4
The ambiguity-inequivalence problem for FTA's is in RP, i.e. the class of problems with randomized polynomial algorithms. $\square$

We have seen that bounding the ambiguity by some fixed constant m yields a class of FTA's with a polynomial equivalence problem. We close this subsection by remarking that, for every fixed $m > 0$, it also can be decided in polynomial time whether or not a given FTA is m-ambiguous.

Theorem 5
For every fixed constant $m > 0$, it can be decided in polynomial time whether or not a given FTA has ambiguity at most m. $\square$

1.3. A Characterization of Finite Ambiguity

In [WeSei86] it is shown that it can be decided in polynomial time whether or not the ambiguity of a finite word automaton is finite. For this a criterion (IDA) is given characterizing an infinite ambiguity. Moreover, this paper proves an upper bound $5^{n/2} \cdot n^n$ for the maximal ambiguity of a finitely ambiguous finite word automaton A having n states. Using an estimation of Baron [Ba88] Kuich slightly improves this upper bound [Kui88]. In [WeSei88] the analysis of finitely ambiguous finite word automata is completed by proving a non-ramification lemma which allows for every word w to construct a word w' of length less than $e^2 \cdot n!$ having the same number of accepting computation paths.

In this subsection we present an extension of the methods of [WeSei86,WeSei88] to finite tree automata.

For $i = 1, 2$ we say A has property (fi) iff A satisfies the corresponding condition:

(f1)   For no $q \in Q$ and x_1–proper tree t there are two different (q,q)-computations of A for t.

(f2)   For no distinct states p, $q \in Q$ and x_1-proper tree t, (p,t,p), (p,t,q), (q,t,q) $\in$ δ.

Observe that the negation of property (f2) is the generalization of the corresponding criterion (IDA) of [WeSei86] for infinite ambiguity of finite word automata. For word automata, property (f1) is implied by property (f2) which is *not* the case for tree automata. The main result of this section is:

Theorem 1

Assume A is a reduced FTA. Then the following four statements are equivalent:

(1)   A has properties (f1) and (f2);

(2)   For every tree $t \in T_\Sigma$, there is a tree t_1 with $\mathrm{depth}(t_1) < e^2 \cdot n!$ such that $da_A(t) = da_A(t_1)$;

(3) $da(A) < 2^{2^{2 \cdot \log(L+1) \cdot n}}$;

(4) $da(A) < \infty$

where n is the number of states of A, e denotes the base of the natural logarithm, and $L > 0$ is the rank of Σ. $\square$

One can construct FTA's that have any of the properties (f1) or (f2) but not the other. Therefore, the characterization given in (1) is irredundant.

The crucial point in the proof of Theorem 1 is to show that statement (1) implies statement (2) or (3). For this, we employ the Non-Ramification Lemma for FTA's.

Assume $t \in T_\Sigma$. To every node $o \in O(t)$ we attach the sets of *attainable, derivable* and *occurring* states respectively:

$$ACC_t(o) := \{q \in Q \mid \exists f \in Q_F : (f,t[x_1/o],q) \in \delta\}$$

$$DER_t(o) := \{q \in Q \mid (q,\sigma_t(o),\varepsilon) \in \delta\}$$

$$OCC_t(o) := ATT_t(o) \cap DER_t(o) .$$

Non-Ramification Lemma for FTA's

Assume there are nodes o_1 and o_1o_2 in $O(t)$ such that $D := OCC_t(o_1) = OCC_t(o_1o_2)$. Consider the factorization $t = u_1u_2u_3$ where $u_1 := t[x_1/o_1]$; $u_2 :- \sigma_t(o_1)[x_1/o_2]$ and $u_3 := \sigma_t(o_1o_2)$. If A has properties (f1) and (f2), then the following holds:

(1)   For every $p \in D$ exactly one (p,q)-computation of A for u_2 exists such that $q \in D$.

(2)   For every $q \in D$ exactly one (p,q)-computation of A for u_2 exists such that $p \in D$. $\square$

The following example shows that the upper bound for the maximal ambiguity of a finitely ambiguous FTA given in Theorem 1 (3) is optimal up to a constant factor in the highest exponent.

Theorem 2

For every $n \geq 3$ and $L \geq 2$ there is a finitely ambiguous FTA $A_{n,L}$ with n states and input signature of rank L such that $da(A_{n,L}) = 2^{2^{\log(L)\cdot(n-2)}}$.

<u>Proof:</u> Define $A_{n,L}$ by $A_{n,L} = (\{1,..,n\},\Sigma,\delta_{n,L},\{1\})$ where $\Sigma_0 := \{\#\}$, $\Sigma_L := \{o\}$ and $\Sigma_m := \emptyset$ otherwise, and $\delta_{n,L} := \{(i,o,(i+1)^L) \mid 1\leq i\leq n-3\} \cup \{n-2\}\times\{o\}\times\{n-1,n\}^L \cup \{(n-1,\#,\varepsilon),(n,\#,\varepsilon)\}$. Then $L(A_{n,L}) = \{\Delta_{n,L}\}$ where $\Delta_{n,L}$ denotes the complete L-ary tree of depth n-2 whose inner nodes are labeled with o and whose leafs are labeled with #. Since $L(A_{n,L})$ is finite, the ambiguity of $A_{n,L}$ is finite, too. There is a bijection between the set of accepting computations of A for $\Delta_{n,L}$ and the set of all words of length L^{n-2} over a two letter alphabet. Therefore, $da(A_{n,L}) = 2^{L^{n-2}}$. $\square$

To complete the investigation of finite ambiguity we show how to construct a polynomial time algorithm which decides finite ambiguity.

Theorem 3

It can be decided in polynomial time whether or not a given FTA is finitely ambiguous.

By Theorem 1 it suffices to find polynomial time algorithms deciding whether or not a reduced FTA A has properties (f1) and (f2). Instead of describing these algorithms explicitly, we reduce the properties (fi) to properties (f'i). The properties (f'1) and (f'2) are described by means of *graphs*, namely G(A), $G(A^2)$ and $G(A^3)$ where G(..) is the *trace graph* corresponding to a given FTA, and A^i denotes the i-th *power* of A.

Opposed to (fi), the new criteria can be easily seen to be decidable in polynomial time. This general method of proving the existence of a polynomial time algorithm by reducing the original problem to some property of trace graphs are used a second time in Subsection 2.1 where we construct a polynomial time algorithm deciding single-valuedness of transducers.

For the ranked alphabet Σ let Σ_B be the (ordinary) alphabet $\Sigma_B := \{ (a,j) \mid m > 0, a \in \Sigma_m, j \in \{1,..,m\}\}$. The set $B(t)$ of *branches* of a tree t is defined by $B(t) := \{\varepsilon\}$ if $t = a \in \Sigma_0$ and $B(t) := \bigcup_{j=1}^{m}(a,j)\cdot B(t_j)$ if $t = a(t_1,..,t_m)$ for some $a \in \Sigma_m$, $m > 0$. Note: the sequence of the second components of the symbols of a branch forms a leaf, whereas the sequence of the first components gives the labels on the path in t from the root of t to this leaf (omitting the label of the leaf itself). A prefix $w = (a_1,j_1)..(a_k,j_k)$ of a branch of t is called *path* in t.

Assume $t \in T_\Sigma$, ϕ is a computation of A for t, and $w = (a_1,j_1)..(a_k,j_k)$ is path in t. The *trace* ϕ_w of ϕ on w is the sequence $\phi_w = (\tau_1,j_1)..(\tau_k,j_k) \in [\delta\times\mathbb{N}]^*$ where $\phi(j_1..j_\kappa) = \tau_\kappa$ for all $\kappa \in \{0,..,k\}$.

Assume $A = (Q,\Sigma,\delta,Q_F)$ is a reduced FTA. The *trace graph* $G(A)$ corresponding to A is the edge-labeled digraph $G(A) = (V,E)$ where the set of vertices $V := Q$, the set of edges E consists of all triples $(q,(\tau,j),q')$ where $\tau = (q,a,q_1...q_m)$ is a q-transition in δ with $q_j = q'$. Observe that the sequences of labels on paths in $G(A)$ describe exactly all possible traces of computations of A.

For defining the k-th power of the FTA A, consider the FTA $A^{(k)} = (Q^k,\Sigma,\delta^k,Q_F^k)$ where $\tau = (<q^{(1)},..,q^{(k)}>,a,<q_1^{(1)},..,q_1^{(k)}>..<q_m^{(1)},..,q_m^{(k)}>) \in \delta^k$ iff $\tau^{(i)} = (q^{(i)},a,q_1^{(i)}..q_m^{(i)}) \in \delta$ for all i. $\tau \in \delta^k$ can be viewed as the k-tuple $\tau = <\tau^{(1)},..,\tau^{(k)}>$ and $\tau^{(i)}$ as its i-th component. Accordingly, every computation ϕ of $A^{(k)}$ for some tree t represents a k-tuple $<\phi^{(1)},..,\phi^{(k)}>$ of computations $\phi^{(i)}$ of A for t; and vice versa. The k-th *power* of A, A^k, is the reduced FTA obtained from $A^{(k)}$ by removing all useless states and transitions.

Finally for $q \in Q$, define $A(q)$ as the reduced FTA obtained from A by employing $\{q\}$ as the set of final states.

A is said to have property (f'1) iff (f'1.1) or (f'1.2) hold:

(f'1.1)   For every edge $(q,(\tau,j),q_j)$ with $\tau = (q,a,q_1...q_m)$ in some strong component of $G(A)$, $da(A(q_{j'})) = 1$ for all $j' \neq j$.

(f'1.2)   For every edge $(<q,q>,(<\tau,\tau'>,j),<q',q'>)$ in some strong component of $G(\Lambda^2)$, $\tau - \tau'$.

There is a polynomial time algorithm to determine the strong components of a digraph. According to Theorem 5 of the last subsection, it can be decided in polynomial time whether or not a given FTA is unambiguous. Hence, it can be decided in polynomial time whether or not an FTA A has property (f'1).

A is said to have property (f'2) iff

(f'2)   there is no path in $G(A^3)$ from $\langle p,p,q \rangle$ to $\langle p,q,q \rangle$ for different states $q \neq p$ in Q.

Reachability in graphs can be decided in polynomial time. Therefore, it also can be decided in polynomial time whether or not an FTA A has property (f'2).

The following proposition relates properties (fi) and (f'i) to each other.

Proposition

Assume A is a reduced FTA. Then the following holds:

(1)   For $i = 1, 2$, if A has property (fi) then A has also property (f'i).

(2)   If A has properties (f'1) and (f'2) then A has also property (f1). If A has property (f'2) then A has also property (f2). $\square$

Now, Theorem 3 follows from Theorem 1 and this proposition. $\square$

2. Bottom-up Finite State Tree Transducers

A bottom-up finite state tree transducer (here FST for short) is a finite state device which produces its output tree while consuming a given input tree in a bottom-up fashion. Since multiple occurrences of variables in patterns are allowed an FST is able to generate several identical copies of images of subtrees. Since some variables can be missing the image of a correctly parsed subtree may be skipped again. In compiler construction finite state transducers are an important tool for manipulating abstract syntax trees [GiSch88]. Formally, FST's can be viewed as one possible generalization of generalized sequential machines (GSM's).

Let X denote the fix denumerable set $\{x_i \mid i \in \mathbb{N}\}$ of variables whereas $X_m := \{x_1,..,x_m\}$. A *bottom-up finite state tree transducer* (FST for short) is a pair $M = (A,T)$. $A = (Q,\Sigma,\delta,Q_F)$ is the FTA *underlying* M whereas $T : \delta \rightarrow T_\Delta(X)$, the *output function* of M, maps every transition $\tau = (q,a,q_1..q_m)$ to its *output pattern* $T(\tau) \in T_\Delta(X_m)$.

Note that an FST according to the definition in [Sei90b] has a transition relation $\delta \subseteq \bigcup_{m \geq 0} Q \times \Sigma_m \times T_\Delta(X_m) \times Q^m$, i.e. the output patterns are part of the transitions. Transducers according to such a definition can be converted into FST's as defined here by allowing "several" transitions $(q,a,q_1...q_m)$ of the underlying finite tree automaton which are distinguished by the different outputs they produce. The extension of the techniques explained here to this slightly more general situation is straight forward and therefore omitted.

The output mapping T of an FST is extended to computations as follows. Assume ϕ is a $(q,q_1..q_k)$-computation of A. If $\phi = x_j$ for some j then $T(\phi) := x_j$. If $\phi = \tau(\phi_1,...,\phi_m)$ then $T(\phi) := T(\tau)[T(\phi_1),..,T(\phi_m)]$. $T(\phi)$ is also called the *output* produced by ϕ.

Observe that by this definition, $T(\phi[\phi_1,..,\phi_k]) = T(\phi)[T(\phi_1),..,T(\phi_k)]$. In abuse of the symbol T we also write $(q,t,s,q_1...q_k) \in T$ provided there is a $(q,q_1...q_k)$-computation ϕ of A for t with $T(\phi) = s$. When dealing with sizes of outputs we abbreviate the functional composition $\omega \circ T$ by Ω.

For some tree $t \in T_\Sigma$, $T_M(t) := \{T(\phi) \mid \phi$ accepting computation of A for t$\}$ denotes the set of outputs of M for t; and $T(M) := \{(t,s) \mid t \in L(A), s \in T_M(t)\}$ is the *translation* defined by M. Two transducers are called *equivalent* iff their translations coincide.

The *size* of M is the sum of the size of its underlying FTA A and the space to represent the output function T. Here, we allow ourselves to use *subtree graphs* whose nodes represent the isomorphy classes of subtrees. This gives us a polynomially sized description of exponentially large trees provided these have only polynomially many non-isomorphic subtrees.

For a given input tree t, $val_M(t) := \#T_M(t)$ denotes the number of different outputs of M for t, whereas the *valuedness* of M is defined by $val(M) := \sup\{val_M(t) \mid t \in T_\Sigma\}$. M is called

- *single-valued*, if $val(M) \le 1$;
- k-*valued*, if $val(M) \le k$;
- *finite-valued*, if $val(M) < \infty$; and
- *infinite-valued*, if $val(M) = \infty$.

In this section we are concerned with the equivalence problem of transducers. Let us first briefly summarize what is known for the word case, i.e. for GSM's. Already very early it was observed that in general the equivalence problem for GSM's in undecidable [Gri68]. However, in 1986 Culik II and Karhumaki prove that the equivalence problem is decidable at least for finite-valued GSM's [CuKa86]. They employ the (confirmed) Ehrenfeucht Conjecture for proving their procedure to be recursive; therefore, no estimation of its complexity is known. In 1988 Weber comes up with a totally different decision procedure (running in double exponential time) which is based on a careful structural analysis of finite-valued GSM's [We88]. In 1989 Weber shows that the equivalence problem is decidable even for finite length-valued GSM's [We89].

Until recently, the knowledge about the equivalence problem and finite-valuedness for tree transducers was comparatively poor. Clearly, since FST's are generalizations of GSM's, equivalence of FST's cannot be decidable either. In 1978 Zachar shows that equivalence is decidable for deterministic FST's [Za78]. In 1980 Engelfrict exhibits a nice generalization of a word lemma by Schützenberger [Sch76] to trees which allows to decide whether or not a given FST is single-valued [En80]. Note that any algorithm which decides single-valuedness can be used to decide equivalence of single-valued FST's.

It turns out that in analogy to the word case [We88], we can construct an algorithm which decides equivalence at least of finite-valued FST's. Again in analogy to the word case, it is based on a precise knowledge about the structural properties of finite-valued FST's.

2.1. Valuedness Bounded by A Constant

In this subsection, we present for every k, a nondeterministic polynomial algorithm which decides whether a given FST has valuedness at least k (Theorem 2). The proof is based on an extension of Engelfriet's Lemma [En80] by means of Ramsey Theory (Theorem 1). Secondly, we present a deterministic polynomial algorithm deciding single-valuedness of FST's (Theorem 5).

We begin by taking a closer look at the monoid $T_\Sigma(x_1)$. Assume $s_1, s_2, t_1, t_2, t'_1, t'_2 \in T_\Sigma(x_1)$. In [En80], Engelfriet points out the following cancellation and factorization properties of $T_\Sigma(x_1)$:

(1)   *Bottom Cancellation:*
        Assume $t_1 \neq t'_1$. Then $s_1 t_1 = s_2 t_1$ and $s_1 t'_1 = s_2 t'_1$ implies $s_1 = s_2$.

(2)   *Top Cancellation:*
        Assume x_1 occurs in s_1. Then $s_1 t_1 = s_1 t_2$ implies $t_1 = t_2$.

(3)   *Factorization:*
        Assume $t_i \neq t'_i$ for $i = 1, 2$. Then $s_1 t_1 = s_2 t_2$ and $s_1 t'_1 = s_2 t'_2$ implies
        $\exists r \in T_\Sigma(x_1): s_1 r = s_2$ or $s_1 = s_2 r$.

Engelfriet employs the properties stated above to derive the following generalization of Schützenberger's lemma for free monoids in [Sch76].

Engelfriet's Lemma
Assume $t_i, t'_i \in T_\Sigma(x_1)$, $i = 0, 1, 2, 3, 4$.
$t_0 .. t_{i-1} \, t_j .. t_4 = t'_0 .. t'_{i-1} \, t'_j .. t'_4$ for all $0 < i < j \leq 4$ implies:
$t_0 t_1 t_2 t_3 t_4 = t'_0 t'_1 t'_2 t'_3 t'_4$. $\square$

Due to the simpler cancellation and factorization properties of a free monoid Schützenberger only needs a subdivision into four factors and three equations as assumptions. Also note, that in [En80] the given lemma is stated with the additional assumption that $t_0 t_2 t_4 = t'_0 t'_2 t'_4$. However, it is not used in the proof. Removing this assumption enables us to use Ramsey Theory [GrRo80] to prove the following extension of the lemma.

Theorem 1
Let $k, n \in \mathbb{N}$, and $t_{\kappa, i} \in T_\Sigma(x_1)$ for $\kappa = 1, .., k$ and $i = 0, .., n$.

(1)   Assume $n \geq 3 \cdot (k^2 - k)!$ and for every $0 < i < j \leq n$ there exist $\kappa < \kappa'$ in
        $\{1, .., k\}$ such that:
        $t_{\kappa, 0} .. t_{\kappa, i-1} \, t_{\kappa, j} .. t_{\kappa, n} = t_{\kappa', 0} .. t_{\kappa', i-1} \, t_{\kappa', j} .. t_{\kappa', n}$.

Then $t_{\kappa_0,0}\cdots t_{\kappa_0,n} = t_{\kappa_1,0}\cdots t_{\kappa_1,n}$ for some $\kappa_0 < \kappa_1$ in $\{1,..,k\}$.

(2)   Assume $n \geq 3\cdot(2k)!$, $t_{0,0},..,t_{0,n}$ is another group of trees in $T_\Sigma(x_1)$ and for every $0 < i < j \leq n$ there exist $\kappa \in \{1,..,k\}$ such that:

$t_{0,0}\cdots t_{0,i-1}\, t_{0,j}\cdots t_{0,n} = t_{\kappa,0}\cdots t_{\kappa,i-1}\, t_{\kappa,j}\cdots t_{\kappa,n}$.

Then $t_{0,0}\cdots t_{0,n} = t_{\kappa_0,0}\cdots t_{\kappa_0,n}$ for some $\kappa_0 \in \{1,..,k\}$. $\square$

Assertion (2) generalizes Engelfriet's Lemma in the same sense as Weber's lemma 2 of [We88] generalizes Schützenberger's lemma. Engelfriet employs his Lemma to prove that single-valuedness of FST's is decidable. The generalization of his Lemma allows us to prove:

Theorem 2

Let $k > 0$ be a fixed constant. Assume M is an FST with n states.

(1) $\mathrm{val}(M) \geq k$ iff there is a tree t of depth less than $3\cdot(k^2-k)!\cdot n^k$ such that $\mathrm{val}_M(t) \geq k$.

(2)   It can be decided in nondeterministic polynomial time whether $\mathrm{val}(M) \geq k$.
     $\square$

One may ask: can one do better, i.e. is k-valuedness complete in coNP? It turns out that, at least for $k = 1$ a *deterministic* polynomial time algorithm can be constructed. For this, let us take a second look at the monoid $T_\Sigma(x_1)$. We observe that the cancellation and factorization properties introduced above can be simplified if all the factors involved contain *at least* one occurrence of the variable x_1.

Assume $s_1,s_2,t_1,t_2 \in T_\Sigma(x_1)$ contain at least one occurrence of x_1. We find:

(1)   *Bottom Cancellation:* $s_1 t_1 = s_2 t_1$ implies $s_1 = s_2$.

(2)   *Top Cancellation:* $s_1 t_1 = s_1 t_2$ implies $t_1 = t_2$.

(3)   *Factorization:* $s_1 t_1 = s_2 t_2$ implies $\exists r : s_1 r = s_2$ or $s_1 = s_2 r$.

Define $\tilde{T}_\Sigma(x_1)$ as the set of all trees of $T_\Sigma(x_1)$ which contain at least one occurrence of x_1. Using the above properties we deduce that $\tilde{T}_\Sigma(x_1)$ is even a *free* monoid!

Call a tree $t \in \tilde{T}_\Sigma(x)$ *irreducible* iff $t \neq x$ and $t = uv$ implies either $u = x$ or $v = x$. For example, $t = a(x,b(x))$ is irreducible whereas $t' = a(b(x),b(x)) = a(x,x)\, b(x)$ is not. Let $I_\Sigma(x)$ denote the set of irreducible trees in $\tilde{T}_\Sigma(x)$. Note that $I_\Sigma(x)$ is infinite.

Theorem 3

(1)   Every tree t in $\tilde{T}_\Sigma(x)$ can be written as a product $t = u_1...u_k$ for irreducible trees $u_1,...,u_k \in I_\Sigma(x)$.

(2)   If $t = u_1...u_k$ and $t = v_1...v_{k'}$ for $u_1,...,u_k,v_1,...,v_{k'} \in I_\Sigma(x)$, then $k = k'$ and $u_\kappa = v_\kappa$ for all κ.

(3)   As a monoid, $\tilde{T}_\Sigma(x)$ is freely generated from $I_\Sigma(x)$, i.e. $\tilde{T}_\Sigma(x) = I_\Sigma(x)^*$.   □

We now present the normal forms necessary for the polynomial test of single-valuedness. As a special convention we introduce a new symbol $\bot$ (i.e. $\bot \notin \Delta$) of rank 0 and define $\Delta_\bot := \Delta \cup \{\bot\}$. We extend the notion of an FST in allowing $\bot$ as an output pattern. However, we only consider FST's (A,T) where an output tree $\bot$ is always substituted for a variable x_j which does not occur in the corresponding output pattern. Therefore, $\bot$ does not occur as the leaf of an output tree $s \neq \bot$ i.e. the output of every $(q,q_1..q_k)$-computation ϕ of A either equals $\bot$ or is in $T_\Delta(X_k)$.

The FST $M = (A,T)$ is called *reduced*, iff (1), (2) and (3) hold:

(1)   A is reduced.

(2)   There is some subset $U(A,T)$ of states such that for every $\tau = (q,a,q_1..q_m) \in \delta$ the following holds:

         if $q \notin U(A,T)$ then $T(\tau) \neq \bot$ and ($q_j \in U(A,T)$ iff x_j does *not* occur in $T(\tau)$);

         if $q \in U(A,T)$ then $T(\tau) = \bot$ and $q_j \in U(A,T)$ for all j.

(3)   If $f \in Q_F$ then $f \neq q_j$ for every transition $(q,a,q_1..q_m) \in \delta$ and every j.

The states in $U(A,T)$ are exactly those which are used by subcomputations ϕ which produce $\bot$. If a computation has reached some state not in $U(A,T)$ we can be sure that the output for the corresponding subcomputation is part of the final output.

It should be noted that in [Sei90b] reducedness does not include property (3) above. However, property (3) can be achieved with little extra computational effort; moreover it simplifies the constructions in the present context.

Secondly, we need a stronger normal form of FST's which allows transitions with output patterns $\neq\bot$ for non-final states q only if the outputs produced by q-computations really depend on the input trees. Assume $M = (A,T)$ is a reduced FST with $A = (Q,\Sigma,\delta,Q_F)$. A state $q \in Q$ is called *constant* iff $\forall$ q-computations ϕ, ϕ': $T(\phi) = T(\phi')$. $Const(A,T)$ denotes the set of constant states of M and $C_T : Const(A,T) \rightarrow T_\Delta \cup \{\bot\}$ the map defined by $C_T(q) := T(\phi)$ for some q-computation ϕ. Observe that always $U(A,T) \subseteq Const(A,T)$. In general, this inclusion is proper; also, $C_T(q)$ possibly may have exponential size. However, the number of *different* subtrees of all the trees $C_T(q)$, $q \in Const(A,T)$, is only polynomial in the size of M. In fact, this is the reason why we are forced to employ graph representations of trees in order to make our algorithm run in polynomial time.

The FST $M = (A,T)$ is called *strongly reduced* iff M is reduced and $U(A,T) = Const(A,T)\backslash Q_F$. One proves:

Theorem 4

(1)   For every FST M there is a reduced FST M_r such that $T(M_r) = T(M)$.
         M_r can be constructed from M in polynomial time.

(2)   For every reduced FST $M = (A,T)$ there is a strongly reduced FST $M_s = (A,T_s)$ such that $T(M) = T(M_s)$

           M_s can be computed in polynomial time. $\square$

Instead of comparing *two* computations w.r.t. *one* output function T it is technically more convenient to consider only one computation but two output functions. Assume $A = (Q,\Sigma,\delta,Q_F)$ is an FTA and $T_1, T_2 : \delta \to T_\Delta(X)$ are maps such that (A,T_i) is an FST for $i = 1, 2$. Then $\Pi = (A,T_1,T_2)$ is called *pairing*. As the size of Π, $|\Pi|$, we simply define $|\Pi| := |(A,T_1)| + |(A,T_2)|$. If (A,T_1) and (A,T_2) are reduced (strongly reduced), then Π is also called reduced (strongly reduced).

We say T_1 is *equivalent* to T_2 (w.r.t. A), $T_1 \equiv T_2$ for short, iff $T(A,T_1) = T(A,T_2)$.

For a (strongly) reduced FST $M = (A,T)$ define the (strongly) reduced pairing $M^2 = (A^2,T_1,T_2)$ where A^2 is the second power of A, and T_i computes the output according to the i-th component of a transition. The following proposition is the formal justification that we may restrict our attention to pairings.

Proposition

Assume $M = (A,T)$ is an FST and $M^2 = (A^2,T_1,T_2)$. Then $\mathrm{val}(M) \le 1$ iff $T_1 \equiv T_2$. $\square$

Assume $\Pi = (A,T_1,T_2)$ is a reduced pairing. We define the following properties (U1), (U2) and (U3):

(U1) $\mathrm{Const}(A,T_1) = \mathrm{Const}(A,T_2)$.

           Observe that in case Π is strongly reduced Property (U1) implies $U(A,T_1) = U(A,T_2)$. Hence for every transition τ the sets of variables occurring in $T_1(\tau)$ and $T_2(\tau)$ agree.

(U2)   Assume $\tau = (q,a,q_1...q_m) \in \delta$, $q_j, q_{j'} \notin U(A,T_1)$ for $j \ne j'$ and $T_i(\tau) = u_i y_i$ where u_i is the maximal prefix of $T_i(\tau)$ in $\tilde{T}_\Delta(*)$. Then there is some $s \in T_\Delta(X_m)$ such that $y_i = s[u_1^{(i)},...,u_m^{(i)}]$ where for every x_j occurring in s, $u_j^{(i)} \in \tilde{T}_\Delta(x_j)$ and $u_j^{(i)} = x_j$ for at least one i.

Property (U3) is described in terms of a subgraph G of the trace graph G(A) of A. G is the maximal subgraph (V,E) of G(A) where $V := (Q\backslash\mathrm{Const}(A,T_i))\cup Q_F$, i.e. V consists of all final states together with all non-constant states. To every node q of G we attach the difference between outputs when reaching this state q during a computation. Formally, the *difference* $\mathrm{diff}(t_1,t_2)$ of trees $t_1, t_2 \in \tilde{T}_\Delta(*)$ is defined as follows. Assume r is the maximal common prefix of t_1 and t_2. Then $\mathrm{diff}(t_1,t_2) = (s_1,s_2)$ iff $t_1 = rs_1$ and $t_2 = rs_2$. Clearly, $\mathrm{diff}(t_1,t_2) = (*,*)$ iff $t_1 = t_2$.

Property (U3) is divided into four assertions. Assertion (1) and (2) give initial and final conditions for the difference of outputs produced along a path. Assertion (4) describes the situation at "branching points" i.e. at nodes o where the output depends on at least two subtrees at o. It states that such nodes are "synchronizing" i.e. both the outputs produced above and below agree. To complete, assertion (3) describes what happens an paths without branching points.

Assume Π has Properties (U1) and (U2). Then Π has Property (U3) iff

(U3)   There is a map diff : $V \to \tilde{T}_\Delta(*)^2$ with:

    (1)   If $q \in Q_F$ then $\mathrm{diff}(q) = (*,*)$.

    (2)   If $\mathrm{diff}(q) = (s_1,s_2)$ and $\tau \in \delta$ is a q-transition with $T_i(\tau) \in T_\Delta$ then $s_1 T_1(\tau) = s_2 T_2(\tau)$.

    Assume $(q,(\tau,j),q') \in E$ and $\mathrm{diff}(q) = (s_1,s_2)$.

    (3)   If $T_i(\tau) \in \tilde{T}_\Delta(x_j)$ and $u_i := T_i(\tau)[*]$ then $\mathrm{diff}(q') = \mathrm{diff}(s_1 u_1, s_2 u_2)$.

    (4)   Assume $T_i(\tau) \notin \tilde{T}_\Delta(x_j)$ and $T_i(\tau) = u_i y_i u'_i$ where $u_i \in \tilde{T}_\Delta(*)$, $u'_i \in \tilde{T}_\Delta(x_j)$ and y_i is the minimal subword containing all variable occurrences $x_{j'}$ with $j' \neq j$. Then $\mathrm{diff}(q') = \mathrm{diff}(u'_1*,u'_2*)$; and $\mathrm{diff}(s_1 u_1, s_2 u_2) = (*,*)$.

Property (U3) is an appropriate generalization of a corresponding property for GSM's characterizing equivalence of two output functions. However, for words (viewed as monadic trees with one special leaf) the situation of (4) never occurs. This observation was exploited by Karhumaki et al. to construct linear sized test sets for regular word languages [KaRy91].

All trees in the image of diff(_) contain at least one occurrence of $*$. They can be spelled by a *finite* set I of irreducible trees which can be computed from Π in polynomial time. Observe again that these trees may have exponential size although they can be represented as a word over I of polynomial length.

Theorem 5

Assume $\Pi = (A,T_1,T_2)$ is a strongly reduced pairing. Then the following two statements are equivalent:

(1) $T_1 \equiv T_2$;

(2) Π has Properties (U1), (U2) and (U3) .

It can be decided in deterministic polynomial time whether or not $T_1 \equiv T_2$. □

Thus, using Theorem 1, the Proposition and Theorem 5 we obtain:

Theorem 6

For every FST $M = (A,T)$ it can be decided in deterministic polynomial time whether or not $\mathrm{val}(M) \leq 1$. □

As a corollary we find:

Corollary 7

Assume $M = (A,T)$ and $M' = (A',T')$ are single-valued FST's such that $L(A) = L(A')$. Then it can be decided in determinisitic polynomial time whether or not M and M' are equivalent. $\square$

This corollary especially shows that equivalence of *deterministic* FST's can be decided in deterministic polynomial time.

2.2. A Characterization of Finite-Valuedness

In this subsection we give a characterization of finite-valued FST's by means of two properties (F1) and (F2). Assume $M = (A,T)$ is a reduced FST with $A = (Q,\Sigma,\delta,Q_F)$.

Property (F1) :
A has property (F1) iff for any x_1-proper $t_1,t_2,t_3 \in T_\Sigma(x_1)$,

$(p,t_1,s_{11},p_1), (p_1,t_2,s_{12},p_1), (p_1,t_3,s_{13},p),$
$(p,t_1,s_{21},p_2), (p_2,t_2,s_{22},p_2), (p_2,t_3,s_{23},q),$
$(q,t_1,s_{31},p_3), (p_3,t_2,s_{32},p_3), (p_3,t_3,s_{33},q) \in T$

implies:
$$s_{11}s_{12}s_{13}\, s_{21}s_{23} = s_{21}s_{22}s_{23}\, s_{31}s_{33} \tag{1}$$

Property (F2) :
A has property (F2) iff for any $\{x_1,x_2\}$-proper $t_1 \in T_\Sigma(\{x_1,x_2\})$, x_1-proper $t_2 \in T_\Sigma(x_1)$ and $t_3 \in T_\Sigma$,

$(p,t_1,s_{11},pp_1), (p_1,t_2,s_{12},p_1), (p_1,t_3,s_{13},\varepsilon),$
$(p,t_1,s_{21},qp_2), (p_2,t_2,s_{22},p_2), (p_2,t_3,s_{23},\varepsilon),$
$(q,t_1,s_{31},qp_3), (p_3,t_2,s_{32},p_3), (p_3,t_3,s_{33},\varepsilon) \in T$

implies:
$$s_{11}[s_{21}[x_1,s_{23}],s_{12}s_{13}] = s_{21}[s_{31}[x_1,s_{33}],s_{22}s_{23}] \tag{2}$$

Property (F1) subsumes all three criteria given in [We90] for finite-valuedness of word transducers. Our additional property (F2) has no analogue in the word case.
An important special case of (F1) is property (F0').

Property (F0') :
A has property (F0') iff for any x_1-proper $t \in T_\Sigma(x_1)$,

$(p,t,s_1,p), (p,t,s_2,q), (p,t,s_3,q), (q,t,s_4,q) \in T$,
$s_1 \neq x_1$ and $\omega(s_2) = \omega(s_3)$

implies:
$$s_2 = s_3 \tag{0'}$$

The following theorem gives a characterization of finite-valued FST's.

Theorem 1

Assume M is a reduced FST. There is a polynomial P independent of M such that the following three statements are equivalent:

(1)   A has properties (F1) and (F2);

(2) $\mathrm{val}(M) < 2^{2^{P(|M|)}}$;

(3) $\mathrm{val}(M) < \infty$. $\square$

Simple examples show that properties (F1) and (F2) are independent. Modifying the example of Subsection 1.3 we find that there are of finite-valued FST's M of arbitrarily big size such that $\mathrm{val}(M) > 2^{2^{c \cdot |M|}}$ for some constant $c>0$.

One possibility to construct a decision procedure for finite-valuedness is to derive an upper bound for the depth of trees t_1,t_2,t_3 which may serve as smallest witnesses for Property (Fi) not to hold. This is done in [Sei90b]. However, a careful extension of the methods in [Sei90c] to derive a deterministic polynomial time algorithm shows:

Theorem 2

It can be decided in deterministic polynomial time whether or not a given FST M is finite-valued. $\square$

The proof relies on a rather tedious series of transformations to the original set {(F1), (F2)} of properties characterizing finite-valued FST's until one ends up in a set of properties which can be decided in polynomial time [Sei90c].

Also, the proof of validity of the implication (3)=>(1) of Theorem 1 in [Sei90b] consists in a sequence of reformulations of the properties (Fj) together with distinctions of various subcases. Especially, the size versions (SFj) of (Fj) are considered. (SFj) is obtained from (Fj) by applying ω to both sides of the conclusion (j). One finds that M as properties (F1) and (F2) iff M has properties (F0'), (SF1) and (SF2). Therefore, it suffices to prove: if $\mathrm{val}(M) < \infty$, then M has properties (F0'), (SF1) and (SF2) .

The proof of the implication (1)=>(2) of Theorem 1 is an induction on the number of strong components of the trace graph G(A) and relies on the following two propositions. Let M = (A,T) be a reduced FST with A = (Q,Σ,δ,Q_F) .

Proposition 1

Assume $t \in T_\Sigma(x_1)$ is x_1-proper, p, q $\in$ Q are in the same strong component of G(A), and ϕ, ϕ' are (p,q)-computations of M for t. If M has property (F0') , then $T(\phi) = T(\phi')$. $\square$

Proposition 2

Assume M has properties (F1) and (F2). Assume $t \in T_\Sigma(x_1)$ is x_1- proper, and $B \subseteq Q$ such that

$\forall p \in B \, \exists q \in B : \exists$ (p,q)-computation of A for t,

$\forall q \in B \, \exists p \in B : \exists$ (p,q)-computation of A for t.

Then for every p, q $\in$ B, the cardinality of the set $\{T(\phi) |\ \phi$ is a (p,q)-computation of A for t} is at most $2^{2^{P(|M|)}}$ for a polynomial P independent of M and t. $\square$

The proof in [We90] of a corresponding result for GSM's is based on the observation that the *difference* between the lengths of possible output words is appropriately bounded. This is no longer true for tree transducers. As a substitute, we prove that the *set* of possible output sizes is finite.

2.3. Equivalence of Finite-valued FST's

This subsection outlines how it can be decided whether the translation of an FST is not included in the translation of a finite-valued FST M' (Theorem 3). Actually, our Theorem 1 of Subsection 2.1., property (F0') and Propositions 1 and 2 of Subsection 2.2. allow to carry over the corresponding constructions of [We88]. Therefore, we proceed as follows. In Theorem 1 we show that this inclusion problem is solvable provided T(M') is a finite union of the translations of single-valued FST's. For this we employ the second extension of Engelfriet's Lemma as stated in Theorem 1 (2) of Subsection 1.1. In Theorem 2 we show that every FST M' which has properties (F1) and (F2) effectively can be decomposed into a finite union of single-valued FST's. Observe that this gives a second proof that properties (F1) and (F2) are necessary for finite-valuedness. Together, Theorems 1 and 2 yield the desired decidability result.

Theorem 1

Assume M = (A,T) is a FST and $M_1,..,M_N$ are single-valued FST's , $M_j = (A_j,T_j)$, such that $L(A) \subseteq L(A_j)$ for all j = 1,..,N . Assume the number of states of any of the automata A_j is bounded by n. Then $T(M) \not\subseteq \bigcup_{j=1}^{N} T(M_j)$ iff there is a pair $(t,s) \in T(M) \setminus \bigcup_{j=1}^{N} T(M_j)$ such that depth(t) < $3 \cdot (2N)! \cdot n^{N+1}$. $\square$

Theorem 2 (Decomposition of finite-valued FST's)

Assume the FST M has properties (F1) and (F2). Then single-valued FST's $M_1,..,M_N$ exist such that

(1) $\qquad N \leq 2^{2^{P(|M|)}}$, and $|M_j| \leq 2^{2^{P(|M|)}}$, j = 1,..,N , for some polynomial P(_) independent of M; and

$$(2) \qquad T(M) = T(M_1) \cup \ldots \cup T(M_N) \ .$$

This decomposition can be found in deterministic time $2^{2^{poly(|M|)}}$. $\square$

The proof of Theorem 2 is based on a "classification" of computations. To every accepting computation ϕ we associate a subset $\Gamma(\phi)$ from a finite set Γ of *specifications* such that $\Gamma(\phi) \cap \Gamma(\phi') \neq \varnothing$ implies $T(\phi) = T(\phi')$ for every two accepting computations ϕ, ϕ' for the same input tree t. Secondly, one shows that it can be decided "online" whether or not an accepting computation admits a given specification, i.e. the set $\Gamma^{-1}(\gamma) := \{ \phi \mid \gamma \in \Gamma(\phi) \}$ is regular.

From Theorems 1 and 2 it is easy to deduce the main theorem of this section:

Theorem 3

Assume M and M' are FST's . If val(M') is finite, then it can be decided in nondeterministic time $2^{2^{2^{poly(|M|+|M'|)}}}$ whether $T(M) \not\sqsubseteq T(M')$. $\square$

As a corollary we obtain:

Corollary 4

Assume M and M' are FST's. If val(M) is finite, then it can be decided in nondeterministic time $2^{2^{2^{poly(|M|+|M'|)}}}$ whether $T(M) \neq T(M')$. $\square$

References

[Aho74]   A.V. Aho, J.E. Hopcroft, J.D. Ullman: The design and analysis of computer algorithms. Addison-Wesley 1974

[Ap76]   T. Apostol: Introduction to analytic number theory. Springer Verlag New York, 1976

[Ba88]   G. Baron: Estimates for bounded automata. Technische Universität Graz und österreichische Computer Gesellschaft, Report 253, part 2, June 1988

[BeReu82]   J. Berstel, C. Reutenauer: Recognizable formal power series on trees. TCS 18 (1982) pp. 115-148

[ChaKoSto81]   A.K. Chandra, D.C. Kozen, L.J. Stockmeyer: Alternation. JACM 28 (1981) pp. 114-133

[CoFr80]   B. Courcelle, P. Franchi-Zannettacchi: Attribute grammars and recursive program schemes, part I. TCS 17 (1982) pp. 163-191

[Cou78]   B. Courcelle: A representation of trees by languages, part II. Theor. Comp. Sci. 7 (1978) pp. 25-55

[CuKa86]   K. Culik II and J. Karhumaki: The equivalence of finite valued transducers (on HDTOL languages) is decidable. TCS 47 (1986), pp. 71-84

[Do70]   J. Doner: Tree acceptors and some of their applications. JCSS 4 (1970) pp. 406-451

[Ei74]   S. Eilenberg: Automata, languages, and machines, Vol. A . Academic Press, New York, 1974

[En80]   J. Engelfriet: Some open questions and recent results on tree transducers and tree languages. In: Formal Language Theory, ed. by R.V. Book, Academic Press 1980, pp. 241-286

[GeStei84]   F. Gecseg, M. Steinby: Tree automata. Akademiai Kiado, Budapest, 1984

[GiSch88]   R. Giegerich, K. Schmal: Code selection techniques: pattern matching, tree parsing and inversion of derivors. Proc. of ESOP 1988, LNCS 300 pp. 245-268

[GrRo80]   R.L. Graham, B.L. Rothschild, J.H. Spencer: Ramsey Theory. John Wiley & Sons, 1980
[Gri68] 16 T. Griffiths: The unsolvability of the equivalence problem for λ-free nondeterministic sequential machines. JACM 15, pp. 409-413

[HaWri60]   G.H. Hardy, E.M. Wright: An introduction to the theory of numbers. Oxford, 4th edition 1960

[KaRy91]   J. Karhumaki, W. Rytter, S. Jarominek: Efficient constructions of test sets for regular and context-free languages. Manuscript 1991

[Kui88]   W. Kuich: Finite automata and ambiguity. Technische Universität Graz und österreichische Computer Gesellschaft, Report 253, June 1988

[MeSto72]   A.R. Meyer, L.J. Stockmeyer: The equivalence problem for regular expressions with squaring requires exponential space. 13th SWAT (1972) pp. 125-129

[Paul78]   W.J. Paul: Komplexitätstheorie. B.G. Teubner Stuttgart 1978

[Ra69]   M.O. Rabin: Decidability of second-order theories and automata on infinite trees. Trans. Amer. Math. Soc. 141 (1969) pp. 1-35

[RoSchoe62]   B. Rosser, L. Schoenfeld: Approximate formulas for some functions of prime numbers. Illinois J. of Mathematics 6 (1962) pp. 64-94

[Sch76]   M. Schützenberger: Sur les relations rationelles entre monoides libres. TCS 3 (1976), pp. 243-259

[Sei89]   H. Seidl: On the finite ambiguity of finite tree automata. Acta Inf. 26, pp. 527-542, 1989

[Sei90a]   H. Seidl: Deciding equivalence of finite tree automata. SIAM J. Comput. 19 (3), pp. 424-437, June 1990

[Sei90b]   Equivalence of finite-valued bottom-up finite state tree transducers is decidable. Proc. CAAP '90, LNCS 431, pp. 269-284, 1990

[Sei90c]   Single-valuedness of bottom-up finite state tree transducers is decidable in polynomial time. Tech. Report 1990

[SteHu81]   R. Stearns, H. Hunt III: On the equivalence and containment problems for unambiguous regular expressions, regular grammars and finite automata. 22th FOCS (1981) pp. 74-81

[SteHu85]   R. Stearns, H. Hunt III: On the equivalence and containment problems for unambiguous regular expressions, regular grammars and finite automata. SIAM J. Comp. 14 (1985) pp. 598-611

[StoMe73]   L.J. Stockmeyer, A.R. Meyer: Word problems requiring exponential time. 5th ACM-STOC pp. 1-9, 1973

[ThaWri68]   J.W. Thatcher, J.B. Wright: Generalized finite automata theory with an application to a decision problem of second order logic. Math. Syst. Th. 2 (1968) pp. 57-81

[Tho84]   W. Thomas: Logical aspects in the study of tree languages. In: 9th Coll. on Trees in Algebra and Programming", B. Courcelle, Ed., Cambridge Univ. Press, 1984, pp. 31-49

[WeSei86]   A. Weber, H. Seidl: On the degree of ambiguity of finite automata. MFCS 1986, Lect. Notes in Comp. Sci. 233, pp. 620-629; to appear in TCS

[WeSei88]   A. Weber, H. Seidl: On finitely generated monoids of matrices with entries in $\mathbb{N}$. Preprint 1988, to appear in RAIRO ITA

[We87]   A. Weber: Über die Mehrdeutigkeit und Wertigkeit von endlichen Automaten und Transducern. Doct. Thesis Frankfurt/Main 1987

[We88]   A. Weber: A decomposition theorem for finite-valued transducers and an application to the equivalence problem. Proc. MFCS 1988 in: LNCS 324, pp. 552-562

[We89]   A. Weber: On the lengths of values in a finite transducer. Proc. of MFCS '89 in: LNCS 379, pp. 523-533, 1989

[We90]   A. Weber: On the valuedness of finite transducers. Acta Informatica 27, pp. 749-780, 1990

[Za78]   Z. Zachar: The solvability of the equivalence problem for deterministic frontier-to-root tree transducers. Acta Cybernetica 4 (1978), pp. 167-177

Tree Automata and Languages
M. Nivat and A. Podelski (editors)

Decidability of the inclusion in monoids generated by tree transformation classes

Zoltán Fülöp   Sándor Vágvölgyi

Research Group on Theory of Automata
Hungarian Academy of Sciences
H-6720 Szeged, Aradi Vértanuk tere 1., Hungary

1   Introduction

In theoretical computer science tree transducers have been studied since the early seventies. In this research area, by a tree we always mean a term (i.e. a polynomial symbol) over a ranked alphabet and by a tree transducer we mean a device, which, by means of its finite number of states and rewrite rules, can transform trees into trees. In this way, each tree transducer induces a relation over terms, which is called a tree transformation and consists of pairs (p, q) of terms such that the tree transducer transforms p into q.

Several types of tree transducers have been studied up to now. First, Rounds [Rou] and Thatcher [Tha1] introduced the concept of the root-to-frontier tree transducer and then, Thatcher [Tha2] defined the notion of the frontier-to-root tree transducer. Using these devices one can model syntax directed translation, which is a wide-spread way of translating high level programming languages, see [Eng5] for a short overview. The names of these devices arise from the fact that a root-to-frontier tree transducer transforms the trees starting form the root and then proceeding towards the frontier, while a frontier-to-root tree transducer works from the frontier towards the root. We note that researchers in tree language theory usually draw trees with root at the top and, in this way, they also use the expressions top-down for root-to-frontier and bottom-up for frontier-to-root, as in the seminal papers [Rou], [Tha1-2], [Eng1] and [Bak]. Later on, to increase the transformational capability, more powerful tree transducers were introduced as well: macro tree transducers [CouFra], [Eng4], [EngVog1], attributed tree transducers [Eng4], [Fül1], [Bar1], high level tree transducers [EngVog4], modular tree transducers [EngVog3] and high level modular tree transducers [Vog2]. These tree transducers induce fairly general tree transformations. For example, attributed tree transducers serve as models for tanslations realized by attribute grammars [Knu], moreover, modular tree transducers compute exactly the class of primitive recursive tree functions [EngVog3].

Nowadays, the theory of tree transducers plays an important role not only in syntax-directed translations but in several other parts of theoretical computer science as well. Instead of giving a long list of applications, we only refer to the theory of term rewrite systems [DerJou], [Hue] (a tree transducer itself is a special term rewrite system) and cite three results concerning ground tree transducers, which were introduced in [DHLT]. A proof in [DHLT] shows that it is decidable whether or not a ground term rewriting system is confluent and, which is even much sronger, it was proved in [DauTis] that the theory of ground term rewriting systems is also decidable. Moreover, in [FülVág8], we gave a fast algorithm for generating a reduced ground term rewriting system from a set of ground term equations, such that this reduced ground term rewriting system provides us with a fast decision algorithm for the word problem of the set of ground term equations.

As we mentioned, a tree transformation induced by a tree transducer is a relation over terms. The composition of tree transformations plays a central role in tree transformation theory and is one of the main concepts in this paper as well. Therefore, before expounding our results, first we recall the notion of the composition and review the most important results concerning it.

Consider the tree transducers A and B (of any type) inducing the tree transformations τ_A and τ_B, respectively. Then, the compositon of $\tau_A \circ \tau_B$ of these two tree transformations is the set of pairs (t, s) of terms such that there is a term r with $(t, r) \in \tau_A$ and $(r, s) \in \tau_B$. Thus the composition of tree transformations is the same as the usual composition of relations. We extend the notion of composition for tree tarnsformation classes (i.e. for classes which consist of tree transformations) in the natural way: the composition $Y \circ Z$ of the tree transformation classes Y and Z is the class of tree transformations $\tau \circ \sigma$ with $\tau \in Y$ and $\sigma \in Z$. We say that the class Y is closed under composition, if $Y \circ Y \subseteq Y$. Moreover, for each natural number n, the nth power of the class Y is defined by $Y^1 = Y$ and $Y^{n+1} = Y^n \circ Y$ for every $n \geq 1$.

In what follows, we shall study tree transformation classes induced by tree transducers. We shall speak about the class of x tree transformations, where x stands for some type of tree transducers, meaning the class of tree transformations which can be induced by tree transducers of type x. The class of root-to-frontier tree transformations is denoted by R, the class of frontier-to-root tree transformations is denoted by F, and the class of homomorphism tree transformations is denoted by H. The prefixes D, L, N and LN in front of R and F stand for the deterministic, the linear, the nondeleting, and the linear, nondeleting subclasses of the classes R and F, respectively. The prefix D may be used together with the other three ones thus, for example, LDR denotes the class of linear, deterministic root-to-frontier tree transformations.

Most results on compositions of tree transformation classes concern one of the following four overlapping areas.

(a) Does a tree transformation class appear as a compositon of its subclasses of simpler structure?

Such connections are usually referred to as decomposition results. Several decomposition results can be found in the papers [Bak], [Eng1-2], [EngFil], [EngVog1], [Füll-3], [FülVág1,4] and in the book [GécSte]. For example, we cite the decompositions $DF = LNDF \circ H$ and $H = NH \circ LH$, see [Eng1] and [FülVág1].

(b) Which tree transformation classes are closed under composition? If a class is not closed under composition, then whether its increasing powers form a proper hierarchy with respect to composition, or whether the hierarchy collapses at some power?

Such problems were investigated in [ArnDau], [Bar2], [DHLT], [Eng2,3,6], [EngVog1,3], [FülVág1,3,4], [GécSte], [VágFül] and [Vog1]. For example, Engelfriet [Eng 6] showed that $F^n \subset F^{n+1}$ and $R^n \subset R^{n+1}$ for every $n \geq 1$. On the other hand, $DF^2 = DF$, i.e., DF is closed under composition, cf. [Eng1].

(c) If the increasing powers form a proper hierarchy under composition, then give a characterization of the n-fold power using another type of tree transducers ([Dau], [EngVog2-4], [Vág]).

(d) Find inclusions and equalities that hold for (compositions of) tree transformation classes. This problem may be considered as both combination and generalization of problems (a) and (b). Such results can be found in [ArnDau], [Bak], [Dau], [Eng1,2], [EngVog1], [Fül1-3], [FülVág1] and [GécSte]. For example, $DR^2 = NDR \circ LH$ was shown in [FülVág1].

It is not hard to observe that from the already verified equalities and inclusions we obtain new ones by applying substitutions. For instance, from the two equalities appearing in (a) and from the equality $NDF = LNDF \circ NH$, which is a special case of the first one in (a), we obtain that $NDF \circ LH = LNDF \circ NH \circ LH = LNDF \circ H = DF$. In a similar way we obtain the inclusion $DR \subset NDR \circ LH$ from the inclusion $DR \subset DR^2$ [Rou] and the equality $DR^2 = NDR \circ LH$ [FülVág1].

Because of the huge amount of already proved relations for classes of tree transformations, one may be interested in generating all valid equalities and inclusions between compositions of tree transformation classes that are taken from a given reservoir set of such classes. More precisely, if we are given a finte set M of tree transformation classes and two expressions of the form

$$Y_1 \circ \ldots \circ Y_m \quad \text{and} \quad Z_1 \circ \ldots \circ Z_n \qquad (*)$$

where $Y_i, Z_j \in M$ for every $1 \leq i \leq m$ and $1 \leq j \leq n$, then we want to decide, *by applying known equalities and inclusions*, which one of the following four mutually excluding statements is true:

(i) $Y_1 \circ \ldots \circ Y_m = Z_1 \circ \ldots \circ Z_n$

(ii) $Y_1 \circ \ldots \circ Y_m \subset Z_1 \circ \ldots \circ Z_n$

(iii) $Z_1 \circ \ldots \circ Z_n \subset Y_1 \circ \ldots \circ Y_m$

(iv) $Y_1 \circ \ldots \circ Y_m$ and $Z_1 \circ \ldots \circ Z_n$ are incomparable.

Consider the following example. Let $M = \{DR\}$. Then, it is obvious that the already known equality $DR^3 = DR^2$ [FülVág1] and inclusion $DR \subset DR^2$ [Rou] are sufficient to decide, for any $m, n \geq 1$ and classes DR^m and DR^n, which one of the statements corresponding to (i), (ii), (iii) and (iv) holds. (Of course, in what follows, we shall be interested in more general sets M.)

The aim of this paper is to provide a method and apply it for more general sets M. The crux of the matter is that we construct objects depending on M using which we are able to decide effectively for arbitrary two expressions of the form $(*)$, which one of the relations (i), (ii), (iii) and (iv) holds. These objects are the following:

(a) a finite Thue system T over M,

(b) a set N of representatives for the Thue congruence $\leftrightarrow_T^*$ generated by T over Σ^*,

(c) the inclusion diagram for the tree transformation classes represented by elements of N,

(d) an algorithm which outputs, for each word $w \in M^*$, the unique representative $u \in N$ of w such that $w \leftrightarrow^*_T u$.

Note that if we choose M to be too general, e.g., if M is the set of all tree transformation classes have been defined in the literature up to now, then our problem becomes so sophisticated that it seems hopeless to construct the above objects. (Later on, in Section 6, we shall detail what we mean by a "too general" M.)

Nonetheless, our method can be used within wide limits. To support this statement, we shall apply it successfully for both choices

$$M = \{ DF, LDF, NDF, LNDF, H, LH, NH \}$$

and

$$M = \{ DR, LDR, NDR, LNDR, H, LH, NH \}.$$

Results concerning the first and the second choice of M were published in [Fül3] and [FülVág1,2,5-7], respectively.

The present paper is organized in the following way. In Section 2 we recall some notations and notions which will be used. Section 3 contains the exact formulation of the problem described above and the method for solving it. In Section 4 we apply this method for the first and then in Section 5 for the second of the above two choices of M. Finally, in Section 6 we summarize the results of the paper and raise some new problems as well.

2   Notions and notations

2.1   Sets and relations

For arbitrary sets A and B, we denote by $A \subseteq B$ that A is a subset of B, by $A \subset B$ that A is a proper subset of B and by $A \not\subseteq B$ that A is not a subset of B. Moreover, $A \bowtie B$ stands for that A and B are incomparable. We shall identify a singleton set $\{ a \}$ with its unique element a.

Let H be a partially ordered set of which elements are sets and let the partial order be the inclusion $\subseteq$. Then, by the inclusion diagram of H, we mean the Hasse diagram [BurSan] of H with respect to the partial order $\subseteq$.

Given two sets A and B, any subset θ of the Cartesian product $A \times B$ is called a relation from A to B. We write $a\theta b$ to mean that $(a, b) \in \theta$. If θ is a relation from A to B, then the set $\{ a \mid a\theta b$ for some $b \in B \}$ is called the domain of θ and is denoted by $dom(\theta)$, while the set $\{ b \in B \mid a\theta b$ for some $a \in A \}$ is the range of θ and is denoted by $ran(\theta)$. We say that θ is total, if $dom(\theta) = A$. For $a \in A$, we introduce the notation

$$\theta(a) = \{ b \mid a\theta b \}$$

and we extend this notation for arbitrary $L \subseteq A$ as follows:

$$\theta(L) = \cup(\, \theta(a) \mid a \in L\,).$$

Moreover, if Y is a class of relations and H is a class of sets, then we put

$$Y(H) = \{\, \theta(L) \mid \theta \in Y,\ L \in H \,\},$$

hence the elements of $Y(H)$ are sets as well.

We now introduce the notion of composition for relations. Let θ be a relation from A to B and let σ be a relation from B to C. Then the composition of θ and σ is the relation $\theta \circ \sigma$ from A to C defined by

$$\theta \circ \sigma = \{\, (a,c) \mid a\theta b \text{ and } b\sigma c \text{ for some } b \in B \,\}.$$

We extend the concept of composition for classes of relations. If Y and Z are classes of relations, then

$$Y \circ Z = \{\, \theta \circ \sigma \mid \theta \in Y \text{ and } \sigma \in Z \,\}.$$

Moreover, let $Y^1 = Y$ and for any $n > 1$, $Y^n = Y \circ Y^{n-1}$. A relation from A to A is also called a relation over A. The identical relation $\{\, (a,a) \mid a \in A \,\}$ over A is denoted by $Id(A)$. Whenever θ is a relation over A, the n-fold composition of θ, denoted by θ^n, is defined by $\theta^0 = Id(A)$ and $\theta^n = \theta \circ \theta^{n-1}$, if $n > 0$. We introduce the notion $\theta^* = \cup(\,\theta^n \mid n \geq 0\,)$, too, that is to say, θ^* is the reflexive, transitive closure of θ.

Let A be an arbitrary set and θ be an equivalence relation over A. A set $N \subset A$ is said to be a set of representatives for θ if, for each $a \in A$, there exists exactly one $b \in N$ such that $a\theta b$.

Finally we introduce the concept of a hierarchy. A sequence of classes $\{\, C_k \mid k \geq 0 \,\}$ is called a hierarchy, if $C_k \subseteq C_{k+1}$ for each $k \geq 0$. A hierarchy is said to be proper, if each inclusion is proper, i.e., $C_k \subset C_{k+1}$, for each $k \geq 0$. The class $\cup(\,C_k \mid k \geq 0\,)$ is the supremum of the hierarchy and is denoted by $\cup C_k$.

2.2 Words, Thue systems and string rewrite systems

An alphabet Σ is a finite nonempty set. The elements of an alphabet Σ are called letters. A word (or string) over Σ is a finite-length Σ-sequence. A typical word can be written as $w = a_1...a_k$, $k \geq 0$ with $a_i \in \Sigma$ for every $1 \leq i \leq k$. We allow $k = 0$, which yields the empty word, which is denoted by λ. The number k is called the length of w, written $l(w)$, and is the number of occurrences of letters of Σ in w. Let Σ^* denote the set of all words over Σ. Let w and z be in Σ^*. Define a binary operation of concatenation where we form the new word wz. Recall that Σ^* is the free monoid generated by Σ under the operation of concatenation with the empty word λ as identity.

Let Σ be an alphabet. A Thue system [Boo2] T over Σ is a finite subset of $\Sigma^* \times \Sigma^*$ and each element (u,v) of T is called a rewriting rule. The Thue congruence generated by T is the reflexive, transitive closure $\leftrightarrow_T^*$ of the relation $\leftrightarrow_T$ defined as follows: for any $w, z \in \Sigma^*$, $w \leftrightarrow_T z$ if and only if there exist $x, y \in \Sigma^*$ and $(u,v) \in T$ such that either $w = xuy$ and $z = xvy$, or $w = xvy$ and $z = xuy$. It is well-known that $\leftrightarrow_T^*$ is the least congruence over Σ^* containing T.

The reduction relation $\rightarrow_T$ over Σ^* is defined as follows: for any $w, z \in \Sigma^*$, $w \rightarrow_T z$ if and only if $w \leftrightarrow_T z$ and $l(w) > l(z)$. A word $w \in \Sigma^*$ is irreducible for T (or T-irreducible) if there is no $z \in \Sigma^*$ such that $w \rightarrow_T z$. The set of all irreducible strings for T is denoted by $IRR(T)$.

Let T be a Thue system over Σ. We say that

(a) T is Church-Rosser if for all $w, z \in \Sigma^*$, if $w \leftrightarrow_T^* z$, then there exists an $x \in \Sigma^*$ such that $w \rightarrow_T^* x$ and $z \rightarrow_T^* x$, and

(b) T is confluent if for all $w, z, x \in \Sigma^*$, if $w \rightarrow_T^* z$ and $w \rightarrow_T^* x$, then there exists a $y \in \Sigma^*$ such that $z \rightarrow_T^* y$ and $x \rightarrow_T^* y$.

It should be clear that for Thue systems, any Church-Rosser system is confluent. However, the possibility of length-preserving rules allows confluent Thue systems that are not Church-Rosser. For instance the system $T = \{ (aa, bb) \}$ over $\{ a, b \}$ is confluent but is not Church-Rosser. If T and T' are Thue systems over Σ such that $\leftrightarrow_T^* = \leftrightarrow_{T'}^*$, then T and T' are said to be equivalent.

A string rewriting system [BKR] S over an alphabet Σ is again a finite subset of $\Sigma^* \times \Sigma^*$. Again, we call the elements of S rewriting rules. The relation $\rightarrow_S^*$ is the reflexive, transitive closure of the relation $\rightarrow_S$ defined by: for $w, z \in \Sigma^*$, $w \rightarrow_S z$ if and only if there exists $x, y \in \Sigma^*$ and $(u, v) \in S$ such that $w = xuy$ and $z = xvy$. We say that z can be derived from w in S, if $w \rightarrow_S^* z$ holds. Notice that the rules of a rewriting system may be applied only in one direction given by the system, while the rules of a Thue system may be applied in either direction. The symmetric, reflexive and transitive closure $\leftrightarrow_S^*$ of $\rightarrow_S$ is a congruence over Σ^*. It is called the Thue congruence generated by S.

Consider a rewriting system S over Σ. We say that S is

(a) noetherian if there are no infinite chains of the form $w_1 \rightarrow_S w_2 \rightarrow_S \ldots$,

(b) Church-Rosser if for every $w, z \in \Sigma^*$, $w \leftrightarrow_S^* z$ implies that $w \rightarrow_S^* x$ and $z \rightarrow_S^* x$ for some $x \in \Sigma^*$, and

(c) confluent if for every $w, z, x \in \Sigma^*$, $w \rightarrow_S^* z$ and $w \rightarrow_S^* x$ imply that $z \rightarrow_S^* y$ and $x \rightarrow_S^* y$ for some $y \in \Sigma^*$.

Note a difference between Thue systems and rewriting systems: a rewriting system S is confluent if and only if it is Church-Rosser, see [Boo2], [Hue], [Jan].

A noetherian and confluent rewriting system is called complete. A word w is called irreducible with respect to S (or S-irreducible) if there is no z such that $w \rightarrow_S z$. The set of irreducible words with respect to S is denoted by $IRR(S)$.

Finally we mention a sufficient condition for S to be noetherian. A weight function is a mapping $\rho : \Sigma \rightarrow \{ 1, 2, \ldots \}$, where for $a \in \Sigma$, $\rho(a)$ is the weight of a. The weight of a word $w \in \Sigma^*$ is defined by: $\rho(w) = 0$ if $w = \lambda$, and $\rho(w) = \rho(v) + \rho(a)$ if $w = va$, for some $v \in \Sigma^*$ and $a \in \Sigma$. For example, if $\rho(a) = 1$ for each $a \in \Sigma$, then $\rho(w) = l(w)$. We say that S is weight reducing if, for each $(u, v) \in S$, $\rho(u) > \rho(v)$ holds. It should be clear that each weight reducing string rewriting system is noetherian as well.

2.3 Trees and tree transformations

A ranked alphabet Σ is an alphabet in which every symbol has a unique rank in the set of nonnegative integers. For any $m \geq 0$, we denote by Σ_m the set of symbols in Σ which have rank m.

For a ranked alphabet Σ and a set H, the set of trees over Σ indexed by H, denoted by $T_\Sigma(H)$, is the smallest set U satisfying the following two conditions:

(i) $H \cup \Sigma_0 \subseteq U$,

(ii) $\sigma(t_1,\ldots,t_m) \in U$ whenever $m > 0$, $\sigma \in \Sigma_m$ and $t_1,\ldots,t_m \in U$.

The set of trees over Σ is $T_\Sigma(\emptyset)$, and we simply write T_Σ for $T_\Sigma(\emptyset)$.

We define the height $height(t)$ of a tree $t \in T_\Sigma$ as follows: if $t \in \Sigma_0$, then $height(t) = 0$ and otherwise, if $t = \sigma(t_1,\ldots,t_m)$ with $\sigma \in \Sigma_m$, $m > 0$, then $height(t) = 1 + \max\{\, height(t_i) \mid 1 \le i \le m \,\}$.

If Σ and Δ are ranked alphabets, then any subset of $T_\Sigma \times T_\Delta$ is a tree transformation from T_Σ to T_Δ. Thus a tree transformation is a relation from T_Σ to T_Δ.

The class of all total, identical tree transformations $Id(T_\Sigma)$ is denoted by I.

We specify a countable set $X = \{\, x_1, x_2, \ldots \,\}$ of symbols called variables and we set $X_m = \{\, x_1, \ldots, x_m \,\}$ for every $m \ge 0$.

We often refer to the set $T_\Sigma(X_m)$, which is abbreviated by $T_{\Sigma,m}$. We distinguish a subset $\hat{T}_{\Sigma,m}$ of $T_{\Sigma,m}$ as follows: a tree $t \in T_{\Sigma,m}$ is in $\hat{T}_{\Sigma,m}$ if and only if each variable in X_m appears exactly once in t and the order of the variables in t is $x_1,\ldots,x_m$. For example, if $\Sigma = \Sigma_0 \cup \Sigma_2$ with $\Sigma_0 = \{\, a \,\}$ and $\Sigma_2 = \{\, \sigma \,\}$, then $\sigma(x_1, \sigma(a, x_1)) \in T_{\Sigma,1}$ but $\sigma(x_1, \sigma(a, x_1)) \notin \hat{T}_{\Sigma,1}$. On the other hand, $\sigma(x_1, \sigma(a, x_2)) \in \hat{T}_{\Sigma,2}$.

Finally, we introduce the notion of tree substitution. Let $m \ge 0$, $t \in T_{\Sigma,m}$ and $a_1,\ldots,a_m \in H$, where H is an arbitrary set. We denote by $t(a_1,\ldots,a_m)$ the tree which is obtained from t by replacing each occurrence of x_i in t by a_i for every $1 \le i \le m$. We obviously have $t(a_1,\ldots,a_m) \in T_\Sigma(H)$.

2.4   Tree transducers

A nice overview of the theory of tree transducers with a unified terminology can be found in [GécSte]. The following definitions are adopted from this book.

A frontier-to-root tree transducer (f transducer for short) is a system $A = (\Sigma, Q, \Delta, R, Q')$, where

(a) Σ and Δ are the input and output ranked alphabets, respectively,

(b) Q is a ranked alphabet, called the set of states, such that $Q = Q_1$ and $Q \cap (\Sigma \cup \Delta \cup X) = \emptyset$,

(c) $Q' \subseteq Q$ is the set of final states,

(d) R is a finite set of rewriting rules of the form $\sigma(q_1(x_1),\ldots,q_m(x_m)) \to q(r)$ with $m \ge 0, \sigma \in \Sigma_m$, $q, q_1,\ldots,q_m \in Q$ and $r \in T_{\Delta,m}$. (Here $\sigma(q_1(x_1),\ldots,\ q_m(x_m))$ is the left-hand side of the rule.)

The tree transducer A can be used to induce a tree transformation from T_Σ to T_Δ in the way described as follows. Define the relation $\Rightarrow_A$, called derivation, over the set $T_\Sigma(Q(T_\Delta))$, where $Q(T_\Delta) = \{\, q(t) \mid q \in Q, t \in T_\Delta \,\}$, so that for any $t, s \in T_\Sigma(Q(T_\Delta))$, $t \Rightarrow_A s$ if and only if the next two conditions hold:

(a) There is a rule of the form $\sigma(q_1(x_1),\ldots,q_m(x_m)) \to q(r)$ in R.

(b) s appears from t by replacing an occurrence of a subtree $q(\sigma(q_1(t_1),\ldots,q_m(t_m)))$ of t by $q(r(t_1,\ldots,t_m))$, where $t_1,\ldots,t_m \in T_\Delta$.

It should be clear that the relation $\Rightarrow_A$ can be interpreted as a method of rewriting terms into terms, hence $\Rightarrow_A^*$ is the iterated application of this method. The tree transformation induced by A is defined as

$$\tau_A = \{\, (t, s) \in T_\Sigma \times T_\Delta \mid t \overset{*}{\underset{A}{\Rightarrow}} q(s) \text{ for some } q \in Q' \,\}.$$

Now we introduce some special types of the f transducer. Let, to this end, $A = (\Sigma, Q, \Delta, R, Q')$ be an f transducer. We say that A is

(a) total if, for each $m \geq 0, \sigma \in \Sigma_m$ and $q_1, \ldots, q_m \in Q$, there is at least one rule in R with left-hand side $\sigma(q_1(x_1), \ldots, q_m(x_m))$;

(b) a deterministic f transducer (df transducer) if there are no two different rules in R with the same left-hand side;

(c) a linear f transducer (lf transducer) if for each rule

$$\sigma(q_1(x_1), \ldots, q_m(x_m)) \to q(r)$$

in R, each of the variables $x_1, \ldots, x_m$ appears at most once in r;

(d) a nondeleting f transducer (nf transducer) if for each rule specified as above, each of the variables $x_1, \ldots, x_m$ appears at least once in r;

(e) a linear nondeleting f transducer (lnf transducer) if A is both linear and nondeleting. This means that for each rule $\sigma(q_1(x_1), \ldots, q_m(x_m)) \to q(r)$ in R, each of the variables $x_1, \ldots, x_m$ appears exactly once in r;

(f) a frontier-to-root homomorphism if Q is a singleton set, $Q = Q'$, moreover if A is total and deterministic.

We shall shall speak about df, ldf, ndf and lndf transducers, of which the last one, for example, stands for the linear nondeleting deterministic frontier-to-root tree transducer.

Note that if A is total, then τ_A is a total tree transformation, while if A is a df transducer, then τ_A is a partial function from T_Σ to T_Δ.

Now let x be a modifier taken from the set { df, ldf, ndf, lndf }. A tree transformation is called an x tree transformation if it can be induced by some x transducer. The classes of all df, ldf, ndf and lndf tree transformations are denoted by DF, LDF, NDF and $LNDF$, respectively. Thus, for example, DF stands for the class of all tree transformations induced by df transducers. Note that I is a proper subclass of any of the classes DF, LDF, NDF and $LNDF$.

Next we provide two examples to illustrate the concept of the f transducer.

Example. Let us consider the nondeleting frontier-to-root homomorphism $A = (\Sigma, Q, \Delta, R, Q')$ defined by $\Sigma = \Sigma_0 \cup \Sigma_1$, $\Sigma_0 = \{\alpha\}$, $\Sigma_1 = \{\beta\}$, $Q = Q' = \{q\}$, $\Delta = \Delta_0 \cup \Delta_2$, $\Delta_0 = \{\alpha\}$, $\Delta_2 = \{\gamma\}$; R consits of the rewriting rules $\alpha \to q(\alpha)$, $\beta(q(x_1)) \to q(\gamma(x_1, x_2))$.

A typical derivation of A looks like as follows.

$$\beta(\beta(\alpha)) \underset{A}{\Rightarrow} \beta(\beta(q(\alpha))) \underset{A}{\Rightarrow} \beta(q(\gamma(\alpha, \alpha))) \underset{A}{\Rightarrow} q(\gamma(\gamma(\alpha, \alpha), \gamma(\alpha, \alpha))).$$

It should be clear that given an input tree of height n over Σ, then A outputs the binary balanced tree of height n over Δ.

Example. The ldf transducer $A = (\Sigma, Q, \Delta, R, Q')$ is defined by $\Sigma = \Sigma_0 \cup \Sigma_2$, $\Sigma_0 = \{\alpha\}$, $\Sigma_2 = \{\gamma\}$, $Q = \{q_o, q_e\}$, $Q' = \{q_o\}$, $\Delta = \Delta_0$, $\Delta_0 = \{\delta, \epsilon\}$, R consists of the rewriting rules $\alpha \to q_o(\delta)$, $\gamma(q_o(x_1), q_o(x_2)) \to q_e(\epsilon)$, $\gamma(q_o(x_1), q_e(x_2)) \to q_o(\delta)$, $\gamma(q_e(x_1), q_o(x_2)) \to q_o(\delta)$ and $\gamma(q_e(x_1), q_e(x_2)) \to q_e(\epsilon)$.

Consider the following derivation of A.

$$\gamma(\alpha, \gamma(\alpha, \alpha)) \underset{A}{\Rightarrow} \gamma(q_o(\delta), \gamma(\alpha, \alpha)) \underset{A}{\Rightarrow} \gamma(q_o(\delta), \gamma(q_o(\delta), \alpha)) \underset{A}{\Rightarrow}$$

$$\gamma(q_o(\delta), \gamma(q_o(\delta), q_o(\delta))) \underset{A}{\Rightarrow} \gamma(q_o(\delta), q_e(\epsilon)) \underset{A}{\Rightarrow} q_o(\delta).$$

It is not hard to see that given an input tree t over Σ, A outputs δ if t contains an odd number of α's and outputs ϵ if t contains an even number of α's. Consequently, the tree transformation induced by A is

$$\tau_A = \{\, (t, \delta) \mid t \in T_\Sigma \text{ and } t \text{ contains an odd number of } \alpha's \,\}.$$

Next we define the concept of the root-to-frontier tree transducer.

The system $A = (\Sigma, Q, \Delta, R, q_0)$ is a root-to-frontier transducer (r transducer for short), where Σ, Q and Δ are the same as in the frontier-to-root case, $q_0 \in Q$ is the initial state and R is a finite set of rewriting rules of the form $q(\sigma(x_1, \ldots, x_m)) \to r$ with $m \geq 0$, $\sigma \in \Sigma_m$, $q \in Q$ and $r \in T_\Delta(Q(X_m))$. (Here $Q(X_m)$ denotes the set $\{\, q(x_i) \mid q \in Q, 1 \leq i \leq m \,\}$.)

The rules in R induce a relation $\Rightarrow_A$, called derivation, over $T_\Delta(Q(T_\Sigma))$ as follows: for $t, s \in T_\Delta(Q(T_\Sigma))$, we write $t \Rightarrow_A s$ if and only if there is a rule $q(\sigma(x_1, \ldots, x_m)) \to r$ in R and s appears from t by replacing a subtree $q(\sigma(t_1, \ldots, t_m))$ of t by $r(t_1, \ldots, t_m)$, where $t_1, \ldots, t_m \in T_\Sigma$. Then the tree transformation induced by A is the relation

$$\tau_A = \{\, (t, s) \in T_\Sigma \times T_\Delta \mid q_0(t) \underset{A}{\overset{*}{\Rightarrow}} s \,\}.$$

Analogously, we define some restrictions on r transducers. Let, to this end, $A = (\Sigma, Q, \Delta, R, Q')$ be an r transducer. We say that A is

(a) total if, for each $m \geq 0, \sigma \in \Sigma_m$ and $q_1, \ldots, q_m \in Q$, there is at least one rule in R with left-hand side $\sigma(q_1(x_1), \ldots, q_m(x_m))$;

(b) a deterministic r transducer (dr transducer) if there are no two different rules in R with the same left-hand side;

(c) a linear r transducer (lr transducer) if for each rule

$$q(\sigma(x_1, \ldots, x_m)) \to r$$

in R, each of the variables $x_1, \ldots, x_m$ appears at most once in r;

(d) a nondeleting r transducer (nr transducer) if for each rule specified as above, each of the variables $x_1, \ldots, x_m$ appears at least once in r.

(e) a linear nondeleting r transducer (lnr transducer) if A is both linear and nondeleting.

(f) a root-to-frontier homomorphism if Q is a singleton set, $Q = \{\, q_0 \,\}$, A is total and is deterministic.

We shall consider dr, ldr, ndr and lndr transducers, of which the last, for example, stands for the linear nondeleting deterministic root-to-frontier tree transducer.

Note that, exactly as in the frontier-to-root case, if A is total, then τ_A is a total tree transformation, while if A is a dr transducer, then τ_A is a partial function from T_Σ to T_Δ.

Now let x be a modifier taken from $\{\, dr, ldr, ndr, lndr \,\}$. A tree transformation is called an x tree transformation if it can be induced by some x transducer. The classes of all dr, ldr, ndr and lndr tree transformations are denoted by DR, LDR, NDR and $LNDR$, respectively. For example, LDR stands for the class of all tree transformations induced by ldr transducers. We again note that I is a proper subclass of any of the classes DR, LDR, NDR and $LNDR$.

Finally we consider the class of homomorphism tree transducers. It is well-known that for each frontier-to-root homomorphism tree transducer A, there exists a root-to-frontier homomorphism tree transducer B such that $\tau_A = \tau_B$, and vice versa, see [GécSte]. Hence we do not distinguish between root-to-frontier and frontier-to-root homomorphism tree transducers and tree transformations induced by them and simply speak about h transducers and h transformations. For this notation we again use the modifiers l, n and ln, thus for instance, by an lh transducer we mean a linear homomorphism tree transducer. The classes of all h, lh, nh and lnh transformations are denoted by H, LH, NH and LNH, respectively. It also holds that I is a proper subclass of any of the classes H, LH, NH and LNH.

We now provide two examples to illustrate the concept of the r transducer.

Example. The ndr transducer $A = (\Sigma, Q, \Delta, R, q_0)$ is defined by $\Sigma = \Sigma_0 \cup \Sigma_1$, $\Sigma_0 = \{\alpha\}$, $\Sigma_1 = \{\beta\}$, $Q = \{q_0, q_1, q_2\}$, $\Delta = \Delta_0 \cup \Delta_2$, $\Delta_0 = \{\alpha, \delta\}$, $\Delta_2 = \{\gamma\}$; R consists of the rewriting rules $q_0(\beta(x_1)) \to \gamma(q_1(x_1), q_2(x_2))$, $q_1(\beta(x_1)) \to \beta(q_1(x_1))$, $q_2(\beta(x_1)) \to \beta(q_2(x_1))$, $q_1(\alpha) \to \alpha$ and $q_2(\alpha) \to \delta$.

Consider the following derivation of A.

$$q_0(\beta(\beta(\alpha))) \underset{A}{\Rightarrow} \gamma(q_1(\beta(\alpha)), q_2(\beta(\alpha))) \underset{A}{\Rightarrow} \gamma(\beta(q_1(\alpha)), q_2(\beta(\alpha))) \underset{A}{\Rightarrow}$$

$$\gamma(\beta(\alpha), q_2(\beta(\alpha))) \underset{A}{\Rightarrow} \gamma(\beta(\alpha), \beta(q_2(\alpha))) \underset{A}{\Rightarrow} \gamma(\beta(\alpha), \beta(\delta)).$$

By induction on the heights of input trees we can easily prove that

$$\tau_A = \{\, (\beta^n(\alpha), \gamma(\beta^{n-1}(\alpha), \beta^{n-1}(\delta))) \mid n = 1, 2, \dots \,\},$$

where $\beta^0(\alpha) = \alpha$ and $\beta^n(\alpha) = \beta(\beta^{n-1}(\alpha))$ if $n > 0$; and similarly for δ.

Example. The ldr transducer $A = (\Sigma, Q, \Delta, R, q_0)$ is defined by $\Sigma = \Sigma_0 \cup \Sigma_1 \cup \Sigma_2$, $\Sigma_0 = \{\alpha\}$, $\Sigma_1 = \{\beta\}$, $\Sigma_2 = \{\gamma\}$, $Q = \{q_0\}$, $\Delta = \Delta_0 = \{\delta\}$; R consists of the rewriting rules $q_0(\beta(x_1)) \to \delta$, $q_0(\gamma(x_1, x_2)) \to \delta$.

It should be clear that $\tau_A = \{\, (t, \delta) \mid t \in T_\Sigma$ and $height(t) > 0 \,\}$.

3 The problem and a general method for the solution

We raise the following problem. Let us be given a finite set M of tree transformation classes (i.e. classes like R, F, their fundamental subclasses or classes of any other type). Give an algorithm that, for every $m, n \geq 0$ and tree transformation classes

$$Y_1 \circ \dots \circ Y_m \quad \text{and} \quad Z_1 \circ \dots \circ Z_n \qquad (*)$$

over M, where $Y_i, Z_j \in M$, for any $1 \leq i \leq m$ and $1 \leq j \leq n$, can decide which one of the following four mutually excluding conditions holds:

(i) $Y_1 \circ \ldots \circ Y_m = Z_1 \circ \ldots \circ Z_n$

(ii) $Y_1 \circ \ldots \circ Y_m \subset Z_1 \circ \ldots \circ Z_n$

(iii) $Z_1 \circ \ldots \circ Z_n \subset Y_1 \circ \ldots \circ Y_m$

(iv) $Y_1 \circ \ldots \circ Y_m \bowtie Z_1 \circ \ldots \circ Z_n.$

Note that, by standard arguments, it is sufficient that the algorithm can decide whether

(v) $Y_1 \circ \ldots \circ Y_m \subseteq Z_1 \circ \ldots \circ Z_n$

holds. However, as we shall see, it is easier to develop an algorithm of which the output is exactly one of the conditions (i)-(iv).

We give a method, by which this algorihm can be constructed provided the set M is not "too general". (After applying our method to concrete choices of M, we explain what we mean by a too general M in Section 6.)

First however we introduce some more notations. We consider the free monoid M^* and, at the same time, the monoid

$$[M] = \{\, Y_1 \circ \ldots \circ Y_m \mid m \geq 0, Y_i \in M \ \text{for} \ 1 \leq i \leq m \,\}$$

generated by M with composition $\circ$. The identity element of $[M]$ is I, the result of the empty composition. We observe that each element of $[M]$ can be represented by an element of M^* which representation can be described by the homomorphism

$$|\ | : M^* \to [M]$$

where $|\ |$ is the unique homomorphism from M^* to $[M]$. Namely, denoting the operation of M^* by $\cdot$, a word $Y_1 \cdot \ldots \cdot Y_m$ in M^* represents the tree transformation class $|\, Y_1 \cdot \ldots \cdot Y_m \,| = Y_1 \circ \ldots \circ Y_m$ in $[M]$. We note that $|\, \lambda \,| = I$. Hence two words $Y_1 \cdot \ldots \cdot Y_m$ and $Z_1 \cdot \ldots \cdot Z_n$ represent the same tree transformation class if and only if $Y_1 \circ \ldots \circ Y_m = Z_1 \circ \ldots \circ Z_n$ which is equivalent to $|\, Y_1 \cdot \ldots \cdot Y_m \,| = |\, Z_1 \cdot \ldots \cdot Z_n \,|$ or, in other words, $Y_1 \cdot \ldots \cdot Y_m \theta Z_1 \cdot \ldots \cdot Z_n$ where θ is the kernel of $|\ |$, i.e., the congruence relation induced by $|\ |$ on M^*.

With these notations in our mind, the method we suggest is to execute the following three tasks.

(a) Give a set of representatives N for θ. (We shall call the elements of N normal forms.)

(b) Give the inclusion diagram of the set $|\, N \,| = \{\, |\, u \,| \mid u \in N \,\}$, i.e., of the set of tree transformation classes represented by normal forms. (Note that having this inclusion diagram, for any given $u, v \in N$, we can read from the diagram, which one of the following conditions holds:

(i') $|\, u \,| \subset |\, v \,|$

(ii') $|\, v \,| \subset |\, u \,|$

(iii') $|\, u \,| = |\, v \,|$

(iv') $|\, u \,| \bowtie |\, v \,|$

(c) Give a finite set $T \subseteq M^* \times M^*$ of generators of θ (i.e. a Thue system T over M such that $\leftrightarrow^*_T = \theta$) and give an algorithm that for any $w \in M^*$, by a suitable sequence of substitutions induced by elements of T, computes the normal form of w, i.e., the unique $u \in N$ for which $w \leftrightarrow^*_T u$.

Now we prove that if we perform (a)-(c), then we have an algorithm that can decide for (*), which one of the conditions (i)-(iv) holds. To this end, take two tree transformation classes $Y_1 \circ \ldots \circ Y_m$ and $Z_1 \circ \ldots \circ Z_n$. First, by the algorithm in (c), compute for the words $Y_1 \cdot \ldots \cdot Y_m$ and $Z_1 \cdot \ldots \cdot Z_n$ the normal forms u and v such that

$$Y_1 \cdot \ldots \cdot Y_m \overset{*}{\underset{T}{\leftrightarrow}} u \quad \text{and} \quad Z_1 \cdot \ldots \cdot Z_n \overset{*}{\underset{T}{\leftrightarrow}} v.$$

Since, by (c), we have $\leftrightarrow^*_T = \theta$, we also have

$$Y_1 \circ \ldots \circ Y_m = \mid u \mid \quad \text{and} \quad Z_1 \circ \ldots \circ Z_n = \mid v \mid.$$

Thus, one of the conditions (i)-(iv) holds for $Y_1 \circ \ldots \circ Y_m$ and $Z_1 \circ \ldots \circ Z_n$ if and only if the corresponding condition of (i')-(iv') holds for $\mid u \mid$ and $\mid v \mid$. Moreover, having the inclusion diagram, by (b), we can read from the diagram which one of (i')-(iv') holds. Hence we obtained the following.

Statement 3.1 *Suppose that we executed the tasks (a)-(c) for M. Then, for any two tree transformation classes (*) over M, it is decidable which one of the conditions (i)-(iv) holds.*

When we applied the above general method to concrete choices of M, it transpired that it is suitable to perform the tasks (a)-(c) in the following five steps.

(I) Give a finite relation $T \subseteq M^* \times M^*$ which is our candidate for the set of generators of θ. (Here we are advised to take into consideration known decomposition results that hold among elements of M.)

(II) Prove that for every $(u, v) \in T$, it holds that $\mid u \mid = \mid v \mid$. (In other case, of course, T cannot be a set of generators of θ. In this way, elements of T represent equalities over $[M]$, hence we shall write $u \doteq v$ rather than (u, v) for $(u, v) \in T$.)

We note that from (II) the inclusion $\leftrightarrow^*_T \subseteq \theta$ follows because, by (II), $T \subseteq \theta$ and $\leftrightarrow^*_T$ is the smallest congruence on T^* containing T.

(III) Give a subset N of M^* which is guessed to be a set of representatives of θ. (Here N should be such that for any $u, v \in N$, $\mid u \mid = \mid v \mid$ if and only if $u = v$; otherwise N cannot be a set of representatives of θ.)

(IV) Give the inclusion diagram of the set $\{ \mid u \mid \mid u \in N \}$, which is, in fact, the set of tree transformation classes represented by the elements of N.

(V) Give an algorithm that for every $w \in M^*$ computes a $u \in N$ such that $w \leftrightarrow^*_T u$.

Next we prove that if we are succesful in steps (I)-(V), then we also performed the tasks (a)-(c).

Lemma 3.2 *Suppose that we have carried out the tasks (I)-(V). Then we have executed (a)-(c).*

Proof. First we show that $\theta \subseteq \leftrightarrow^*_T$ also holds. Therefore, let $w, w' \in M^*$ be such that $w\theta w'$. Then construct, by (V), the normal forms u and u' for which $w \leftrightarrow^*_T u$ and $w' \leftrightarrow^*_T u'$. Since $\leftrightarrow^*_T \subseteq \theta$, we also have $w\theta u$ and $w'\theta u'$ from which $u\theta u'$ follows. Then, by (III), $u = u'$ and thus $w \leftrightarrow^*_T u = u' \leftrightarrow^*_T w'$. Consequently we obtain that $\leftrightarrow^*_T = \theta$, which together with (V) yields (c). Moreover, by (III) and (V), it also follows that N is really a set of representatives of θ, hence we also have (a). Finally by (IV) we have (b). $\qquad\square$

For the sake of a better understanding we perform the steps (I)-(V) for the previous example $M = \{DR\}$.

(I') Let $T = \{DR^3 \doteq DR^2\}$. (Of course our motive is the equality $DR^3 = DR^2$ which was proved to be valid in [FülVág1].)

(II') We have nothing to do, since in [FülVág1] it was shown that $DR^3 = DR^2$.

(III') Let $N = \{\lambda, DR, DR^2\}$.

(IV') It is obvious that $I \subset DR$. Moreover it was proved in [Rou], that $DR \subset DR^2$. Thus, the inclusion diagram of the set $\{\,|\,u\,|\,|\,u \in N\,\}$ is

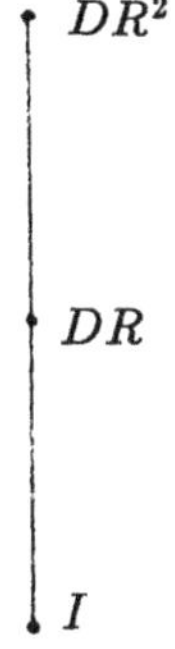

(V') The algorithm that computes for every $w \in M^*$ a $u \in N$ with $w \leftrightarrow^*_T u$ is as follows. Let $m \geq 0$ and $w = DR^m$.

(i)   If $m = 0$, then let $u = \lambda$.

(ii) If $m = 1$, then let $u = DR$.

(iii) If $m \geq 2$, then (by $m - 2$ applications of the reduction $DR^3 \doteq DR^2$) we have $u = DR^2$.

Obviously, in each of the above cases we have $w \leftrightarrow^*_T u$.

After these, by Statement 3.1, if we are given two arbitrary tree transformation classes

$$DR^m \text{ and } DR^n$$

with $m, n \geq 0$, then we can easily decide which one of the conditions corresponding to (i)-(iv) holds between these classes.

In the next two sections we carry out (I)-(V) for two more general choices of the set M.

Finally we consider the possibility of a more convenient computation of the normal forms. For this we need the following theorems concerning Thue systems and string rewriting systems.

Theorem 3.3 ([Jan],[Boo2]) *A Thue system T (noetherian string rewriting system S) is Church-Rosser if and only if each class of the congruence $\leftrightarrow_T^*$ ($\leftrightarrow_S^*$) contains exactly one T-irreducible (S-irreducible) element.*

Theorem 3.4 ([Boo1],[BooODu]) *For any Thue system T (weight reducing string rewrite system S) a linear time algorithm can be given that for every word w computes a T-irreducble (S-irreducible) element u such that $w \leftrightarrow_T^* u$ ($w \leftrightarrow_S^* u$).*

Note that if T (S) appearing in Theorem 3.4. is even Church-Rosser, then, by Theorem 3.3, u is the unique irreducible element in the congruence class containing w.

Hence, it would be desirable if our Thue system T given in (I) were Church-Rosser and if it satisfied $N = IRR(T)$, where N is the set of normal forms appearing in (III). If this were the case, then we would not need to give the algorithm appearing in (V), for instead of it we could use the linear time algorithm described in Theorem 3.4, that for each $w \in M^*$ computes the normal form $u \in N$ of w (since u is the unique T-irreducible elment in the $\leftrightarrow_T^*$ class of w due to the equality $N = IRR(T)$).

Unfortunately, T does not have the above properties in general. Still, we can avoid the specification and then the use of the algorithm (V) if, for T, we give a string rewriting system S which is

(i) weight reducing,

(ii) equivalent to T (in the sense $(\leftrightarrow_T^* = \leftrightarrow_S^*)$ and

(iii) for which $N = IRR(S)$.

Really if we have such an S, then S is Church-Rosser (because in every congruence class of $\leftrightarrow_S^*$ there is exactly one S-irreducible element). Moreover, the algorithm in Theorem 3.4 computes, for every word $w \in M^*$, the unique normalform $u \in N$ of w (because $N = IRR(S)$).

In the next two sections, in both cases, after applying our method to concrete choices of M, we shall give a string rewriting system S that have the above three properties.

4   The frontier-to-root case

In this section we apply the method described in the previous section to the case when M consists of the class of df transformations and some of its more important fundamental subclasses.

Namely, we take

$$M = \{DF, LDF, NDF, LNDF, H, LH, NH\},$$

the class LNH is not taken into M since, as it is not difficult to see, $LNH \circ Y = Y \circ LNH = Y$ for every $Y \in M$. In the rest of the section we perform the steps (I)-(V) of Section 3 to the above choice of M; the notations $[M]$, $|\ |$, θ and T stand for the same as in the previous section.

(I) Let $T \subseteq M^* \times M^*$ consist of the following 13 elements, where the elements (u, v) of T are written in the form $u \doteq v$.

(1) $NH \cdot LH \doteq H$

(2) $LH \cdot NH \doteq H$

(3) $NH \cdot NH \doteq NH$

(4) $LH \cdot LH \doteq LH$

(5) $LNDF \cdot LNDF \doteq LNDF$

(6) $LNDF \cdot LH \doteq LDF$

(7) $LNDF \cdot NH \doteq NDF$

(8) $LNDF \cdot H \doteq DF$

(9) $NH \cdot NDF \doteq NDF$

(10) $LH \cdot LDF \doteq LDF$

(11) $NH \cdot LDF \doteq DF$

(12) $LDF \cdot LNDF \doteq LDF$

(13) $NDF \cdot LNDF \doteq NDF$

(II) This step is in fact the proof of the following theorem.

Theorem 4.1 ([Fül3]) *For every $u \doteq v \in T$, it holds that $|\ u\ | = |\ v\ |$.*

Proof. (Outline.) The theorem states that the elements of T are valid equations over $[M]$, for example equation (8) means that $LNDF \circ H = DF$. However, many elements of T were already known to be valid in the literature. Thus the validity of (8) follows from the proofs of Lemma 4.1 of [Eng 1] and of Theorem 3.3 on page 158 of [GécSte], while (6) and (7) are special cases of (8). Moreover, (1) and (2) were proved in [FülVág1].

Many others, namely (3), (4), (5), (9) and (10) can be proved by the following technique. We observe that any of these formal equations has the form $Y \cdot Z \doteq Z$, where $Y, Z \in M$. We have to prove that $Y \circ Z = Z$. First we show that $Y \circ Z \subseteq Z$. For this, we take the y transducer $A = (\Sigma, P, \Delta, R, P')$ and the z transducer $B = (\Delta, Q, \Omega, R', Q')$. (For instance, for equation (10), A is an lh and B is an ldf transducer.) We construct the df transducer $C = (\Sigma, P \times Q, \Omega, R'', P' \times Q')$, called the compositon of A and B, such that R'' is defined by the following two conditions:

(i) for every $\sigma \in \Sigma_0$, $p \in P$ and $q \in Q$, the rule $\sigma \to \langle p, q \rangle(r)$ is in R'' if and only if there is a rule $\sigma \to p(r')$ in R such that $r' \Rightarrow_B^* q(r)$;

(ii) for every $m \geq 1$, $\sigma \in \Sigma_m$, $p, p_1, \ldots p_m \in P$ and $q, q_1, \ldots, q_m \in Q$, the rule $\sigma(\langle p_1, q_1 \rangle(x_1), \ldots, \langle p_m, q_m \rangle(x_m)) \to \langle p, q \rangle(r)$ is in R'' if and only if there is a rule of the form $\sigma(p_1(x_1), \ldots, p_m(x_m)) \to p(r')$ in R such that $r'(q_1(x_1), \ldots q_m(x_m)) \Rightarrow_B^* q(r)$.

It is not difficult to prove, see [Bak] for example, that for this df transducer C, $\tau_C = \tau_A \circ \tau_B$ holds. Moreover C inherits the properties both A and B have, for example if both A and B are linear (nondeleting, of one state etc.), then C is also linear (nondeleting, of one state etc.). Thus $Y \circ Z \subseteq Z$ follows. (Note that $\tau_C = \tau_A \circ \tau_B$ does not hold in general if A and B are nondeterministic f transducers.)

To prove the conversed inclusion $Z \subseteq Y \circ Z$, we only note that for any tree transformation $\tau \in Z$ we have $\tau = \tau' \circ \tau$, where τ' is a suitable identical tree transformation which is, due to $I \subseteq Y$, always in Y.

The validity of equations (12) and (13) can be proved similarly since both of them have the form $Z \cdot Y \doteq Z$, with $Y, Z \in M$.

Thus the only relatively new decomposition result is the validity of (11), which was shown in [Fül3].	□

(III) Now we give a candidate for the set of representatives of θ. Let

$$N = M \cup \{ \lambda, H \cdot NDF, H \cdot LNDF, NH \cdot LNDF, LH \cdot LNDF \}.$$

We state that N is a set of representatives for θ. (As we already saw in Lemma 3.2, it will follow that N is in fact a set of representatives of θ only after performing steps (IV) and (V).)

(IV) Next we give the inclusion diagram of the tree transformation classes represented by the elements of N.

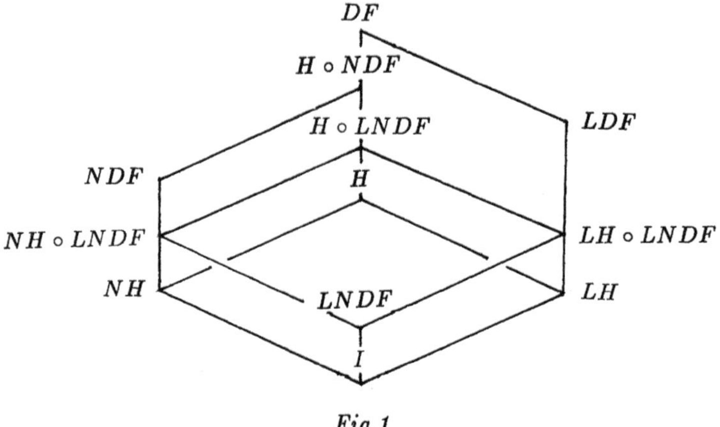

Fig.1

Theorem 4.2 ([Fül3]) *The diagram in Figure 1 is the iclusion diagram of the set* $\{ \mid u \mid \mid u \in N \}$ *of tree transformation classes.*

We note, it can be seen from the diagram that different elements of N represent different tree transformation classes, more exactly for any $u, v \in N$, $\mid u \mid = \mid v \mid$ if and only if $u = v$.

(V) In the last step we have to give an algorithm which, for every $w \in M^*$, computes its normalform, i.e., the unique $u \in N$ such that $w \leftrightarrow_T^* u$.

Theorem 4.3 ([Fül3]) *For every* $w \in M^*$, *there can effectively be given a* $u \in N$ *such that* $w \leftrightarrow_T^* u$.

Proof. (Outline.) First we give further 6 formal equations, that can be derived from elements of T. We have

$$(14)\ H \cdot NH \doteq H \qquad\qquad (17)\ LH \cdot H \doteq H$$

$$(15)\ NH \cdot H \doteq H \qquad\qquad (18)\ LH \cdot NH \doteq NH \cdot LH$$

$$(16)\ H \cdot LH \doteq H \qquad\qquad (19)\ H \cdot H \doteq H$$

	DF	NDF	LDF	LNDF	H	NH	LH
DF	DF 8, 11, 12, 11, 8, 19, 8	DF 7, 11, 12, 11, 8, 14, 8	DF 8, 1, 10, 11, 8, 5, 8	DF 11, 12, 11	DF 8, 19, 8	DF 8, 14, 8	DF 8, 16, 8
NDF	DF 8, 13, 7, 15, 8	NDF 7, 9, 7, 5, 7	DF 7, 11, 8, 5, 8	NDF 13	DF 7, 15, 8	NDF 7, 3, 7	DF 7, 1, 8
LDF	DF 8, 12, 6, 17, 8	DF 7, 12, 6, 2, 8	LDF 6, 12, 6 4, 6	LDF 12	DF 6, 17, 8	DF 6, 2, 8	LDF 6, 4, 6
LNDF	DF 8, 5, 8	NDF 7, 5, 7	LDF 6, 5, 6	LNDF 5	DF 8	NDF 7	LDF 6
H	DF 11, 4, 1, 10, 11	H · NDF	DF 1, 10, 11	H · LNDF	H 19	H 14	H 16
NH	DF 11, 3, 11	NDF 9	DF 11	NH · LNDF	H 15	NH 3	H 1
LH	DF 11, 2, 1 10, 11	H · NDF 9, 2	LDF 10	LH · LNDF	H 17	H 2	LH 4
H · NDF	DF 8, 13, 7, 15, 8, 11, 14, 1, 10, 11	H · NDF 7, 9, 7, 5, 7	DF 7, 11, 8, 5, 8, 11, 14, 1, 10, 11	H · NDF 13	DF 7, 15, 8, 11, 14, 1, 10, 11	H · NDF 7, 3, 7	DF 7, 1, 8, 11, 14, 1, 10, 11
H · LNDF	DF 8, 5, 8, 11, 14, 1, 10, 11	H · NDF 7, 5, 7	DF 6, 5, 6, 1, 10, 11	H · LNDF 5	DF 8, 11, 14, 1, 10, 11	H · NDF 7	DF 6, 1, 10, 11
NH · LNDF	DF 8, 5, 8, 11, 3, 11	NDF 7, 5, 7, 9	DF 6, 5, 6, 11	NH · LNDF 5	DF 8, 11, 3, 11	NDF 7, 9	DF 6, 11
LH · LNDF	DF 8, 5, 8, 11, 2, 1, 10, 11	H · NDF 7, 5, 7, 9, 2	LDF 6, 5, 6, 10	LH · LNDF 5	DF 8, 11, 2, 1, 10, 11	H · NDF 7, 9, 2	LDF 6, 10

Table 1.

and, for example, (14) can be obtained as $H \cdot NH \doteq LH \cdot NH \cdot NH \doteq LH \cdot NH \doteq H$, applying (2), (3) and (2) of T, respectively.

Next we construct Table 1. This Cayley-like table contains the following information. To each element v of N (except λ) there belongs a row, and to each element Y of M there belongs a column of the table. The entry determined by v and Y contains a $u \in N$ such that $v \cdot Y \leftrightarrow^*_T u$. Moereover, if $v \cdot Y = u$, then there is nothing else in the entry, as in the case $v = H$ and $Y = NDF$, but if $v \cdot Y \neq u$, then the entry contains the numbers of the equations of (1)-(19) by which we have $v \cdot Y \leftrightarrow^*_T u$. For example, in case of $v = NH \cdot LNDF$ and $Y = LH$ we have $v \cdot Y = NH \cdot LNDF \cdot LH \leftrightarrow^*_T DF = u$ such that $NH \cdot LNDF \cdot LH \leftrightarrow_T NH \cdot LDF$, by (6), and $NH \cdot LDF \leftrightarrow_T DF$, by (11).

Now we prove our theorem by induction on $l(w)$ as follows.

If $l(w) = 0$, meaning that $w = \lambda$, then we have $u = \lambda$.

If $l(w) > 0$, that is $w = x \cdot Y$ for some $x \in M^*$ and $Y \in M$, then construct, by induction hypothesis, the $v \in N$ such that $x \leftrightarrow^*_T v$, and then determine $u \in N$, by our table as above, such that $v \cdot Y \leftrightarrow^*_T u$. We obtain that $w = x \cdot Y \leftrightarrow^*_T v \cdot Y \leftrightarrow^*_T u$. $\square$

With this, we performed (I)-(V) and thus, by Lemma 3.2, we executed (a)-(c) of Section 3 for this choice of M. Hence we can apply Statement 3.1.

Finally, as an example, we apply the algorithm we developed to determine which one of the conditions corresponding to (i)-(iv) of Section 3 holds for the tree transformation classes

$$LNDF \circ NH \circ LDF \quad \text{and} \quad NH \circ LNDF \circ NDF.$$

We obtain, applying Table 1 in the way described above, that $LNDF \cdot NH \cdot LDF \leftrightarrow^*_T DF$ and that $NH \cdot LNDF \cdot NDF \leftrightarrow^*_T NDF$. Further, by the diagram in Fig. 1, we have $NDF \subset DF$, hence we also have

$$NH \circ LNDF \circ NDF \subset LNDF \circ NH \circ LDF.$$

Finally, we give a string rewriting system S over M with the properties described at the end of Section 3. First, however, we note that our T does not have the Church-Rosser property. In fact it is easy to see that $NDF \cdot NH \leftrightarrow^*_T LNDF \cdot NH$ but there exists no $z \in M^*$ for which $NDF \cdot NH \to^*_T z$ and $LNDF \cdot NH \to^*_T z$. However, we have the following result.

Theorem 4.4 *A weight reducing string rewriting system S over M can be given such that $\leftrightarrow^*_T = \leftrightarrow^*_S$ and that $N = IRR(S)$.*

Proof. (Outline.) Let S be the set of the pairs $(v \cdot Y, u)$, where $v \in N$, $Y \in M$ and u is in the entry determined by v and Y of Tabie 1, provided that $v \cdot Y \neq u$. Thus the case $v = H$ and $Y = LNDF$ does not incduce a rule, since $v \cdot Y = u$ in Table 1. We observe that S is "almost length reducing", the only exception is the rule $(LH \cdot NDL, H \cdot NDL)$. But if we put $\rho(LH) = 2$ and $\rho(Y) = 1$ for any other $Y \in M$, then S is weight reducing with respect to this weight function ρ. Since $T \subseteq S$, we have $\leftrightarrow^*_T \subseteq \leftrightarrow^*_S$. On the other hand any element of S can be derived from elements of T as we saw in the proof of Theorem 4.3, that is to say, $S \subseteq \leftrightarrow^*_T$. Hence we also have $\leftrightarrow^*_S \subseteq \leftrightarrow^*_T$. The only thing remained to prove is that $N = IRR(S)$. $\square$

By the above theorem, instead of the algorithm given in (V), we can apply the algorithm appearing in Theorem 3.4 to compute the normal form u of any word $w \in M^*$.

5 The root-to-frontier case

In this section we put

$$M = \{\, DR, LDR, NDR, LNDR, H, LH, NH \,\},$$

i.e., this choice of M is the root-to-frontier counterpart of that of Section 4. We perform (I)-(V) of Section 3 to this M.

(I) Now let $T \in M^* \times M^*$ be the set of the following formal equations:

(1) $DR \cdot NDR \doteq DR$	(11) $LNDR \cdot LDR \doteq LDR^2$
(2) $DR \cdot NH \doteq DR$	(12) $LNDR \cdot LH \doteq LDR^2$
(3) $NDR^2 \doteq NDR$	(13) $NH \cdot NDR \doteq NDR$
(4) $NDR \cdot LNDR \doteq NDR$	(14) $NH \cdot LDR \doteq DR$
(5) $NDR \cdot NH \doteq NDR$	(15) $NH^2 \doteq NH$
(6) $NDR \cdot LH \doteq DR^2$	(16) $NH \cdot LH \doteq H$
(7) $LDR \cdot LNDR \doteq LDR$	(17) $LH \cdot LDR \doteq LDR$
(8) $LNDR \cdot DR \doteq DR^2$	(18) $LH^2 \doteq LH$
(9) $LNDR \cdot NDR \doteq NDR$	(19) $LH \cdot NH \doteq H$
(10) $LNDR^2 \doteq LNDR$	

(II) Next we prove that elements of T represent valid equalities over M.

Theorem 5.1 ([FülVág1]) *For every $u \doteq v \in T$, it holds that $\mid u \mid = \mid v \mid$.*

Proof. (Outline.) Similarly to the frontier-to-root case, it is useful to introduce the concept of the composition of dr transducers. For this, let $A = (\Sigma, P, \Delta, R, p_0)$ and $B = (\Delta, Q, \Omega, R', q_0)$ be dr transducers. We define the dr transducer $C = (\Sigma, P \times Q, \Omega, R'', \langle p_0, q_0 \rangle)$, where R'' is the smallest set U of rules for which the following two conditions hold:

(i) if there is a rule $p(\sigma) \to t$ in R with $p \in P$, $\sigma \in \Sigma_0$ and $t \in T_\Delta$ and $q(t) \Rightarrow_B^* r$ for some $q \in Q$ and $r \in T_\Omega$, then let the rule $\langle p, q \rangle(\sigma) \to r$ be in U,

(ii) if there is a rule

$$p(\sigma(x_1, \ldots, x_m)) \to t(p_1(x_{i_1}), \ldots, p_n(x_{i_n})),$$

where $m \geq 1$, $n \geq 0$, $\sigma \in \Sigma_m$, $p, p_1, \ldots, p_n \in P$ and $t \in \hat{T}_{\Delta, n}$, in R, and we have

$$q(t) \overset{*}{\underset{B}{\Rightarrow}} r(q_{11}(x_1), \ldots, q_{1v_1}(x_1), \ldots, q_{n1}(x_n), \ldots, q_{nv_n}(x_n))$$

for some $q \in Q$, $v_j \geq 0$, $q_{j1}, \ldots, q_{jv_j} \in Q$ $(1 \leq j \leq n)$ and $r \in \hat{T}_{\Omega, v}$ with $v = v_1 + \ldots + v_n$; then let the rule

$$\langle p, q \rangle(\sigma(x_1, \ldots, x_m)) \to r(\langle q_{11}, p_1 \rangle(x_{i_1}), \ldots, \langle q_{nv_n}, p_n \rangle(x_{i_n}))$$

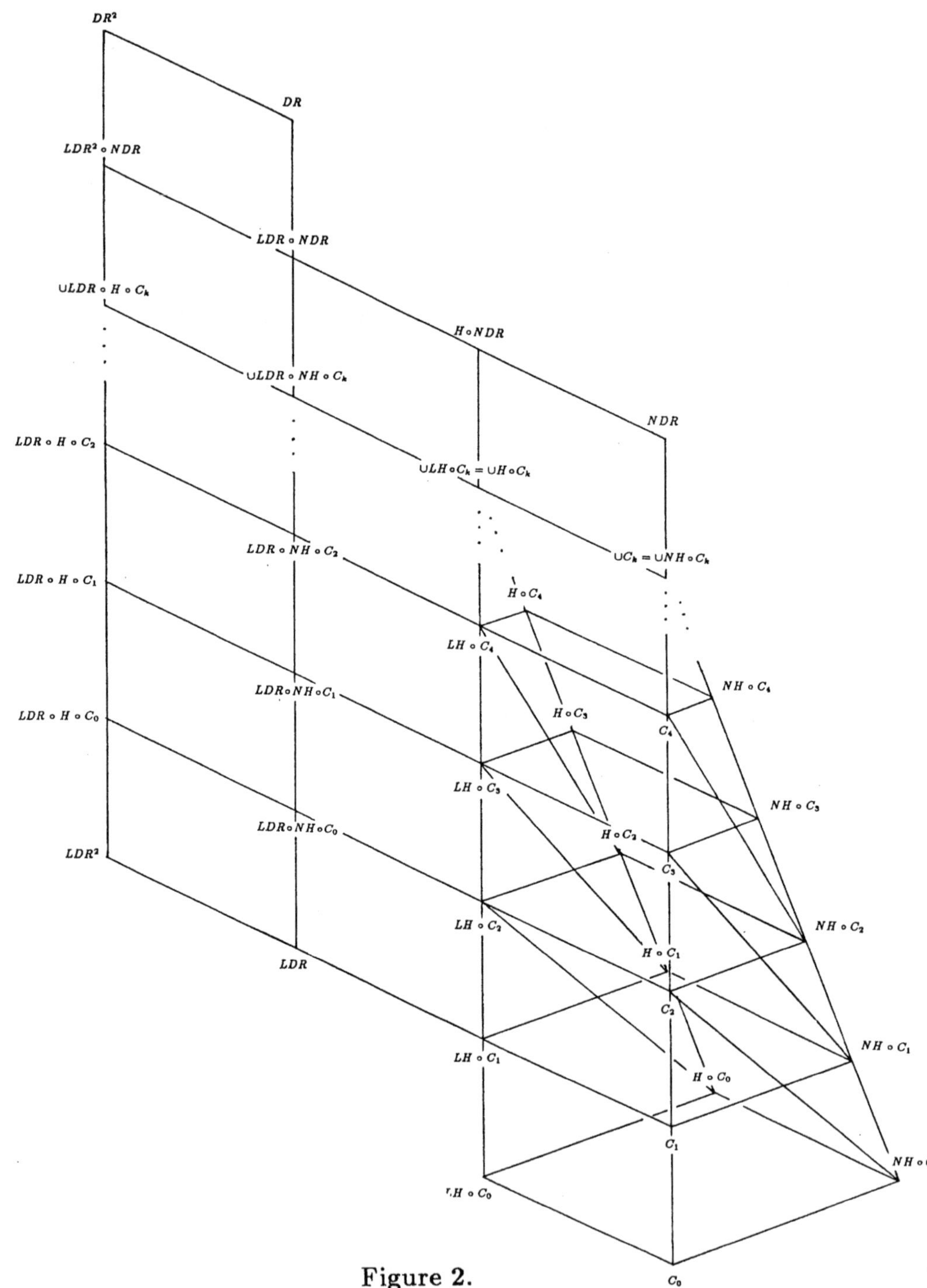

Figure 2.

be in U. For this dr transducer C, it holds that $\tau_A \circ \tau_B \subseteq \tau_C$, which is weaker than in the frontier-to-root case, for a proof see [Bak]. Moreover, the following three statements hold:

(a) if A is total or B is nondeleting, then $\tau_A \circ \tau_B = \tau_C$;

(b) if both A and B have property x, where x can be any of the modifiers total, one-state, linear and nondeleting, then C also have property x;

(c) for $\iota_A = Id(dom(\tau_A))$ we have $\iota_A \circ \tau_C = \tau_A \circ \tau_B$.

Now we consider the proof of our theorem. Using the above composition technique for dr transducers and its properties (a), (b), we can prove that (1)-(5), (7), (9), (10), (13), (15), (17) and (18) are valid in $[M]$. Further, the proof of (14) can be found in [Bak] and from that proof it transpires that (16) is a special case of (14). Finally, we proved in [FülVág1], that (6), (8), (11), (12) and (19) represent valid equalities over $[M]$. We note that property (c) of the composition technique described above can be used in the proof of (8). $\qquad\square$

(III) Next we give the set N of representatives for θ. We define two sets

$$N_1 = \{\, DR, LDR, NDR, LNDR, DR^2, LDR \cdot NDR, LDR^2,$$

$$H \cdot NDR, LDR^2 \cdot NDR \,\}$$

and

$$N_2 = \{\, H, NH, LH, LDR \cdot NH, LDR \cdot H \,\}.$$

Moreover we define z_k for every $k \geq 0$ as follows. Let $z_0 = \lambda$, let $z_{k+1} = z_k \cdot LNDR$ if k is an even and let $z_{k+1} = z_k \cdot NH$ if k is an odd integer. Then we put

$$N = N_1 \cup N_2 \cup \{\, z_k \mid k \geq 0 \,\} \cup \{\, z \cdot z_k \mid z \in N_2 \text{ and } k \geq 1 \,\}.$$

Theorem 5.2 ([FülVág5]) *The diagram in Fig. 2 is the inclusion diagram of the set* $\{\, |\, u\, | \mid u \in N \,\}$.

We note that our diagram in Fig. 2 slightly differs from the corresponding diagram in [FülVág5]. Namely, in that paper we wrote $LDR \circ H$ for $LNDR \circ H$. However, it was proved in [FülVág1] that $LDR \circ H = LNDR \circ H$, which we will see in (V), too. Moreover, we note that the suprema of the 6 hierarchies appearing in the diagram are also indicated apart from that they do not belong to the set $\{\, |\, u\, | \mid u \in N \,\}$. This is because we also proved in [FülVág5] that these suprema are properly contained in the classes above them, respectively.

(V) Finally we give the algorithm that computes the normal forms.

Theorem 5.3 ([FülVág6]) *For every $w \in M^*$, there can effectively be given a $u \in N$ such that $w \leftrightarrow_T^* u$.*

Proof. (Outline.) The proof can be given in the same way as in the frontier-to-root case. Here we also need some further formal equations that can be derived from T. Namely we have

	DR	NDR	LDR	$LNDR$	H	NH	LH
DR	DR^2	DR 1	DR^2 2, 14	DR 1, 4, 1	DR^2 19, 30, 2	DR 2	DR^2 30
LDR	DR^2 35	$LDR \cdot NDR$	LDR^2	LDR 7	$LDR \cdot H$	$LDR \cdot NH$	LDR^2 34
NDR	DR^2 33	NDR 3	DR^2 32	NDR 4	DR^2 31	NDR 5	DR^2 6
$z_1 = LNDR$	DR^2 8	NDR 9	LDR^2 11	$LNDR$ 10	$LDR \cdot H$ 19, 12, 34, 19	z_2	LDR^2 12
DR^2	DR^2 28	DR^2 1	DR^2 2, 14, 28	DR^2 26	DR^2 16, 2, 30, 28	DR^2 2	DR^2 30, 28
$LDR \cdot NDR$	DR^2 33, 35, 28	$LDR \cdot NDR$ 3	DR^2 32, 35, 28	$LDR \cdot NDR$ 4	DR^2 31, 35, 28	$LDR \cdot NDR$ 5	DR^2 6, 35, 28
LDR^2	DR^2 35, 35, 28	$LDR^2 \cdot NDR$	LDR^2 29	LDR^2 7	$LDR \cdot H$ 34, 23	$LDR \cdot H$ 34, 19	LDR^2 34, 29
$H \cdot NDR$	DR^2 33, 14, 20, 27	$H \cdot NDR$ 3	DR^2 32, 14, 20, 27	$H \cdot NDR$ 4	DR^2 31, 14, 20, 27	$H \cdot NDR$ 5	DR^2 6, 14, 20, 27
$LDR^2 \cdot NDR$	DR^2 33, 35, 35, 28, 28	$LDR^2 \cdot NDR$ 3	DR^2 32, 35, 35, 28, 28	$LDR^2 \cdot NDR$ 4	DR^2 31, 35, 35, 28, 28	$LDR^2 \cdot NDR$ 5	DR^2 6, 35, 35, 28, 28
H	DR 14, 20, 27	$H \cdot NDR$	DR 27	$H \cdot z_1$	H 24	H 20	H 22
NH	DR 14, 15, 14	NDR 13	DR 14	$NH \cdot z_1$	H 21	NH 15	H 16
LH	DR 14, 19, 27	$H \cdot NDR$ 13, 19	LDR 17	$LH \cdot z_1$	H 23	H 19	LH 18
$LDR \cdot NH$	DR^2 14, 15, 14, 35	$LDR \cdot NDR$ 13	DR^2 14, 35	$LDR \cdot NH \cdot z_1$	$LDR \cdot H$ 21	$LDR \cdot NH$ 15	$LDR \cdot H$ 16
$LDR \cdot H$	DR^2 14, 20, 27, 35	$LDR^2 \cdot NDR$ 19, 34, 13	DR^2 27, 35	$LDR \cdot H \cdot z_1$	$LDR \cdot H$ 24	$LDR \cdot H$ 20	$LDR \cdot H$ 22
z_2	DR^2 14, 15, 14, 8	NDR 13, 9	DR^2 14, 8	z_3	$LDR \cdot H$ 21, 19, 12, 34, 19	z_2 15	$LDR \cdot H$ 16, 19, 12, 34, 19
$z_{2k-1}, k \geq 2$	DR^2 8, 14, 15, 28	NDR 9, 13	DR^2 2, 8, 11, 14, 15, 28	z_{2k-1} 10	DR^2 2, 8, 12 14, 15, 19, 28	z_{2k}	DR^2 2, 8, 12 14, 15, 28
$z_{2k}, k \geq 2$	DR^2 8, 14, 15, 28	NDR 13, 9	DR^2 8, 14, 15, 28	z_{2k+1}	DR^2 2, 8, 12, 14 15, 19, 21, 28	z_{2k} 15	DR^2 2, 8, 12, 14, 15 16, 19, 28, 30, 34

Table 2.

(20) $H \cdot NH \doteq H$ (28) $DR^3 \doteq DR^2$

(21) $NH \cdot H \doteq H$ (29) $LDR^3 \doteq LDR^2$

(22) $H \cdot LH \doteq H$ (30) $DR \cdot LH \doteq DR^2$

(23) $LH \cdot H \doteq H$ (31) $NDR \cdot H \doteq DR^2$

(24) $H \cdot H \doteq H$ (32) $NDR \cdot LDR \doteq DR^2$

(25) $LH \cdot NH \doteq NH \cdot LH$ (33) $NDR \cdot DR \doteq DR^2$

(26) $DR \cdot LNDR \doteq DR$ (34) $LDR \cdot LH \doteq LDR^2$

(27) $H \cdot LDR \doteq DR$ (35) $LDR \cdot DR \doteq DR^2$

Then we construct Table 2 which can be used similarly to the frontier-to-root case. The only significant difference is that now the table should have an infinite number of rows because of the fact that in this case N is an infinte set. Fortunately we need not consider rows of the form $z \cdot z_k$ with $z \in N_2$ and $k \geq 1$, and succeeded in reducing the infinitely many rows z_k $(k \geq 1)$ to four ones, in fact to the rows $z_1, z_2, \{\, z_{2k-1} \mid k \geq 2 \,\}$ and $\{\, z_{2k} \mid k \geq 2 \,\}$. Thus for example we have $z_{11} \cdot LDR \leftrightarrow_T^* DR^2$ and the formal equations (2), (8), (11), (14), (15) and (28) can be used to obtain DR^2 from $z_{11} \cdot LDR$. Hence we can compute the normal form of any word of the form $z_k \cdot Y$ where $k \geq 0$ and $Y \in M$. The normal form of a word of the form $z \cdot z_k \cdot Y$ with $z \in N_2$, $k \geq 1$ and $Y \in M$ can be computed as follows. First we compute, in the way described above, the $v \in N$ such that $z_k \cdot Y \leftrightarrow_T^* v$ and then compute, by the application of Table 2 $l(v)$ times, $u \in N$ such that $z \cdot v \leftrightarrow_T^* u$. Then we have $z \cdot z_k \cdot Y \leftrightarrow_T^* z \cdot v \leftrightarrow_T^* u$.

Finally, the normal form of an arbitrary word $w \in M^*$ can be computed by recursion on $l(w)$, exactly as in the frontier-to-root case. $\square$

We succeeded in (I)-(V) for this choice of M. As an example, we apply the algorithm we developed to determine the inclusion relation between the tree transformation classes

$$LNDR \circ LH \circ NH \circ LNDR \quad \text{and} \quad LH \circ LNDR \circ H.$$

By the algorithm in Theorem 5.3 we have $LNDR \cdot LH \leftrightarrow_T LDR^2$ and $LDR^2 \cdot NH \leftrightarrow_T^* LDR \cdot H$. Moreover, we see that $LDR \cdot H \cdot LNDR = LDR \cdot H \cdot z_1$ is already a normal form hence $LNDR \cdot LH \cdot NH \cdot LNDR \leftrightarrow_T^* LDR \cdot H \cdot z_1$. As for the other word $LH \cdot LNDR \cdot H$, we have that $LH \cdot LNDR$ is a normal form and that $LNDR \cdot H \leftrightarrow_T^* LDR \cdot H$. Then, again by Table 2, $LH \cdot LDR \leftrightarrow_T^* LDR$ and we obtain that $LDR \cdot H$ is already a normal form. Hence $LH \cdot LNDR \cdot H \leftrightarrow_T^* LDR \cdot H$. Then the inclusion diagram in Fig. 2 shows that $LDR \circ H \subset LDR \circ H \circ z_1$, which in fact means that

$$LH \circ LNDR \circ H \subset LNDR \circ LH \circ NH \circ LNDR.$$

We note that our T given in (I) does not have the Church-Rosser property. Really, it is even not confluent because we have $NDR \cdot NH \cdot LDR \rightarrow_T NDR \cdot LDR$ and $NDR \cdot NH \cdot LDR \rightarrow_T NDR \cdot DR$ but there exists no $x \in M^*$ such that $NDR \cdot LDR \rightarrow_T^* x$ and $NDR \cdot DR \rightarrow_T^* x$. However T is equivalent to a weight reducing string rewriting system S, for which $N = IRR(S)$.

Theorem 5.4 ([FülVág7]) *There can be given a weight reducing string rewriting system S over M such that $\leftrightarrow_T^* = \leftrightarrow_S^*$ and $N = IRR(S)$.*

Proof. (Outline.) Analogously to the frontier-to-root case we give S using Table 2. Let S be the set of all pairs of the form $(u \cdot Y, v)$ where u belongs to a row of Table 2 except the last three rows, $Y \in M$ and v is the entry of Table 2 determined by u and Y. Moreover we require that $u \cdot Y \neq v$. Then $T \subseteq S$ and thus $\leftrightarrow_T^* \subseteq \leftrightarrow_S^*$. Conversely, any elements of S can be proved from elements of T, that is $S \subseteq \leftrightarrow_T^*$, hence $\leftrightarrow_S^* \subseteq \leftrightarrow_T^*$, too.

Moreover we choose the weight function ρ determined by $\rho(DR) = 1$, $\rho(LDR) = 2$, $\rho(NDR) = 3$, $\rho(LNDR) = 4$, $\rho(H) = 5$, $\rho(LH) = 6$ and $\rho(NH) = 7$. It is an easy exercise to show that S is weight reducing under ρ, while the exact proof of $N = IRR(S)$ can be found in [FülVág7]. $\qquad\square$

In the light of the above theorem and the notes at the end of Section 3 we can again apply the algorithm appearing in Theorem 3.4 to compute the normal form $u \in N$ of an arbitrary word $w \in M^*$.

6 Conclusion

During the research summarized in this paper we developed a method by which, given a finite set M of tree transformation classes, an algorithm can be constructed that decides which one of the possible four relations $\subset$, $\supset$, $=$ and $\bowtie$ holds between two arbitrarily given elements of the monoid $[M]$ generated by M with composition. However, it may happen that the application of our method to certain concrete choices of M will not be successful. This is because there are several crucial points in which we may fail during the application of the method, i.e., during the execution of steps (I)-(V) described in Section 3. For example, we do not know in advance if θ has a finite set of generators or even if it has, we may not be able to give this set. Another critical step is (IV) since in many cases it may be very difficult to state the inclusion between two elements C and D of $\{ \mid u \mid \mid u \in N \}$; even if we know that, let us say, $C \subseteq D$, it is often not trivial at all whether the inclusion is proper or not. We certainly would face these problems if, for example, we chose M as the union of the sets considered in Sections 4 and 5, not to speak about the choice of the nondeterministic versions of these sets. We thought of the above difficulties when we mentioned in Section 3 that M should not be "too general".

However, we believe that the results of Sections 4 and 5 are sufficient arguments to convince the reader that our method is applicable for a relatively wide range of tree transformation classes. We think that it would be worth-while to consider further choices of M that, for example, contain the simple but very useful class of relabeling tree transformations [GécSte], [Eng1]. Another, probably fruitful choice of M could be the set that consists of the five classes of look-ahead transformations considered in [FülVág4]. In that paper, many decomposition results were obtained for these classes, and in [FülVág9] a hierarchy theorem was proved for iterated deterministic top-down look-ahead. These findings could be the first step towards giving the set of generators for θ.

References

[ArnDau] Arnold, A. and Dauchet, M., Morphismes et bimorphismes d'arbes, Theoret. Comput Sci. 20 (1982) 33-93.

[Bak] Baker, B. S., Composition of top-down and bottom-up tree transducers, Inform. and Control 41 (1979) 186-213.

[Bar1] Bartha M., An algebraic definition of attributed transformations, in Proc. of the 1981 FCT Conference (Szeged, Hungary, Aug. 24-28), Lecture Notes in Computer Science, vol. 117, Springer-Verlag, Berlin, 1981.

[Bar2] Bartha M., Linear deterministic attributed transformations, Acta Cybernet. 6 (1983) 125-147.

[BKR] Benninghofen, B., Kemmerich, S. and Richter, M. M., Systems of reductions, Lecture Notes in Computer Science vol. 277, Springer-Verlag 1987, Berlin.

[Boo1] Book, R. V., Confluent and other types of Thue systems, J. Assoc. Comput. Mach. 29 (1982) 171-182.

[Boo2] Book, R. V., Thue systems and the Church-Rosser property: replacement systems, specification of formal languages and presentations of monoids, in Progress in combinatorics on words (L.Cummings ed.), 1-38, Academic Press, 1983, New York.

[BooODu] Book, V. and O'Dunlaing, P., Testing for the Church-Rosser Property, Theoret. Comput. Sci. 16 (1981) 223-229.

[Bursan] Burris, S. and Sankappanavar, H. P., A Course in Universal Algebra, Springer-Verlag, New-York, 1981.

[CouFra] Courcelle, B. and Franchi-Zannettacci, P., Attribute grammars and recursive program schemes, Part I and II, Theoret. Comput. Sci. 17 (1982) 163-191, 235-257.

[Dau] Dauchet, M., Transductions de forets, bimorphisms de magmoides, Thése de doctorat, University of Lille, France, 1977.

[DauTis] Dauchet, M. and Tison S., The theory of ground rewrite systems is decidable, Technical report Nr. 182, University of Lille, France, 1990.

[DerJou] Dershowitz, N. and Jounnaud, J. P., Rewrite systems, in Handbook of Theoretical Computer Science, North-Holland, 1989, Amsterdam.

[DHLT] Dauchet, M., Heuillard, T., Lescanne, P. and Tison, S., Decidability of the Confluence of Finite Ground Term Rewrite Systems and of Other Related Term Rewrite Systems, Inform. and Computation, 88 (1990) 187-201.

[Eng1]   Engelfriet, J., Bottom-up and top-down tree transformations - a comparison, Mah. Systems Theory 9 (1975) 198-231.

[Eng2]   Engelfriet, J., Top-down trre transducers with regular look-ahead, Math. Systems Theory 10 (1977) 289-303.

[Eng3]   Engelfriet, J., On trec transducers for partial functions, Inform. Process. Lett. 7 (1978) 170-172.

[Eng4]   Engelfriet, J., Some open questions and recent results on tree transducers and tree languages, in "Formal Language Theory; Perspectives and Open Problems" (R. V. Book ed.) Academic Press, New York, 1980.

[Eng5]   Engelfriet, J., Tree transducers and syntax-directed semantics, in Proc. of the 7th CAAP, Lille, 1982, 82-107.

[Eng6]   Engelfriet, J., Three hierarchies of transducers, Math. Systems Theory 15 (1982) 95-125.

[EngFil]   Engelfriet, J. and File, G., The formal power of one-visit attribute grammars, Acta Inform. 16 (1981) 275-302.

[EngVog1] Engelfriet, J. and Vogler, H., Macro tree transducers, J. Comput. System Sci. 31 (1985) 71-146.

[EngVog2] Engelfriet, J. and Vogler, H., Pushdown machines for the macro tree transducer, Theoret. Comput. Sci. 42 (1986) 251-368.

[EngVog3] Engelfriet, J. and Vogler, H., Modular tree transducers, Theoret. Comput. Sci. 78 (1991) 267-303.

[EngVog4] Engelfriet, J. and Vogler, H., High level tree transducers and iterated pushdown transducers, Acta Inform. 26 (1988) 131-192.

[Fül1]   Fülöp, Z., On attributed tree transducers, Acta Cybernet. 5 (1981) 261-279.

[Fül2]   Fülöp, Z., Decomposition results concerning K-visit attributed tree transducers, Acta Cybernet. 6 (1983) 163-171.

[Fül3]   Fülöp, Z., A complete description for a monoid of deterministic bottom-up tree transformation classes, Theoret. Comput. Sci. 88 (1991) 253-268.

[FülVág1] Fülöp, Z. and Vágvölgyi, S., Results on compositions of deterministic root-to-frontier tree transfomations, Acta Cybernet. 8 (1987) 49-61.

[FülVág2] Fülöp, Z. and Vágvölgyi, S., On ranges of compositions of deterministic root-to-frontier tree transformations, Acta Cybernet. 8 (1988) 259-266.

[FülVág3] Fülöp, Z. and Vágvölgyi, S., Top-down tree transducers with determinstic top-down look-ahead, Inform. Process. Lett. 33 (1989/90) 3-5.

[FülVág4] Fülöp, Z. and Vágvölgyi, S., Variants of top-down tree transducers with look-ahead, Math. Systems Theory 21 (1989) 125-145.

[FülVág5] Fülöp, Z. and Vágvölgyi, S., A complete classification of deterministic root-to-frontier tree transformation classes, Theoret. Comput. Sci. 81 (1991) 1-15.

[FülVág6] Fülöp, Z. and Vágvölgyi, S., A finite presentation for a monoid of tree transformation classes, in Proc. 2nd Confernce on Automata, Languages and Programming Systems (F. Gécseg and I. Peák eds.), Salgótarján, 1988.

[FülVág7] Fülöp, Z. and Vágvölgyi, S., A complete term rewriting system for a monoid of tree transformation classes, Inform. and Computation 86 (1990) 195-212.

[FülVág8] Fülöp, Z. and Vágvölgyi, S., Ground term rewriting rules for the word problem of a ground term equation system, Bulletin of the EATCS 45 (1991) 186-201.

[FülVág9] Fülöp, Z. and Vágvölgyi, S., Iterated deterministic top-down look-ahead, in Proc. of the 1989 FCT Conference (Szeged Aug. 21-25), Lecture Notes in Computer Science, vol. 380, Springer-Verlag, Berlin, 1989.

[GécSte] Gécseg, F. and Steinby, M., Tree automata, Akadémiai Kiadó, 1984, Budapest.

[Hue] Huet, G., Confluent reductions: Abstract properties and applications to term rewriting systems, Journal of the ACM 27 (1980) 797-821.

[Jan] Jantzen, M., Confluent String Rewriting, Springer-Verlag, 1988, Berlin-Heidelberg-New York.

[Knu] Knuth, D. E., Semantics of context-free languages, Math. Systems Theory 2 (1968) 127-145.

[Rou] Rounds, W. C., Mappings and grammars on trees, Math. Systems Theory 4 (1970) 257-287.

[Tha1] Thatcher, J. W., Generalized^2 sequential machine maps, J. Comput. System Sci. 4 (1970) 339-367.

[Tha2] Thatcher, J. W., Tree automata: an informal survey, in Currents in the Theory of Computing (A. V. Aho ed.) Prentice-Hall, 1973, Englewood Cliffs.

[Vág] Vágvölgyi, S., On compositions of root-to-frontier tree transformations, Acta Cybernet. 7 (1986) 443-480.

[VágFül] Vágvölgyi, S. and Fülöp Z., An infinite hierarchy of tree transformations in the class NDR, Acta Cybernet. 8(1987) 153-168.

[Vog1] Vogler, H., Basic tree transducers, J. Comput. System Sci. 34 (1987) 87-128.

[Vog2]   Vogler, H., High level modular tree transducers, in " 1978-1988, Ten years IIG, Institutes for Information Processing" (H. J. Pongratz and W. Schinnerl eds.) Report Nr. 260, Institutes for Information Processing, Graz University of Technology.

Tree Automata and Languages
M. Nivat and A. Podelski (editors)
© 1992 Elsevier Science Publishers B.V. All rights reserved.

Tree-Adjoining Grammars and Lexicalized Grammars*

Aravind K. Joshi and Yves Schabes

Department of Computer and Information Science
University of Pennsylvania
Philadelphia, PA 19104-6389, USA

1   Introduction

In this paper, we will describe a tree generating system called tree-adjoining grammar
(TAG) and state some of the recent results about TAGs. The work on TAGs is motivated
by linguistic considerations. However, a number of formal results have been established
for TAGs, which we believe, would be of interest to researchers in tree grammars and
tree automata. After giving a short introduction to TAG, we briefly state these results
concerning both the properties of the string sets and tree sets (Section 2). We will
also describe the notion of lexicalization of grammars (Section 3) and investigate the
relationship of lexicalization to context-free grammars (CFGs) and TAGs (Section 4).

The notion of lexicalization is linguistically very significant. Formally, this notion
leads to a tree representation for CFGs which may be regarded as a *stronger* version of
the Greibach Normal Form for CFGs, in the sense that the structures are preserved and
not just the string sets.

The motivations for the study of tree-adjoining grammars (TAG) are of linguistic and
formal nature. The elementary objects manipulated by a TAG are trees, i.e., structured
objects and not strings. Using structured objects as the elementary objects of a formalism,
it is possible to construct formalisms whose properties relate directly to the strong generative capacity (structural description) which is more relevant to linguistic descriptions
than the weak generative capacity (set of strings).

TAG is a tree-generating system rather than a string generating system. The set of
trees derived in a TAG constitute the object language. Hence, in order to describe the
derivation of a tree in the object language, it is necessary to talk about derivation 'trees'
for the object language trees. These derivation trees are important both syntactically

---

*This work was partially supported by NSF grants DCR-84-10413, ARO Grant DAAL03-87-0031, and
DARPA Grant N0014-85-K0018.
We are grateful to Anne Abeillé, Bruno Courcelle, Anthony Kroch, Mitch Marcus, Fernando Pereira,
Stuart Shieber, Mark Steedman and Marilyn Walker for providing valuable comments.

and semantically. It has also turned out that some other formalisms which are weakly equivalent to TAGs are similar to each other in terms of the properties of the derivation 'trees' of these formalisms (Weir, 1988; Joshi et al., forthcoming 1991).

Another important linguistic motivation for TAGs is that TAGs allow factoring recursion from the statement of linguistic constraints (dependencies), thus making these constraints strictly local, and thereby simplifying linguistic description (Kroch and Joshi, 1985).

Lexicalization of grammar formalism is also one of the key motivations, both linguistic and formal. Most current linguistic theories give lexical accounts of several phenomena that used to be considered purely syntactic. The information put in the lexicon is thereby increased in both amount and complexity[1].

On the formal side, lexicalization allows us to associate each elementary structure in a grammar with a lexical item (terminal symbol in the context of formal grammars). The well-known Greibach Normal Form (CNF) for CFG is a kind of lexicalization, however it is a *weak* lexicalization in a certain sense as it does not preserve structures of the original grammar. Our tree based approach to lexicalization allows us to achieve lexicalization while preserving structures, which is linguistically very significant.

2   Tree-Adjoining Grammars

TAGs were introduced by Joshi, Levy and Takahashi (1975) and Joshi (1985). For more details on the original definition of TAGs, we refer the reader to (Joshi, 1987; Kroch and Joshi, 1985). It is known that tree-adjoining languages (TALs) generate some strictly context-sensitive languages and fall in the class of the so-called 'mildly context-sensitive languages' (Joshi et al., forthcoming 1991). TALs properly contain context-free languages and are properly contained by indexed languages.

Although the original definition of TAGs did not include substitution as a combining operation, it can be easily shown that the addition of substitution does not affect the formal properties of TAGs.

We first give an overview of TAG and then we study the lexicalization process.

Definition 1 (tree-adjoining grammar)
A tree-adjoining grammar (TAG) consists of a quintuple (Σ, NT, I, A, S), where

(i) Σ is a finite set of terminal symbols;
(ii) NT is a finite set of non-terminal symbols[2]: $\Sigma \cap NT = \emptyset$;
(iii) S is a distinguished non-terminal symbol: $S \in NT$;
(iv) I is a finite set of finite trees, called *initial trees,* characterized as follows (see tree on the left in Figure 1):

   • interior nodes are labeled by non-terminal symbols;

---

[1]Some of the linguistic formalisms illustrating the increased use of lexical information are, lexical rules in LFG (Kaplan and Bresnan, 1983), GPSG (Gazdar et al., 1985), HPSG (Pollard and Sag, 1987), Combinatory Categorial Grammars (Steedman, 1987), Karttunen's version of Categorial Grammar (Karttunen, 1986), some versions of GB theory (Chomsky, 1981)), and Lexicon-Grammars (Gross, 1984).

[2]We use lower-case letters for terminal symbols and upper-case letters for non-terminal symbols.

- the nodes on the frontier of initial trees are labeled by terminals or non-terminals; non-terminal symbols on the frontier of the trees in I are marked for substitution; by convention, we annotate nodes to be substituted with a down arrow ($\downarrow$);

(v) A is a finite set of finite trees, called *auxiliary trees*, characterized as follows (see tree on the right in Figure 1):

  - interior nodes are labeled by non-terminal symbols;
  - the nodes on the frontier of auxiliary trees are labeled by terminal symbols or non-terminal symbols. Non-terminal symbol on the frontier of the trees in A are marked for substitution except for one node, called the *foot node*; by convention, we annotate the foot node with an asterisk ($*$); the label of the foot node must be identical to the label of the root node.

In *lexicalized TAG*, at least one terminal symbol (the anchor) must appear at the frontier of all initial or auxiliary trees.

The trees in $I \cup A$ are called *elementary trees*. We call an elementary tree an X-type elementary tree if its root is labeled by the non-terminal X.

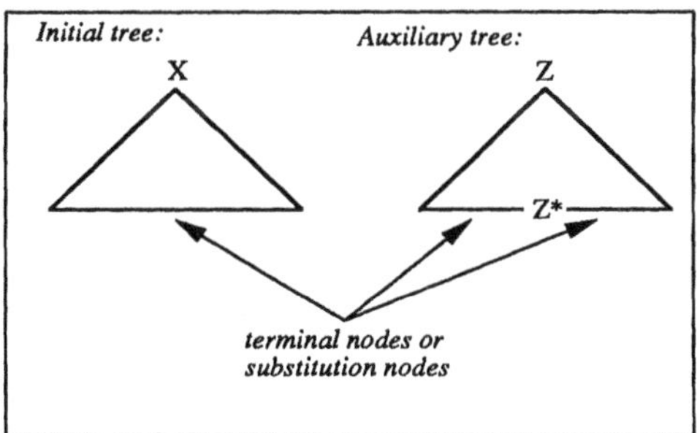

Figure 1: Schematic initial and auxiliary trees.

A tree built by composition of two other trees is called a *derived tree*.

We now define the two composition operations that TAG uses: *adjoining* and *substitution*.

Adjoining builds a new tree from an auxiliary tree β and a tree α (α is any tree, initial, auxiliary or derived). Let α be a tree containing a non-substitution node n labeled by X and let β be an auxiliary tree whose root node is also labeled by X. The resulting tree, γ, obtained by adjoining β to α at node n (see top two illustrations in Figure 2) is built as follows:

- the sub-tree of α dominated by n, call it t, is excised, leaving a copy of n behind.

- the auxiliary tree β is attached at the copy of n and its root node is identified with the copy of n.

- the sub-tree t is attached to the foot node of β and the root node of t (i.e. n) is identified with the foot node of β.

The top two illustrations in Figure 2 illustrate how adjoining works. The auxiliary tree β_1 is adjoined on the VP node in the tree α_2. α_1 is the resulting tree.

Substitution takes only place on non-terminal nodes of the frontier of a tree (see bottom two illustrations in Figure 2). An example of substitution is given in the fourth illustration (from the top) in Figure 2. By convention, the nodes on which substitution is allowed are marked by a down arrow ($\downarrow$). When substitution occurs on a node n, the node is replaced by the tree to be substituted. When a node is marked for substitution, only trees derived from initial trees can be substituted for it.

By definition, any adjunction on a node marked for substitution is disallowed. For example, no adjunction can be performed on any NP node in the tree α_2. Of course, adjunction is possible on the root node of the tree substituted for the substitution node.

2.1   Adjoining Constraints

In the system that we have described so far, an auxiliary tree β can be adjoined on a node n if the label of n is identical to the label of the root node of the auxiliary tree β and if n is labeled by a non-terminal symbol not annotated for substitution. It is convenient for linguistic description to have more precision for specifying which auxiliary trees can be adjoined at a given node. This is exactly what is achieved by *constraints on adjunction* (Joshi, 1987). In a TAG $G = (\Sigma, NT, I, A, S)$, one can, for each node of an elementary tree (on which adjoining is allowed), specify one of the following three constraints on adjunction:

- *Selective Adjunction ($SA(T)$, for short)*: only members of a set $T \subseteq A$ of auxiliary trees can be adjoined on the given node. The adjunction of an auxiliary is not mandatory on the given node.

- *Null Adjunction (NA for short)*: it disallows any adjunction on the given node.[3]

- *Obligatory Adjunction ($OA(T)$, for short)*: an auxiliary tree member of the set $T \subseteq A$ must be adjoined on the given node. In this case, the adjunction of an auxiliary tree is mandatory. OA is used as a notational shorthand for $OA(A)$.

If there are no substitution nodes in the elementary trees and if there are no constraints on adjoining, then we have the 'pure' (old) Tree Adjoining Grammar (TAG) as described in (Joshi et al., 1975).

The operation of substitution and the constraints on adjoining are both needed for linguistic reasons. Constraints on adjoining are also needed for formal reasons in order to obtain some closure properties.

---

[3]Null adjunction constraint corresponds to a selective adjunction constraint for which the set of auxiliary trees T is empty: $NA = SA(\emptyset)$

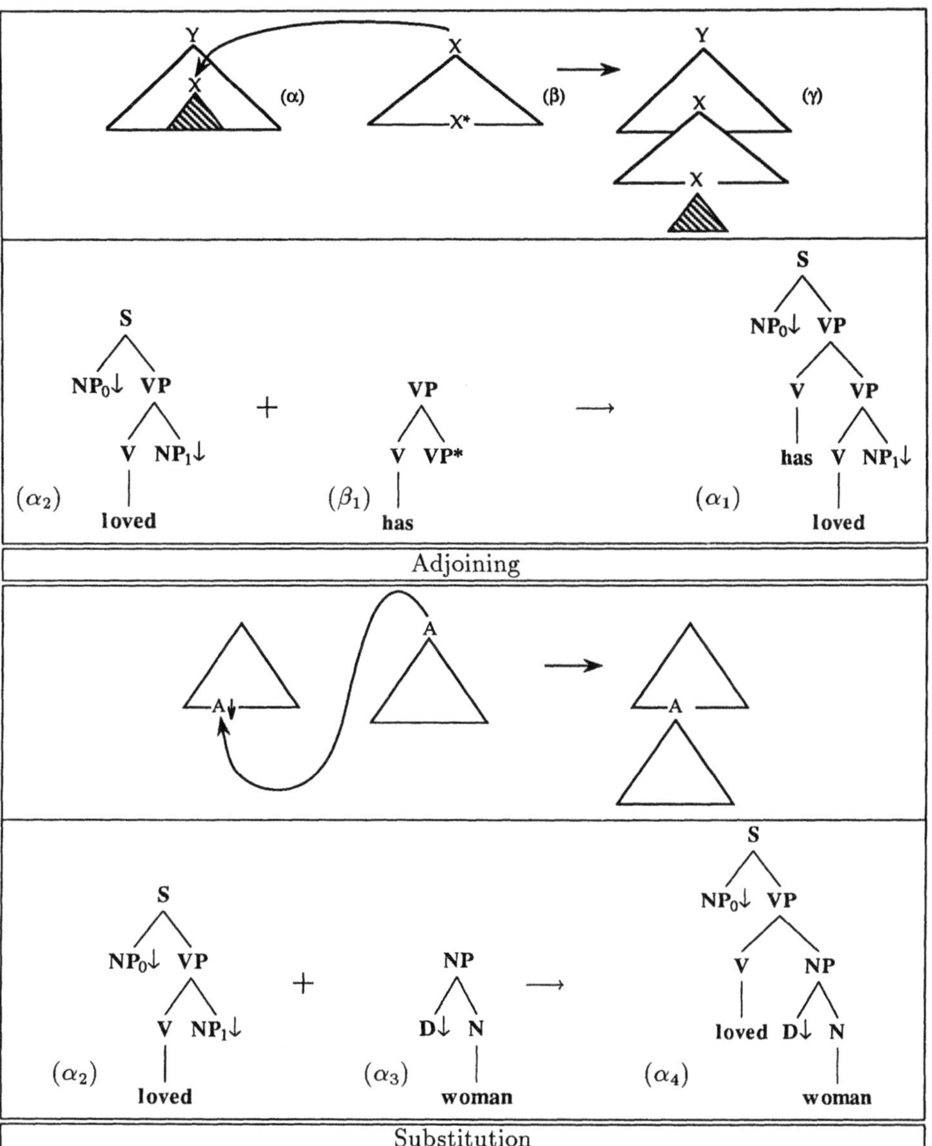

Figure 2: Combining operations: adjoining and substitution

2.2   Derivation in TAG

We now define by an example the notion of derivation in a TAG. Unlike CFGs, the tree obtained by derivation (the *derived tree*) does not give enough information to determine how it was constructed. The *derivation tree* is an object that specifies uniquely how a derived tree was constructed. Both operations, adjunction and substitution, are considered in a TAG derivation. Take for example the derived tree α_5 in Figure 3; α_5 yields the sentence *yesterday a man saw Mary*. It has been built with the elementary trees shown in Figure 4.

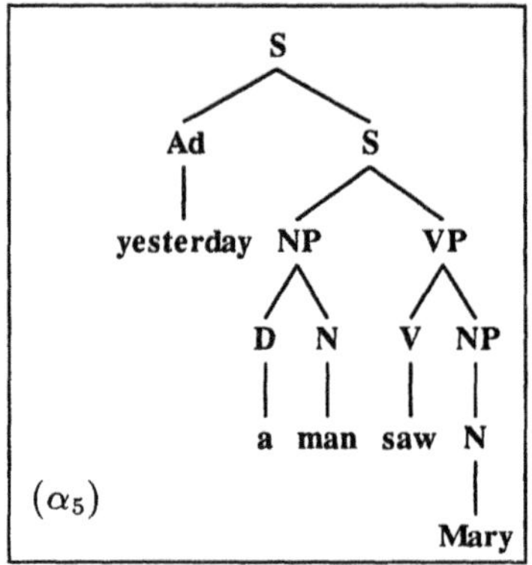

Figure 3: Derived tree for: *yesterday a man saw Mary.*

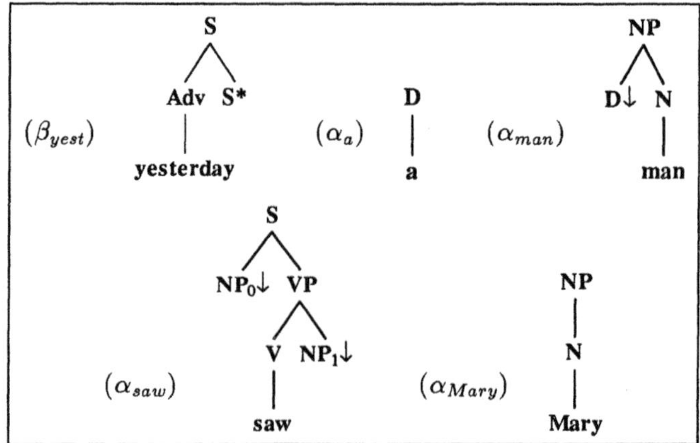

Figure 4: Some elementary trees.

The root of a derivation tree for TAGs is labeled by an S-type initial tree. All other nodes in the derivation tree are labeled by auxiliary trees in the case of adjunction or initial trees in the case of substitution. A tree address is associated with each node (except the root node) in the derivation tree. This tree address is the address of the node in the parent tree to which the adjunction or substitution has been performed. We use the following convention: trees that are adjoined to their parent tree are linked by an unbroken line to their parent, and trees that are substituted are linked by a dashed line.[4] Since by definition, adjunction can only occur at a particular node one time, all the children of a node in the derivation tree will have distinct addresses associated with them.

The derivation tree in Figure 5 specifies how the derived tree α_5 pictured in Figure 3 was obtained.

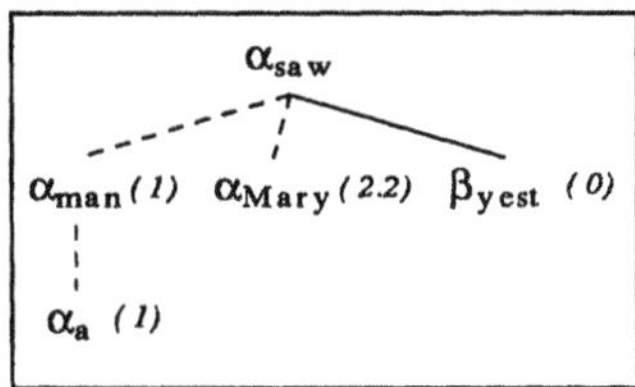

Figure 5: Derivation tree for *Yesterday a man saw Mary*.

This derivation tree (Figure 5) should be interpreted as follows: α_a is substituted in the tree α_{man} at the node of address 1 (D), α_{man} is substituted in the tree α_{saw} at address 1 (NP_0) in α_{man}, α_{Mary} is substituted in the tree α_{saw} at node $2 \cdot 2$ (NP_1) and the tree β_{yest} is adjoined in the tree α_{saw} at node 0 (S).

The order in which the derivation tree is interpreted has no impact on the resulting derived tree.

2.3 Some properties of the string languages and tree sets

We summarize some of the well known properties of tree-adjoining grammar's string languages and of the tree sets.

The *tree set* of a TAG is defined as the set of completed[5] initial trees derived from some S-rooted initial tree:

$$T_G = \{t \mid t \text{ is 'derived' from some S-rooted initial tree}\}$$

The string language of a TAG, $L(G)$, is then defined as the set of yields of all the trees in the tree set:

$$L_G = \{w \mid w \text{ is the yield of some } t \text{ in } T_G\}$$

Adjunction is more powerful than substitution and it generates some context-sensitive languages (see Joshi [1985] for more details). [6]

---

[4] We will use Gorn addresses as tree addresses: 0 is the address of the root node, k is the address of the kth child of the root node, and $p \cdot q$ is the address of the qth child of the node at address p.

[5] We say that an initial tree is *completed* if there is no substitution nodes on the frontier of it.

[6] Adjunction can simulate substitution with respect to the weak generative capacity. It is also possible to encode a context-free grammar with auxiliary trees using adjunction only. However, although the languages correspond, the possible encoding does not directly reflect the tree set of original context-free grammar since this encoding uses adjunction.

Some well known properties of the string languages follow:

- context-free languages are strictly included in tree-adjoining languages, which themselves are strictly included in indexed languages;
$$CFL \subset TAL \subset Indexed\ Languages \subset CSL$$

- TALs are semilinear;

- All closure properties of context-free languages also hold for tree-adjoining languages. In fact, TALs are a full abstract family of languages (full AFLs).

- a variant of the push-down automaton called embedded push-down automaton (EPDA) (Vijay-Shanker, 1987) characterizes exactly the set of tree-adjoining languages, just as push-down automaton characterizes CFLs.

- there is a pumping lemma for tree-adjoining languages.

- tree-adjoining languages can be parsed in polynomial time, in the worst case in $O(n^6)$ time.

Some well know properties of the tree sets of tree-adjoining grammars follow:

- the tree sets of recognizable sets (regular tree sets)(Thatcher, 1971) are strictly included in the tree sets of tree-adjoining grammars, $T(TAG)$;
$$recognizable\ sets \subset T(TAG)$$

- the set of paths of all the trees in the tree set of a given TAG, $\mathcal{P}(T(G))$, is a context-free language;
$$\mathcal{P}(T(G))\ is\ a\ CFL$$

- the tree sets of TAG are equivalent to the tree sets of linear indexed languages. Hence, linear versions of Schimpf-Gallier tree automaton (Schimpf and Gallier, 1985) are equivalent to $T(TAG)$;

- for every TAG, G, the tree set of G, $T(G)$, is recognizable in polynomial time, in the worst case in $O(n^3)$-time, where n is the number of nodes in a tree $t \in T(G)$.

We now give two examples to illustrate some properties of tree-adjoining grammars.

Example 1 Consider the following TAG $G_1 = (\{a, e, b\}, \{S\}, \{\alpha_6\}, \{\beta_2\}, S)$

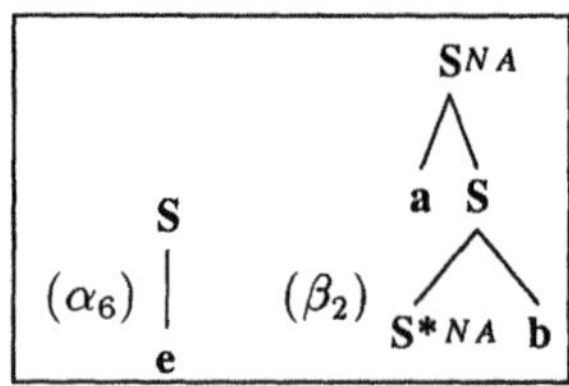

G_1 generates the language $L_1 = \{a^n e b^n | n \geq 1\}$. For example, in Figure 6, α_7 has been obtained by adjoining β_2 on the root node of α_6 and α_8 has been obtained by adjoining β_2 on the node at address 2 in in the tree α_7.

Although L_1 is a context-free language, G_1 generates cross serial dependencies. For example, α_7 generates, $a_1 e b_1$, and α_8 generates, $a_1 a_2 e b_2 b_1$. It can be shown that $T(G_1)$ is not recognizable.

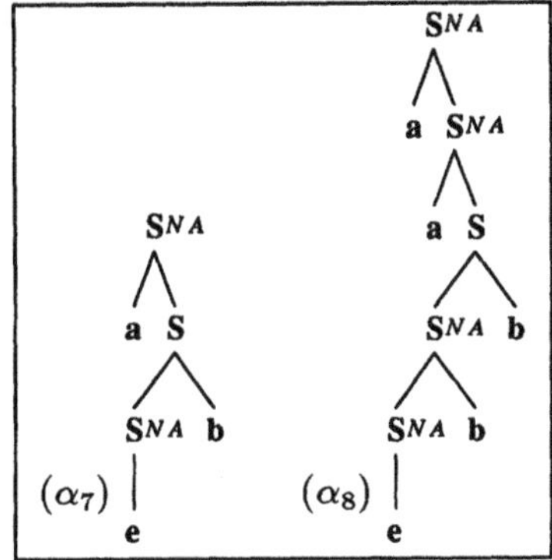

Figure 6: Some derived trees of G_1

Example 2 Consider the following TAG, $G_2 = (\{a, b, c, d, e\}, \{S\}, \{\alpha_6\}, \{\beta_3\}, S)$

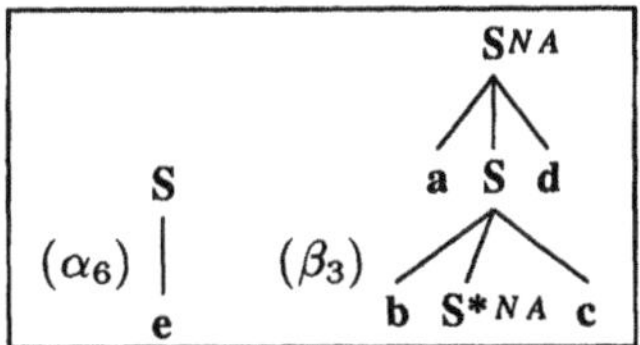

G_2 generates the context-sensitive language $L_2 = \{a^n b^n e c^n d^n | n \geq 1\}$. For example, in Figure 7, α_9 has been obtained by adjoining β_3 on the root node of α_6 and α_{10} has been obtained by adjoining β_3 on the node at address 2 in in the tree α_9.

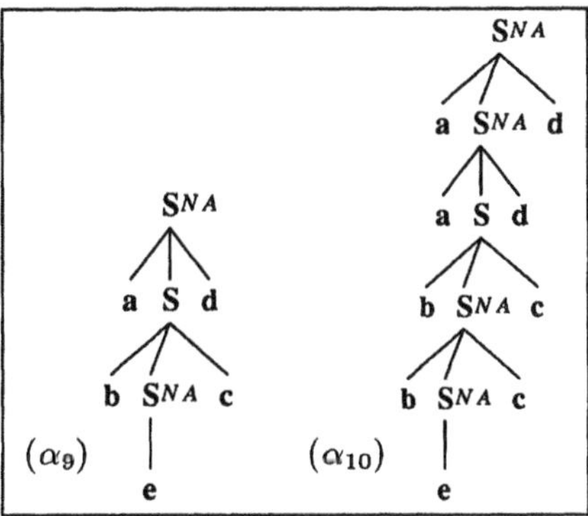

Figure 7: Some derived trees of G_2

One can show that $\{a^n b^n c^n d^n e^n | n \geq 1\}$ is not a tree-adjoining language.

We have seen in Section 2.2 that the derivation in TAG is also a tree. For a given

TAG, G, it can be shown that the set of derivations trees of G, $D(G)$, is recognizable (in fact a local set). Many different grammar formalisms have been shown to be equivalent to TAG, their derivation trees are also local sets.

3   Lexicalized Grammars

We define a notion of "lexicalized grammars" that is of both linguistic and formal significance. We then show how TAG arises in the processes of lexicalizing context-free grammars.

In this "lexicalized grammar" approach (Schabes et al., 1988; Schabes, 1990), each elementary structure is systematically associated with a lexical item called the *anchor*. By 'lexicalized' we mean that in each structure there is a lexical item that is realized. The 'grammar' consists of a lexicon where each lexical item is associated with a finite number of structures for which that item is the anchor. There are operations which tell us how these structures are composed. A grammar of this form will be said to be 'lexicalized'.

Definition 2 (Lexicalized Grammar) A grammar is 'lexicalized' if it consists of:

- a finite set of structures each associated with a lexical item; each lexical item will be called the *anchor* of the corresponding structure;

- an operation or operations for composing the structures.

We require that the anchor must be an overt (i.e. not the empty string) lexical item.

The *lexicon* consists of a finite set of structures each associated with an anchor. The structures defined by the lexicon are called *elementary structures*. Structures built by combination of others are called *derived structures*.

As part of our definition of lexicalized grammars, we require that the structures be of finite size. We also require that the combining operations combine a finite set of structures into a finite number of structures. We will consider operations that combine two structures to form one derived structure.

Other constraints can be put on the operations. For examples, the operations could be restricted not to copy, erase or restructure unbounded components of their arguments. We could also impose that the operations yield languages of constant growth (Joshi [1985]). The operations that we will use have these properties.

Categorial Grammars (Lambek, 1958; Steedman, 1987) are lexicalized according to our definition since each basic category has a lexical item associated with it.

As in Categorial Grammars, we say that the *category* of a word is the entire structure it selects. If a structure is associated with an anchor, we say that the entire structure is the *category* structure of the anchor.

We also use the term 'lexicalized' when speaking about structures. We say that a structure is *lexicalized* if there is at least one overt lexical item that appears in it. If more than one lexical item appears, either one lexical item is designated as the anchor or a subset of the lexical items local to the structure are designated as *multi-component anchor*. A grammar consisting of only lexicalized structures is of course lexicalized.

For example, the following structures are lexicalized according to our definition:[7]

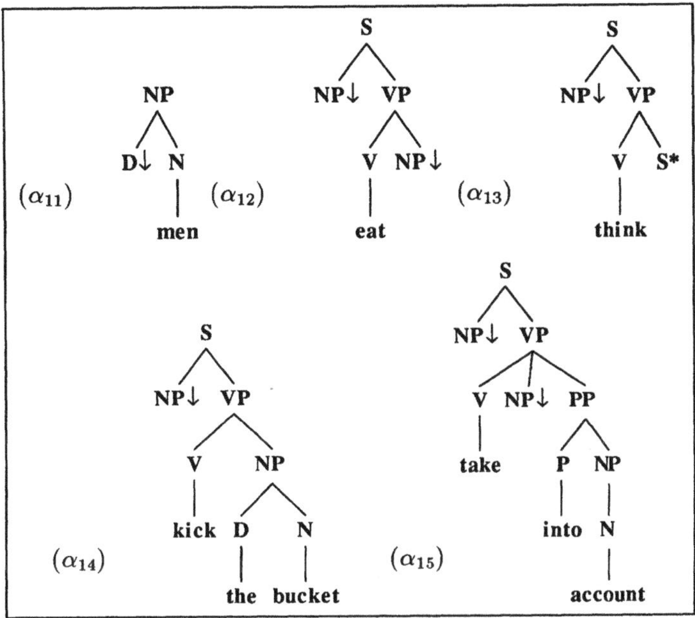

Some simple properties follow immediately from the definition of lexicalized grammars.

Proposition 1 *Lexicalized grammars are finitely ambiguous.*

A grammar is said to be finitely ambiguous if there is no sentence of finite length that can be analyzed in an infinite number of ways.

The fact that lexicalized grammars are finitely ambiguous can be seen by considering an arbitrary sentence of finite length. The set of structures anchored by the words in the input sentence consists of a set of structures necessary to analyze the sentence; while any other structure introduces lexical items not present in the input string. Since the set of selected structures is finite, these structures can be combined in finitely many ways (since each tree is associated with at least one lexical item and since structures can combine to produce finitely many structures). Therefore lexicalized grammars are finitely ambiguous.

Since a sentence of finite length can only be finitely ambiguous,the search space used for analysis is finite. Therefore, the recognition problem for lexicalized grammars is decidable

Proposition 2 *It is decidable whether or not a string is accepted by a lexicalized grammar.*[8]

---

[7]The interpretation of the annotations on these structures is not relevant now and it will be given later.

[8]Assuming that one can compute the result of the combination operations.

Having stated the basic definition of lexicalized grammars and also some simple properties, we now turn our attention to one of the major issues: can context-free grammars be lexicalized?

Not every grammar is in a lexicalized form. Given a grammar G stated in a formalism, we will try find another grammar G_{lex} (not necessarily stated in the same formalism) that generates the same language and also the same tree set as G and for which the lexicalized property holds. We refer to this process as lexicalization of a grammar.

Definition 3 (Lexicalization) We say that a formalism F can be lexicalized by another formalism F', if for any finitely ambiguous grammar G in F there is a grammar G' in F' such that G' is a lexicalized grammar and such that G and G' generate the same tree set (and a fortiori the same language).

The next section discusses what it means to lexicalize a grammar. We will investigate the conditions under which such a 'lexicalization' is possible for CFGs and tree-adjoining grammars (TAGs). We present a method to lexicalize grammars such as CFGs, while keeping the rules in their full generality. We then show how a lexicalized grammar naturally follows from the extended domain of locality of TAGs.

4 'Lexicalization' of CFGs

Our definition of lexicalized grammars implies their being finitely ambiguous. Therefore a necessary condition of lexicalization of a CFG is that it is finitely ambiguous. As a consequence, recursive chain rules obtained by derivation (such as $X \overset{*}{\Rightarrow} X$) or elementary (such as $X \to X$) are disallowed since they generate infinitely ambiguous branches without introducing lexical items.

In general, a CFG will not be in lexicalized form. For example a rule of the form, $S \to NP\ VP$ or $S \to S\ S$, is not lexicalized since no lexical item appears on the right hand side of the rule.

A lexicalized CFG would be one for which each production rule has a terminal symbol on its right hand side. These constitute the structures associated with the lexical anchors. The combining operation is the standard substitution operation.[9]

Lexicalization of CFG that is achieved by transforming it into an equivalent Greibach Normal Form CFG, can be regarded as a *weak* lexicalization, because it does not give us the same set of trees as the original CFG.[10] Our notion of lexicalization can be regarded as *strong* lexicalization.

In the next sections, we propose to extend the domain of locality of context-free grammar in order to make lexical item appear local to the production rules. The domain of

---

[9]Variables are independently substituted by substitution. This standard (or first order) substitution contrasts with more powerful versions of substitution which allow to substitute multiple occurrences of the same variable by the same term.

[10]To our knowledge, there is no known tree transducer that transforms the derivation of a context-free grammar transformed in Greibach normal form to its original derivation. We strongly suspect that such a tree transducer can very probably be found. However, the transducer will probably be quite complex and will need to work on distant pieces since leaves of the derivation tree must be put back to higher positions in the tree.

locality of a CFG is extended by using a tree rewriting system that uses only substitution. We will see that in general, CFGs cannot be lexicalized using substitution alone, even if the domain of locality is extended to trees. Furthermore, in the cases where a CFG could be lexicalized by extending the domain of locality and using substitution alone, we will show that, in general, there is not enough freedom to choose the anchor of each structure. This is important because we want the choice of the anchor for a given structure to be determined on purely linguistic grounds. We will then show how the operation of adjunction enables us to freely 'lexicalize' CFGs.

4.1   Substitution and Lexicalization of CFGs

We already know that we need to assume that the given CFG is finitely ambiguous in order to be able to lexicalized it. We propose to extend the domain of locality of CFGs to make lexical items appear as part of the elementary structures by using a grammar on trees that uses substitution as combining operation. This tree-substitution grammar consists of a set of trees that are not restricted to be of depth one (rules of context-free grammars can be thought as trees of depth one) combined with substitution.[11]

A finite set of elementary trees that can be combined with substitution define a tree-based system that we will call a *tree substitution grammar*.

Definition 4 (Tree-Substitution Grammar)
A Tree-Substitution Grammar (TSG) consists of a quadruple (Σ, NT, I, S), where

(i) Σ is a finite set of terminal symbols;
(ii) NT is a finite set of non-terminal symbols[12]: $\Sigma \cap NT = \emptyset$;
(iii) S is a distinguished non-terminal symbol: $S \in NT$;
(iv) I is a finite set of finite trees whose interior nodes are labeled by non-terminal symbols and whose frontier nodes are labeled by terminal or non- terminal symbols. All non-terminal symbols on the frontier of the trees in I are marked for substitution. The trees in I are called *initial* trees.

We say that a tree is *derived* if it has been built from some initial tree in which initial or derived trees were substituted. A tree will be said of *type X* if its root is labeled by X. A tree is considered *completed* if its frontier is to be made only of nodes labeled by terminal symbols.

Whenever the string of labels of the nodes on the frontier of an X-type initial tree t_X is $\alpha \in (\Sigma \cup NT)^*$, we will write: $Fr(t_X) = \alpha$.

As for TAG, the derivation in TSG is stated in the form of a tree called the *derivation tree* (see Section 2.2).

It is easy to see that the set of languages generated by this tree rewriting system is exactly the same set as context-free languages.

We now come back to the problem of lexicalizing context-free grammars. One can try to lexicalize finitely ambiguous CFGs by using tree-substitution grammars. However we will exhibit a counter example that shows that, in the general case, finitely ambiguous

---

[11] We assume here first order substitution meaning that all substitutions are independent.
[12] We use lower-case letters for terminal symbols and upper-case letters for non-terminal symbols.

CFGs cannot be lexicalized with a tree system that uses substitution as the only combining operation.

Proposition 3   *Finitely ambiguous context-free grammars cannot be lexicalized with a tree-substitution grammar.*

Proof[13] of Proposition 3
We show this proposition by a contradiction. Suppose that finitely ambiguous CFGs can be lexicalized with TSG. Then the following CFG can be lexicalized:[14]

Example 3 (counter example)

$$S \to S\ S$$
$$S \to a$$

Suppose there were a lexicalized TSG G generating the same tree set as the one generated by the above grammar. Any derivation in G must start from some initial tree. Take an arbitrary initial tree t in G. Since G is a lexicalized version of the above context-free grammar, there is a node n on the frontier of t labeled by a. Since substitution can only take place on the frontier of a tree, the distance between n and the root node of t is constant in any derived tree from t. And this is the case for any initial tree t (of which there are only finitely many). This implies that in any derived tree from G there is at least one branch of bounded length from the root node to a node labeled by a (that branch cannot further expand). However in the derivation trees defined by the context-free grammar given above, a can occur arbitrarily far away from the root node of the derivation. Contradiction.

□

The CFG given in Example 3 cannot be lexicalized with a TSG. The difficulty is due to the fact that TSGs do not permit the distance between two nodes in the same initial tree to increase.

For example, one might think that the following TSG is a lexicalized version of the above grammar:

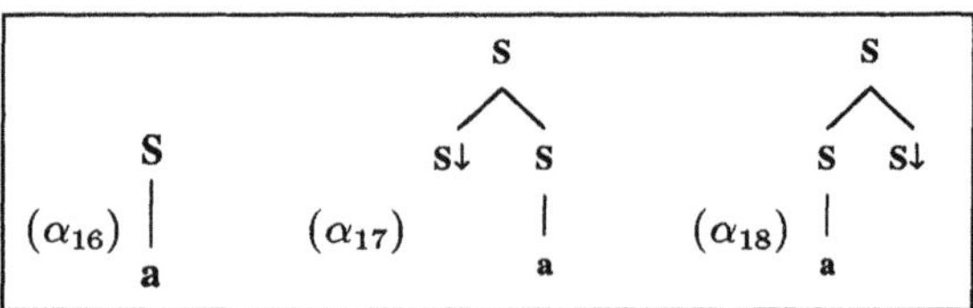

---

[13]The underlying idea behind this proof was suggested to us by Stuart Shieber.

[14]This example was pointed out to us by Fernando Pereira.

However, this lexicalized TSG does not generate all the trees generated by the context-free grammar; for example the following tree (α_{19}) cannot be generated by the above TSG:

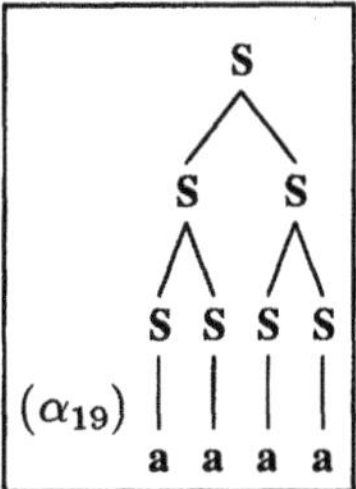

We now turn to a less formal observation. Even if some CFGs can be lexicalized by using TSG, the choice of the lexical items that emerge as the anchor may be too restrictive, for example, the choice may not be linguistically motivated.

Consider the following example:

Example 4

$$
\begin{aligned}
S &\to NP\ VP \\
VP &\to adv\ VP \\
VP &\to v \\
NP &\to n
\end{aligned}
$$

The grammar can be lexicalized as follows:

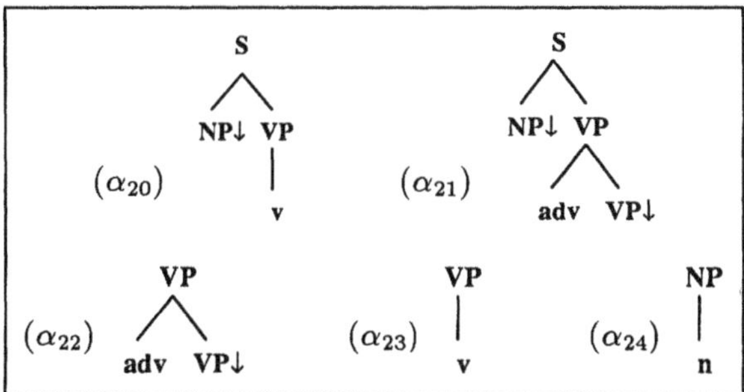

This tree-substitution grammar generates exactly the same set of trees as in Example 4, however, in this lexicalization one is forced to choose *adv* (or *n*) as the anchor of a structure rooted by S (α_{21}), and it cannot be avoided. This choice is not linguistically motivated. If one tried not to have an S-type initial tree anchored by n or by *adv*, recursion on the VP node would be inhibited.

For example, the grammar written below:

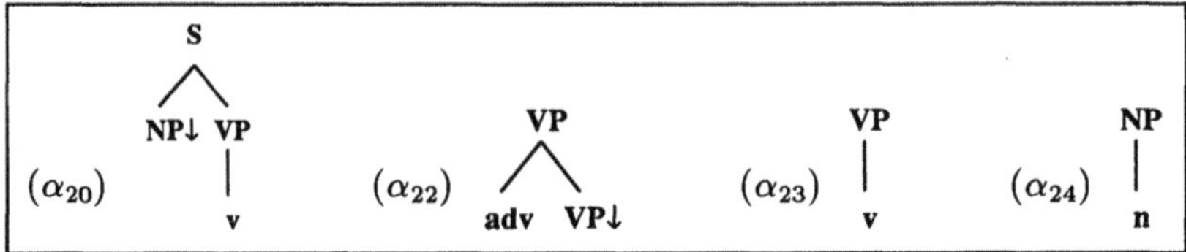

does not generate the tree α_{25}:

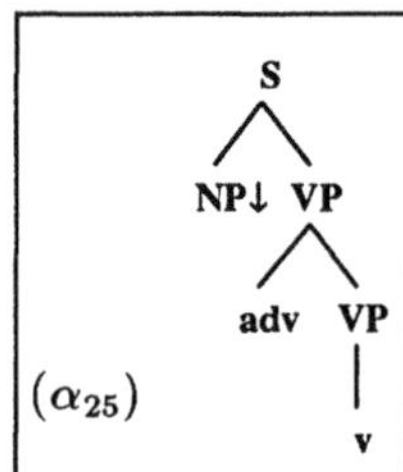

This example shows that even when it is possible to lexicalize a CFG, substitution (TSG) alone does not allow us to freely choose the lexical anchors. Substitution alone forces us to make choices of anchors that might not be linguistically (syntactically or semantically) justified. From the proof of proposition 3 we conclude that a tree based system that can lexicalize context-free grammars must permit the distance between two nodes in the same tree to be increased during a derivation. In the next section, we suggest the use of an additional operation when defining a tree-based system in which one tries to lexicalize CFGs.

4.2   Lexicalization of CFGs with TAGs

Another combining operation is needed to lexicalize finitely ambiguous CFGs. As the previous examples suggest us, we need an operation that is capable of inserting a tree inside another one. We suggest using adjunction as an additional combining operation. A tree-based system that uses substitution and adjunction coincides with a tree-adjoining grammar (TAG).

We first show that the CFGs in examples 3 and 4 for which TSG failed can be lexicalized within TAGs.

Example 5 Example 3 could not be lexicalized with TSG. It can be lexicalized by using adjunction as follows: [15]

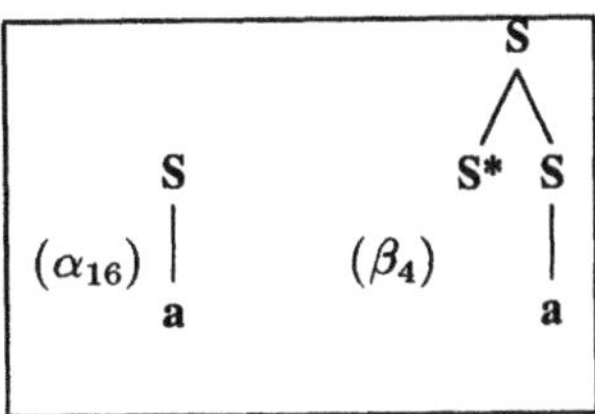

The auxiliary tree β_4 can now be inserted by adjunction inside the derived trees.
For example, the following derived trees can be derived by successive adjunction of β_4:

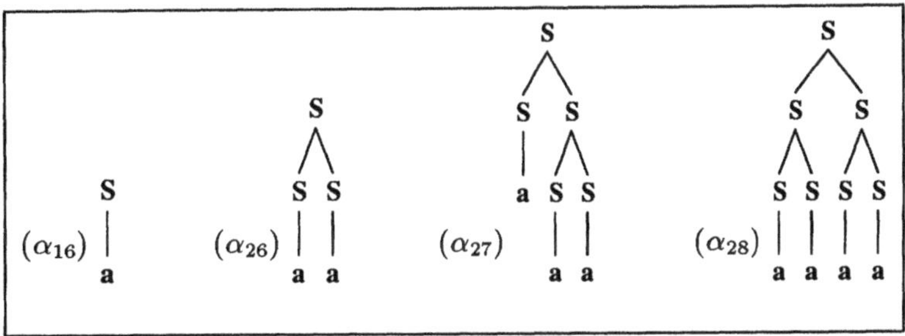

Example 6 The CFG given in Example 4 can be lexicalized by using adjunction and one can choose the anchor freely:[16]

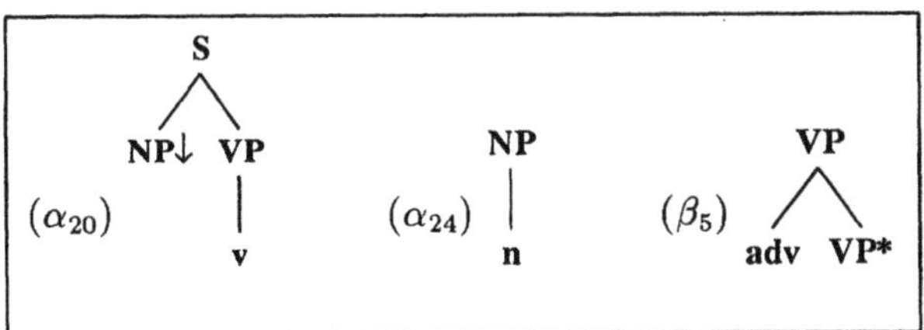

The auxiliary tree β_5 can be inserted in α_{20} at the VP node by adjunction. Using adjunction one is thus able to choose the appropriate lexical item as anchor. The following trees (α_{29} and α_{30}) can be derived by substitution of α_{24} into α_{20} for the NP node and

---

[15] a is taken as the lexical anchor of both the initial tree α_{16} and the auxiliary tree β_4.

[16] We chose v as the lexical anchor of α_{20} but, formally, we could have chosen n instead.

by adjunction of β_5 on the VP node in α_{20}:

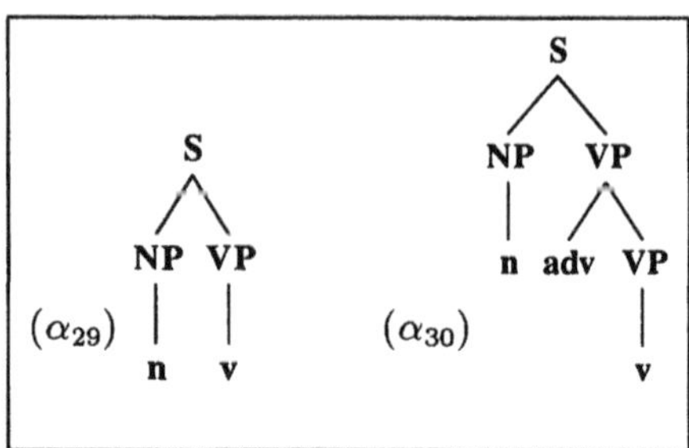

We are now ready to prove the main result: any finitely ambiguous context-free grammar can be lexicalized within tree-adjoining grammars; furthermore adjunction is the only operation needed. Substitution as an additional operation enables one to lexicalize CFGs in a more compact way.

Proposition 4 *If $G = (\Sigma, NT, P, S)$ is a finitely ambiguous CFG which does not generate the empty string, then there is a lexicalized tree-adjoining grammar $G_{lex} = (\Sigma, NT, I, A, S)$ generating the same language and tree set as G. Furthermore G_{lex} can be chosen to have no substitution nodes in any elementary trees.*

We give a constructive proof of this proposition. Given an arbitrary CFG, G, we construct a lexicalized TAG, G_{lex}, that generates the same language and tree set as G. The construction is not optimal with respect to time or the number of trees but it does satisfy the requirements.

The idea is to separate the recursive part of the grammar G from the non-recursive part. The non-recursive part generates a finite number of trees, and we will take those trees as initial TAG trees. Whenever there is in G a recursion of the form $B \overset{*}{\Rightarrow} \alpha B \beta$, we will create an B-type auxiliary tree in which α and β are expanded in all possible ways by the non-recursive part of the grammar. Since the grammar is finitely ambiguous and since $\lambda \notin L(G)$, we are guaranteed that $\alpha\beta$ derives some lexical item within the non-recursive part of the grammar. The proof follows.

Proof of Proposition 4

Let $G = (\Sigma, NT, P, S)$ be a finitely ambiguous context-free grammar s.t. $\lambda \notin L(G)$. We say that $B \in NT$ is a *recursive symbol* if and only if $\exists \alpha, \beta \in (\Sigma \cup NT)^*$ s.t. $B \overset{*}{\Rightarrow} \alpha B \beta$. We say that a production rule $B \to \delta$ is recursive whenever B is recursive.

The set of production rules of G can be partitioned into two sets: the set of recursive production rules, say $R \subseteq P$, and the set of non-recursive production rules, say $NR \subseteq P$; $R \cup NR = P$ and $R \cap NR = \emptyset$. In order to determine whether a production is recursive, given G, we construct a directed graph $\mathcal{G}$ whose nodes are labeled by non-terminal symbols and whose arcs are labeled by production rules. There is an arc labeled by $p \in P$ from a node labeled by B to a node labeled by C whenever p is of the form $B \to \alpha C \beta$, where $\alpha, \beta \in (\Sigma \cup NT)^*$. Then, a symbol B is recursive if the node labeled by B in $\mathcal{G}$ belongs to a cycle. A production is recursive if there is an arc labeled by the production which belongs to a cycle.

Let $L(NR) = \{w|S\overset{*}{\Rightarrow}w$ using only production rules in $NR\}$. $L(NR)$ is a finite set. Since $\lambda \notin L(G)$, $\lambda \notin L(NR)$. Let I be the set of all derivation trees defined by $L(NR)$. I is a finite set of trees; the trees in I have at least one terminal symbol on the frontier since the empty string is not part of the language. I will be the set of initial trees of the lexicalized TAG G_{lex}.

We then form a base of minimal cycles of $\mathcal{G}$. Classical algorithms on graphs gives us methods to find a finite set of so-called 'base-cycles' such that any cycle is a combination of those cycles and such that they do not have any sub-cycle. Let $\{c_1 \cdots c_k\}$ be a base of cycles of $\mathcal{G}$ (each c_i is a cycle of $\mathcal{G}$).

We initialize the set of auxiliary trees of G_{lex} to the empty set, i.e. $A := \emptyset$. We repeat the following procedure for all cycles c_i in the base until no more trees can be added to A.

For all nodes n_i in c_i, let B_i be the label of n_i,
    According to c_i, $B_i\overset{*}{\Rightarrow}\alpha_i B_i\beta_i$,
    If B_i is the label of a node in a tree in $I \cup A$ then
        for all derivations $\alpha_i\overset{*}{\Rightarrow}w_i \in \Sigma^\star$, $\beta_i\overset{*}{\Rightarrow}z_i \in \Sigma^\star$
       that use only non-recursive production rules
           add to A the auxiliary tree corresponding to all derivations:
           $B_i\overset{*}{\Rightarrow}\alpha_i B_i\beta_i\overset{*}{\Rightarrow}w_i B_i z_i$ where the node labeled B_i on the frontier is the foot node.

In this procedure, we are guaranteed that the auxiliary trees have at least one lexical item on the frontier, because $\alpha_i\beta_i$ must always derive some terminal symbol otherwise, the derivation $B_i\overset{*}{\Rightarrow}\alpha_i B_i\beta_i$ would derive a rule of the form $B_i\overset{*}{\Rightarrow}B_i$ and the grammar would be infinitely ambiguous.

It is clear that G_{lex} generates exactly the same tree set as G. Furthermore G_{lex} is lexicalized.

$\square$

We just showed that adjunction is sufficient to lexicalize context-free grammars. However, the use of substitution as an additional operation to adjunction enables one to lexicalize a grammar with a more compact TAG.

5   Closure of TAGs under Lexicalization

In the previous section, we showed that context-free grammars can be lexicalized within tree-adjoining grammars. Only adjunction is necessary but the addition of substitution gives us the possibility to have a more compact representation of the lexicalized grammar. We now ask ourselves if TAGs are closed under lexicalization: given a finitely ambiguous TAG, G, ($\lambda \notin L(G)$), is there a lexicalized TAG, G_{lex}, which generates the same language and the same tree set as G? The answer is yes. We therefore establish that TAGs are closed under lexicalization. The following proposition holds:

Proposition 5 (TAGs are closed under lexicalization)
If G is a finitely ambiguous TAG that uses substitution and adjunction as combining operation, s.t. $\lambda \notin L(G)$, then there exists a lexicalized TAG G_{lex} which generates the same language and the same tree set as G.

The proof of this proposition is similar to the proof of proposition 4 and we only give a sketch of it. It consists of separating the recursive part of the grammar from the non-recursive part. The recursive part of the language is represented in G_{lex} by auxiliary trees. Since G is finitely ambiguous, those auxiliary trees will have at least one terminal symbol on the frontier. The non-recursive part of the grammar is encoded as initial trees. Since the empty string is not generated, those initial trees have at least one terminal symbol on the frontier. In order to determine whether an elementary tree is recursive, given G, we construct a directed graph $\mathcal{G}$ whose nodes are labeled by elementary trees and whose arcs are labeled by tree addresses. There is an arc labeled by ad from a node labeled by β to a node labeled by α whenever β can operate (by adjunction or substitution) at address ad in α. Then, an elementary tree δ is recursive if the node labeled by δ in $\mathcal{G}$ belongs to a cycle. The construction of the lexicalized TAG is then similar to the one proposed for proposition 4.

6 Conclusion[17]

The elementary objects manipulated by a tree-adjoining grammar are trees, i.e., structured objects and not strings. The properties of TAGs relate directly to the strong generative capacity (structural description) which is more relevant to linguistic descriptions than the weak generative capacity (set of strings). The tree sets of TAGs are not recognizable sets but are equivalent to the tree sets of linear indexed languages. Hence, tree-adjoining grammars generate some context-sensitive languages. However, tree-adjoining languages are strictly contained in the class of indexed languages.

The lexicalization of grammar formalisms is of linguistic and formal interest. We have taken the point of view that rules should not be separated totally from their lexical realization. In this "lexicalized" approach, each elementary structure is systematically associated with a lexical anchor. These structures specify extended domains of locality (as compared to Context Free Grammars) over which constraints can be stated.

The process of lexicalization of context-free rules forces us to use operations for combining structures that make the formalism fall in the class of mildly context sensitive languages. Substitution and adjunction give us the freedom to lexicalize CFGs. Elementary structures of extended domain of locality, when they are combined with substitution and adjunction, yield Lexicalized TAGs. TAGs were so far introduced as an independent formal system. We have shown that they derive from the lexicalization process of context-free grammars. We also have shown that TAGs are closed under lexicalization.

It is still an open problem whether or not adjoining is the 'minimal' operation needed for lexicalizing CFGs, i.e., whether there exists a tree-gluing operation say Φ such that substitution and Φ can lexicalize any CFG, and such that the tree sets of the tree system with substitution and Φ are properly contained in the tree sets of TAG.

Bibliography

Anne Abeillé, Kathleen M. Bishop, Sharon Cote, and Yves Schabes. 1990. A lexicalized tree adjoining grammar for english. Technical Report MS-CIS-90-24, Department of Computer and Information Science, University of Pennsylvania.

Anne Abeillé. 1988. Parsing french with tree adjoining grammar: some linguistic accounts. In *Proceedings of the 12th International Conference on Computational Linguistics (COLING'88)*, Budapest, August.

N. Chomsky. 1981. *Lectures on Government and Binding*. Foris, Dordrecht.

G. Gazdar, E. Klein, G. K. Pullum, and I. A. Sag. 1985. *Generalized Phrase Structure Grammars*. Blackwell Publishing, Oxford. Also published by Harvard University Press, Cambridge, MA.

---

[17]There are several important papers about TAGs describing their linguistic and formal properties. Some of these are: Joshi (1987), Joshi, Vijay-Shanker and Weir (forthcoming 1991), Vijay-Shanker (1987), Weir (1988), Schabes (1990; 1991), Schabes and Joshi (1988; 1989), Kroch (1987), Kroch and Joshi (1985), Abeillé, Bishop, Cote and Schabes (1990), Abeillé (1988). A reader interested in TAGs will find these papers very useful.

Maurice Gross. 1984. Lexicon-grammar and the syntactic analysis of french. In *Proceedings of the 10th International Conference on Computational Linguistics (COLING'84)*, Stanford, 2-6 July.

Aravind K. Joshi, L. S. Levy, and M. Takahashi. 1975. Tree adjunct grammars. *Journal of Computer and System Sciences*, 10(1).

Aravind K. Joshi, K. Vijay-Shanker, and David Weir. forthcoming, 1991. The convergence of mildly context-sensitive grammatical formalisms. In Peter Sells, Stuart Shieber, and Tom Wasow, editors, *Foundational Issues in Natual Language Processing*. MIT Press, Cambridge MA.

Aravind K. Joshi. 1985. How much context-sensitivity is necessary for characterizing structural descriptions—Tree Adjoining Grammars. In D. Dowty, L. Karttunen, and A. Zwicky, editors, *Natural Language Processing—Theoretical, Computational and Psychological Perspectives*. Cambridge University Press, New York. Originally presented in a Workshop on Natural Language Parsing at Ohio State University, Columbus, Ohio, May 1983.

Aravind K. Joshi. 1987. An Introduction to Tree Adjoining Grammars. In A. Manaster-Ramer, editor, *Mathematics of Language*. John Benjamins, Amsterdam.

R. Kaplan and J. Bresnan. 1983. Lexical-functional grammar: A formal system for grammatical representation. In J. Bresnan, editor, *The Mental Representation of Grammatical Relations*. MIT Press, Cambridge MA.

Lauri Karttunen. 1986. Radical lexicalism. Technical Report CSLI-86-68, CSLI, Stanford University. Also in *Alternative Conceptions of Phrase Structure*, University of Chicago Press, Baltin, M. and Kroch A., Chicago, 1989.

Anthony Kroch and Aravind K. Joshi. 1985. Linguistic relevance of tree adjoining grammars. Technical Report MS-CIS-85-18, Department of Computer and Information Science, University of Pennsylvania, April.

Anthony Kroch. 1987. Unbounded dependencies and subjacency in a tree adjoining grammar. In A. Manaster-Ramer, editor, *Mathematics of Language*. John Benjamins, Amsterdam.

Joachim Lambek. 1958. The mathematics of sentence structure. *American Mathematical Monthly*, 65:154–170.

Carl Pollard and Ivan A. Sag. 1987. *Information-Based Syntax and Semantics. Vol 1: Fundamentals*. CSLI.

Yves Schabes and Aravind K. Joshi. 1988. An Earley-type parsing algorithm for Tree Adjoining Grammars. In *26th Meeting of the Association for Computational Linguistics (ACL'88)*, Buffalo, June.

Yves Schabes and Aravind K. Joshi. 1989. The relevance of lexicalization to parsing. In *Proceedings of the International Workshop on Parsing Technologies*, Pittsburgh, August. To also appear under the title *Parsing with Lexicalized Tree adjoining Grammar* in *Current Issues in Parsing Technologies*, MIT Press.

Yves Schabes, Anne Abeillé, and Aravind K. Joshi. 1988. Parsing strategies with 'lexicalized' grammars: Application to tree adjoining grammars. In *Proceedings of the 12th International Conference on Computational Linguistics (COLING'88)*, Budapest, Hungary, August.

Yves Schabes. 1990. *Mathematical and Computational Aspects of Lexicalized Grammars*. Ph.D. thesis, University of Pennsylvania, Philadelphia, PA, August. Available as technical report (MS-CIS-90-48, LINC LAB179) from the Department of Computer Science.

Yves Schabes. 1991. The valid prefix property and left to right parsing of tree-adjoining grammar. In *Proceedings of the second International Workshop on Parsing Technologies*, Cancun, Mexico, February.

K. M. Schimpf and J. H. Gallier. 1985. Tree pushdown automata. *Journal of Computer and System Sciences*, 30:25–39.

Mark Steedman. 1987. Combinatory grammars and parasitic gaps. *Natural Language and Linguistic Theory*, 5:403–439.

J. W. Thatcher. 1971. Characterizing derivations trees of context free grammars through a generalization of finite automata theory. *Journal of Computer and System Sciences*, 5:365–396.

K. Vijay-Shanker. 1987. *A Study of Tree Adjoining Grammars*. Ph.D. thesis, Department of Computer and Information Science, University of Pennsylvania.

David J. Weir. 1988. *Characterizing Mildly Context-Sensitive Grammar Formalisms*. Ph.D. thesis, Department of Computer and Information Science, University of Pennsylvania.

Tree Automata and Languages
M. Nivat and A. Podelski (editors)
1992 Elsevier Science Publishers B.V.

A Short Proof of the Factorization Forest Theorem

Imre Simon[1]

Instituto de Matemática e Estatística
Universidade de São Paulo
05508 São Paulo, SP, Brasil

e-mail: isimon@ime.usp.br

1   Introduction

A short proof of the Factorization Forest Theorem is given. More precisely it is proved that
every morphism $f\colon A^+ \to S$, with S a finite semigroup, admits a Ramseyan factorization
forest of finite height. The proof is based on the Krohn-Rhodes decomposition Lemma
and yields an exponential upper bound on the height of the factorization forest obtained.
This contrasts with the linear bound obtained by the original (more complicated) proof.

2   Factorization forests of finite height

We shall need some definitions and results from [8]. For an alphabet A we denote the free
semigroup generated by A either by A^+, as usual, or by $\mathcal{F}(A)$. In the second notation
the elements of $\mathcal{F}(A)$ will be represented as $(a_1, a_2, \ldots, a_p)$, where $a_i \in A$.
A *factorization forest* $F = (X, d)$ *over* A consists of a subset X of A^+ together with a
function $d\colon X \to \mathcal{F}(X)$ such that for every $x \in X$, $d(x) = (x_1, x_2, \ldots, x_p)$ implies that
$x = x_1 x_2 \cdots x_p$. In other words, $d(x)$ is a factorization of x whose factors belong to X.
The name factorization forest as well as the terminology used is inspired by the following
construction. Given F we associate to each $x \in X$ a rooted ordered tree $T(x)$ whose
vertices are labeled by elements of X. If $: d(x) := 1$ then $T(x)$ consists just of the root
labeled x. If $d(x) = (x_1, x_2, \ldots, x_p)$, with $p > 1$, then the root of $T(x)$ has degree p and a
copy of $T(x_i)$ is associated to the i-th direct descendant of the root.
The elements of X will be called *vertices*. The *degree* of $x \in X$ is 0 if $: d(x) := 1$
and is $: d(x) :$ otherwise. The *external set* of F is the set of vertices of degree 0. Let
$d(x) = (x_1, \ldots, x_p)$; the *height* of x, denoted $h(x)$, is defined by

$$h(x) = \begin{cases} 0 & \text{if } x \text{ is external,} \\ 1 + \max\{\, h(x_i) \mid 1 \le i \le p \,\} & \text{otherwise.} \end{cases}$$

The *height* of F is $h(F) = \sup\{\, h(x) \mid x \in X \,\}$.

Let $f: A^+ \to S$ be a semigroup morphism. Factorization forest F is *Ramseyan mod f* if for every x of degree $p \geq 3$, $d(x) = (x_1, x_2, \ldots, x_p)$ implies that there exists an idempotent $e \in S$ such that $e = f(x) = f(x_1) = \cdots = f(x_p)$. Factorization forest F is *alphabetical* if every external vertex of F is a letter in A. Let F be a factorization forest over A; we say that *f admits F* if F is alphabetical with vertex set A^+. We say that *f admits a Ramseyan factorization forest* if it admits a factorization forest F over A, which is Ramseyan mod f.

We recall the main result of [8].

Theorem 1 *Every morphism $f: A^+ \to S$, from a free semigroup to a finite one, admits a Ramseyan factorization forest of height at most $9 : S :$.*

Let S be a finite semigroup. We call the *height of S* the least integer $H(S)$ such that every morphism $f: A^+ \to S$, from a free semigroup to S admits a Ramseyan factorization forest of height at most $H(S)$. We emphasize that we did not restrict the alphabet A to be finite in the above definition. Theorem 1 guarantees that for every finite semigroup S, $H(S) \leq 9 : S :$. The (linear) precision of this result puts heavy constraints on the proof which results quite complicated. Our aim in this paper is to prove that for every finite semigroup S, $H(S) \leq 2^{:S:+1} - 2$. Even though this result is much weaker than the previous one its interest lies in the fact that its proof is much shorter and thus it might throw some light on the nature of the Factorization Forest Theorem. The proof relies on the Krohn-Rhodes Decomposition Lemma.

3 The short proof

Initially we state three Lemmata whose proofs we defer to the next section.

Lemma 2 *The height of a finite group G is at most $3 : G :$.*

Lemma 3 *The height of a finite cyclic semigroup S is at most $3 : S :$.*

Lemma 4 *Let S be a finite semigroup, such that $S = V \cup T$, where V is a proper left ideal of S and T is a proper subsemigroup of S. Then, $H(S) \leq H(V) + H(T) + 2$.*

The following Proposition is due to Krohn and Rhodes [4] and has been used as the basic decomposition tool in the original Proof of the Krohn-Rhodes Prime Decomposition Theorem. It has many variants; some of these can be found in [6, Proposition 2.5 or Lemma 7.2.7] and it has been used in many contexts. The original proof of this result was obtained by using the Local Structure Theory of finite semigroups. An elegant, short and elementary proof, found by A. H. Clifford, appears in [5, Lemma 3.7, page 105]. Recall that a semigroup is *left simple* if it does not contain proper left ideals.

Proposition 5 *Let S be a finite semigroup. Then either*

1. *S is left simple,*

2. *S is cyclic,*

3. *there exist a proper left ideal V of S and a proper subsemigroup T of S such that $S = V \cup T$.*

We are ready to prove our main result.

Theorem 6 *The height of a finite semigroup S is at most $2^{:S:+1} - 2$.*

Proof. We proceed by induction on the cardinal of S. If $: S := 1$ then the height of S is 1. Assume that $: S :> 1$. Initially we consider 4 cases:

- If S is a group then we are done in view of Lemma 2.

- If S is a cyclic semigroup then we are done in view of Lemma 3.

- If S is the union of a proper left ideal V and a proper subsemigroup T then from the induction hypothesis we conclude that $H(V) \leq 2^{:V:+1} - 2$ and $H(T) \leq 2^{:T:+1} - 2$. From Lemma 4, $H(S) \leq H(V) + H(T) + 2 \leq 2^{:V:+1} + 2^{:T:+1} - 2 \leq 2^{:S:+1} - 2$, since $1 \leq: V :,: T :<: S :.$

- If S is the union of a proper right ideal V and a proper subsemigroup T then an argument dual to the prevoius one shows that $H(S) \leq 2^{:S:+1} - 2$.

Assume now that neither of the four cases holds. Then, by Proposition 5, S is left simple. Also, by the dual of Proposition 5, S is right simple. But then S is a group [2, page 6] and it satisfies the first case. A contradicton which establishes the theorem. ∎

4 Proof of the Lemmata

The proof of the case of groups is inspired by the McNaughton-Yamada proof of Kleene's Theorem.

Proof of Lemma 2. Let G have n elements, and let $g_1, g_2, \ldots g_n$ be an enumeration of them. We define a linear order on G by putting $g_i \leq g_j$ if and only if $i \leq j$. Let $f : A^+ \to G$ be a semigroup morphism for which we shall construct a Ramseyan alphabetical factorization forest of height at most $3n$.
For $0 \leq k \leq n$ and $1 \leq i, j \leq n$ we define

$$X_{i,j}^{(k)} = \{\, u \in f^{-1}(g_i^{-1}g_j) \mid g_i f(v) \leq g_k, \text{ for all } v, w \in A^+, u = vw \,\}.$$

We observe that for every i and k, $X_{i,i}^{(k)} \subseteq f^{-1}(e)$, where e is the identity of G. Also, a standard argument in Automaton Theory [3, page 175] shows that, for every i, j, k, $X_{i,j}^{(k)} \subseteq A$ if $k = 0$ and, for $k \geq 1$,

$$X_{i,j}^{(k)} = X_{i,j}^{(k-1)} \cup X_{i,k}^{(k-1)}(X_{k,k}^{(k-1)})^* X_{k,j}^{(k-1)}. \tag{1}$$

Now we claim that for every k, there exists an alphabetical factorization forest $F_k = (X^{(k)}, d_k)$ of height at most $3k$, with vertex set

$$X^{(k)} = \bigcup_{1 \leq i,j \leq n} X_{i,j}^{(k)},$$

which is Ramseyan modulo f. We proceed by induction; for $k = 0$ we recall that $X^{(k)} \subseteq A$, hence forest F_0 of height 0 trivially exists. Assume that $k > 0$. Forest F_k factorizes all words in F_{k-1} exactly as forest F_{k-1} does. The words $x \in X^{(k)} - X^{(k-1)}$ are factorized as follows:

- If $x \in X_{k,k}^{(k)}$ then using (1) we conclude that $x \in (X_{k,k}^{(k-1)})^* \subseteq X_{k,k}^{(k)}$ and we put $d_k(x) = (x_1, x_2, \ldots, x_p)$, where each $x_l \in X_{k,k}^{(k-1)}$. This factorization of x is unique, furthermore the hypothesis $x \notin X^{(k-1)}$ implies that $p \geq 2$. Since $f(X_{k,k}^{(k-1)}) = e$ we conclude that the Ramseyan condition is satisfied by $d_k(x)$. Finally, we observe that $h_k(x) \leq 1 + 3(k-1) = 3k - 2$.

- Assume now that $x \in X^{(k)} - X^{(k-1)} - X_{k,k}^{(k)}$ is such that $x \in X_{k,j}^{(k)}$, for some $j \neq k$. According (1), $x = uv$, with $u \in (X_{k,k}^{(k-1)})^*$ and $v \in X_{k,j}^{(k-1)}$. Note that neither u nor v can be empty. We put $d_k(x) = (u, v)$ and using the previous case and the induction hypothesis we conclude that $h_k(x) \leq 3k - 1$ in this case.

- Assume now that $x \in X^{(k)} - X^{(k-1)} - (\cup_j X_{k,j}^{(k)})$. There exists then $1 \leq i, j \leq n$ such that $x \in X_{i,j}^{(k)}$, for some $i \neq k$. According (1) $x = uv$, with $u \in X_{i,k}^{(k-1)}$ and $v \in (X_{k,k}^{(k-1)})^* X_{k,j}^{(k-1)}$. It follows that $v \in X_{k,j}^{(k)}$. Note that neither u nor v can be empty. We put $d_k(x) = (u, v)$ and using the previous cases and the induction hypothesis we conclude that $h_k(x) \leq 3k$ in this case.

The proof of the Lemma is concluded by the observation that $X^{(n)} = A^+$. ∎

Proof of Lemma 3. Let n be the cardinal of S. From the classification of finite cyclic semigroups we conclude that there exist $s \in S$ and positive integers r and m such that $n = r + m - 1$, $s^{r+m} = s^r$ and $S = \{s, s^2, \ldots, s^{r+m-1}\}$. Furthermore, $K = \{s^r, s^{r+1}, \ldots, s^{r+m-1}\}$ is a cyclic group of cardinal m.
Let $f \colon A^+ \to S$ be a semigroup morphism for which we shall construct a Ramseyan alphabetical factorization forest $F = (A^+, d)$ of height at most $3n$. Let

$$X = \{u \in A^+ \mid f(x) \notin K\},$$

$$Y = \{u \in A^+ \mid u = va, \text{ with } v \in X \cup \{1\}, a \in A \text{ and } f(u) \in K\}.$$

Clearly, the length of words in $X \cup Y$ is bounded by r. Next we observe that $A^+ = X \cup Y^+ \cup Y^+ X$. Now we define F as follows:

- For letters $x \in A$ we put $d(x) = (x)$ and for words $x \in (X \cup Y) - A$ we define $d(x) = (v, a)$, where $x = va$, with $v \in A^+$ and $a \in A$. Clearly, every word in $X \cup Y$ has height at most $r \leq n$.

- Now we shall define the decomposition of words $x \in Y^+ - Y$. Initially we take a bijection $\beta \colon B \to Y$, where B is a new alphabet. The function $f\beta$ has a unique extension to a morphism $f' \colon B^+ \to K$. From Lemma 2 we can take an alphabetical factorization forest $F' = (B^+, d')$ of height at most $3m$ which is Ramseyan modulo f'. If $r = 1$ then $Y = A$, hence this forest can be taken for F and we are done. Next we note that for every $x \in Y^+ - Y$ there exists a unique factorization $x =

$x_1 x_2 \cdots x_p$, for some $p \geq 2$ and $x_i \in Y$. This factorization defines the word $u = b_1 b_2 \cdots b_p \in B^+$ with height $h'(u)$ in F'. If $d'(u) = (u_1, u_2, \ldots, u_k)$ then we put $f(x) = (\beta(u_1), \beta(u_2), \ldots, \beta(u_k))$. It is easy to see that this decomposition satisfies the Ramseyan condition. Furthermore, by induction on the height $h'(u)$ of u one proves that the height $h(x)$ of x will be at most $h'(u) + r \leq 3m + r \leq 3n$.

- It remains to consider the words $x \in Y^+ X$. These have a factorization $x = uv$, with $u \in Y^+$ and $v \in X$. If we put $d(x) = (u, v)$ the height of x will satisfy $h(x) \leq 1 + \max\{ h(u), h(v) \}$. From the previous cases we conclude that $h(x) \leq 1 + 3m + r \leq 3n$

The proof of Lemma 3 is complete. ∎

Proof of Lemma 4. Let $f : A^+ \to S$ be a semigroup morphism for which we shall construct a Ramseyan alphabetical factorization forest of height at most $H(V) + H(T) + 2$. Initially we observe that V being a left ideal of S it is also a subsemigroup of S. Let $f' : B \to V$ be a bijection, where B is a new alphabet. We also denote by f' the unique extension of f' to a semigroup morphism $f' : B^+ \to V$. From the hypothesis we can take an alphabetical factorization forest $F' = (B^+, d')$ of height at most $H(V)$ which is Ramseyan modulo f'. Let $C = \{ a \in A \mid f(a) \in T \}$ and let $f'' : C^+ \to T$ be the restriction of f to C^+. From the hypothesis we can take an alphabetical factorization forest $F'' = (C^+, d'')$ of height at most $H(T)$ which is Ramseyan modulo f''. Let $D = A - C$. Then, $f(D) \subseteq V$. On the other hand, since $A = C \cup D$, it follows that

$$A^+ = C^+ \cup (D \cup C^+ D)^+ \cup (D \cup C^+ D)^+ C^+.$$

Guided by the above identity we shall construct the alphabetical factorization forest $F = (A^+, d)$ of height at most $H(V) + H(T) + 2$ which shall be Ramseyan modulo f. Let $x \in A^+$.

- If $x \in C^+$ then we put $d(x) = d''(x)$ and since every segment of x also belongs to C^+ it is easy to see that d satisfies the Ramseyan condition and that $h(x) = h''(x) \leq H(T)$.

- If $x \in D \cup C^+ D$ then we put $d(x) = (x)$ if $x \in D$ or $d(x) = (u, v)$, where $u \in C^+$ and $v \in D$. Using the previous case we have that $h(x) \leq 1 + H(T)$.

- If $x \in (D \cup C^+ D)^+ - (D \cup C^+ D)$ then x has a unique (because C and D are disjoint) factorization $x = u_1 u_2 \cdots u_k$, with $u_i \in D \cup C^+ D$ and $k \geq 2$. Since $f(D) \subseteq V$ and V is a left ideal in S, it follows that for each i, $f(u_i) \in V$; hence there exists $b_i \in B$ such that $f'(b_i) = f(u_i)$. Let $y = b_1 b_2 \cdots b_k$ and let $d'(y) = (y_1, y_2, \ldots, y_p)$. This determines a factorization $x = x_1 x_2 \cdots x_p$ of x, such that each $x_i \in (D \cup C^+ D)^+$ and $f(x_i) = f'(y_i)$. We put $d(x) = (x_1, x_2, \ldots, x_p)$. Now, $f(x_i) = f'(y_i)$ implies that $d(x)$ has the Ramseyan property since $d'(y)$ has it. On the other hand, by inducton on $h'(y)$ and using the previous case one sees that $h(x) \leq 1 + h'(y) + H(T)$; hence, $h(x) \leq 1 + H(V) + H(T)$.

- Remains to consider the case when $x \in (D \cup C^+ D)^+ C^+$. In this case $x = uv$ for some $u \in (D \cup C^+ D)^+$ and $v \in C^+$. We put $d(x) = (u, v)$ and using the previous cases we conclude that $h(x) \leq 2 + H(V) + H(T)$.

The proof of Lemma 4 is complete.   ■

NOTES: In 1978 the author, unaware of [1], rediscovered the theorem of T. C. Brown on locally finite semigroups. The proof just given is a paraphrasing of the author's proof of Brown's result, which appears in [7]. Even though published only now, this is the first proof the author obtained, in 1986, for the Factorization Forest Theorem.

References

[1] T. C. Brown. An interesting combinatorial method in the theory of locally finite semigroups. *Pacific J. Math.*, 36:285–289, 1971.

[2] A. H. Clifford and G. B. Preston. *The Algebraic Theory of Semigroups, Vol I, second edition.* American Mathematical Society, Providence, R.I., 1964.

[3] S. Eilenberg. *Automata, Languages, and Machines, Volume A.* Academic Press, New York, NY, 1974.

[4] K. Krohn and J. Rhodes. Algebraic theory of machines i. prime decomposition theorem of finite semigrooups and machines. *Trans. Amer. Math. Soc.*, 116:450–464, 1965.

[5] K. Krohn, J. Rhodes, and B. Tilson. The prime decomposition theorem of the algebraic theory of machines. In M. A. Arbib, editor, *Algebraic Theory of Machines, Languages and Semigroups*, pages 81–125, Academic Press, New York, NY, 1968.

[6] G. Lallement. *Semigroups and Combinatorial Applications.* John Wiley & Sons, New York, NY, 1979.

[7] I. Simon. Caracterização de conjuntos racionais limitados. 1978. Tese de Livre-Docência, Instituto de Matemática e Estatística da Universidade de São Paulo.

[8] I. Simon. Factorization forests of finite height. *Theoretical Comput. Sci.*, 72:65–94, 1990.

Tree Automata and Languages
M. Nivat and A. Podelski (editors)
© 1992 Elsevier Science Publishers B.V. All rights reserved.

439

Unification Procedures In Automated Deduction Methods Based on Matings: A Survey

Jean H. Gallier

Digital PRL, 85 avenue Victor-Hugo, 92563 Rueil-Malmaison Cedex, France, and
Department of Computer and Information Science, University of Pennsylvania,
Philadelphia, PA 19104, USA

Abstract

Unification procedures arising in methods for automated theorem proving based on
matings are surveyed. We begin by reviewing some fundamentals of automated deduction, including the Skolem form and the Skolem-Herbrand-Gödel theorem. Next,
the method of matings for first-order languages without equality due to Andrews and
Bibel is presented. Standard unification is described in terms of transformations on
systems (following the approach of Martelli and Montanari, anticipated by Herbrand).
Some fast unification algorithms are also sketched, in particular, a unification closure
algorithm inspired by Paterson and Wegman's method. The method of matings is then
extended to languages with equality. This extention leads naturally to a generalization
of standard unification called rigid E-unification (due to Gallier, Narendran, Plaisted,
and Snyder). The main properties of rigid E-unification, decidability, NP-completeness,
and finiteness of complete sets, are discussed.

1 Introduction

Unification is a very general computational paradigm that plays an important role in
many different areas of symbolic computation. For example, unification plays a central
role in

- Automated Deduction (First-order logic with or without equality, higher-order
  logic);

- Logic Programming (Prolog, λ-Prolog);

- Constraint-based Programming;

- Type Inferencing (ML, ML^+, etc.);

- Knowledge-Base Systems, Feature structures; and

- Computational Linguistics (Unification grammars).

In this survey, we shall focus on unification problems arising in methods for automated theorem proving based on matings. This covers at least the kind of unification arising in resolution (Robinson [61], Plotkin [58]), matings (Andrews, Bibel [4, 11, 12, 13]), equational matings (Gallier, Plaisted Narendran, Raatz, Snyder [28, 27]), and ET-proofs (Miller, Pfenning [57, 51, 53]). Clearly, many other important parts of unification theory are left out, and we apologize for this. In particular, we will not cover the classification theory of the Siekmann school (for example, [62, 69, 15, 16]), the many unification procedures for special theories (AC, etc., see Siekmann [62]), the combination of unification procedures (for example, Yelick [71], Schmidt-Schauss [65], Boudet, Jouannaud, and Schmidt-Schauss [14]) order-sorted unification (for example, Meseguer, Goguen, and Smolka [50], and Isakowitz [37]), semi-unification (see [41] for references), unification applied to type-inferencing (for example, Milner [54], Kfoury, Tiuryn, and Urzyczyn, [40,42], and Remy [59, 60]), unification in computational linguistics (for example, Shieber [63]), and unification in feature structures (for example, Aït-Kaci [2, 3]). Fortunately for the uninitiated reader, there are other very good survey papers covering significant parts of the above topics: the survey by Siekmann [62], the survey by Knight [43], and the survey by Kirchner and Jouannaud [38]. The most elementary and having the broadest coverage is probably Knight's survey. Our account has a narrower focus, but it also sketches some of the proof techniques, which is usually missing from the other surveys. The topics that we will cover are:

- Standard unification;

- Rigid E-unification (E a finite set of first-order equations), a decidable form of E-unification recently introduced in the framework of Andrews and Bibel's method of matings [4, 6, 11, 12, 13].

These unification problems will be tackled using the *method of transformations* on term systems, already anticipated in Herbrand's thesis [32] (1930), and revived very effectively by Martelli and Montanari [49] for standard unification. In a nutshell, the method is as follows:

A unification problem is gradually transformed into one whose solution is (almost) obvious.

This approach is an instance of a very old method in mathematics, but a fairly recent trend in computer science, namely, the specification of procedures and algorithms in terms of inference rules. There are a number of significant advantages to this method. (1) A clean separation of logic and control is achieved. (2) The correctness of the procedure obtained this way is often easier to establish, and irrelevant implementation issues are avoided. (3) The actual design of algorithms from the procedure specified by rules can be viewed as an optimization process.

Another benefit of this approach to the design of algorithms is that one often gains a deeper understanding of the problem being solved, and one understands more easily the differences between algorithms solving a same problem. The effectiveness of this method for tackling unification problems was first shown by Martelli and Montanari [49] (although, as we said earlier, it was anticipated by Herbrand [32]). Similarly, Bachmair,

Dershowitz, Hsiang, and Plaisted [8, 7, 9, 10] showed how to describe and study Knuth-Bendix completion procedures [44] in terms of proof rules. Presented in terms of proof rules, completion procedures are more transparent, and their correctness proofs are significantly simplified. Many other examples of the effectiveness of the method of proof rules can be easily found (for example, in type inference problems, and Gentzen-style automated deduction, see Gallier [22] for the latter). Jouannaud and Kirchner [38] also emphasize the proof rules method, and provide many more examples of its use.

Although in this paper the perspective is to discuss unification in terms of transformations (proof rules), we should not forget about the history of unification theory, and the major turning points.[1] Undoubtedly, the invention of the resolution method and of the first unification algorithm by Alan Robinson in the early sixties [61] (1965) marks the beginning of a new era. In the post Robinson era, we encounter Plotkin's seminal paper on building-in equational theories [58] (1972), in which E-unification is introduced, and then Gérard Huet's thesis [35] (1976) (with Huet [34] as a precursor). Huet's thesis (1976) makes a major contribution to the theory of higher-order unification, in that it shows that a restricted form of unification, preunification, is sufficient for most theorem-proving applications. The first practical higher-order (pre)unification algorithm is defined and proved correct. Huet also gives a quasi-linear unification algorithm for standard unification, and an algorithm for unifying infinite rational trees. In the more recent past, in our perspective, we would like to mention Martelli and Montanari's paper showing the effectiveness of the method of transformations [49] (1982), and Claude Kirchner's thesis [39] in which the method of transformations is systematically applied to E-unification. We also would like to mention that most of the results on E-unification and higher-order unification discussed in this paper originate from Wayne Snyder's thesis [66] (1988). A more comprehensive presentation of these results will appear in Snyder [68].

Unification theory is a very active field of research, and it is virtually impossible to keep track of all the papers that have appeared on this subject. As evidence that unification is a very active field of research, two special issues of the *Journal of Symbolic Computation* are devoted to unification theory (Part I in Vol. 7(3 & 4), and Part II in Vol. 8(1 & 2), both published in 1989). It is our hope that this paper will inspire other researchers to work in this area.

The paper is organized as follows. Section 2 provides a review of background material relevant to unification. The notion of Skolem form and the Skolem-Herbrand-Gödel theorem are reviewed in Section 3. The method of matings for languages without equality is presented in Section 4. Section 5 is devoted to a presentation of standard unification using the method of transformations on terms systems. Fast unification methods are discussed in Section 6. The method of equational matings, a generalization of matings to languages with equality, is presented in Section 7. Section 8 is devoted to rigid E-unification, an extension of standard unification arising in the framework of equational matings. We give an overview of results of Gallier, Narendran, Plaisted,

---

[1] The following list is by no means exclusive, and only reflects the perspective on unification adopted in this paper.

and Snyder [27], showing among other things that rigid E-unification is decidable and NP-complete. Directions for further research are discussed in section 9.

2 Algebraic Background

We begin with a brief review of algebraic background material. The purpose of this section is to establish the notation and the terminology used throughout this paper. As much as possible, we follow Huet [36] and Gallier [22].

Definition 2.1 Let $\longrightarrow \subseteq A \times A$ be a binary relation on a set A. The *converse* (or *inverse*) of the relation $\longrightarrow$ is the relation denoted as $\longrightarrow^{-1}$ or $\longleftarrow$, defined such that $u \longleftarrow v$ iff $v \longrightarrow u$. The symmetric closure of $\longrightarrow$, denoted by $\longleftrightarrow$, is the relation $\longrightarrow \cup \longleftarrow$. The transitive closure, reflexive and transitive closure, and the reflexive, symmetric, and transitive closure of $\longrightarrow$ are denoted respectively by $\xrightarrow{+}$, $\xrightarrow{*}$, and $\xleftrightarrow{*}$.

Definition 2.2 A relation $\longrightarrow$ on a set A is *Noetherian* or *well founded* iff there are no infinite sequences $\langle a_0, \ldots, a_n, a_{n+1}, \ldots \rangle$ of elements in A such that $a_n \longrightarrow a_{n+1}$ for all $n \geq 0$.

Definition 2.3 A *preorder* $\preceq$ on a set A is a binary relation $\preceq \subseteq A \times A$ that is reflexive and transitive. A *partial order* $\preceq$ on a set A is a preorder that is also antisymmetric. The converse of a preorder (or partial order) $\preceq$ is denoted as $\succeq$. A *strict ordering* (or *strict order*) $\prec$ on a set A is a transitive and irreflexive relation. Given a preorder (or partial order) $\preceq$ on a set A, the strict ordering $\prec$ associated with $\preceq$ is defined such that $s \prec t$ iff $s \preceq t$ and $t \not\preceq s$. Conversely, given a strict ordering $\prec$, the partial ordering $\preceq$ associated with $\prec$ is defined such that $s \preceq t$ iff $s \prec t$ or $s = t$. The converse of a strict ordering $\prec$ is denoted as $\succ$. Given a preorder (or partial order) $\preceq$, we say that $\preceq$ is well founded iff $\succ$ is well founded.

Definition 2.4 Let $\longrightarrow \subseteq A \times A$ be a binary relation on a set A. We say that $\longrightarrow$ is *locally confluent* iff for all $a, a_1, a_2 \in A$, if $a \longrightarrow a_1$ and $a \longrightarrow a_2$, then there is some $a_3 \in A$ such that $a_1 \xrightarrow{*} a_3$ and $a_2 \xrightarrow{*} a_3$. We say that $\longrightarrow$ is *confluent* iff for all $a, a_1, a_2 \in A$, if $a \xrightarrow{*} a_1$ and $a \xrightarrow{*} a_2$, then there is some $a_3 \in A$ such that $a_1 \xrightarrow{*} a_3$ and $a_2 \xrightarrow{*} a_3$. We say that $\longrightarrow$ is *Church-Rosser* iff for all $a_1, a_2 \in A$, if $a_1 \xleftrightarrow{*} a_2$, then there is some $a_3 \in A$ such that $a_1 \xrightarrow{*} a_3$ and $a_2 \xrightarrow{*} a_3$. We say that $a \in A$ is *irreducible* iff there is no $b \in A$ such that $a \longrightarrow b$. It is well known (Huet [36]) that a Noetherian relation is confluent iff it is locally confluent and that a relation is confluent iff it is Church-Rosser. A relation $\longrightarrow$ is *canonical* iff it is Noetherian and confluent. Given a canonical relation $\longrightarrow$, it is well known that every $a \in A$ reduces to a unique irreducible element $a{\downarrow} \in A$ called the *normal form of a*, and that $a \xleftrightarrow{*} b$ iff $a{\downarrow} = b{\downarrow}$ (Huet [36]).

Definition 2.5 Terms are built up inductively from a *ranked alphabet* (or *signature*) Σ of constant and function symbols, and a countably infinite set $\mathcal{X}$ of *variables*. For simplicity of exposition, we assume that Σ is a one-sorted ranked alphabet, i.e., that there is a *rank function* $r\colon \Sigma \to \mathbf{N}$ assigning a *rank* (or *arity*) $r(f)$ to every symbol $f \in \Sigma$ ($\mathbf{N}$ denotes the set of natural numbers). We let $\Sigma_n = \{f \in \Sigma \mid r(f) = n\}$. Symbols in Σ_0 (of rank zero) are called *constants*.

Definition 2.6 We let $T_\Sigma(\mathcal{X})$ denote the set of terms built up inductively from Σ and $\mathcal{X}$. Thus, $T_\Sigma(\mathcal{X})$ is the smallest set with the following properties:

- $x \in T_\Sigma(\mathcal{X})$, for every $x \in \mathcal{X}$;

- $c \in T_\Sigma(\mathcal{X})$, for every $c \in \Sigma_0$;

- $f(t_1, \ldots, t_n) \in T_\Sigma(\mathcal{X})$, for every $f \in \Sigma_n$ and all $t_1, \ldots, t_n \in T_\Sigma(\mathcal{X})$.

Given a term $t \in T_\Sigma(\mathcal{X})$, we let $Var(t)$ be the set of variables occurring in t. A term t is a *ground term* iff $Var(t) = \emptyset$.

It is well known that $T_\Sigma(\mathcal{X})$ is the term algebra freely generated by $\mathcal{X}$, and this allows us to define substitutions.

Definition 2.7 A *substitution* is a function $\varphi\colon \mathcal{X} \to T_\Sigma(\mathcal{X})$ such that $\varphi(x) \neq x$ for only finitely many $x \in \mathcal{X}$. The set $D(\varphi) = \{x \in \mathcal{X} \mid \varphi(x) \neq x\}$ is the *domain* of φ, and the set $I(\varphi) = \bigcup_{x \in D(\varphi)} Var(\varphi(x))$ is the *set of variables introduced* by φ.

A substitution $\varphi\colon \mathcal{X} \to T_\Sigma(\mathcal{X})$ with domain $D(\varphi) = \{x_1, \ldots, x_n\}$ and such that $\varphi(x_i) = t_i$ for $i = 1, \ldots, n$, is denoted as $[t_1/x_1, \ldots, t_n/x_n]$. Since $T_\Sigma(\mathcal{X})$ is freely generated by $\mathcal{X}$, every substitution $\varphi\colon \mathcal{X} \to T_\Sigma(\mathcal{X})$ has a unique homomorphic extension $\widehat{\varphi}\colon T_\Sigma(\mathcal{X}) \to T_\Sigma(\mathcal{X})$. For every term $t \in T_\Sigma(\mathcal{X})$, we denote $\widehat{\varphi}(t)$ as $t[\varphi]$ or even as $\varphi(t)$ (with an intentional identification of φ and $\widehat{\varphi}$).

Definition 2.8 Given two substitutions φ and ψ, their *composition* denoted $\varphi\,;\psi$ is the substitution defined such that $\varphi\,;\psi(x) = \widehat{\psi}(\varphi(x))$ for all $x \in \mathcal{X}$. Thus, note that $\varphi\,;\psi = \varphi \circ \widehat{\psi}$, but not $\varphi \circ \psi$, where $\circ$ denotes the composition of functions (written in diagram order). A substitution φ is *idempotent* iff $\varphi\,;\varphi = \varphi$. It is easily seen that a substitution φ is idempotent iff $I(\varphi) \cap D(\varphi) = \emptyset$. A substitution φ is a *renaming* iff $\varphi(x)$ is a variable for every $x \in D(\varphi)$, and φ is injective over its domain. Given a set V of variables and a substitution φ, the *restriction of φ to V* is the substitution denoted $\varphi|_V$ defined such that, $\varphi|_V(x) = \varphi(x)$ for all $x \in V$, and $\varphi|_V(x) = x$ for all $x \notin V$.

There will be occasions where it is necessary to replace a subterm of a given term with another term. We can make this operation precise by defining the concept of a tree address originally due to Gorn.

Definition 2.9 Given a term $t \in T_\Sigma(\mathcal{X})$, the set $Tadd(t)$ of *tree addresses* in t is a set of strings of positive natural numbers defined as follows (where ϵ denotes the null string):

- $Tadd(x) = \{\epsilon\}$, for every $x \in \mathcal{X}$;

- $Tadd(c) = \{\epsilon\}$, for every $c \in \Sigma_0$;

- $Tadd(f(t_1, \ldots, t_n)) = \{\epsilon\} \cup \{iw \mid w \in Tadd(t_i),\ 1 \leq i \leq n\}$.

Definition 2.10 Given any $\beta \in Tadd(t)$, the *subtree* rooted at β in t is denoted as t/β. Given $t_1, t_2 \in T_\Sigma(\mathcal{X})$ and $\beta \in Tadd(t_1)$, the tree $t_1[\beta \leftarrow t_2]$ obtained by replacing the subtree rooted at β in t_1 with t_2 can be easily defined.

Definition 2.11 Let $\longrightarrow$ be a binary relation $\longrightarrow \subseteq T_\Sigma(\mathcal{X}) \times T_\Sigma(\mathcal{X})$. (i) The relation $\longrightarrow$ is *monotonic* (or *stable under the algebra structure*) iff for every two terms s, t and every function symbol $f \in \Sigma$, if $s \longrightarrow t$ then $f(\ldots, s, \ldots) \longrightarrow f(\ldots, t, \ldots)$.

(ii) The relation $\longrightarrow$ is *stable* (under substitution) if $s \longrightarrow t$ implies $s[\sigma] \longrightarrow t[\sigma]$ for every substitution σ.

Definition 2.12 A strict ordering $\prec$ has the *subterm property* iff $s \prec f(\ldots, s, \ldots)$ for every term $f(\ldots, s, \ldots)$. A *simplification ordering* $\prec$ is a strict ordering that is monotonic and has the subterm property (since we are considering symbols having a fixed rank, the deletion property is superfluous, as noted in Dershowitz [20]). A *reduction ordering* $\prec$ is a strict ordering that is monotonic, stable (under substitution), and such that $\succ$ is well founded. With a slight abuse of language, we will also say that the converse $\succ$ of a strict ordering $\prec$ is a simplification ordering (or a reduction ordering). It is shown in Dershowitz [20] that there are simplification orderings that are total on ground terms.

Definition 2.13 A *set of rewrite rules* is a binary relation $R \subseteq T_\Sigma(\mathcal{X}) \times T_\Sigma(\mathcal{X})$ such that $Var(r) \subseteq Var(l)$ whenever $\langle l, r \rangle \in R$. A rewrite rule $\langle l, r \rangle \in R$ is usually denoted as $l \to r$. A rewrite rule $s \to t$ is a *variant* of a rewrite rule $u \to v \in R$ iff there is some renaming ρ with domain $Var(u) \cup Var(v)$ such that $s = u[\rho]$ and $t = v[\rho]$.

Let $R \subseteq T_\Sigma(\mathcal{X}) \times T_\Sigma(\mathcal{X})$ be a set of rewrite rules.

Definition 2.14 The relation $\longrightarrow_R$ over $T_\Sigma(\mathcal{X})$ is defined as the smallest stable and monotonic relation that contains R. This is the *rewrite relation* associated with R. This relation is defined explicitly as follows: Given any two terms $t_1, t_2 \in T_\Sigma(\mathcal{X})$, then

$$t_1 \longrightarrow_R t_2$$

iff there is some variant $l \to r$ of some rule in R, some tree address β in t_1, and some substitution σ, such that

$$t_1/\beta = l[\sigma], \quad \text{and} \quad t_2 = t_1[\beta \leftarrow r[\sigma]].$$

The concept of an equation is similar to that of a rewrite rule, but equations can be used oriented forward or backward, and the restriction $Var(r) \subseteq Var(l)$ is dropped.

Definition 2.15 A *set of equations* is a binary relation $E \subseteq T_\Sigma(\mathcal{X}) \times T_\Sigma(\mathcal{X})$. An equation $\langle l, r \rangle \in E$ is usually denoted as $l \doteq r$, to emphasize the difference with a rewrite rule. An equation $s \doteq t$ is a *variant* of an equation $u \doteq v \in E$ iff there is some renaming ρ with domain $Var(u) \cup Var(v)$ such that $s = u[\rho]$ and $t = v[\rho]$.

Definition 2.16 The relation $\longleftrightarrow_E$ over $T_\Sigma(\mathcal{X})$ is defined as the smallest symmetric relation containing E that is stable, and monotonic. This relation is defined explicitly as follows: Given any two terms $t_1, t_2 \in T_\Sigma(\mathcal{X})$, then

$$t_1 \longleftrightarrow_E t_2$$

iff there is some variant $l \doteq r$ of some equation in $E \cup E^{-1}$, some tree address β in t_1, and some substitution σ, such that

$$t_1/\beta = l[\sigma], \quad \text{and} \quad t_2 = t_1[\beta \leftarrow r[\sigma]].$$

Note that an equation can be used oriented forward or backward, since E^{-1} consists of all $r \doteq l$ such that $l \doteq r \in E$.

Definition 2.17 The reflexive and transitive closure of $\longrightarrow_R$ is denoted as $\overset{*}{\longrightarrow}_R$, and the reflexive and transitive closure of $\longleftrightarrow_E$ as $\overset{*}{\longleftrightarrow}_E$. Sometimes, $\overset{*}{\longleftrightarrow}_E$ is denoted as $=_E$. It is easily seen that $\overset{*}{\longleftrightarrow}_E$ is an equivalence relation. In fact, $\overset{*}{\longleftrightarrow}_E$ is the smallest congruence containing E that is stable under substitution, and $\overset{*}{\longleftrightarrow}_E = (\longrightarrow_E \cup \longrightarrow_E^{-1})^*$. It can be shown that $E \models u \doteq v$ iff $u \overset{*}{\longleftrightarrow}_E v$ (a form of Birkhoff's completeness theorem). A set R of rewrite rules is called Noetherian, confluent, Church-Rosser, or canonical, iff the relation $\longrightarrow_R$ has the corresponding property.

3 The Skolem-Herbrand-Gödel Theorem

An automated deduction method is a procedure for checking whether an arbitrary formula is provable. In this survey, we are restricting ourselves to first-order logic, for which, by Gödel's completeness theorem, we know that a formula is provable iff it is valid. Thus, the problem is equivalent to designing a procedure for checking whether an arbitrary formula is valid. This problem is also called the *validity problem*. The main difficulty in automated deduction is to deal with the quantifiers. To be more precise, the difficulty is to design procedures that handle the quantifiers efficiently. For a quantifier-free formula, also called a proposition, it can be shown that the validity problem is decidable. For propositions without equality, this is fairly easy to show, and there are a number of algorithmic methods for deciding the validity problem: the truth-table method, the resolution method, the matings method, Davis and Putnam's method, etc. (see Gallier [22], or Manna [48]). For quantifier-free formulae with equality, the validity problem is also decidable, but this is harder to prove. Decidability can be established using the "congruence closure method", and an algorithm using congruence closure can be designed (Kozen [45,46], Nelson and Oppen [55], Downey, Sethi, and Tarjan [21]). In the general case of quantifed formulae, Church [18] (1936) proved that the validity problem is undecidable . A particularly simple proof of this important result was later given by Floyd (see Manna [48], page 105-106).

Two important theorems help in dealing with quantifiers. The first theorem, essentially due to Skolem, shows that the validity problem can be reduced to the validity problem for formulae containing only one kind of quantifiers (say $\forall$). The second one, known as the Skolem-Herbrand-Gödel theorem, shows that the validity of a quantified formula can be reduced to the validity of a *quantifier-free formula*, modulo guessing some (ground) substitutions. Without digressing excessively, we would like to warn the readers of a confusion often made between too different important theorems, Herbrand's theorem, and the Skolem-Herbrand-Gödel theorem. The first theorem, *Herbrand's theorem* [32] (1930), is about the *provability* of first-order formulae, and its (meta)proof does not appeal to the semantics of first-order logic at all. Furthermore, Herbrand's theorem also yields some information on the length of proofs. Historically, Herbrand's theorem was proved in 1930, before the Skolem-Herbrand-Gödel theorem (which, according to Peter Andrews, was apparently only formulated in the early fifties by Quine). The second theorem, the *Skolem-Herbrand-Gödel theorem* (see Andrews [4, 5] or Gallier [22]), is about *unsatisfiability*, a semantic notion, and its (meta)proof can be presented essentially as a semantic argument (a certain model is constructed). Furthermore, the Skolem-Herbrand-Gödel theorem does not yield any information on the length of proofs. Basically, the Skolem-Herbrand-Gödel theorem combines results of Skolem [64] (1928) and Gödel [30, 31] (1930), from his proof of the completeness theorem, but since it is definitely a "semantic version" of Herbrand's theorem, it is appropriate to refer to it by the concatenation of the three names! Finally, the (meta)proof of Herbrand's theorem is significantly harder than the (meta)proof of the Skolem-Herbrand-Gödel theorem, but it yields more information, namely, some complexity-theoretic information about the length of proofs. For our purposes, the Skolem-Herbrand-Gödel theorem is all we need.

What Skolem (essentially) showed is that given a quantified (first-order) formula A, one can associate two formulae A_{vff} and A_{sff} with the following properties:

(1) The formula A_{vff} contains only *existensial* quantifiers, and A is valid iff A_{vff} is valid.

(2) The formula A_{sff} contains only *universal* quantifiers, and A is satisfiable iff A_{sff} is satisfiable.

Following Goldfarb (see [32]), we call A_{vff} the *validity functional form* of A. Intuitively speaking, it is obtained from A by "eliminating" universal quantifiers. Dually, we call A_{sff} the *satisfiability functional form* of A. Intuitively speaking, it is obtained from A by "eliminating" existential quantifiers. Now, recall that we say that a formula A is unsatisfiable iff it is not satisfied in any structure, and that A is valid iff $\neg A$ is unsatisfiable. Thus, (2) can be equivalently stated as A is unsatisfiable iff A_{sff} is unsatisfiable.

Oddly, early researchers in the field of automated deduction have shown a preference for A_{sff}, the *satisfiability functional form*, often referred to as the *Skolem form* of A. This is perhaps because the main property of A_{sff} (A is satisfiable iff A_{sff} is satisfiable) can be intuitively justified by an appeal to the axiom of choice. The consequence of this bias is that most automated deduction methods are traditionally presented as *refutation* methods. This means that in order to show that A is valid, we attempt to show that $\neg A$ is unsatisfiable, that is, we try to show that the Skolem form $(\neg A)_{sff}$ of $\neg A$, is unsatisfiable. We have an unfortunate first step which consists in negating what we are trying to prove! We believe that tradition is worth fighting when it is silly, and it would certainly make more sense to present automated deduction methods, the resolution method in particular, in their positive version, as *proving* methods, rather than as *refutation* (negative) methods. The drawback of such a choice is that one has to constantly translate traditional refutation methods into their positive form, in order to compare them with other (positive) proving methods. Consequently, mostly for ease of comparison with other methods, we will follow the tradition, not without some guilt feelings, and present our methods as refutation methods. Therefore, we will be using the *satisfiability functional form* A_{sff}, often called the *Skolem form* of A, and use the version of the Skolem-Herbrand-Gödel theorem dealing with unsatisfiability, rather than validity.

Roughly, the Skolem-Herbrand-Gödel theorem asserts that a formula A is unsatisfiable iff some conjunction of substitution instances of subformulae of the Skolem form of A is unsatisfiable. The crucial idea is that the unsatisfiability of a *quantified* formula A is reduced to the unsatisfiability of some *quantifier-free* formula (obtainable from A). The price of this reduction is that one needs to "guess" some substitutions and which subformulae of A need to be instantiated with these substitutions.

It is possible to define the Skolem form of an arbitrary formula and state a version of the Skolem-Herbrand-Gödel theorem for arbitrary formulae. However, in the fully general case, one needs to deal with negative and positive occurrences of subformulae, which complicates matters and obscures the main point of the theorem. We can give a simpler version of the Skolem-Herbrand-Gödel theorem for formulae in a special form, the *negation normal form* (for short, *nnf*), where negation only applies to atomic formulae. Since every formula is equivalent to another formula in *nnf*, there is no loss of generality. Furthermore, contrary to other normal formal forms, such as prenex form, the *nnf* of a formula is linear in the size of the original formula. The following example should give a crisper idea of what we are talking about.

Example 3.1 Let $A = \exists x \forall y (P(y) \supset P(x))$. In order to prove that A is valid, we will attempt to prove that $\neg A$ is unsatisfiable.

Step 1: Compute $\neg A$. We have

$$\neg A = \forall x \exists y (P(y) \wedge \neg P(x)).$$

Step 2: Compute the Skolem form $\forall x B_0$ of $\neg A$. We have

$$\forall x B_0 = \forall x (P(f(x)) \wedge \neg P(x)).$$

Step 3: Find a conjunction of (ground) instances of B_0 which is unsatisfiable. Observe that

$$C = (P(f(a)) \wedge \neg P(a)) \wedge (P(f(f(a))) \wedge \neg P(f(a)))$$

is unsatisfiable.

One should note that no substitution σ makes $\sigma(B_0) = \sigma(P(f(x)) \wedge \neg P(x))$ unsatisfiable. A systematic way to find a conjunction of (ground) instances of B_0 which is unsatisfiable is to duplicate B_0 and try again. After duplication (with renaming of the second conjunct), we have

$$B_1 = (P(f(x)) \wedge \neg P(x)) \wedge (P(f(y)) \wedge \neg P(y)).$$

The key point is that unsatisfiability will be achieved if we can find **mated pairs** of literals, that is, pairs of literals of opposite signs. In B_1, the literals $P(f(x))$ and $\neg P(y)$ form a mated pair. If we apply the substitution $[a/x, f(a)/y]$, we get $P(f(a))$ and $\neg P(f(a))$. What happens is that $P(f(x))$ and $P(y)$ are **unified** by the substitution $[a/x, f(a)/y]$.

This is a general phenomenon, and it is at the heart of the method of matings of Bibel and Andrews [4, 6, 11, 12, 13]. The crucial observation due to Andrews and Bibel is that a quantifier-free formula (in *nnf*) is unsatisfiable iff certain sets of literals occurring in A (called *vertical paths*) are unsatisfiable. Matings come up as a convenient method for checking that vertical paths are unsatisfiable. Roughly speaking, a *mating* is a set of pairs of literals of opposite signs (mated pairs) such that all these (unsigned) pairs are globally unified by some substitution. The importance of matings stems from the fact that a quantifier-free formula A has a mating iff there is a ground substitution θ such that $\theta(A)$ is unsatisfiable. Thus, we see where unification comes into the picture, at least in the case of formulae without equality. Things are more complicated when formulae contain equality. In this case, we are naturally led to more general forms of unification, E-unification and rigid E-unification. We now proceed with a more rigorous presentation of the concept of Skolem form and of the Skolem-Herbrand-Gödel theorem. First, we recall the formal definition of the negation normal form.

Definition 3.2 Formulae in *negation normal form* (for short, in *nnf*) are defined inductively as follows. A formula A is in *nnf* iff either

(1) A is an atomic formula or the negation $\neg B$ of an atomic formula, or

(2) $A = (B \vee C)$, where B and C are in *nnf*, or

(3) $A = (B \wedge C)$, where B and C are in *nnf*, or

(4) $A = \forall x B$, where B is in *nnf*, or

(5) $A = \exists x B$, where B is in *nnf*.

Lemma 3.3 *For every formula A, one can construct a formula B in nnf such that $A \equiv B$ is valid.*

From now on, we will be dealing only with *rectified formulae*, that is, formulae in which no variable occurs both free and bound, and distinct occurrences of quantifiers bound distinct variables. It is easy to show that for every formula A, one can construct a rectified formula B equivalent to A. We now give an algorithm to compute the Skolem form of a formula in *nnf*. First, it is necessary to compute the universal scope of a subformula.

Definition 3.4 Given a (rectified) formula A in *nnf*, the set $US(A)$ of pairs $\langle B, L \rangle$ where B is a subformula of A and L is a sequence of variables, is defined inductively as follows:

$$US_0 = \{\langle A, \langle \rangle \rangle\};$$
$$\begin{aligned}
US_{k+1} = US_k &\cup \{\langle C, L \rangle, \langle D, L \rangle \mid \langle B, L \rangle \in US_k, \\
&\quad B \text{ is of the form } (C \wedge D) \text{ or } (C \vee D)\} \\
&\cup \{\langle C, L \rangle \mid \langle \exists x C, L \rangle \in US_k\} \\
&\cup \{\langle C, \langle y_1, \ldots, y_m, x \rangle \rangle \mid \langle \forall x C, \langle y_1, \ldots, y_m \rangle \rangle \in US_k\}.
\end{aligned}$$

For every subformula B of A, the sequence L of variables such that $\langle B, L \rangle$ belongs to $US(A) = \bigcup US_k$ is the *universal scope* of B.

Example 3.5 Let

$$A = \forall x (P(a) \vee \exists y (Q(y) \wedge \forall z (P(y, z) \vee \exists u Q(x, u)))) \vee \exists w Q(a, w).$$

Then,

$$\langle \exists y (Q(y) \wedge \forall z (P(y, z) \vee \exists u Q(x, u))), \langle x \rangle \rangle,$$
$$\langle \exists u Q(x, u), \langle x, z \rangle \rangle, \text{ and}$$
$$\langle \exists w Q(a, w), \langle \rangle \rangle$$

define the universal scope of the subformulae of A of the form $\exists x B$.

Definition 3.6 Given a rectified sentence[2] A in *nnf*, the *Skolem form* (or *Skolem normal form*) of A is defined recursively as follows. Let A' be any subformula of A:

(i) If A' is either an atomic formula B or the negation $\neg B$ of an atomic formula B, then $SK(A') = A'$.

(ii) If A' is of the form $(B * C)$, where $* \in \{\vee, \wedge\}$, then $SK(A') = (SK(B) * SK(C))$.

(iii) If A' is of the form $\forall x B$, then $SK(A') = \forall x SK(B)$.

(iv) If A' is of the form $\exists x B$, then if $\langle y_1, \ldots, y_m \rangle$ is the universal scope of $\exists x B$ (that is, the sequence of variables such that $\langle \exists x B, \langle y_1, \ldots, y_m \rangle \rangle \in US(A)$) then

    (a) If $m > 0$, create a new *Skolem function symbol* $f_{A'}$ of rank m and let $SK(A') = SK(B[f_{A'}(y_1, \ldots, y_m)/x])$.

    (b) If $m = 0$, create a new *Skolem constant* $f_{A'}$ and let $SK(A') = SK(B[f_{A'}/x])$.

Observe that since the sentence A is rectified, all subformulae A' of the form $\exists x B$ are distinct, and since the Skolem symbols are indexed by the subformulae A', they are also distinct.

---

[2] Recall that a *sentence* is a formula without any free variables, and it is also called a *closed formula*.

Example 3.7 Let

$$A = \forall x(P(a) \vee \exists y(Q(y) \wedge \forall z(P(y,z) \vee \exists u Q(x,u)))) \vee \exists w Q(a,w).$$

$SK(\exists w Q(a,w)) = Q(a,c),$
$SK(\exists u Q(x,u)) = Q(x,f(x,z)),$
$SK(\exists y(Q(y) \wedge \forall z(P(y,z) \vee \exists u Q(x,u)))) =$
$$(Q(g(x)) \wedge \forall z(P(g(x),z) \vee Q(x,f(x,z))))), \quad \text{and}$$
$SK(A) = \forall x(P(a) \vee (Q(g(x)) \wedge \forall z(P(g(x),z) \vee Q(x,f(x,z))))) \vee Q(a,c).$

The main property of Skolem forms is given in the following lemma.

Lemma 3.8 *Let* **L** *be a first-order language with or without equality. Let A be a rectified* **L**-*sentence in nnf, and let B be its Skolem normal form. The sentence A is satisfiable iff its Skolem form B is satisfiable.*

Proof. Let C be any subformula of A. We show that the following properties hold:

(a) For every structure **A** such that all function, predicate, and constant symbols in the Skolem form $SK(C)$ of C receive an interpretation, for every assignment s, if $\mathbf{A} \models SK(C)[s]$ then $\mathbf{A} \models C[s]$.

(b) For every structure **A** such that exactly all function, predicate, and constant symbols in C receive an interpretation, for every assignment s (with range A), if $\mathbf{A} \models C[s]$ then there is an expansion **B** of **A** such that $\mathbf{B} \models SK(C)[s]$.

The proof is by induction on the size of subformulae of A. Details can be found in Gallier [22]. $\square$

Warning: In general, a formula A and its Skolem form $SK(A)$ are **not** equivalent. For example, $\exists x P(x)$ and $P(a)$ are not equivalent.

The Skolem-Herbrand-Gödel theorem can be stated in a very concise form if we introduce the notion of a *compound instance* due to Andrews. Recall that a *literal* is either an atomic formula or the negation of an atomic formula.

Definition 3.9 Let A be a rectified sentence in *nnf* and let B its Skolem form. The set of *compound instances* (for short, *c-instances*) of B is defined inductively as follows:

(i) If B is a literal, then B is its only c-instance;

(ii) If B is of the form $(C * D)$, where $* \in \{\vee, \wedge\}$, for any c-instance H of C and c-instance K of D, then $(H * K)$ is a c-instance of B;

(iii) If B is of the form $\forall x C$, for any k closed terms $t_1,\ldots,t_k$, if H_i is a c-instance of $C[t_i/x]$ for $i = 1,\ldots,k$, then $H_1 \wedge \ldots \wedge H_k$ is a c-instance of B.

Example 3.10 Let

$$B = \forall x(P(x) \vee \forall y Q(y, f(x))) \wedge (\neg P(a) \wedge (\neg Q(a, f(a)) \vee \neg Q(b, f(a)))).$$

Then,

$$(P(a) \vee (Q(a, f(a)) \wedge Q(b, f(a)))) \wedge (\neg P(a) \wedge (\neg Q(a, f(a)) \vee \neg Q(b, f(a))))$$

is a c-instance of B.

Note that c-instances are quantifier free. We are now ready to state a version of the Skolem-Herbrand-Gödel theorem due to Andrews [4] (1981), but first, a minor technicality has to be taken care of. If the first-order language under consideration does not have any constants, the theorem fails. For example, the formula $\forall x \forall y (P(x) \wedge \neg P(y))$ is unsatisfiable, but if the language has no constants, we cannot find a ground substitution instance of $P(x) \wedge \neg P(y)$ that is unsatisfiable. To avoid this problem, we will assume that if any first-order language $\mathbf{L}$ does not have constants, the special constant $\#$ is added to it.

Theorem 3.11 *Let $\mathbf{L}$ be a first-order language with or without equality. Given any rectified sentence A in nnf, if B is the Skolem form of A, then A is unsatisfiable if and only if some compound instance C of B is unsatisfiable.*

Remark: There is an algorithm for deciding whether a c-instance is unsatisfiable if equality is absent, but in case equality is present, such an algorithm is much less trivial. Such an algorithm based on congruence closure exists.

In view of lemma 3.8, if we are interested in deciding unsatisfiability, we can restrict our attention to universal sentences (that is, sentences containing only universal quantifiers). Then, theorem 3.11 can be sated as follows:

Corollary 3.12 *Given a universal sentence A in nnf, A is unsatisfiable if and only if some compound instance C of A is unsatisfiable.*

Lemma 3.12 is the theoretical basis of many refutation procedures, in particular the resolution method, and the method of matings. In the next section, we look at the method of matings, as presented by Andrews [4]. The same method was also investigated by Bibel [11, 12, 13] under the name of "connection method", and in fact, probably predates the method of matings.

4 The Method of Matings

If one wants to write a procedure based on lemma 3.12, the first problem to solve is to find a way of generating compound instances nicely. Andrews proposed a convenient notion, the notion of *amplification* [4].

Definition 4.1 Given a universal formula A, C is obtained from B by *quantifier duplication* iff C results from B by replacing some subformula $\forall x M$ of B by $(\forall x M \wedge \forall x M)$.

If $C_1 \Rightarrow C_2, \ldots, C_{n-1} \Rightarrow C_n$, with $B = C_1$, $C = C_n$, and C_{i+1} is obtained from C_i by quantifier duplication, $1 \leq i < n$, then C is obtained from B by some *sequence of quantifier duplications*.

If $A \Rightarrow^* B$ by some sequence of quantifier duplications, C is a rectified sentence equivalent to B, and D obtained from C by deleting the quantifiers in C, then D is an *amplification* of A.

The following lemma shows that every compound instance arises from some amplification.

Lemma 4.2 *Let* **L** *be a first-order language with or without equality. Given a universal sentence A in nnf, C is a c-instance of A iff there is some amplification D of A and some (ground) substitution θ such that $C = \theta(D)$.*

Form theorem 3.11 and lemma 4.2, we have the following variant of the Skolem-Herbrand-Gödel theorem.

Theorem 4.3 *Let* **L** *be a first-order language with or without equality. Given a universal sentence A in nnf, A is unsatisfiable iff there is some amplification D of A and some (ground) substitution σ such that $\sigma(D)$ is unsatisfiable.*

The next step towards automated deduction is to find a method for deciding whether, given a quantifier-free formula D, there is some susbtitution σ such that $\sigma(D)$ is undecidable. We first solve this problem for the case of first-order languages **without equality**. We present the method of *vertical paths*, a variant of the disjunctive normal form.

Definition 4.4 Let A be a quantifier-free formula in *nnf*. The set $vp(A)$ of *vertical paths* in A is the set of sets of literals defined inductively as follows:

If A is a literal, then $vp(A) = \{\{A\}\}$;

If $A = (B \wedge C)$, then $vp(A) = \{\pi_1 \cup \pi_2 \mid \pi_1 \in vp(B), \pi_2 \in vp(C)\}$;

If $A = (B \vee C)$, then $vp(A) = vp(B) \cup vp(C)$.

The fundamental property of vertical paths in given in the following lemma.

Lemma 4.5 *Let* **L** *be a first-order language with or without equality. A quantifier-free formula A in nnf is unsatisfiable iff every vertical path in A is unsatisfiable.*

Let us now look more closely at vertical paths, and see what it means for a vertical path to be unsatisfiable. This is where the assumption that equality does not occur simplifies matters drastically. Given a literal L, if $L = A$ where A is a positive atom, then $\neg L = \neg A$, else if $L = \neg A$ where A is a positive atom, then $\neg L = A$. We say that L and $\neg L$ are *complementary*.

- For languages **without** equality, a vertical path $\{L_1, \ldots, L_m\}$ is unsatisfiable iff two of the literals L_i, L_j are complementary, that is, $L_i = \neg L_j$.

- If the formula A is of the form $\sigma(D)$, this means that there are literals $\sigma(L_i)$ and $\neg\sigma(L_j)$ such that
$$\sigma(L_i) = \sigma(L_j).$$

A substitution such that $\sigma(L_i) = \sigma(L_j)$ is called a *unifier* of L_i and L_j. Thus, we see that looking for an automated deduction procedure based on the Skolem-Herbrand-Gödel theorem leads to unification. It also leads to *matings*, which are convenient for checking that vertical paths are unsatisfiable.

Definition 4.6 Given a quantifier-free formula A in *nnf*, a *mating* for A is a pair $\mathcal{M} = \langle MS, \sigma \rangle$, where

(1) MS is a set of pairs of literals of opposite sign (in A), and

(2) σ is a substitution such that, for every pair $(L, \neg L') \in MS$,

$$\sigma(L) = \sigma(L').$$

A mating is *p-acceptable* iff every vertical path $\pi \in vp(A)$ contains some mated pair $(L, \neg L') \in MS$.

The following lemma is the bridge between theorem 4.3 and lemma 4.5.

Lemma 4.7 *Given a quantifier-free formula A in nnf, the following properties hold:*

(1) Given a substitution θ, if $\theta(A)$ is unsatisfiable, then there is a p-acceptable mating $\mathcal{M}$ for A.

(2) If $\mathcal{M}$ is a p-acceptable mating for A with associated substitution $\sigma_\mathcal{M}$, then $\sigma_\mathcal{M}(A)$ is unsatisfiable.

The completeness and soundness for the method of matings is an immediate consequence of theorem 4.3, lemma 4.5, and lemma 4.7.

Theorem 4.8 *Given a universal sentence A in nnf, A is unsatisfiable iff some amplification D of A has a p-acceptable mating.*

Let us work out an example in detail to illustrate theorem 4.8.

Example 4.9 (Due to Andrews [4]) Let A be the formula

$$\exists x \forall y (Px \equiv Py) \supset (\exists x Px \equiv \forall y Py).$$

We want to prove that A is valid. First, we negate A and eliminate $\equiv$ and $\supset$:

$$\exists x \forall y [(\neg Px \vee Py) \wedge (\neg Py \vee Px)] \wedge [(\exists x Px \wedge \exists y \neg Py) \vee (\forall y Py \wedge \forall x \neg Px)]$$

Next, we skolemize:

$$\forall y[(\neg Pc \vee Py) \wedge (\neg Py \vee Pc)] \wedge [(Pd \wedge \neg Pe) \vee (\forall z Pz \wedge \forall x \neg Px)]$$

Next, we amplify (duplicate quantifiers):

$$\forall y[(\neg Pc \vee Py) \wedge (\neg Py \vee Pc)] \wedge \forall y[(\neg Pc \vee Py) \wedge (\neg Py \vee Pc)]$$
$$\wedge [(Pd \wedge \neg Pe) \vee (\forall z Pz \wedge \forall x \neg Px)]$$

Rectify variables:

$$\forall y[(\neg Pc \vee Py) \wedge (\neg Py \vee Pc)] \wedge \forall w[(\neg Pc \vee Pw) \wedge (\neg Pw \vee Pc)]$$
$$\wedge [(Pd \wedge \neg Pe) \vee (\forall z Pz \wedge \forall x \neg Px)]$$

Delete Quantifiers:

$$[(\neg Pc \vee Py) \wedge (\neg Py \vee Pc)] \wedge [(\neg Pc \vee Pw) \wedge (\neg Pw \vee Pc)]$$
$$\wedge [(Pd \wedge \neg Pe) \vee (Pz \wedge \forall x \neg Px)]$$

Vertical paths displayed in "matrix form":

$$\begin{bmatrix} \neg Pc & \vee & Py \\ \neg Py & \vee & Pc \\ \neg Pc & \vee & Pw \\ \neg Pw & \vee & Pc \\ \begin{bmatrix} Pd \\ \neg Pe \end{bmatrix} & \vee & \begin{bmatrix} Pz \\ \neg Px \end{bmatrix} \end{bmatrix}$$

There are 32 vertical paths.

The substitution $\theta = [d/x,\ d/y,\ c/z,\ e/w]$ "mates" all the vertical paths:

$$\begin{bmatrix} \neg Pc & \vee & Pd \\ \neg Pd & \vee & Pc \\ \neg Pc & \vee & Pe \\ \neg Pe & \vee & Pc \\ \begin{bmatrix} Pd \\ \neg Pe \end{bmatrix} & \vee & \begin{bmatrix} Pc \\ \neg Pd \end{bmatrix} \end{bmatrix}$$

A p-acceptable mating is given below:

$$\{\langle \neg Pc, Pz \rangle,\ \langle \neg Py, Pd \rangle,\ \langle \neg Pc, Pc \rangle,\ \langle Py, \neg Px \rangle,\ \langle \neg Pe, Pw \rangle\}$$

The substitution θ unifies (in fact, in an mgu) of the set of pairs:

$$\{\langle Pc, Pz \rangle,\ \langle Py, Pd \rangle,\ \langle Pc, Pc \rangle,\ \langle Py, Px \rangle,\ \langle Pe, Pw \rangle\}$$

A naive procedure implementing the method of matings is given below.

Definition 4.10 (A Procedure for Finding Matings)

Let A_0 be a universal sentence in *nnf*. The formula A_0 evolves in steps called *quantifier duplication steps*.

Let A be the evolving formula

Let $\widehat{A}$ be obtained from A by deleting the quantifiers (an *amplification* of A_0).

Initially, $A := A_0$.

1. : Construct $vp(\widehat{A})$, the set of sets of literals called *vertical paths*.

2. : Find whether there is a substitution σ such that for every vertical path $\pi \in vp(\widehat{A})$, $\sigma(\pi)$ is unsatisfiable. If step 2 succeeds, go to step 4. Otherwise, go to step 3.

3. : Choose some universal subformula $\forall x B$ of A, and replace it by $(\forall x B \wedge \forall x B)$. Then, rectify variables in this new formula, obtaining A'. Let $A := A'$ (*quantifier duplication step*). Go back to step 1.

4. : Stop, A_0 is unsatisfiable (and so are $\widehat{A}$ and A).

If A_0 is unsatisfiable, this procedure stops when it succeeds in finding some substitution closing all vertical paths in step 2. For languages without equality, we can use *unification* to check whether a set of pairs can be mated.

We now briefly discuss some optimizations of the naive procedure. To trim the search space, we can use a *connection graph* (Kowalski). Given a mating $\mathcal{M}$ for A, literals M and $\neg N$ are *potential mates* w.r.t. $\mathcal{M}$ iff some vertical path contains both M and $\neg N$ and $\sigma_{\mathcal{M}}(M)$ and $\sigma_{\mathcal{M}}(N)$ are unifiable (i.e., there is a substitution such that $\theta(\sigma_{\mathcal{M}}(M)) = \theta(\sigma_{\mathcal{M}}(N))$). The *connection graph* for A and $\mathcal{M}$ has the literals of A as nodes, and there is an edge from M to $\neg N$ iff M and $\neg N$ are potential mates w.r.t. $\mathcal{M}$. When building a mating, we can choose a pair of potential mates and add it to the current mating.

The way of eliminating $\equiv$ can have a great influence on the number of vertical paths. $A \equiv B$ can be transformed to

$$(\neg A \vee B) \wedge (A \vee \neg B), \quad \text{or} \quad (A \wedge B) \vee (\neg A \wedge \neg B).$$

Example 4.11 (revisited) Let A be the formula

$$\exists x \forall y (Px \equiv Py) \supset (\exists x Px \equiv \forall y Py).$$

Negate and eliminate $\equiv$ and $\supset$:

$$\exists x \forall y [(Px \wedge Py) \vee (\neg Px \wedge \neg Py)] \wedge [(\exists x Px \wedge \exists y \neg Py) \vee (\forall y Py \wedge \forall x \neg Px)]$$

Skolemize:

$$\forall y [(Pc \wedge Py) \vee (\neg Pc \wedge \neg Py)] \wedge [(Pd \wedge \neg Pe) \vee (\forall z Pz \wedge \forall x \neg Px)]$$

Duplicate quantifier and rectify:

$$\forall y[(Pc \wedge Py) \vee (\neg Pc \wedge \neg Py)] \wedge \forall w[(Pc \wedge Pw) \vee (\neg Pc \wedge \neg Pw)]$$
$$\wedge [(Pd \wedge \neg Pe) \vee (\forall z Pz \wedge \forall x \neg Px)]$$

Delete quantifiers, and display in matrix form:

$$\begin{bmatrix} \begin{bmatrix} Pc \\ Py \end{bmatrix} \vee \begin{bmatrix} \neg Pc \\ \neg Py \end{bmatrix} \\[2ex] \begin{bmatrix} Pc \\ Pw \end{bmatrix} \vee \begin{bmatrix} \neg Pc \\ \neg Pw \end{bmatrix} \\[2ex] \begin{bmatrix} Pd \\ \neg Pe \end{bmatrix} \vee \begin{bmatrix} Pz \\ \neg Px \end{bmatrix} \end{bmatrix}$$

There are 8 vertical paths instead of 32.

We now briefly discuss some ways of reducing the number of vertical paths. Observe that

$$M = \left[(L_1 \wedge P_1) \vee \ldots \vee (L_n \wedge P_n)\right] \wedge \left[(\neg L'_1 \wedge \ldots \wedge \neg L'_n \wedge Q) \vee R\right]$$

is equivalent to

$$N = \left[(L_1 \wedge P_1) \vee \ldots \vee (L_n \wedge P_n)\right] \wedge R,$$

when $L_i = L'_i$.

If A is a sentence containing M and A^* is the result of substituting N for M in A,

$$\begin{bmatrix} \begin{bmatrix} L_1 \\ P_1 \end{bmatrix} \vee \quad \ldots \quad \vee \begin{bmatrix} L_n \\ P_n \end{bmatrix} \\[3ex] \begin{bmatrix} \neg L_1 \\ \vdots \\ \neg L_n \\ Q \end{bmatrix} \vee R \end{bmatrix}$$

occurs in A, and

$$\begin{bmatrix} \begin{bmatrix} L_1 \\ P_1 \end{bmatrix} \vee \ldots \vee \begin{bmatrix} L_n \\ P_n \end{bmatrix} \\[2ex] R \end{bmatrix}$$

occurs in A^*.

If $\mathcal{M}$ is a mating for A such that each pair $\langle L_i, \neg L'_i \rangle$ or $\langle L'_i, \neg L_i \rangle$ is in $\mathcal{M}$, there is a mating $\mathcal{M}^*$ associated with $\sigma_{\mathcal{M}}(A^*)$.

Another kind of simplification due to Prawitz is as follows: $(L \vee Q) \wedge (\neg L \vee R)$ simplifies to $(L \wedge R) \vee (\neg L \wedge Q)$. Also, we can try to use symmetries in matings to minimize the search space. For more details, the reader is referred to Andrews [4].

5 Standard Unification

We saw in the previous section how (standard) unification arises naturally in the context of the method of matings. Historically, unification was brought to the fore as a seminal component of automated deduction systems by Robinson in 1964, and has been studied by numerous researchers since that time. In this section we present an abstract view of unification as a set of non-deterministic rules for transforming a unification problem into an explicit representation of its solution, if such exists. This elegant approach is due to Martelli and Montanari [49], but was in fact implicit in Herbrand's thesis (1930).[3] For a very good historical account and technical details on standard unification, the reader is referred to Knight's survey article [43], Jouannaud and Kirchner's survey article [38], and Lassez, Maher, and Marriot [47].

It is natural to define a unification problem as a set $\{\langle u_1, v_1\rangle, \ldots, \langle u_n, v_n\rangle\}$ of ordered pairs of terms. This is fine, but it turns out that it is more convenient to allow repetitions of pairs $\langle u_i, v_i\rangle$, and to allow $\langle u_i, v_i\rangle$ to be unordered. Thus, we adopt the following definition of a unification problem.

Definition 5.1 A *term pair* or just a *pair* is a multiset of two terms, denoted, by $\langle u, v\rangle$. A *term system* (or *system*) is a (finite) multiset of term pairs. In denoting term systems, we will often drop the curly brackets and simply write $\langle u_1, v_1\rangle, \ldots, \langle u_n, v_n\rangle$.

The reason why multisets are more convenient than sets is that they lead to a simpler statement of the transformations. This shows up in two ways. Firstly, since multiset union is not idempotent (contrary to set union), when we write $S \cup \{\langle u, v\rangle\}$, we mean the multiset consisting of all pairs in S distinct from $\langle u, v\rangle$, and of the $m + 1$ pairs $\langle u, v\rangle$, where m is the number of occurrences of $\langle u, v\rangle$ in S. Thus, in a transformation $S \cup \{\langle u, v\rangle\} \implies S \cup R$, where S and R are multisets of (unordered pairs) and $\langle u, v\rangle$ does not belong to R, there are fewer occurrences of the pair $\langle u, v\rangle$ on the right-hand side of the transformations than there are on the left-hand side. If S, R were interpreted as sets, and $\cup$ as set union, we would have to stipulate that $\langle u, v\rangle$ does not belong to S, or explicitly remove it from S on the right-hand side. Secondly, if pairs $\langle u, v\rangle$ are considered ordered, then a special transformation switching pairs $\langle u, x\rangle$ to $\langle x, u\rangle$ is needed when x is a variable but u is not. If we treat $\langle u, v\rangle$ as unordered, such a transformation is unnecessary. We can now state the *(standard) unification problem*:

Definition 5.2 Given a term system $S = \langle u_1, v_1\rangle, \ldots, \langle u_n, v_n\rangle$, the (standard) unification problem is to find some (all) substitution(s) σ s.t. $u_i[\sigma] = v_i[\sigma]$ for every i, $1 \le i \le n$.

We let $U(S)$ denote the set of all unifiers of S.

Example 5.3 The substitution $\sigma = [a/x, \ a/y]$ is a unifier of the pair

$$\langle f(x, g(a, y)), \ f(x, g(y, x))\rangle.$$

---

[3] It is remarkable that in his thesis, Herbrand gave all the steps of a (nondeterministic) unification algorithm based on transformations on systems of equations. These transformations are given at the end of the section on property A, page 148 of Herbrand [32].

In fact, σ is the only unifier of this pair.

Example 5.4 For every term t, the substitution $\sigma = [a/x, a/y, t/z]$ is a unifier of the pair $\langle f(z, g(a, y)), f(z, g(y, x)) \rangle$.

Observe that in example 5.4, there is an infinite number of unifiers. This leads us to the following question.

Question: How do we compare unifiers?

Answer: Define a preorder $\leq$ on substitutions.

Definition 5.5 Let V be a set of variables. We write $\sigma = \theta[V]$ iff $\sigma(x) = \theta(x)$ for all $x \in V$, and we write $\sigma \leq \theta[V]$ (σ *is more general than* θ *over* V) iff there exists a substitution η such that $\theta = \sigma\,;\eta[V]$.

The intuitive idea behind these definitions is that σ is more general than θ (over V) when each $\theta(x)$ can be obtained from $\sigma(x)$ by instantiating some of the variables occurring in $\sigma(x)$. The reason for relativizing the definition of $\leq$ to a set V of variables is technical. For one thing, some of the results are incorrect if V is left out. Also, in theorem proving applications, it is often desirable to compare substitutions with respect to a "protected set" of variables V. A crucial concept in unification theory is that of a *most general unifier*.

Definition 5.6 Given a term system S and a finite set V of "protected" variables, a substitution σ is a *most general unifier for S away from V* (or *mgu away from V*) iff:

(i) $D(\sigma) \subseteq Var(S)$ and $I(\sigma) \cap (V \cup D(\sigma)) = \emptyset$;

(ii) σ is a unifier of S, i.e. $\sigma \in U(S)$;

(iii) For every unifier θ of S, $\sigma \leq \theta[Var(S)]$.

Note that condition (i) implies that σ is idempotent. When V is not significant, we just call σ an *mgu*. A number of questions now arise naturally.

Some Questions:

- 1. Is $\leq$ well-founded?

- 2. Given S, can we decide whether S is unifiable?

- 3. Given S, if S is unifiable, is there a *mgu*?

The answer to all questions is YES.

That $\leq$ is well-founded is shown in Huet [35]. The decidability of standard unification is implicit in Herbrand's thesis [32] (1930), and it is also settled by Robinson [61] (1965), who gives the first algorithm to find *mgu*'s.

We will now show that questions (1)-(3) have a positive answer. A key point is the similarity between solving a unification problem and solving a system of linear equations:

$$u_1 = v_1$$
$$\vdots$$
$$u_n = v_n$$

and

$$x_1 = a_{1\,1}x_1 + \cdots + a_{1\,n}x_n$$
$$\vdots$$
$$x_m = a_{m\,1}x_1 + \cdots + a_{m\,n}x_n.$$

However, there are some major differences.

- In unification, we cannot assume that the algebraic structure is a field.

- The u_i may not be variables.

- We may have $u_i = v_j$ for $i \neq j$.

Nevertheless, the basic idea of *variable elimination* (as in Gaussian elimination) applies. If u_i is a variable that does not occur in v_i, we can *substitute v_i for u_i* in the rest of the system, and preserve the set of solutions:

$$u_1[v_i/u_i] = v_1[v_i/u_i]$$
$$\vdots$$
$$u_i = v_i$$
$$\vdots$$
$$u_n[v_i/u_i] = v_n[v_i/u_i].$$

This leads to the idea of the *method of transformations* on term systems:

Attempt to transform S into a system S' which is obviously solved.

One of the critical issues is to decide what we mean by a solved system. Quite obviously, a solved system is one that should represent a unifying substitution. Since we are dealing with multisets of unordered pairs, we have to be a little careful in formalizing this idea.

Definition 5.7 A term pair $\langle u, v \rangle$ is in *solved form* in a system S iff either u or v is a variable, say x, and this variable x does not occur anywhere else in S; in particular, if $x = u$ then $x \notin Var(v)$ (and similarly if $x = v$ then $x \notin Var(u)$). The variable x is called a *solved variable*. A system is in solved form if all its pairs are in solved form; a variable is *unsolved* if it occurs in S but is not solved.

Note that a solved form system is always a *set* of solved pairs. Also, note that in a solved pair $\langle u, v \rangle$, it is possible that both u and v are variables, and in this case, it may be that only one of the two is solved in S, or that both are solved in S. Thus, ignoring the order in term pairs, a system is *solved* iff it is of the form

$$S = \langle x_1, v_1 \rangle, \ldots, \langle x_n, v_n \rangle,$$

where $x_1, \ldots, x_n$ are **distinct variables**, and $x_i \notin Var(v_j)$ for all i, j, $1 \leq i, j \leq n$. A system in solved form defines essentially a unique substitution as shown in the definition below.

Definition 5.8 Given a system S in solved form, we define the substitution σ_S as follows: if $S = \langle x_1, v_1 \rangle, \ldots, \langle x_n, v_n \rangle$, then $\sigma_S = [v_1/x_1, \ldots, v_n/x_n]$.

Actually, the above definition is ambiguous, and this is the one place where we might regret our definition of a term system where pairs $\langle u, v \rangle$ are unordered. Let us explain where the difficulty lies. There is no problem with a solved pair $\langle u, v \rangle$ in which **only one** of u, v, say u, is a solved variable, because then the substitution component must be $[v/u]$. But when **both** u and v are solved variables, we can use $[u/v]$ or $[v/u]$ interchangeably as a substitution component. Thus, σ_S is not uniquely defined. However, note for any two σ'_S and σ''_S obtained from S, there is a renaming permutation ρ such that $\sigma'_S = \sigma''_S; \rho$ (where ρ is determined by the pairs $\langle u, v \rangle$ where both u and v are solved in S). Thus, σ_S is uniquely defined, modulo some inessential renaming permutation. The important fact is that σ_S is an idempotent *mgu* of S, as we now show.

Lemma 5.9 *Let* $S = \langle x_1, t_1 \rangle, \ldots, \langle x_n, t_n \rangle$ *be in solved form, where the* $x_1, \ldots, x_n$ *are solved variables. If* $\sigma = [t_1/x_1, \ldots, t_n/x_n]$, *then* σ *is an idempotent mgu of* S. *Furthermore, for any unifier* θ *of* S, *we have* $\theta = \sigma; \theta$.

Proof. We simply observe that for any θ, $\theta(x_i) = \theta(t_i) = \theta(\sigma(x_i))$ for $1 \leq i \leq n$, and $\theta(x) = \theta(\sigma(x))$ otherwise. Clearly σ is an *mgu*, and since $D(\sigma) \cap I(\sigma) = \emptyset$ by the definition of solved forms, it is idempotent. $\square$

The next question is to find sets of transformations for solving unification problems. The following properties of such a set $\mathcal{T}$ of transformations are desirable:

1. (Soundness) Whenever $S \overset{*}{\Longrightarrow}_{\mathcal{T}} S'$, then $U(S') \subseteq U(S)$.

2. (Completeness) For any unifier θ of S, there is some solved S' s.t. $S \overset{*}{\Longrightarrow}_{\mathcal{T}} S'$, and $\sigma_{S'} \leq_E \theta[Var(S)]$.

When $\mathcal{T}$ satisfies (1) and (2), we say that $\mathcal{T}$ *is a complete set of transformations*. We also want the transformations to be as *deterministic as possible*, to reduce the search space. The set of transformations given in the next definition is a variant of the Herbrand–Martelli–Montanari transformations.

Definition 5.10 (The Set of Transformations ST) Let S be any term system (possibly empty), and u, v two terms. The set T consists of the following transformations:

$$\{\langle u, u\rangle\} \cup S \Longrightarrow S \qquad\qquad (triv)$$

$$\{\langle f(u_1,\ldots,u_k), f(v_1,\ldots,v_k)\rangle\} \cup S \Longrightarrow \{\langle u_1, v_1\rangle,\ldots,\langle u_k, v_k\rangle\} \cup S \qquad (dec)$$

$$\{\langle x, v\rangle\} \cup S \Longrightarrow \{\langle x, v\rangle\} \cup S[v/x], \qquad\qquad (vel)$$

where x is a variable s.t. $\langle x, v\rangle$ is **not** solved in $\{\langle x, v\rangle\} \cup S$, and $x \notin Var(v)$.

It should be noted that in transformation (vel), one should not relax the condition "$\langle x, v\rangle$ is **not** solved" to "x is not solved". If this were allowed, one could eliminate the variable x even when v is also a solved variable, and this could have the effect that v could then become unsolved again, leading to an infinite (cyclic) sequence of transformations.

The set ST is a complete set of transformations for standard unification. The basic idea of the transformations is to transform the original problem into a solved form which represents its own solution.

Example 5.11

$$\langle f(x, g(a, y)),\ f(x, g(y, x))\rangle$$
$$\Longrightarrow_{dec} \langle x, x\rangle,\ \langle g(a, y), g(y, x)\rangle$$
$$\Longrightarrow_{triv} \langle g(a, y), g(y, x)\rangle$$
$$\Longrightarrow_{dec} \langle a, y\rangle,\ \langle y, x\rangle$$
$$\Longrightarrow_{vel} \langle a, y\rangle,\ \langle a, x\rangle\,.$$

The reader can immediately verify that the substitution $[a/y, a/x]$ is a unifier of the original system (in fact, it is an mgu). The sense in which these transformations preserve the logically invariant properties of a unification problem is shown in the next lemma.

Lemma 5.12 *Let the set of all standard unifiers of a system S be denoted by $U(S)$. If $S \Longrightarrow S'$ using any transformation from ST, then $U(S) = U(S')$.*

Proof. The only difficulty concerns (vel). Suppose $\{\langle x, v\rangle\} \cup S \Longrightarrow_{vel} \{\langle x, v\rangle\} \cup \sigma(S)$ with $\sigma = [v/x]$. For any substitution θ, if $\theta(x) = \theta(v)$, then $\theta = \sigma;\theta$, since $\sigma;\theta$ differs from θ only at x, but $\theta(x) = \theta(v) = \sigma;\theta(x)$. Thus,

$$\theta \in U(\{\langle x, v\rangle\} \cup S)$$
$$\text{iff }\ \theta(x) = \theta(v)\ \text{ and }\ \theta \in U(S)$$
$$\text{iff }\ \theta(x) = \theta(v)\ \text{ and }\ \sigma;\theta \in U(S)$$
$$\text{iff }\ \theta(x) = \theta(v)\ \text{ and }\ \theta \in U(\sigma(S))$$
$$\text{iff }\ \theta \in U(\{\langle x, v\rangle\} \cup \sigma(S)). \quad \square$$

The point here is that the most important feature of a unification problem—its set of solutions—is preserved under these transformations, and hence we are justified in our method of attempting to transform such problems into a trivial (solved) form in which the existence of an *mgu* is evident.

We may now show the soundness and completeness of these transformations following [49].

Theorem 5.13 (Soundness) *If $S \overset{*}{\Longrightarrow} S'$ with S' in solved form, then $\sigma_{S'} \in U(S)$.*

Proof. Using the previous lemma and a trivial induction on the length of transformation sequences, we see that $U(S) = U(S')$, and so clearly $\sigma_{S'} \in U(S)$. □

Theorem 5.14 (Completeness) *Every sequence of transformations*

$$S = S_0 \Longrightarrow S_1 \Longrightarrow S_2 \Longrightarrow \ldots$$

must eventually terminate. Furthermore, S is unifiable iff every system S' derivable from S is in solved form, and for every $\theta \in U(S)$, $\sigma_{S'} \leq \theta$.

Proof. We first show that every transformation sequence terminates. For any system S, let us define a complexity measure $\mu(S) = \langle n, m \rangle$, where n is the number of *unsolved* variables in the system, and m is the sum of the sizes of all the terms in the system. Then the lexicographic ordering on $\langle n, m \rangle$ is well-founded, and each transformation produces a new system with a measure strictly smaller under this ordering: (*triv*) and (*dec*) must decrease m and can not increase n, and (*vel*) must decrease n.

Therefore the relation $\Longrightarrow$ is well-founded, and every transformation sequence must end in some system to which no transformation applies. Suppose a given sequence ends in a system S'. Now $\theta \in U(S)$ implies by lemma 5.12 that $\theta \in U(S')$, and so S' can contain no pairs of the form $\langle f(t_1, \ldots, t_n), g(t'_1, \ldots, t'_m) \rangle$ or of the form $\langle x, t \rangle$ with $x \in Var(t)$. But since no transformation applies, all pairs in S' must be in solved form. Finally, since $\theta \in U(S')$, by lemma 5.9 we must have $\sigma_{S'} \leq \theta$. □

Putting these two theorems together, we have that the set $\mathcal{ST}$ can always find an *mgu* for a unifiable system of terms; as remarked in [49], this abstract formulation can be used to model many different unification algorithms, by simply specifying data structures and a control strategy.

The first published unification algorithm due to Robinson ([61]) can run in exponential time. Since Robinson's seminal discovery, several polynomial-time algorithms for standard unification have been given, including one by Robinson himself. Among them, we single out a quasi-linear algorithm due to Huet [35] (1976), a linear-time algorithm due to Paterson and Wegman [56] (1978), and quasi-linear and linear algorithms due to Martelli and Montanari [49] (1982). For an excellent account of standard unification, the reader is referred to Knight's survey article [43], and to Jouannaud and Kirchner's survey article [38]. Martelli and Montanari's important contribution ([49]), perhaps even more than their algorithm itself,[4] is to have demonstrated with perfect

---

[4] Their algorithm is not so different from Paterson and Wegman's algorithm.

clarity that the method of transformations is remarkably well suited for tackling unification problems. In some sense, Martelli and Montanari revived Herbrand's approach, which led to new important work by Kirchner and others. In the next section, we sketch some versions of fast unification algorithms.

6 Fast (Standard) Unification

If one looks closely at the set of transformations ST, one realizes that the complexity of unification algorithms is related to explicit variable elimination. Thus, a main concern in designing fast unification algorithms is to avoid explicit variable elimination. We first give the intuition behind a fast unification algorithm due to Martelli and Montanari [49].

Starting from a system S, suppose we only apply decomposition and deletion of trivial rules If no failure takes place and there are still pairs left, we must reach a system S' such that every pair is of the form $\langle x, v \rangle$, where x is a variable.

We can group all pairs sharing some common element to form equivalence classes. Thus, S' can be viewed as a partition, where every class is of the form

$$\{x_1, \ldots, x_k, t_1, \ldots, t_m\},$$

where $x_1, \ldots, x_k$ are variables ($k > 0$) and $t_1, \ldots, t_m$ are nonvariable terms ($m \geq 0$).

Clearly, S' is unifiable only if for every class, all nonvariable terms have the same root symbol. The other basic idea is to **analyze dependencies among variables** (analogy with solving systems of linear equations).

A precedence relation on classes can be defined as follows:

$C < C'$ iff C' contains some variable x that occurs in some nonvariable term t in C.

Intuitively, the class of C *needs* the value of the variable x. The following is easily shown.

Lemma 6.1 *If $<^+$ is reflexive, then S' is not unifiable.*

From now on, we are dealing with sets of the form

$$\{x_1, \ldots, x_k, t_1, \ldots, t_m\},$$

that Martelli and Montanari call *multiequations*, and write in the form

$$\{x_1, \ldots, x_k\} := \{t_1, \ldots, t_m\}.$$

Since eliminating duplicate terms may be costly, they allow both sides to be multisets. If $<^+$ is acyclic (irreflexive), roughly speaking, Martelli and Montanari do the following:

Pick some class $\{x_1, \ldots, x_k, t_1, \ldots, t_m\}$ where $m > 0$ and the t_i are not constants. Since all the t_i have the same root symbol, say f, form the new system in which the above class is replaced by the sets

$$\{x_1, \ldots, x_k, f(y_1, \ldots, y_n)\}, \{y_1, t_1/1, \ldots, t_m/1\}, \ldots, \{y_n, t_1/n, \ldots, t_m/n\}.$$

Again, group blocks together to form a partition, and check for acyclicity of the new $<^+$. Continue this process until failure, or no new classes are formed. If the last $<^+$ obtained is acyclic, we can form a *mgu* in *triangular form*, by using any total ordering of the classes extending $<^+$. Formally, a triangular form is defined as follows.

Definition 6.2 Given an idempotent substitution σ (i.e., $D(\sigma) \cap I(\sigma) = \emptyset$) with domain $D(\sigma) = \{x_1, \ldots, x_k\}$, a *triangular form for σ* is a finite set T of pairs $\langle x, t \rangle$ where $x \in D(\sigma)$ and t is a term, such that this set T can be sorted (possibly in more than one way) into a sequence $\langle \langle x_1, t_1 \rangle, \ldots, \langle x_k, t_k \rangle \rangle$ satisfying the following properties: for every i, $1 \leq i \leq k$,

(1) $\{x_1, \ldots, x_i\} \cap Var(t_i) = \emptyset$, and

(2) $\sigma = [t_1/x_1]; \ldots ; [t_k/x_k]$.

The set of variables $\{x_1, \ldots, x_k\}$ is called the *domain* of T. Note that in particular $x_i \notin Var(t_i)$ for every i, $1 \leq i \leq k$, but variables in the set $\{x_{i+1}, \ldots, x_k\}$ may occur in $t_1, \ldots, t_i$. It is easily seen that σ is an (idempotent) *mgu* of the term system T.

Example 6.3 Consider the substitution $\sigma = [f(f(x_3, x_3), f(x_3, x_3))/x_1, f(x_3, x_3)/x_2]$. The system $T = \{\langle x_1, f(x_2, x_2) \rangle, \langle x_2, f(x_3, x_3) \rangle\}$ is a triangular form of σ since it can be ordered as $\langle \langle x_1, f(x_2, x_2) \rangle, \langle x_2, f(x_3, x_3) \rangle \rangle$ and $\sigma = [f(x_2, x_2)/x_1]; [f(x_3, x_3)/x_2]$.

It turns out that we have computed a certain relation on terms, a *unification closure*. We now define this concept, due to Paterson and Wegman [56] (1978), and give a fast algorithm based on it.

Definition 6.4 Let Σ be a finite ranked alphabet (signature). Consider a finite graph G whose nodes are labeled with symbols in Σ or variables in $\mathcal{X}$. Let $\Lambda : V \to \Sigma \cup \mathcal{X}$ be the labeling function.

If $\Lambda(u)$ is a constant or a variable, then u is a terminal node; If $\Lambda(u)$ is a function symbol of rank k, then u has k immediate successors $u[1], \ldots, u[k]$.

An equivalence relation R on a graph G is a *unification closure* iff, for every pair (u, v) of nodes in V^2, whenever uRv then:

(1) Either $\Lambda(u) = \Lambda(v)$, or one of $\Lambda(u)$, $\Lambda(v)$ is a variable;

(2) If $\Lambda(u) = \Lambda(v)$ and $r(\Lambda(u)) = n$, then for every i, $1 \leq i \leq n$, $u[i]Rv[i]$.

Graphically, if u and v are two nodes labeled with the same symbol f of rank n, if $u[1], \ldots, u[n]$ are the successors of u and $v[1], \ldots, v[n]$ are the successors of v,

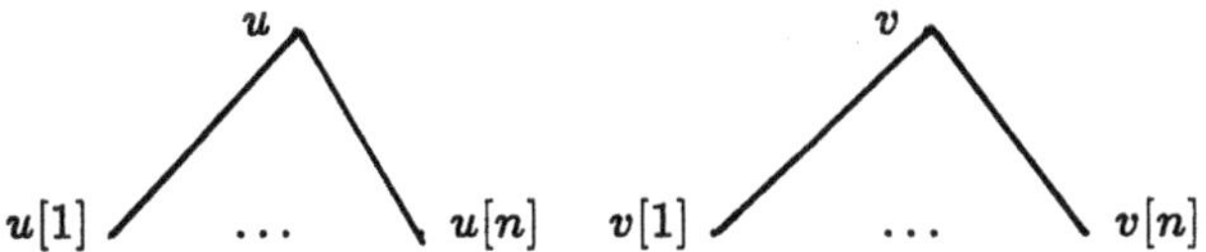

if u and v are equivalent then $u[i]$ and $v[i]$ are equivalent for all $i, 1 \leq i \leq n$. We have a kind of forward closure. The following lemma is easily shown.

Lemma 6.5 *There is an algorithm which, given any arbitrary relation R_0 on a finite graph G, decides whether the smallest unification closure containing a relation R_0 on G exists, and if so computes it.*

In order to test whether a system $S = \{\langle u_1, v_1 \rangle, \ldots, \langle u_m, v_m \rangle\}$ is unifiable, we can compute the unification closure of the relation S on the graph G_S constructed as follows:

(i) The set set of nodes of G_S is the set of all subterms of terms in S.

(ii) Every subterm that is either a constant or a variable is a terminal node labeled with that symbol.

(iii) For every subterm of the form $f s_1 \ldots s_k$, the label is f, and there is an edge from $f s_1 \ldots s_k$ to s_i, for each $i, 1 \leq i \leq k$.

Let R be the least unification closure containing the relation S on the graph G_S, if it exits. A new graph G_S/R can be constructed as follows:

(i) The nodes of G_S/R are the equivalence classes of R.

(ii) There is an edge from a class C to a class C' iff there is an edge in G_S from some node s in class C to some node t in class C'.

The following lemma is essentially due to Paterson and Wegman [56] (1978).

Lemma 6.6 *The system S is unifiable iff the unification closure R exists and the graph G_S/R is acyclic.*

When S is unifiable, let $\langle C_1, \ldots, C_n \rangle$ be the sequence of all equivalence classes containing some variable, ordered such that, if there is a path from C_i to C_j, then $i < j$.

For each $i, 1 \leq i \leq n$, if C_i contains some nonvariable term, let t_i be any such term, else let t_i be any variable in C_i.

Let

$$\sigma_i = [t_i/z_1, \ldots, t_i/z_k],$$

where $\{z_1, \ldots, z_k\}$ is the set of variables in C_i, and let $\sigma = \sigma_1 ; \ldots ; \sigma_n$. Then σ is a most general unifier of S. The substitution σ has a triangular representation.

Example 6.7 Consider the pair

$$\langle f(f(x_2, x_2), f(x_3, x_3)), f(x_1, x_2)\rangle.$$

The nontrivial classes of the unification closure containing variables are

$$\{x_1, f(x_2, x_2)\}, \{x_2, f(x_3, x_3)\}.$$

The first class precedes the second. A triangular form of the *mgu* σ is

$$\sigma = [f(x_2, x_2)/x_1]; [f(x_3, x_3)/x_2].$$

Note that $\sigma = [f(f(x_3, x_3), f(x_3, x_3))/x_1, f(x_3, x_3)/x_2]$.

We now give a simple fast algorithm, assuming for simplicity that every symbol in Σ is either a constant or binary. This algorithm is basically Ravi Sethi's algorithm, in [1].

```
procedure unif(u, v : node; var R : partition; var flag : bool);
  var s, t : node; flag1, flag2 : bool;
  begin
    flag1 := false; flag2 := false;
    s := find(R, u); t := find(R, v);
    if s = t then
      flag := true
    else
      if nonvar(s) and nonvar(t) and root(s) = root(t) then
        union(R, s, t);
        unif(left(s), left(t), R, flag1);
        unif(right(s), right(t), R, flag2);
        flag := flag1 and flag2
      else
        if var(s) or var(t) then
          union(R, s, t);
          flag := true
        else
          flag := false
        endif
      endif
    endif;
    if flag and acyclic(R) then
      printsubst(R)
    else
      failure(R)
    endif
  end
```

Using Tarjan's fast version of *union* and *find*, the algorithm runs in $O(n\alpha(n))$, where $\alpha(n)$ is a sort of inverse of Ackermann's function that grows extremely slowly.

We conclude this section with some comments on Paterson and Wegman's fast algorithm [56] (1978). Paterson and Wegman's algorithm is a unification closure algorithm, and it uses the concept of a "root class". A *root class* is an equivalence class of nodes containing only terms corresponding to nodes of the DAG that have no parents. It is immediately seen that if a system S is unifiable, then every class that is mimimal in the ordering defined such that $C < C'$ iff there is a path from C to C' in the DAG, is a root class.

Paterson and Wegman's algorithm processes root classes first. A root class has the property that no further nodes can be added to it as the result of computing a unification closure, because propagation proceeds from parent to children. Thus, once a root class has been processed, it can be deleted.

It should be noted that Paterson and Wegman's original algorithm contains bugs. The bugs were reported and fixed by De Champeaux [19]. One of the bugs is a trivial typo. The other bug is more subtle. In the Paterson-Wegman's algorithm, the equivalence of two elements is represented by the existence of a special (undirected) edge between these two elements (not to be confused with the edges of the DAG). The problem is that multiple edges can be created by the algorithm, but this is not taken into account by the algorithm.

We now come back to the method of matings in the general case of languages with equality.

7 Equational Matings

In this section, we show that the method of matings for languages **without** equality presented in section 4 can be generalized to languages **with** equality. This generalization will lead us to a decidable form of unification extending standard unification and called *rigid E-unification*. The generalized method of matings was first presented in Gallier, Raatz, and Snyder [23] (1987), where it was conjectured that rigid E-unification is decidable. Several months later, Gallier, Narendran, Plaisted, and Snyder proved that rigid E-unification is NP-complete and that finite complete sets of rigid E-unifiers always exist. These results were announced (without complete proofs) at LICS'88 [25]. Full details and proofs appear in Gallier, Narendran, Plaisted, and Snyder [27]. A detailed presentation of the method of equational matings is given in [23, 26], or [28].

First, it is important to note that lemma 3.8, theorem 3.11, theorem 4.3, and lemma 4.5, also hold for languages **with** equality. The main difference with the case of languages without equality, is that the criterion for checking whether a vertical path is unsatisfiable is more complicated, and involves some equality reasoning.

A criterion for the unsatisfiability of a conjunction of literals based on the concept of congruence closure is known. In order to explain this criterion, it is convenient to represent every atomic formula as an equation. This can be done by adding to our language (which already contains the special sort *bool*) the constant $\top$ of sort *bool*, interpreted as **true**. Then, every atomic formula $Pt_1 \ldots t_n$ of sort *bool* can be expressed as the equation $(Pt_1 \ldots t_n \doteq \top)$. Hence, we can assume that all atomic formulae are equations. The notations $Pt_1 \ldots t_n$ and $(Pt_1 \ldots t_n \doteq \top)$ will be used interchangeably for atomic formulae of sort *bool*.

Given a vertical path π, we can arrange the literals in π by grouping positive and negative literals together, to form a conjunction C_π of the form

$$(s_1 \doteq t_1) \wedge \ldots \wedge (s_m \doteq t_m) \wedge \neg(s_1' \doteq t_1') \wedge \ldots \wedge \neg(s_n' \doteq t_n').$$

The congruence closure method defined below enables us to decide whether conjunctions of the above form are satisfiable or not.

Definition 7.1 (Congruence closure) Let $TERMS(\pi)$ be the set of all subterms of terms in π. Construct the labeled directed graph G_π as follows:

- The set of Nodes of G_π is $TERMS(\pi)$.

- The node $f(t_1, \ldots, t_n)$ is labeled with f.

- For each node $f(t_1, \ldots, t_n)$, there is an edge from $f(t_1, \ldots, t_n)$ to each t_i.

A relation $\simeq$ on the set of nodes of G_π is *G-congruential* iff, for any two nodes $f(s_1, \ldots, s_n)$ and $f(t_1, \ldots, t_n)$, if $s_i \simeq t_i, 1 \leq i \leq n$, then $f(s_1, \ldots, s_n) \simeq f(t_1, \ldots, t_n)$.

Given a vertical path

$$\pi = \{(s_1 \doteq t_1), \ldots, (s_m \doteq t_m), \neg(s_1' \doteq t_1'), \ldots, \neg(s_n' \doteq t_n')\},$$

let $E = \{(s_1 \doteq t_1), \ldots, (s_m \doteq t_m)\}$. The following results can be shown (see Kozen [45], Nelson and Oppen [55], or Gallier [22]).

Lemma 7.2 *There is a smallest G-congruential equivalence relation on G_π containing E. It is called the congruence closure of E, and it is denoted as $\stackrel{*}{\cong}_E$.*

Lemma 7.3 *A vertical path π is unsatisfiable iff $s_j' \stackrel{*}{\cong}_E t_j'$ for some $j, 1 \leq j \leq n$.*

The congruence closure $\stackrel{*}{\cong}_E$ can be computed in polynomial time (Kozen [45,46], Nelson and Oppen [55], Downey, Sethi, and Tarjan [21]).

Now, recall that theorem 4.3 states that a universal sentence A in *nnf* is unsatisfiable iff there is some amplification D of A and some (ground) substitution σ such that $\sigma(D)$ is unsatisfiable. Since $\sigma(D)$ is quantifier-free, by lemma 4.5, $\sigma(D)$ is unsatifiable iff all vertical paths in $\sigma(D)$ are unsatisfiable. In view of lemma 7.3, we can now give the criterion stating that vertical paths in $\sigma(D)$ are unsatisfiable:

Given any $\pi = \{(s_1 \doteq t_1), \ldots, (s_m \doteq t_m), \neg(s_1' \doteq t_1'), \ldots, \neg(s_n' \doteq t_n')\}$ in $vp(D)$,[5] there is some i $(1 \leq i \leq n)$, such that $\{\sigma(s_1 \doteq t_1), \ldots, \sigma(s_m \doteq t_m), \neg\sigma(s_i' \doteq t_i')\}$ is unsatisfiable.

---

[5] **Warning:** $\pi \in vp(D)$, not $\pi \in vp(\sigma(D))$.

Since the set $\{\sigma(s_1 \doteq t_1), \ldots, \sigma(s_m \doteq t_m), \neg\sigma(s'_i \doteq t'_i)\}$ consists of quantifier-free formulae, it is unsatisfiable iff $\sigma(s'_i \doteq t'_i)$ is provable from $\{\sigma(s_1 \doteq t_1), \ldots, \sigma(s_m \doteq t_m)\}$, treated as a set of *ground equations*. By lemma 7.3, this is equivalent to saying that

$\sigma(s'_i)$ and $\sigma(t'_i)$ are congruent modulo the congruence closure associated with the set of equations $\{\sigma(s_1 \doteq t_1), \ldots, \sigma(s_m \doteq t_m)\}$.

The definition of an equational mating is motivated by the above observation. It is designed so that we have a criterion expressed in terms of vertical paths for testing whether given a quantifier-free formula D, there is some substitution σ such that $\sigma(D)$ is unsatisfiable (see lemma 7.5).

Definition 7.4 Let A be a quantifier-free formula in nnf. An *equational mating* $\mathcal{M}$ for A is a pair $\langle MS, \sigma \rangle$, where MS is a set of sets of literals called *mated sets* and σ is a substitution, such that, each mated set is a subset of some vertical path $\pi \in vp(A)$ and is of the form

$$\{(s_1 \doteq t_1), \ldots, (s_m \doteq t_m), \neg(s \doteq t)\} \subseteq \pi,$$

where $m \geq 0,$[6] and, for every mated set $\{(s_1 \doteq t_1), \ldots, (s_m \doteq t_m), \neg(s \doteq t)\} \in MS$, the set of literals $\{\sigma(s_1 \doteq t_1), \ldots, \sigma(s_m \doteq t_m), \neg\sigma(s \doteq t)\}$ is unsatisfiable. The substitution associated with the mating $\mathcal{M}$ is also denoted as $\sigma_{\mathcal{M}}$. We also commit a slight abuse of language (and notation) and say that a mated set belongs to $\mathcal{M}$.

An equational mating $\mathcal{M}$ is a *refutation mating* iff $\sigma_{\mathcal{M}}(A)$ is unsatisfiable.

An equational mating $\mathcal{M}$ is *path acceptable*[7] (for short, *p-acceptable*), iff, for every path $\pi \in vp(A)$, there is some mated set $\{(s_1 \doteq t_1), \ldots, (s_m \doteq t_m), \neg(s \doteq t)\} \in \mathcal{M}$, such that

$$\{(s_1 \doteq t_1), \ldots, (s_m \doteq t_m), \neg(s \doteq t)\} \subseteq \pi.$$

A number of remarks are in order:

(1) Given the substitution σ, the mating condition can be tested using the congruence closure method. The difficulty is to decide whether or not the substitution σ exists.

(2) Given a family MS of mated sets, let $\vec{E} = (E_S)_{S \in MS}$ be the family of sets of equations of the form $E_S = \{(s_1 \doteq t_1), \ldots, (s_m \doteq t_m)\}$ and $S = \{\langle s, t \rangle \mid S \in MS\}$ the set of pairs where E_S and $\langle s, t \rangle$ are associated with the mated sets $S = \{(s_1 \doteq t_1), \ldots, (s_m \doteq t_m), \neg(s \doteq t)\} \in MS$. Observe that $\mathcal{M} = \langle MS, \sigma \rangle$ is a mating iff σ is a solution of the following problem:

Problem 1: Given $\vec{E} = \{E_i \mid 1 \leq i \leq n\}$ a family of n finite sets of equations and $S = \{\langle u_i, v_i \rangle \mid 1 \leq i \leq n\}$ a set of n pairs of terms, is there a substitution θ such that, treating each set $\theta(E_i)$ as a set of *ground* equations (i.e. holding the variables in $\theta(E_i)$ "rigid"), $\theta(u_i)$ and $\theta(v_i)$ are provably equal from $\theta(E_i)$ for $i = 1, \ldots, n$?

---

[6] The case $m = 0$ is indeed possible when $\sigma(s) = \sigma(t)$, i.e., when σ is a unifier of s and t.

[7] A path acceptable mating is also called a *spanning mating* by Miller [52].

Equivalently, is there a substitution θ such that $\theta(u_i)$ and $\theta(v_i)$ can be shown congruent from $\theta(E_i)$ by the congruence closure method for $i = 1,\ldots,n$?

Problem 1 is a unification problem more general than standard unification. A substitution θ solving the above problem is called a *rigid $\vec{E}$-unifier of S*, and a pair $\langle \vec{E}, S \rangle$ such that S has some rigid $\vec{E}$-unifier is called an *equational premating*. This key observation is used in searching for the substitutions associated with matings. They are the rigid $\vec{E}$-unifiers of S.

The following lemma is a straightforward generalization of a lemma 4.7 to languages with equality.

Lemma 7.5 *Given a quantifier-free formula A in nnf, the following properties hold:*

(1) Given a substitution θ, if $\theta(A)$ is unsatisfiable, then there is a p-acceptable equational mating $\mathcal{M}$ for A.

(2) A p-acceptable equational mating $\mathcal{M}$ for A is a refutation mating for A, i.e. $\sigma_{\mathcal{M}}(A)$ is unsatisfiable.

Corollary 7.6 *Given a quantifier-free formula A in nnf, there is a substitution θ such that $\theta(A)$ is unsatisfiable iff there is a p-acceptable equational mating $\mathcal{M}$ for A.* $\square$

As in section 4, the completeness and soundness for the method of equational matings is an immediate consequence of theorem 4.3, lemma 4.5, and lemma 7.5.

Theorem 7.7 *Given a universal sentence A in nnf, A is unsatisfiable iff some amplification D of A has a p-acceptable equational mating.*

Let us give some examples illustrating the use of theorem 7.7. From this point on, for the sake of brevity, we will use interchangeably the terms equational mating and mating.

Example 7.8 Consider the following Horn formula A, where x, y, z denote variables:

$$(a \doteq b) \,\wedge$$
$$((f^3 x \doteq x) \vee \neg(fx \doteq fb)) \,\wedge$$
$$(Qa \vee \neg(f^3 a \doteq a)) \,\wedge$$
$$((f^5 y \doteq y) \vee \neg Qy) \,\wedge$$
$$(Ra \vee \neg(fa \doteq a) \vee \neg Pfa) \,\wedge$$
$$\neg Rfz \,\wedge$$
$$Pa$$

There are 24 vertical paths in A. Let $\theta = [a/x, a/y, a/z]$. The substitution θ closes all the paths in $\theta(A)$, which is easy to see for the 21 vertical paths containing the sets of literals $\{(f^3 a \doteq a), \neg(f^3 a \doteq a)\}$, $\{Qa, \neg Qa\}$, and $\{(a \doteq b), \neg(fa \doteq fb)\}$. A p-acceptable mating for A is given by θ and the following set of 6 sets of literals:

$$\{\{(f^3 x \doteq x), \neg(f^3 a \doteq a)\},$$
$$\{Qa, \neg Qy\},$$
$$\{(a \doteq b), \neg(fx \doteq fb)\},$$
$$\{(f^5 y \doteq y), (f^3 x \doteq x), Ra, \neg Rfz\},$$
$$\{(f^5 y \doteq y), (f^3 x \doteq x), \neg(fa \doteq a)\},$$
$$\{(f^5 y \doteq y), (f^3 x \doteq x), Pa, \neg Pfa\}\}.$$

The above set is a mating because $(fa \doteq a)$ is equationally provable from $(f^3 a \doteq a)$ and $(f^5 a \doteq a)$. Indeed, $(f^3 a \doteq a)$ implies $(f^4 a \doteq fa)$, which implies $(f^5 a \doteq f^2 a)$, which, by transitivity, implies $(f^2 a \doteq a)$. In turn, $(f^2 a \doteq a)$ implies $(f^3 a \doteq fa)$, and by one more application of transitivity, this implies $(fa \doteq a)$. According to lemma 7.5, $\theta(A)$ is unsatisfiable.

Example 7.9 Let A be the following (equational) sentence:

$$\forall x \forall y \forall z (*(x, *(y, z)) \doteq *(*(x, y), z))) \wedge \tag{1}$$
$$\forall u (*(u, 1) \doteq u) \wedge \tag{2}$$
$$\forall v (*(1, v) \doteq v) \wedge \tag{3}$$
$$\forall w (*(w, w) \doteq 1) \wedge \tag{4}$$
$$\neg (*(a, b) \doteq *(b, a)). \tag{5}$$

The first three equations are the axioms for monoids (a binary operation $*$ which is associative and has an identity element 1), the fourth equation asserts that the square of every element is the identity, and the fifth asserts the negation of the commutativity of $*$ (A is the result of a Skolemization). The unsatisfiability of A asserts that any monoid such that the square of every element is the identity is commutative.

Consider the following amplification D of A in the left column and the set MS consisting of one set of literals in the right column:

$$
\begin{aligned}
D = \quad & (*(u_1, 1) \doteq u_1) & MS = \{\{ & (*(u_1, 1) \doteq u_1), \\
\wedge \ & (*(w_1, w_1) \doteq 1) & & (*(w_1, w_1) \doteq 1), \\
\wedge \ & (*(x_1, *(y_1, z_1)) \doteq *(*(x_1, y_1), z_1))) & & (*(x_1, *(y_1, z_1)) \doteq *(*(x_1, y_1), z_1))), \\
\wedge \ & (*(x_2, *(y_2, z_2)) \doteq *(*(x_2, y_2), z_2))) & & (*(x_2, *(y_2, z_2)) \doteq *(*(x_2, y_2), z_2))), \\
\wedge \ & (*(w_2, w_2) \doteq 1) & & (*(w_2, w_2) \doteq 1), \\
\wedge \ & (*(1, v_1) \doteq v_1) & & (*(1, v_1) \doteq v_1), \\
\wedge \ & (*(x_3, *(y_3, z_3)) \doteq *(*(x_3, y_3), z_3))) & & (*(x_3, *(y_3, z_3)) \doteq *(*(x_3, y_3), z_3))), \\
\wedge \ & (*(x_4, *(y_4, z_4)) \doteq *(*(x_4, y_4), z_4))) & & (*(x_4, *(y_4, z_4)) \doteq *(*(x_4, y_4), z_4))), \\
\wedge \ & (*(w_3, w_3) \doteq 1) & & (*(w_3, w_3) \doteq 1), \\
\wedge \ & \neg (*(a, b) \doteq *(b, a)). & & \neg (*(a, b) \doteq *(b, a))\}\}.
\end{aligned}
$$

Let θ be the substitution

$$[a/u_1, (a * b)/w_1, a/x_1, (a * b)/y_1, (a * b)/z_1,$$
$$a/x_2, a/y_2, b/z_2, a/w_2, b/v_1,$$
$$b/x_3, (a * b)/y_3, b/z_3, a/x_4, b/y_4, b/z_4, b/w_3].$$

We claim that $\langle MS, \theta \rangle$ is a mating for D. For simplicity of notation let us adopt infix notation, and denote $*(s,t)$ as $s * t$. Then, we have:

$$
\begin{aligned}
a * b &= \{a * 1\} * b & \text{by (2)} \\
&= \{a * [(a * b) * (a * b)]\} * b & \text{by (4)} \\
&= \{[a * (a * b)] * (a * b)\} * b & \text{by (1)} \\
&= \{[(a * a) * b] * (a * b)\} * b & \text{by (1)} \\
&= \{[1 * b] * (a * b)\} * b & \text{by (4)} \\
&= \{b * (a * b)\} * b & \text{by (3)} \\
&= b * \{(a * b) * b\} & \text{by (1)} \\
&= b * \{a * (b * b)\} & \text{by (1)} \\
&= b * \{a * 1\} & \text{by (4)} \\
&= b * a, & \text{by (2)}
\end{aligned}
$$

which shows that $\langle MS, \theta \rangle$ is a p-acceptable equational mating for D (there is a single vertical path in D).

8 Rigid E-Unification

In the second remark following definition 7.4, it was noted that a new form of unification, "rigid E-unification", arises naturally in extending Andrews and Bibel's theorem proving method of matings [4, 6, 11, 12, 13], to first-order languages with equality. What was noted in this remark, is that $\mathcal{M} = \langle MS, \sigma \rangle$ is a mating iff σ is a solution of the following problem:

Problem 1: Given $\vec{E} = \{E_i \mid 1 \leq i \leq n\}$ a family of n finite sets of equations and $S = \{\langle u_i, v_i \rangle \mid 1 \leq i \leq n\}$ a set of n pairs of terms, is there a substitution θ such that, treating each set $\theta(E_i)$ as a set of *ground* equations (i.e. holding the variables in $\theta(E_i)$ "rigid"), $\theta(u_i)$ and $\theta(v_i)$ are provably equal from $\theta(E_i)$ for $i = 1, \ldots, n$?

Equivalently, is there a substitution θ such that $\theta(u_i)$ and $\theta(v_i)$ can be shown congruent from $\theta(E_i)$ by the congruence closure method for $i = 1, \ldots, n$?

Actually, it turns out that problem 1 reduces to the following simpler problem:

Problem 2: Given a finite set $E = \{u_1 \doteq v_1, \ldots, u_n \doteq v_n\}$ of equations and a pair $\langle u, v \rangle$ of terms, is there a substitution θ such that, treating $\theta(E)$ as a set of ground equations, $\theta(u) \overset{*}{\longleftrightarrow}_{\theta(E)} \theta(v)$, that is, $\theta(u)$ and $\theta(v)$ are congruent modulo $\theta(E)$ by congruence closure (Kozen [45], Nelson and Oppen [55])?

The substitution θ is called a *rigid E-unifier of u and v.*

Example 8.1 Let $E = \{fa \doteq a, \; ggx \doteq fa\}$, and $\langle u, v \rangle = \langle gggx, x \rangle$. Then, the substitution $\theta = [ga/x]$ is a rigid E-unifier of u and v. Indeed, $\theta(E) = \{fa \doteq a, \; ggga \doteq$

$fa\}$, and $\theta(gggx)$ and $\theta(x)$ are congruent modulo $\theta(E)$, since

$$\theta(gggx) = ggggga \longrightarrow gfa \qquad\qquad \text{using } ggga \doteq fa$$
$$\longrightarrow ga = \theta(x) \qquad\qquad \text{using } fa \doteq a.$$

Note that θ is not the only rigid E-unifier of u and v. For example, $[gfa/x]$ or more generally $[gf^n a/x]$ is a rigid E-unifier of u and v. However, θ is more general than all of these rigid E-unifiers (in a sense to be made precise later).

The importance of rigid E-unification stems from the fact that it is decidable, and in fact NP-complete, see Gallier, Narendran, Plaisted, and Snyder [27]. Remarkably, it can also be shown that there is always a finite set of most general rigid E-unifiers called a complete set of rigid E-unifiers [27].

It is interesting to observe that the notion of rigid E-unification arises by *bounding* the resources, in this case, the number of available instances of equations in E. In order to understand more clearly the concept of rigid E-unification, let us recall what (unrestricted) E-unification is. We are given a set of equations $E = \{u_1 \doteq v_1, \ldots, u_n \doteq v_n\}$, and (for simplicity) a pair of terms $\langle u, v \rangle$. The problem is to decide whether is there a substitution θ s.t. $\theta(u) \xleftrightarrow{\;*\;}_E \theta(v)$.

Note that there is *no bound* on the number of instances of equations in E that can be used in the proof that $\theta(u) \xleftrightarrow{\;*\;}_E \theta(v)$. Going back to definition 2.16, we observe that $\theta(u) \xleftrightarrow{\;*\;}_E \theta(v)$ iff is there a *multiset* of equations (from E)

$$\left\{ \binom{u_1' \doteq v_1'}{n_1}, \ldots, \binom{u_m' \doteq v_m'}{n_m} \right\}$$

and m sets of substitutions $\{\sigma_{j,1}, \ldots, \sigma_{j,n_j}\}$, s.t., letting

$$E' = \{\sigma_{j,k}(u_j' \doteq v_j') \mid 1 \leq j \leq m,\ 1 \leq k \leq n_j\},$$

we have $\theta(u) \xleftrightarrow{\;*\;}_{E'} \theta(v)$, considering E' as **ground**. Basically, the restriction imposed by rigid E-unification is that $n_1 = \ldots = n_m = 1$, i.e., at most a single instance of each equation in E can be used. In fact, these instances $\theta(u_1 \doteq v_1), \ldots, \theta(u_n \doteq v_n)$ must arise from the substitution θ itself. Also, once these instances have been created, the remaining variables (if any) are considered rigid, that is, treated as constants, so that it is not possible to instantiate these instances. Thus, rigid E-unification and Girard's linear logic [29] share the same spirit. Since the resources are bounded, it is not too surprising that rigid E-unification is decidable, but it is not obvious at all that the problem is in NP. The special case of rigid E-unification where E is a set of ground equations has been investigated by Kozen who has shown that this problem is NP-complete (Kozen, [45,46]). Thus, rigid E-unification is NP-hard. We also showed that it is in NP, hence NP-complete.

Our plan for the rest of this section is to define precisely what complete sets of rigid E-unifiers are, and to sketch the decision procedure. The definitions of a rigid E-unifier, the preorder $\leq_E$, and complete sets of rigid E-unifiers, will parallel those given for E-unification, but equations are considered as ground in equational proofs. It will be convenient to write $u \cong_E^* v$ to express that $u \xleftrightarrow{\;*\;}_E v$, treating the equations in E as *ground equations*.

Definition 8.2 Let $E = \{(s_1 \doteq t_1), \ldots, (s_m \doteq t_m)\}$ be a finite set of equations, and let $Var(E) = \bigcup_{(s \doteq t) \in E} Var(s \doteq t)$ denote the set of variables occurring in E.[8] Given a substitution θ, we let $\theta(E) = \{\theta(s_i \doteq t_i) \mid s_i \doteq t_i \in E,\ \theta(s_i) \neq \theta(t_i)\}$. Given any two terms u and v,[9] a substitution θ is a *rigid unifier of u and v modulo E* (for short, a *rigid E-unifier of u and v*) iff

$\theta(u) \stackrel{*}{\cong}_{\theta(E)} \theta(v)$, that is, $\theta(u)$ and $\theta(v)$ are congruent modulo the set $\theta(E)$ considered as a set of *ground* equations.

The following example should help grasping the notion of rigid E-unification. The problem is to show that if $x \cdot x = 1$ in a monoid, then the monoid is commutative.

Example 8.3

$$
\begin{aligned}
E = \{ & u_1 \cdot 1 \doteq u_1 \\
& w_1 \cdot w_1 \doteq 1 \\
x_1 \cdot (y_1 \cdot z_1) &\doteq (x_1 \cdot y_1) \cdot z_1 \\
x_2 \cdot (y_2 \cdot z_2) &\doteq (x_2 \cdot y_2) \cdot z_2 \\
& w_2 \cdot w_2 \doteq 1 \\
& 1 \cdot v_1 \doteq v_1 \\
x_3 \cdot (y_3 \cdot z_3) &\doteq (x_3 \cdot y_3) \cdot z_3 \\
x_4 \cdot (y_4 \cdot z_4) &\doteq (x_4 \cdot y_4) \cdot z_4 \\
& w_3 \cdot w_3 \doteq 1 \}.
\end{aligned}
$$

$$\langle u, v \rangle = \langle a \cdot b,\ b \cdot a \rangle.$$

The reader can verify that θ below is a rigid E-unifier:

$$
\begin{aligned}
\theta = [& a/u_1,\ a/x_1,\ a/x_2,\ a/y_2,\ a/w_2,\ a/x_4, \\
& b/z_2,\ b/v_1,\ b/x_3,\ b/z_3,\ b/y_4,\ b/z_4,\ b/w_3, \\
& a \cdot b/w_1,\ a \cdot b/y_1,\ a \cdot b/z_1,\ a \cdot b/y_3].
\end{aligned}
$$

Definition 8.4 Let E be a (finite) set of equations, and W a (finite) set of variables. For any two substitutions σ and θ, $\sigma =_E \theta[W]$ iff $\sigma(x) \stackrel{*}{\cong}_E \theta(x)$ for every $x \in W$. The relation $\sqsubseteq_E$ is defined as follows. For any two substitutions σ and θ, $\sigma \sqsubseteq_E \theta[W]$ iff $\sigma =_{\theta(E)} \theta[W]$. The set W is omitted when $W = \mathcal{X}$ (where $\mathcal{X}$ is the set of variables), and similarly E is omitted when $E = \emptyset$.

Intuitively speaking, $\sigma \sqsubseteq_E \theta$ iff σ can be generated from θ using the equations in $\theta(E)$. Clearly, $\sqsubseteq_E$ is reflexive. However, it is not symmetric as shown by the following example.

---

[8] It is possible that equations have variables in common.

[9] It is possible that u and v have variables in common with the equations in E.

Example 8.5 Let $E = \{fx \doteq x\}$, $\sigma = [fa/x]$ and $\theta = [a/x]$. Then $\theta(E) = \{fa \doteq a\}$ and $\sigma(x) = fa \stackrel{*}{\cong}_{\theta(E)} a = \theta(x)$, and so $\sigma \sqsubseteq_E \theta$. On the other hand $\sigma(E) = \{ffa \doteq fa\}$, but a and fa are not congruent from $\{ffa \doteq fa\}$. Thus $\theta \sqsubseteq_E \sigma$ *does not* hold.

It is not difficult to show that $\sqsubseteq_E$ is also transitive. We also need an extension of $\sqsubseteq_E$ defined as follows.

Definition 8.6 Let E be a (finite) set of equations, and W a (finite) set of variables. The relation $\leq_E$ is defined as follows: for any two substitutions σ and θ, $\sigma \leq_E \theta[W]$ iff $\sigma\,;\eta \sqsubseteq_E \theta[W]$ for some substitution η (that is, $\sigma\,;\eta =_{\theta(E)} \theta[W]$ for some η).

Intuitively speaking, $\sigma \leq_E \theta$ iff σ is more general than some substitution that can be generated from θ using $\theta(E)$. Clearly, $\leq_E$ is reflexive. The transitivity of $\leq_E$ is also shown easily. When $\sigma \leq_E \theta[W]$, we say that σ *is (rigid) more general than θ over* W. It can be shown that if σ is a rigid E-unifier of u and v and $\sigma \leq_E \theta$, then θ is a rigid E-unifier of u and v. The converse is false. Finally, the crucial concept of a complete set of rigid E-unifiers can be defined.

Definition 8.7 Given a (finite) set E of equations, for any two terms u and v, letting $V = Var(u) \cup Var(v) \cup Var(E)$, a set U of substitutions is a *complete set of rigid E-unifiers for u and v* iff: For every $\sigma \in U$,

(i) $D(\sigma) \subseteq V$ and $D(\sigma) \cap I(\sigma) = \emptyset$ (idempotence),

(ii) σ is a rigid E-unifier of u and v,

(iii) For every rigid E-unifier θ of u and v, there is some $\sigma \in U$, such that, $\sigma \leq_E \theta[V]$.

Suppose we want to find a rigid E-unifier θ of u and v. There is an algorithm using transformations for finding rigid E-unifiers. Roughly, the idea is to use a form of unfailing completion procedure (Knuth and Bendix [44], Huet [36], Bachmair [8], Bachmair, Dershowitz, and Plaisted [9], Bachmair, Dershowitz, and Hsiang [10]). In order to clarify the differences between our method and unfailing completion, especially for readers unfamiliar with this method, we briefly describe the use of unfailing completion as a refutation procedure. For more details, the reader is referred to Bachmair [8].

Let E be a set of equations, and $\succ$ a reduction ordering total on ground terms. The central concept is that of E being *ground Church-Rosser w.r.t.* $\succ$. The crucial observation is that every ground instance $\sigma(l) \doteq \sigma(r)$ of an equation $l \doteq r \in E$ is orientable w.r.t. $\succ$, since $\succ$ is total on ground terms. Let $E^{\succ}$ be the set of all instances $\sigma(l) \doteq \sigma(r)$ of equations $l \doteq r \in E \cup E^{-1}$ with $\sigma(l) \succ \sigma(r)$ (the set of *orientable instances*). We say that E is *ground Church-Rosser w.r.t.* $\succ$ iff for every two ground terms u, v, if $u \stackrel{*}{\longleftrightarrow}_E v$, then there is some ground term w such that $u \stackrel{*}{\longrightarrow}_{E^{\succ}} w$ and $w \stackrel{*}{\longleftarrow}_{E^{\succ}} v$. Such a proof is called a *rewrite proof*.

An unfailing completion procedure attempts to produce a set E^{∞} equivalent to E and such that E^{∞} is ground Church-Rosser w.r.t. $\succ$. In other words, every ground equation provable from E has a rewrite proof in E^{∞}. The main mechanism involved is the computation of critical pairs. Given two equations $l_1 \doteq r_1$ and $l_2 \doteq r_2$ where l_2 is unifiable with a subterm l_1/β of l_1 which is not a variable, the pair $\langle \sigma(l_1[\beta \leftarrow r_2]), \sigma(r_1)\rangle$ where σ is a *mgu* of l_1/β and l_2 is a *critical pair*.

If we wish to use an unfailing completion procedure as a refutation procedure, we add two new constants T and F and a new binary function symbol *eq* to our language. In order to prove that $E \models u \doteq v$ for a ground equation $u \doteq v$, we apply the unfailing completion procedure to the set $E \cup \{eq(u,v) \doteq F, \ eq(z,z) \doteq T\}$, where z is a new variable. It can be shown that $E \models u \doteq v$ iff the unfailing completion procedure generates the equation $F \doteq T$. Basically, given any proof of $F \doteq T$, the unfailing completion procedure extends E until a rewrite proof is obtained. It can be shown that unfailing completion is a complete refutation procedure, but of course, it is not a decision procedure. It should also be noted that when unfailing completion is used as a refutation procedure, E^∞ is actually never generated. It is generated "by need", until $F \doteq T$ turns up.

We now come back to our situation. Without loss of generality, it can be assumed that we have a rigid E-unifier θ of T and F such that $\theta(E)$ is ground. In this case, equations in $\theta(E)$ are orientable instances. The crucial new idea is that in trying to obtain a rewrite proof of $F \doteq T$, we still compute critical pairs, but we **never rename variables**. If l_2 is equal to l_1/β, then we get a critical pair essentially by simplification. Otherwise, some variable in l_1 or in l_2 gets bound to a term *not* containing this variable. Thus the total number of variables in E keeps decreasing. Therefore, after a polynomial number of steps (in fact, the number of variables in E) we must stop or fail. So we get membership in NP. Oversimplifying a bit, we can say that our method is a form of lazy unfailing completion with no renaming of variables.

However, there are some significant departures from traditional Knuth-Bendix completion procedures, and this is for two reasons. The first reason is that we must ensure termination of the method. The second is that we want to show that the problem is in NP, and this forces us to be much more concerned about efficiency.

Our method can be described in terms of a single transformation on triples of the form $\langle S, \mathcal{E}, \mathcal{O}\rangle$, where S is a unifiable set of pairs, $\mathcal{E}$ is a set of equations, and $\mathcal{O}$ is something that will be needed for technical reasons and can be ignored for the present. Starting with an initial triple $\langle S_0, \mathcal{E}_0, \mathcal{O}_0\rangle$ initialized using E and u, v (except for $\mathcal{O}$ that must be guessed), if the number of variables in E is m, one considers sequences of transformations

$$\langle S_0, \mathcal{E}_0, \mathcal{O}_0\rangle \Rightarrow^+ \langle S_k, \mathcal{E}_k, \mathcal{O}_k\rangle$$

consisting of at most $k \leq m$ steps. It will be shown that u and v have some rigid E-unifier iff there is some sequence of steps as above such that the special equation $F \doteq T$ is in $\mathcal{E}_k$ and S_k is unifiable. Then, the most general unifier of S_k is a rigid E-unifier of u and v.

Roughly speaking, $\mathcal{E}_{k+1}$ is obtained by overlapping equations in $\mathcal{E}_k$ (forming critical pairs), as in unfailing Knuth-Bendix completion procedures, except that no renaming of variables takes place. In order to show that the number of steps can be bounded by m, it is necessary to show that some measure decreases every time an overlap occurs, and there are two difficulties. First, the overlap of two equations may involve the identity substitution when some equation simplifies another one. In this case, the number of variables does not decrease, and no other obvious measure decreases. Second, it is more difficult to handle overlap at variable occurrences than it is in the traditional case, because we are not allowed to form new instances of equations.

The first difficulty can be handled by using a special procedure for reducing a set of (ground) equations. Such a procedure is presented in Gallier et al. [24] and runs in polynomial time (see also [67]). Actually, one also needs a total simplification ordering $\prec$ on ground terms, and a way of orienting equations containing variables, which is the purpose of the mysterious component $\mathcal{O}$. The second difficulty is overcome by noticing that one only needs to consider ground substitutions, that the ordering $\prec$ (on ground terms) can be extended to ground substitutions, and that given any rigid E-unifier θ of u and v, there is always a least rigid E-unifier σ (w.r.t $\prec$) that is equivalent to θ (in a sense to be made precise).

Other complications arise in proving that the method is in NP, in particular, we found it necessary to represent most general unifiers (mgu's) by their triangular form as in Martelli and Montanari [49]. This concept has already been defined in definition 6.2.

The triangular form $T = \{\langle x_1, t_1 \rangle, \dots, \langle x_k, t_k \rangle\}$ of a substitution σ also defines a substitution, namely $\sigma_T = [t_1/x_1, \dots, t_k/x_k]$. This substitution is usually different from σ and not idempotent as can be seen from example 6.3. However, this substitution plays a crucial role in our decision procedure because of the following property.

Lemma 8.8 *Given a triangular form* $T = \{\langle x_1, t_1 \rangle, \dots, \langle x_k, t_k \rangle\}$ *for a substitution* σ *and the associated substitution* $\sigma_T = [t_1/x_1, \dots, t_k/x_k]$, *for every unifier* θ *of* T, $\theta = \sigma_T \, ; \theta$.

An other important observation about σ_T is that even though it is usually not idempotent, at least one variable in $\{x_1, \dots, x_k\}$ does not belong to $I(\sigma_T)$ (otherwise, condition (1) of the triangular form fails). We will assume that a procedure TU is available, which, given any unifiable term system S, returns a triangular form for an idempotent mgu of S, denoted by $TU(S)$. When S consists of a single pair $\langle u, v \rangle$, $TU(S)$ is also denoted by $TU(u, v)$.

One of the major components of the decision procedure for rigid E-unification is a procedure for creating a reduced set of rewrite rules equivalent to a given (finite) set of ground equations. Given a set R of rewrite rules, we say that R *is rigid reduced* iff

(1) No lefthand side of any rewrite rule $l \to r \in R$ is reducible by any rewrite rule in $R - \{l \to r\}$ treated as a ground rule;

(2) No righthand side of any rewrite rule $l \to r \in R$ is reducible by any rewrite rule in R treated as a ground rule.

A procedure for creating a rigid reduced set of rewrite rules equivalent to a given (finite) set of rewrite rules was first presented in Gallier et al. [24] and runs in polynomial time. However, due to the possibility that variables may occur in the equations, we have to make some changes to this procedure. Roughly speaking, given a "guess" $\mathcal{O}$ (a preorder which we call an *order assignment*) of the ordering among all subterms of the terms in a set of equations E, we can run the reduction procedure R on E and $\mathcal{O}$ to produce a reduced rewrite system $R(E, \mathcal{O})$ equivalent to E, and whose orientation is dictated by the preorder $\mathcal{O}$. The precise definition of an order assignment $\mathcal{O}$ is too involved to be reproduced here, but this is not essential anyway. All we need to know is that we have an algorithm R such that, given a set E of equations and an order-assignment $\mathcal{O}$, a rigid-reduced set of rewrite rules $R(E, \mathcal{O})$ is returned, the rules in $R(E, \mathcal{O})$ being oriented by $\mathcal{O}$. We are now ready to define a procedure for finding rigid E-unifiers.

This method uses the reduction procedure just discussed, and a single transformation on certain systems defined next. First, the following definition is needed.

Definition 8.9 Given a set E of equations and some equation $l \doteq r$, the set of equations obtained from E by deleting $l \doteq r$ and $r \doteq l$ from E is denoted by $(E - \{l \doteq r\})^{\dagger}$. Formally, we let $(E - \{l \doteq r\})^{\dagger} = \{u \doteq v \mid u \doteq v \in E,\ u \doteq v \neq l \doteq r,\ \text{and}\ u \doteq v \neq r \doteq l\}$.

Definition 8.10 Let $\prec$ be a total simplification ordering on ground terms. We shall be considering finite sets of equations of the form $\mathcal{E} = \mathcal{E}_{\Sigma} \cup \{eq(u, v) \doteq F,\ eq(z, z) \doteq T\}$,[10] where $\mathcal{E}_{\Sigma}$ is a set of equations over $T_{\Sigma}(\mathcal{X})$, and $u, v \in T_{\Sigma}(\mathcal{X})$. We define a transformation on systems of the form $\langle S, \mathcal{E}, \mathcal{O} \rangle$, where S is a term system, $\mathcal{E}$ a set of equations as above, and $\mathcal{O}$ an order assignment:

$$\langle S_0,\ \mathcal{E}_0,\ \mathcal{O}_0 \rangle \Rightarrow \langle S_1,\ \mathcal{E}_1,\ \mathcal{O}_1 \rangle,$$

where $l_1 \doteq r_1,\ l_2 \doteq r_2 \in \mathcal{E}_0 \cup \mathcal{E}_0^{-1}$, either l_1/β is *not* a variable or the equation $l_2 \doteq r_2$ is degenerate,[11] $l_1/\beta \neq l_2$, $TU(l_1/\beta, l_2)$ represents an *mgu* of l_1/β and l_2 in triangular form,[12] $\sigma = [t_1/x_1, \ldots, t_p/x_p]$ where $TU(l_1/\beta, l_2) = \{\langle x_1, t_1 \rangle, \ldots, \langle x_p, t_p \rangle\}$,

$$\mathcal{E}_1' = \sigma((\mathcal{E}_0 - \{l_1 \doteq r_1\})^{\dagger} \cup \{l_1[\beta \leftarrow r_2] \doteq r_1\}),$$

$\mathcal{O}_1$ is an order assignment on $\mathcal{E}_1'$ compatible with $\mathcal{O}_0$, $S_1 = S_0 \cup TU(l_1/\beta, l_2)$, and $\mathcal{E}_1 = R(\mathcal{E}_1', \mathcal{O}_1)$.

Observe that $\sigma(l_1[\beta \leftarrow r_2] \doteq r_1)$ looks like a critical pair of equations in $\mathcal{E}_0 \cup \mathcal{E}_0^{-1}$, but it is not. This is because a critical pair is formed by applying the *mgu* of l_1/β and l_2 to $l_1[\beta \leftarrow r_2] \doteq r_1$, but $[t_1/x_1, \ldots, t_p/x_p]$ is usually not a *mgu* of l_1/β and l_2. It is the composition $[t_1/x_1]; \ldots; [t_p/x_p]$ that is a *mgu* of l_1/β and l_2. The reason for not applying the *mgu* is that by repeated applications of this step, exponential size terms could be formed, and it would not be clear that the decision procedure is in NP. We have chosen an approach of "lazy" (or delayed) unification. Also note that we use the rigid reduced system $R(\mathcal{E}_1', \mathcal{O}_1)$ rather than $\mathcal{E}_1'$, and so, a transformation step is defined only if R does not fail. The method for finding E-unifiers is then is the following.

---

[10] eq, T, F are some new symbols not occurring in E, u, v.

[11]   An equation $x \doteq v$ is degenerate if x is a variable and $x \notin Var(v)$.

[12]   Note that we are requiring that l_1/β and l_2 have a *nontrivial* unifier. The triangular form of *mgus* is important for the NP-completeness of this method.

Definition 8.11 (Method) Let $E_{u,v} = E \cup \{eq(u,v) \doteq F, \ eq(z,z) \doteq T\}$, $\mathcal{O}_0$ an order assignment on $E_{u,v}$, $\mathcal{S}_0 = \emptyset$, $\mathcal{E}_0 = R(E_{u,v}, \mathcal{O}_0)$, m the total number of variables in $\mathcal{E}_0$, and $V = Var(E) \cup Var(u,v)$. For any sequence

$$\langle \mathcal{S}_0, \mathcal{E}_0, \mathcal{O}_0 \rangle \Rightarrow^+ \langle \mathcal{S}_k, \mathcal{E}_k, \mathcal{O}_k \rangle$$

consisting of at most m transformation steps, if $\mathcal{S}_k$ is unifiable and $k \leq m$ is the first integer in the sequence such that $F \doteq T \in \mathcal{E}_k$, return the substitution $\theta_{\mathcal{S}_k}|_V$, where $\theta_{\mathcal{S}_k}$ is the *mgu* of $\mathcal{S}_k$ (over $T_\Sigma(\mathcal{X})$).

Example 8.12 Let E be the set of equations $E = \{fa \doteq a, \ ggx \doteq fa\}$, and $\langle u,v \rangle = \langle gggx, x \rangle$. We have

$$E_{u,v} = \{fa \doteq a, \ ggx \doteq fa, \ eq(gggx, x) \doteq F, \ eq(z,z) \doteq T\}.$$

The congruence closure Π of $E_{u,v}$ has three nontrivial classes $\{a, fa, ggx\}$, $\{eq(gggx, x), F\}$, and $\{eq(z,z), T\}$. Let $\mathcal{O}_0$ be the order assignment on $E_{u,v}$ such that

$$T \prec_{\mathcal{O}_0} eq(gggx, x),$$
$$F \prec_{\mathcal{O}_0} eq(z, z),$$
$$a \prec_{\mathcal{O}_0} fa \prec_{\mathcal{O}_0} ggx,$$

the least elements of classes being ordered in the order of listing of the classes. We have $\mathcal{S}_0 = \emptyset$, and the reduced system $\mathcal{E}_0 = R(E_{u,v}, \mathcal{O}_0)$ is

$$\mathcal{E}_0 = \{fa \doteq a, \ ggx \doteq a, \ eq(ga, x) \doteq F, \ eq(z,z) \doteq T\}.$$

Note that there is an overlap between $eq(ga, x) \doteq F$ and $eq(z,z) \doteq T$ at address ϵ in $eq(ga, x)$, and we obtain the triangular system $\{\langle x, ga \rangle, \langle z, ga \rangle\}$ and the new equation $F \doteq T$. Thus, we have

$$\langle \mathcal{S}_0, \mathcal{E}_0, \mathcal{O}_0 \rangle \Rightarrow \langle \mathcal{S}_1, \mathcal{E}_1, \mathcal{O}_1 \rangle,$$

where $\mathcal{S}_1 = \{\langle x, ga \rangle, \langle z, ga \rangle\}$

$$\mathcal{E}_1' = \{fa \doteq a, \ ggga \doteq a, \ eq(ga, ga) \doteq F, \ F \doteq T\},$$

and $\mathcal{O}_1$ is the restriction of $\mathcal{O}_0$ to the subterms in $\mathcal{E}_1'$. After reducing $\mathcal{E}_1'$, we have

$$\mathcal{E}_1 = \{fa \doteq a, \ ggga \doteq a, \ eq(ga, ga) \doteq T, \ F \doteq T\}.$$

Since $F \doteq T \in \mathcal{E}_1$ and $\mathcal{S}_1$ is unifiable, the restriction $[ga/x]$ of the *mgu* $[ga/x, ga/z]$ of $\mathcal{S}_1$ to $Var(E) \cup Var(u,v) = \{x\}$ is a rigid E-unifier of $gggx$ and x.

The following major results are proved in Gallier, Narendran, Plaisted, and Snyder [27].

Theorem 8.13 *The procedure given by definition 8.11 is a decision procedure for rigid E-unification. Furthermore, it belongs to NP.*

The soundness and completeness of the method are subsumed by the following result.

Theorem 8.14 *Let E be a set of equations over $T_\Sigma(\mathcal{X})$, u, v two terms in $T_\Sigma(\mathcal{X})$, m the number of variables in $E \cup \{u, v\}$, and $V = Var(E) \cup Var(u, v)$. There is a finite complete set of rigid E-unifiers for u and v given by the set*

$$\{\theta_{\mathcal{S}_k}|_V \mid \langle \mathcal{S}_0, \mathcal{E}_0, \mathcal{O}_0 \rangle \Rightarrow^+ \langle \mathcal{S}_k, \mathcal{E}_k, \mathcal{O}_k \rangle,\ k \leq m\},$$

for any order assignment $\mathcal{O}_0$ on $E_{u,v}$, with $\mathcal{S}_0 = \emptyset$, $\mathcal{E}_0 = R(E_{u,v}, \mathcal{O}_0)$, and where $\mathcal{S}_k$ is unifiable, $F \doteq T \in \mathcal{E}_k$, $F \doteq T \notin \mathcal{E}_i$ for all i, $0 \leq i < k$, and $\theta_{\mathcal{S}_k}$ is the mgu of $\mathcal{S}_k$ over $T_\Sigma(\mathcal{X})$.

Thus, we note another major difference between general E-unification and rigid E-unification. In rigid E-unification, there is always a finite complete set of (rigid) E-unifiers.

The above results have been improved by Isakowitz [37] and by Choi and Gallier [17]. Isakowitz has shown that order assignments can be dispensed with if a different reduction procedure is used. Isakowitz also studied the extension of rigid E-unification to order-sorted logic, and proved results analogous to those presented here for some subclasses of equations. Choi and Gallier have obtained a more direct proof of the NP-completeness of rigid E-unification that also avoids order assignments. This new proof is more algebraic and uses some key ideas from Kozen [45].

9 Conclusion and Directions For Further Research

We surveyed two methods for automated theorem proving, the method of matings for languages without equality, and the method of equational matings, for languages with equality. We also surveyed various unification procedures associated with theorem proving methods based on matings. These include standard unification and rigid E-unification. The crucial property of these unification methods is that they are decidable. However, their complexity is very different: standard unificatin can be performed in linear-time, but rigid E-unification is NP-complete.

An area of research that remains wide open is the study of efficient implementations of equational matings and rigid E-unification. This is a difficult problem, since rigid E-unification is NP-complete, and one needs to isolate interesting special classes of formulae for which tractable algorithms can be found. On a more theoretical level, it would be interesting to study the generalization of linear logic including equations. We conjecture that this will lead naturally to rigid E-unification. Finally, investigating whether the method of matings can be generalized either to order-sorted logic or to higher-order logic, remains to be done.

10 References

[1] Aho, A., Sethi, R., and Ullman, J. *Compilers, Principles, Techniques, and Tools,* Addison Wesley (1986).

[2] Aït-Kaci, H. A lattice theoretic approach to computation based on a calculus of partially ordered type structures, Ph.D. thesis. Department of Computer and Information Science, University of Pensylvania, PA (1984).

[3] Aït-Kaci, H. An algebraic semantics approach to the effective resolution of type equations. *Theoretical Computer Science* 45, pp. 293-351 (1986).

[4] Andrews, P. Theorem Proving via General Matings. *J.ACM* 28(2), 193-214, 1981.

[5] Andrews, P. *An Introduction to Mathematical Logic and Type Theory: To Truth Through Proof.* Academic Press, New York, 1986.

[6] Andrews, P.B., D. Miller, E. Cohen, F. Pfenning, "Automating Higher-Order Logic," *Contemporary Mathematics* 29, 169-192, 1984.

[7] Bachmair, L., *Proof Methods for Equational Theories,* Ph.D thesis, University of Illinois, Urbana Champaign, Illinois (1987).

[8] Bachmair, L., *Canonical Equational Proofs,* Research Notes in Theoretical Computer Science, Wiley and Sons, 1989.

[9] Bachmair, L., Dershowitz, N., and Plaisted, D., "Completion without Failure," *Resolution of Equations in Algebraic Structures,* Vol. 2, Aït-Kaci and Nivat, editors, Academic Press, 1-30 (1989).

[10] Bachmair, L., Dershowitz, N., and Hsiang, J., "Orderings for Equational Proofs," In *Proc. Symp. Logic in Computer Science,* Boston, Mass. (1986) 346-357.

[11] Bibel, W. Tautology Testing With a Generalized Matrix Reduction Method, *TCS* 8, pp. 31-44, 1979.

[12] Bibel, W. On Matrices With Connections, *J.ACM* 28, pp. 633-645, 1981.

[13] Bibel, W. *Automated Theorem Proving.* Friedr. Vieweg & Sohn, Braunschweig, 1982.

[14] Boudet, A., Jouannaud, J.-P., and Schmidt-Schauss, M. Unification in Boolean Rings and Abelian Groups. *Journal of Symbolic Computation* 8(5), pp. 449-478 (1989).

[15] Bürckert, H., Herold, A., and Schmidt-Schauss, M. On equational theories, unification, and (un)decidability. Special issue on Unification, Part II, *Journal of Symbolic Computation* 8(1 & 2), 3-50 (1989).

[16] Bürckert, H., Matching–A Special Case of Unification? *Journal of Symbolic Computation* 8(5), pp. 523-536 (1989).

[17] Choi, J., and Gallier, J.H. A simple algebraic proof of the NP-completeness of rigid E-unification, in preparation (1990).

[18]   Church, A., "A note on the Entscheidungsproblem", *JSL* 1 (1936) 40-41, corrections, 101-102.

[19]   De Champeaux, D., "About the Paterson-Wegman linear unification," *Journal of Computer and System Sciences*, 32(1) (1986) 79-90.

[20]   Dershowitz, N,. "Termination of Rewriting," Journal of Symbolic Computation 3 (1987) 69-116.

[21]   Downey, Peter J., Sethi, Ravi, and Tarjan, Endre R. "Variations on the Common Subexpressions Problem." *J.ACM* 27(4), 758-771, 1980.

[22]   Gallier, J.H. *Logic for Computer Science: Foundations of Automatic Theorem Proving*, Harper and Row, New York (1986).

[23]   Gallier, J.H., Raatz, S., and Snyder, W., "Theorem Proving using Rigid E-Unification: Equational Matings," *LICS'87*, Ithaca, New York (1987) 338-346.

[24]   Gallier, J.H., Narendran, P., Plaisted, D., Raatz, S., and Snyder, W., "Finding canonical rewriting systems equivalent to a finite set of ground equations in polynomial time," submitted to *J.ACM* (1987).

[25]   Gallier, J.H., Narendran, P., Plaisted, D., and Snyder, W., "Rigid E-unification is NP-complete," *LICS'88*, Edinburgh, Scotland, July 5-8, 1988, 218-227.

[26]   Gallier, J.H., Raatz, S, and Snyder, W. Rigid E-Unification and its Applications to Equational Matings. *Resolution of Equations in Algebraic Structures*, Vol. 1, Aït-Kaci and Nivat, editors, Academic Press, 151-216 (1989).

[27]   Gallier, J.H., Narendran, P., Plaisted, D., and Snyder, W. Rigid E-Unification: NP-completeness and Applications to Theorem Proving. Special issue of *Information and Computation* 87(1/2), 129-195 (1990).

[28]   Gallier, J.H., Narendran, P., Raatz, S., and Snyder, W. Theorem Proving Using Equational Matings and Rigid E-Unification. To appear in *J.ACM*, pp. 62 (1990).

[29]   Girard, J.Y., "Linear Logic," *Theoretical Computer Science* 50:1 (1987) 1-102.

[30]   Gödel, Kurt. Die Vollstandigkeit der Axiome des Logischen Funktionenkalküls, *Monatsh. Math. Phys.* 37, 349-360 (1930), translated in Gödel [31], 44-123 (1986).

[31]   Gödel, Kurt. *Collected Works, Vol. I, Publications 1929-1936*, Edited by S. Feferman, J. Dawson, S. Kleene, G. Moore, R. Solovay, and J. van Heijenoort, Oxford University Press (1986).

[32]   Herbrand, J., "Sur la Théorie de la Démonstration," in *Logical Writings*, W.D. Goldfarb, ed., Harvard University Press (1971).

[33]   Hindley, J., and Seldin, J., *Introduction to Combinators and Lambda Calculus*, Cambridge University Press (1986).

[34] Huet, G. *Constrained Resolution: A Complete Method for Higher-Order Logic*, Ph.D. thesis, Case Western Reserve University (1972).

[35] Huet, G., *Résolution d'Equations dans les Langages d'Ordre* $1, 2, \ldots, \omega$, Thèse d'Etat, Université de Paris VII (1976).

[36] Huet, G., "Confluent Reductions: Abstract Properties and Applications to Term Rewriting Systems," JACM 27:4 (1980) 797-821.

[37] Isakowitz, T. Theorem Proving Methods For Order-Sorted Logic, Ph.D. thesis. Department of Computer and Information Science, University of Pensylvania (1989).

[38] Jouannaud, J.-P., and Kirchner, C. Solving Equations in Abstract Algebras: A Rule-Based Survey of Unification. Technical Report, University of Paris Sud (1989).

[39] Kirchner, C., *Méthodes et Outils de Conception Systematique d'Algorithmes d'Unification dans les Theories Equationnelles*, Thèse d'Etat, Université de Nancy I (1985).

[40] Kfoury, A.J., J. Tiuryn, and P. Urzyczyn, "An analysis of ML typability", submitted (a section of this paper will appear in the proceedings of CAAP 1990 under the title "ML typability is DEXPTIME-complete").

[41] Kfoury, A.J., J. Tiuryn, and P. Urzyczyn, "The undecidability of the semi-unification problem", Proceedings of STOC (1990).

[42] Kfoury, A.J., and J. Tiuryn, "Type Reconstruction in Finite-Rank Fragments of the Polymorphic Lambda Calculus," LICS'90, Philadelpha, PA.

[43] Knight. K. "A Multidisciplinary Survey," *ACM Computing Surveys*, Vol. 21, No. 1, pp. 93-124 (1989).

[44] Knuth, D.E. and Bendix, P.B., "Simple Word Problems in Univeral Algebras," in *Computational Problems in Abstract Algebra*, Leech, J., ed., Pergamon Press (1970).

[45] Kozen, D., "Complexity of Finitely Presented Algebras," Technical Report TR 76-294, Department of Computer Science, Cornell University, Ithaca, New York (1976).

[46] Kozen, D., "Positive First-Order Logic is NP-Complete," IBM Journal of Research and Development, 25:4 (1981) 327-332.

[47] Lassez, J.-L., Maher, M., and Marriot, K. Unification Revisited. *Foundations of Deductive Databases and Logic Programming*, J. Minker, editor, Morgan-Kaufman, pp. 587-625 (1988).

[48] Manna, Zohar. *Mathematical Theory of Computation*. McGraw-Hill (1974).

[49] Martelli, A., Montanari, U., "An Efficient Unification Algorithm," ACM Transactions on Programming Languages and Systems, 4:2 (1982) 258-282.

[50] Meseguer, J., Goguen, J. A., and Smolka, G. Order-Sorted Unification. *Journal of Symbolic Computation* 8(4), pp. 383-413 (1989).

[51] Miller, D., *Proofs in Higher-Order Logic*, Ph.D. thesis, Carnegie-Mellon University, 1983.

[52] Miller, D. A. Expansion Trees and Their Conversion to Natural Deduction Proofs. In *7th International Conference on Automated Deduction, Napa, CA*, edited by R.E. Shostak, L.N.C.S, No. 170, New York: Springer Verlag, 1984.

[53] Miller, D. A compact Representation of Proofs. *Studia Logica* 4/87, pp. 347-370 (1987).

[54] Milner, R. A theory of type polymorphism in programming. *J. Comput. Sys. Sci.* 17, pp. 348-375 (1978).

[55] Nelson G. and Oppen, D. C. Fast Decision Procedures Based on Congruence Closure. *J. ACM* 27(2), 356-364, 1980.

[56] Paterson, M.S., Wegman, M.N., "Linear Unification," Journal of Computer and System Sciences, 16 (1978) 158-167.

[57] Pfenning, F., *Proof Transformations in Higher-Order Logic*, Ph.D. thesis, Department of Mathematics, Carnegie Mellon University, Pittsburgh, Pa. (1987).

[58] Plotkin, G., "Building in Equational Theories," Machine Intelligence 7 (1972) 73-90.

[59] Rémy, Didier, "Algèbres Touffues. Application au typage polymorphique des objects enregistrements dans les languages fonctionnels." Thèse, Université Paris VII, 1990.

[60] Rémy, Didier, "Records and variants as a natural extension in ML," *Sixteenth ACM Annual Symposium on Principles of Programming Languages*, Austin Texas, 1989.

[61] Robinson, J.A., "A Machine Oriented Logic Based on the Resolution Principle," JACM 12 (1965) 23-41.

[62] Siekmann, J. H. Unification Theory, Special Issue on Unification, Part I, *Journal of Symbolic Computation* 7(3 & 4), pp. 207-274 (1989).

[63] Shieber, S. *An Introduction to Unification-Based Approaches to Grammar.* CSLI Lecture Notes Series, Center for the study of Language and Information, Stanford, CA (1986).

[64] Skolem, Thoralf. Über die Mathematische Logik, *Norsk matematisk tidsskrift* 10, 125-142 (1928). translated in van Heijenoort [70], 508-524 (1967).

[65] Schmidt-Schauss, M. Unification in a Combination of Arbitrary Disjoint Equational Theories. Special issue on Unification, Part II, *Journal of Symbolic Computation* 8(1 & 2), 51-99 (1989).

[66] Snyder, W. Complete Sets of Transformations for General Unification, Ph.D. thesis. Department of Computer and Information Science, University of Pensylvania, PA (1988).

[67] Snyder, W., "Efficient Ground Completion: A Fast Algorithm for Generating Reduced Ground Rewriting Systems from a Set of Ground Equations," RTA'89, Chapel Hill, NC (journal version submitted for publication).

[68] Snyder, W. *The Theory of General Unification*. Birkhauser Boston, Inc. (in preparation).

[69] Szabo, P. *Unifikationstheorie erster Ordnung*, Ph.D. thesis, Universität Karlsruhe (1982).

[70] van Heijenoort, Jean (editor). *From Frege to Gödel. A Source Book in Mathematical Logic, 1879-1931*, Harvard University Press (1967).

[71] Yelick, K. Unification in combinations of collapse-free regular theories. *Journal of Symbolic Computation* 3(1 & 2), pp. 153-182 (1987).